WORLD *of* ANATOMY AND PHYSIOLOGY

WORLD *of* ANATOMY AND PHYSIOLOGY

K. Lee Lerner and Brenda Wilmoth Lerner, *Editors*

Volume 2

K-Z

General Index

GALE®

THOMSON

GALE

Detroit • New York • San Diego • San Francisco • Cleveland • New Haven, Conn. • Waterville, Maine • London • Munich

World of Anatomy and Physiology
K. Lee Lerner and Brenda Wilmoth Lerner

Project Editor
Kimberley A. McGrath

Editorial
Mark Springer

Permissions
Kim Davis

Imaging and Multimedia
Randy Bassett, Mary K. Grimes, Lezlie Light, Dan Newell, David G. Oblender, Robyn V. Young

Product Design
Cindy Baldwin, Michelle DiMercurio, Michael Logusz

Manufacturing
Wendy Blurton, Evi Seoud

LIBRARY OF CONGRESS CATALOG-IN-PUBLICATION DATA

World of anatomy and physiology / K. Lee Lerner and Brenda Wilmoth Lerner, editors.
  p. cm.
Includes bibliographical references and index.
ISBN 0-7876-5684-4 (set : hardcover)
1. Physiology—Encyclopedias. 2. Anatomy—Encyclopedias. I. Lerner, K. Lee. II. Lerner, Brenda Wilmoth.
QP11.W67 2002
612'.003—dc21   2002005517

ISBN 0-7876-5684-4 (set), ISBN 0-7876-5685-2 (v.1), ISBN 0-7876-5686-0 (v.2)

Printed in the United States of America
10 9 8 7 6 5 4 3 2 1

CONTENTS

VOLUME 2: K-Z

V

W

Y

Z

INTRODUCTION

World of Anatomy and Physiology is devoted to the study of the intricate relationships of form and function within the human body. Although anatomy is the most ancient of all medical studies, the early years of the twenty-first century are an exciting time to undertake the study of the structure and function of the human body. Around the world, thousands of dedicated research and clinical science specialists provide a constant stream of insights into the most intimate aspects of our anatomy and physiology.

Recent rapid progress in cell biology allows, for the first time, insight into the fundamental mechanisms of development. Never before in human history has information moved so rapidly from the laboratory to the clinical setting. The development of reproductive medicine and gene therapy promises to propel us into a new and revolutionary era of biotechnology and biomedical science.

World of Anatomy and Physiology is a collection of 650 entries on topics covering a range of interests—from biographies of the pioneers of anatomy and physiology to explanations of the latest developments and advances in embryology and developmental biology. Despite the complexities of terminology and advanced knowledge of biochemistry needed to fully explore some of the topics in physiology, every effort has been made to set forth entries in everyday language and to provide accurate and generous explanations of the most important terms. The editors intend *World of Anatomy and Physiology* for a wide range of readers. Accordingly, *World of Anatomy and Physiology* articles are designed to instruct, challenge, and excite less-experienced students, while providing a solid foundation and reference for more advanced students.

The very essence of the attributes that define life is the province of two complementary branches of science, anatomy and physiology. All of the sub-disciplines of medical science are built on the fundamental foundations of anatomy and physiology.

Anatomy studies the structure of body parts and their relationships to one another, while physiology concerns the function of the body's structural machinery and how all the body parts work to carry out life-sustaining activities. Questions such as how the living body is able to see, hear, keep warm, or digest food are at the core of anatomical and physiological research.

As our most accomplished athletes and artists prove, the human body has the ability to move gracefully, to lift an arm accurately, and perform many various tasks. The anatomist and physiologist observe and question how body's nervous system, muscles, and joints perform these tasks. To date, the investigation of such questions has depended largely on anatomical studies, which involved examination of anatomical structure by dissection. Technological advances in microcopy and imaging now allow anatomists to view living body and cellular structure on the molecular scale.

Advances in molecular biology allow researchers to investigate a broad range of phenomena, including the physiochemical processes taking place in the cells and tissues of the body, the electrical events underlying the actions of the nervous system, and the feedback mechanisms that allow fine control of complex physiological processes.

With the rapid expansion of scientific knowledge, various disciplines have split off from the parent disciplines of anatomy and physiology. Biochemistry, the study of chemical processes in cells and tissues was probably the first discipline to diverge, then also biophysics, which deals with physical processes within cells. More recently, neurophysiology became established as a separate and specialized area of study. Quite often, anatomical or physiological research is carried out in departments of medicine, where basic scientific knowledge is required for the understanding of disease. The reverse situation is also common: basic research in anatomy or physiology reveals how a disease occurs. For example, basic research into the pancreas and the hormone insulin led to the discovery of successful treatment for the disease diabetes mellitus. Finally, anatomists and physiologists have long been interested in the science underlying such things as athletic performance, the function of the respiratory and nervous systems under anesthesia, the effects of low or high barometric pressures, lack of oxygen, and so on. This all comes under the heading of applied anatomy and physiology.

In all these aspects of the study of anatomy and physiology, it is important to appreciate the role of mechanisms that control bodily functions. For example, we cannot divorce the study of muscular contraction from the feed-back systems that control the structural elements of muscle. All muscles contain structures that signal back to the central nervous system the muscle length and the degree of contractile force. At the spinal level, reflex actions occur to control the performance as precisely as possible in terms of length and force. Without these sophisticated control systems, it becomes very difficult to use muscles properly, precisely, or accurately.

Similar mechanisms exist to control arterial blood pressure. Obviously, if blood pressure is too high, there will be an increased likelihood of rupture of blood vessels and consequent hemorrhage as occurs in strokes. If blood pressure falls too low, the blood supply to the brain will be impaired and consciousness may be lost. When one considers the fact that the human body contains trillions of cells in nearly constant activity, and that remarkably little goes wrong with it, one appreciates what a marvelous system it is. Walter Cannon, an American physiologist of the early twentieth century, spoke of the "wisdom of the body" and he coined the term "homeostasis" to describe its ability to maintain relatively stable internal conditions even though the outside world changes continuously. Although the literal translation of homeostasis is "unchanging," the term does not actually mean a static, or unchanging, state in the body. Rather, it indicates a dynamic state of equilibrium, or a balance, in which the internal conditions vary, but always within relatively narrow limits. In general, the body is in homeostasis when its needs are adequately met and it is functioning smoothly.

Maintaining homeostasis is much more complicated than it appears at first glance. Virtually every organ plays a role in maintaining the constancy of the internal environment. Adequate blood levels of vital nutrients must be continuously present, and heart activity and blood pressure must be constantly monitored and adjusted so that the blood is propelled to all body tissues. Metabolic wastes must not be allowed to accumulate and the body temperature must be precisely controlled. A wide variety of chemical, thermal, and neural factors act and interact in complex ways—sometimes helping and sometimes hindering the body as it works to maintain its "steady rudder" to steer carefully between the dangers of anatomical and physiological extremes.

A system of communication is essential for homeostasis to function. Communication is accomplished chiefly by the nervous and endocrine systems, which use electrical and chemical mechanisms to transmit impulses, information, and instructions. All homeostatic control mechanisms have at least three interdependent control mechanisms: receptors, control centers and effectors. Most homeostatic control systems function as negative feedback mechanisms, where the output of the system shuts off the original stimulus or reduces its intensity. In positive feedback systems, the response enhances the original stimulus so that the activity, or output, is accelerated.

Blood clotting is an example of a process controlled by positive feedback mechanisms.

Although *World of Anatomy and Physiology* concentrates on topics in classical anatomy and physiology, the editors have tried to provide insight into important areas of developmental and reproductive biology. Almost daily, new discoveries extend our understanding of reproductive biology and embryology. At an equally rapid pace, biotechnologies emerge to expand applications of those discoveries. The pace of change and innovation is challenging to all publications. For example, during the writing of *World of Anatomy and Physiology*, researchers announced the creation of cloned human embryos that grew to the six-cell stage. Days before going to press, researchers at the Whitehead Institute for Biomedical Research announced results in the British journal *Nature* that indicated that fully differentiated adult cells can be used to form clones.

Accordingly, *World of Anatomy and Physiology* has attempted to incorporate references and basic explanations of the latest findings and applications. Although certainly not a substitute for in-depth study of important topics such as stem cell research or cloning, the editors hope to provide students and readers with the basic information and insights that will enable a greater understanding of the news and stimulate critical thinking regarding current events in biomedicine.

In the Classical world, the Greek philosopher Aristotle (384–322 B.C.) raised fundamental questions about form, function, reproduction, and development. The quest for insight and answers into the embryological development of humans continues to fascinate and challenge modern scientists. We hope that *World of Anatomy and Physiology* inspires a new generation of scientists who will join in the exciting quest to unlock the remaining secrets of life. It is our modest wish that this book provide valuable information to students and readers regarding topics that play an increasingly prominent role in our civic debates, and a fundamentally important and intimate role in our everyday lives.

K. Lee Lerner and Brenda Wilmoth Lerner, editors
New Orleans
February, 2002

How to Use the Book

Students who are new to the study of anatomy and physiology should start their studies with a careful reading of the article titled, "Anatomical nomenclature," and with an inspection of the article's accompanying diagrams. Over the centuries, anatomists developed a standard nomenclature, or method of naming anatomical structures. In order to standardize nomenclature, anatomical terms relate to the *standard anatomical position*. When the human body is in the standard anatomical position it is upright, erect on two legs, facing frontward, with the arms at the sides each rotated so that the palms of the hands turn forward. Based upon this position, the terms superior, inferior, anterior, posterior and a number of other terms gain clarity and precision. The editors recommend that the reader bookmark the "Anatomical nomenclature" article so quick reference may be made during the reading of other articles. With a little practice and patience, seemingly daunting anatomical nomenclature becomes understandable and readers will achieve comfort with its use.

Although we have attempted to include a number of stimulating photos and informative diagrams *World of Anatomy and Physiology* is not an atlas of anatomy. The deepest understanding of anatomical form requires an atlas that can visually reinforce descriptive writings. There are a number of excellent anatomical atlases available to students and readers that will greatly enhance the reading of *World of Anatomy and Physiology* articles. In particular, the editors recommend the McMinn and Hutchings classic work titled *"Color Atlas of Human Anatomy* and Marjorie England's *Color Atlas of Life Before Birth*, both published by Year Book Medical Publishers, Chicago.

The articles in the book are meant to be understandable to anyone with a curiosity about topics in anatomy or physiology. Cross-references to related articles, definitions, and biographies in this collection are indicated by **bold-faced type**, and these cross-references will help explain and expand the individual entries. Although far from containing a comprehensive collection of topics related to genetics, *World of Anatomy and Physiology* carries specifically selected topical entries that directly impact topics in classical anatomy and physiology. For those readers interested in genetics, the editors recommend Gale's *World of Genetics* as an accompanying reference. For those readers interested in specific genetic disorders, the editors highly recommend the *Gale Encyclopedia of Genetic Disorders*.

This first edition of *World of Anatomy and Physiology* has been designed with ready reference in mind:

- **Entries are arranged alphabetically**, rather than by chronology or scientific field. In addition to classical topics, *World of Anatomy and Physiology* contains many articles addressing the impact of advances in anatomy and physiology on history, ethics, and society.
- **Bold-faced terms** direct reader to related entries.
- **"See also" references** at the end of entries alert the reader to related entries not specifically mentioned in the body of the text.
- A **Sources Consulted** section lists the most worthwhile print material and web sites we encountered in the compilation of this volume. It is there for the inspired reader who wants more information on the people and discoveries covered in this volume.
- The **Historical Chronology** includes many of the significant events in the advancement of anatomy and physiology. The most current entries date from just days before *World of Anatomy and Physiology* went to press.
- A **comprehensive General Index** guides the reader to topics and persons mentioned in the book. Bolded page references refer the reader to the term's full entry.

Although there is an important and fundamental link between the composition and shape of biological molecules and their functions in biological systems, a detailed understanding of biochemistry is neither assumed or required for *World of Anatomy and Physiology*. Accordingly, students and other readers should not be intimidated or deterred by the complex names of biochemical molecules (especially the names for particular proteins, enzymes, etc.). Where necessary, sufficient information regarding chemical structure is provided. If desired, more information can easily be obtained from any basic chemistry or biochemistry reference.

Advisory Board

In compiling this edition, we have been fortunate to rely upon the expertise and contributions of the following scholars who served as academic and contributing advisors for *World of Anatomy and Physiology*, and to them we would like to express our sincere appreciation for their efforts to ensure that *World of Anatomy and Physiology* contains the most accurate and timely information possible:

Robert G. Best, Ph.D.
Director, Division of Genetics, Department of Obstetrics and Gynecology
University of South Carolina School of Medicine
Columbia, South Carolina

Antonio Farina, M.D., Ph.D.
Visiting Professor, Department of Pathology and Laboratory Medicine
Brown University School of Medicine
Providence, Rhode Island
and
Department of Embryology, Obstetrics, and Gynecology
University of Bologna
Bologna, Italy

Brian D. Hoyle, Ph.D.
Microbiologist
Nova Scotia, Canada

Eric v.d. Luft, Ph.D., M.L.S.
Curator of Historical Collections
SUNY Upstate Medical University
Syracuse, New York

Danila Morano, M.D.
Department of Embryology, Obstetrics, and Gynecology
University of Bologna
Bologna, Italy,

Judyth Sassoon, Ph.D., ARCS
Department of Biology & Biochemistry
University of Bath
Bath, England

Constance K. Stein, Ph.D.
Director of Cytogenetics, Assistant Director of Molecular Diagnostics
SUNY Upstate Medical University
Syracuse, New York

Acknowledgments

It has been our privilege and honor to work with the following contributing writers and scientists who represent scholarship in anatomy and physiology spanning five continents: Mary Brown; Sherri Chasin Calvo; Bryan Cobb, M.S., *Ph.D. (Department of Genetics, University of Alabama, Birmingham); Sandra Galeotti, M.S.; Larry Gillman, Ph.D.; Brook Hall, Ph.D.; Nicole LeBrasseur, Ph.D.; Adrienne Wilmoth Lerner, M.A.* (Graduate Student, Department of History, Vanderbilt University); Agnieszka Lichanska, Ph.D.; Jill Liske, M.Ed.; Kelli Miller; Lissa Rotundo; Tabitha Sparks, Ph.D.; Susan Thorpe-Vargas, Ph.D.; and David Tulloch, Ph.D.

(*) Anticipated by date of publication

Many of the academic advisors for *World of Anatomy and Physiology*, along with others, authored specially commissioned articles within their field of expertise. The editors would like to specifically acknowledge the following special contributions:

Antonio Farina, M.D., Ph.D.
Ethical issues in embryological research

Danila Morano, M.D.
Pregnancy, maternal physiological and anatomical changes

Judyth Sassoon, Ph.D.
Biochemistry

Constance K. Stein, Ph.D.
Stem cells

The editors would like to extend special thanks Dr. Judyth Sassoon for her contributions to the introduction to *World of Anatomy and Physiology*. The editors also wish to acknowledge Dr. Eric v.d. Luft for his diligent and extensive research related to his compilation of the "History of anatomy and physiology" series of articles.

The editors gratefully acknowledge the assistance of Ms. Robyn Young and the Gale Imaging and Multimedia team for their guidance through the complexities and difficulties related to graphics. Last, but certainly not least, the editors thank Ms. Kimberley McGrath, whose continued wit and guidance kept *World of Anatomy and Physiology* a labor of love.

The editors lovingly dedicate this book to Mary Josephine Wilmoth.

K

KANDEL, ERIC R. (1929-)

Austrian-born American neurobiologist

Eric Kandel received the Nobel Prize in physiology or medicine in 2000 for his research on the neurological mechanisms of memory. With colleagues **Arvid Carlsson** and **Paul Greengard**, Kandel discovered the connections between varying synaptic functions and the physiology of learning and memory.

Born in Vienna, Austria, Kandel immigrated to the United States in 1939. In college, Kandel majored in literature and history. Opting for a career in the sciences, Kandel then studied medicine at New York University, earning his degree in 1956. After post-doctorial work in both the United States and France, Kandel held professorships at Harvard and New York University before moving his research to Columbia University. There he embarked on his prize-winning research and was founding director of the Center for Neurobiology and Behavior.

Over the course of two decades, Kandel researched **brain** function and learning on the cellular and molecular level. He centered his studies on the mechanisms of learning by observing the manner in which parts of the brain learn from, and adapt and respond to certain stimuli. Using the three basic modes of learning (sensitization, habituation, and conditioning), Kandel discovered that the brain reacted to the stimuli and showed evidence of actual physical change, as well manifesting behavioral change. Kandel identified that stimuli created changes in behavior by altering neural connections.

Furthering the idea of the connection between synaptic change and learning. Kandel endeavored to identify physiological differences between the creating of long-term and short-term memory. In his experiments on sea slugs, Kandel discovered that the brain's synapses produce different reactions to various stimuli. The nature of these chemical reactions dictates whether the experience is stored in short-term or long-term memory. In other words, varying levels of learning and memory are the result of subtlety different processes in the brain. While at first glance sea slugs may seem too distant a specimen to bear relevance on human medicine, such studies are foundational to developing research **Alzheimer's disease**, Parkinson's disease, and traumatic head injuries. Increased understandings of the specific functions of **neurotransmitters**, the chemical processes of synapses and their role in memory, is the essential cornerstone for pharmacological and surgical treatment of neurological disorders and injuries.

See also Brain, intellectual functions

KATZ, BERNARD (1911-)

German-born English physician

Renowned as a skilled experimentalist at a young age, Bernard Katz's research concerned the nature of nerve transmissions, especially those that cause the stimulation of muscles. He discovered the existence of tiny packages, or "quanta" of neurotransmitter molecules that are responsible for many neural phenomena. For this discovery, he was awarded a share of the 1970 Nobel Prize in physiology or medicine with biochemical pharmacologist **Julius Axelrod** and physiologist **Ulf Euler**.

Katz was born in Leipzig, Germany, in 1911. His father was Max Katz and his mother, the former Eugenie Rabinowitz. He completed his high school education at Leipzig's Albert Gymnasium in 1929 and then embarked on a study of medicine at the University of Leipzig. In 1933, Katz was awarded the university's Siegfried Garten Prize for research in physiology. A year later he received his medical degree.

Katz's future prospects in Germany in 1934 were not promising. Adolf Hitler had been named chancellor of the German Reich a year earlier and had already begun his purge of Jewish intellectuals. Katz decided to leave his native land, and was able to obtain a postdoctoral fellowship at the

University of London. There he continued his physiological research under physiologist Archibald Vivian Hill, Nobel Prize winner in physiology or medicine in 1922 for his discovery of the thermodynamics of muscle activity.

As Katz was awarded his Ph.D. in physiology in 1938, the outbreak of war in Europe was imminent. He left England for Australia where he became Beit Memorial Research Fellow at Sydney Hospital. In 1942, the war in the Pacific had also broken out, and Katz joined the Royal Australian Air Force, where he served as a radar officer. At the war's conclusion, he returned to London and joined the staff at University College, London. He was appointed assistant director of research in the Biophysics Research Unit and Henry Head Research Fellow in 1946. Four years later he became a Reader in Physiology and, in 1952, was chosen to be head of biophysics, a post he held until his retirement in 1978.

Katz first studied the nature of nerve transmissions in the early 1940s, in conjunction with physiologist John C. Eccles and neurophysiologist Stephen Kuffler. The problem on which this team worked was the mechanism by which **neurons** (nerve cells) stimulate muscle cells. At the time, it was still not clear whether this process was purely electrical or whether it involved a chemical component.

At University College in 1950, Katz began the work for which he was later awarded a Nobel Prize, the study of electrical and chemical changes at the end of a neuron. Some years earlier, physiologists and pharmacologists Henry Hallett Dale and **Otto Loewi** had demonstrated that neurons release the chemical acetylcholine at their terminal end. Acetylcholine eventually was recognized as the first of a group of chemical compounds known as **neurotransmitters**, a name that identifies their function of transmitting neural messages across the synapse (the point at which a nervous impulse passes from one neuron to another or from a nerve to a muscle or other **tissue**) between cells. In the early 1950s, Katz made two important discoveries about the release of acetylcholine by neurons. First, he found that this release occurs continuously and spontaneously, even if the neuron is at rest, although at much lower levels than during excitation. Second, he discovered that acetylcholine is released in discrete particles that contain a few hundred or few thousand molecules of the chemical, but whose size is always some integral multiple of the smallest package observed. These "packets" of neurotransmitters are, then, similar to quanta of excitatory units. Some years later, these packets—now called vesicles—were actually observed within neurons by means of electron microscopy.

Working with a variety of associates, Katz continued to study the nature of nerve transmission within a neuron and between neurons. In 1967, for example, he and Ricardo Miledi found that the release of acetylcholine is mediated by a **calcium** ion. Two years later, he and his colleagues were able to show that the electrical potential at the terminal of an axon (the long tubular extension of the neuron cell body that transmits nerve impulses away from the cell body) can be precisely and quantitatively associated with the electrochemical potential of the number of acetylcholine molecules found within a vesicle. It was for these discoveries and his work on "the humoral transmitters in the nerve terminals and the mecha-

nism for their storage, release, and inactivation" that he received a share of the 1970 Nobel Prize in physiology or medicine.

In addition to the Nobel Prize, Katz has been awarded the Baly Medal of the Royal College of Physicians and the Copley Medal of the Royal Society, both in 1967. He has also served as Herter Lecturer at Johns Hopkins University in 1958, Dunham Lecturer at Harvard in 1961, and Croonian Lecturer at the Royal Society, also in 1961. He was knighted by Queen Elizabeth II in 1969. Katz married Marguerite Penly in 1945. The couple have two sons.

See also Nerve impulses and conduction of impulses; Nervous system overview

KENDALL, EDWARD C. (1886-1972)
American biochemist

Edward C. Kendall is best remembered as a pioneer in the discovery and isolation of several important **hormones**. As a young scientist he isolated the hormone thyroxine from the thyroid glands of cattle; today, thyroxine is produced synthetically and used in the treatment of thyroid disorders. Later, he isolated six hormones produced by the adrenal cortex. One of these was cortisone, which proved to be a breakthrough in the treatment of rheumatoid arthritis. Kendall's work led to the 1950 Nobel Prize in medicine and physiology, which he shared with colleagues Philip S. Hench and **Tadeus Reichstein**.

Edward Calvin Kendall was born in South Norwalk, Connecticut, the youngest of three children. His father, George Stanley Kendall, was a dentist, and his mother was Eva Frances (Abbott). Kendall showed a curious nature early on, and when he entered Columbia University in 1904 he chose chemistry as his primary area of study. He earned his bachelor of science degree in 1908, his master's degree in 1909, and his Ph.D. in chemistry in 1910—all from Columbia.

Upon his graduation, he accepted a position with the pharmaceutical firm Parke, Davis, and Company in Detroit. After four months at Parke, Davis, Kendall left and returned to New York. He soon found a position at St. Luke's Hospital. It was at St. Luke's where Kendall began his work on isolating thyroxine. Nearly twenty years earlier, the German chemist Eugen Baumann had discovered high concentrations of iodine in the thyroid gland. Scientists were later able to obtain a protein called thyroglobulin; Kendall's aim was to isolate the active compound in this protein.

He was able to purify the protein, and early experiments with patients at St. Luke's proved successful. But while the physicians at St. Luke's were eager to find new ways to treat patients, their emphasis on actual research was not as strong as Kendall thought it should be. He left St. Luke's at the end of 1913, and headed west to the Mayo Clinic, where over the next four decades he did his most important work.

By the end of 1914, Kendall had isolated thyroxine. This breakthrough discovery eventually led to synthetic production of the substance, which in turn led to more effective treatment of thyroid disorders. For his work he was awarded the Chandler Prize by his alma mater in 1925. Kendall also isolated the peptide glutathione from yeast and determined its structure.

Now Kendall was ready to tackle the challenge of isolating hormones from the adrenal gland. During the 1930s, he managed to isolate more than two dozen hormones, or corticoids (so called because they came from the cortex, or outer section, of the gland). The six most important hormones were each assigned a letter A through F. Compound E—cortisone—turned out to be the most significant of these.

Compound E was not easy to synthesize. Kendall worked for several years with a substance obtained from cattle **bile** and was finally successful in producing a small amount of the compound late in 1946. Kendall's research got a boost from the United States government, which gave top medical priority to the investigation of cortisone during World War II. This was prompted in part by rumors (later proven untrue) that German scientists had been extracting adrenal gland extract from Argentine cattle and giving it to Nazi pilots to boost their strength (they were supposed to be able to fly planes at heights up to 40,000 ft. [12,192 m]). The U.S. Office of Scientific Research and Development (with which Kendall served as a civilian during the war) gave him support, and the pharmaceutical firm Merck and Company sent a scientist to help him complete the synthesis.

Research by Kendall's colleague Hench showed that cortisone might be useful in the treatment of rheumatoid arthritis. Actual experimentation with patients began in 1948, and the results were dramatic. Rheumatoid arthritis is a painful condition that causes severe **pain** and swelling in the **joints**; cortisone, though not a cure, was able to control the symptoms. It also controlled symptoms in some skin diseases and **eye** disorders. Reichstein, working independently of Kendall and Hench, also synthesized cortisone in Switzerland. It was for their work and research with cortisone that the three men were awarded the Nobel Prize. Kendall was also awarded several honorary degrees, including one from Columbia.

In 1951, Kendall accepted a position as visiting professor at Princeton University, where he remained for the rest of his life. Among his other awards were the American Public Health Association's Lasker Award and the American Medical Association's Scientific Achievement Award. He was a member of several organizations, and in addition to his book *Thyroxine* and his memoirs, he wrote articles for numerous scientific publications. He served as president of the American Society of Biological Chemists from 1925 to 1926 and the Endocrine Society from 1930 to 1931.

Kendall married Rebecca Kennedy in 1915 and eventually had three sons and a daughter. Kendall died in New Jersey in 1972, and was buried in Rochester, Minnesota, home of the Mayo Clinic.

See also Adrenal glands and hormones; Thyroid histophysiology and hormones

KHORANA, HAR GOBIND (1922-   )
Indian-born American biochemist

Har Gobind Khorana, an organic chemist who specialized in the study of proteins and nucleic acids, shared the Nobel Prize in physiology or medicine with **Robert W. Holley** and **Marshall W. Nirenberg** in 1968 for discoveries related to the **genetic code** and its function in **protein synthesis**. In addition to developing methods for investigating the structure of the nucleic acids, Khorana introduced many of the techniques that allowed scientists to decipher the genetic code and show how **ribonucleic acid (RNA)** can specify the structure of proteins. Four years after winning the Nobel Prize, Khorana succeeded in synthesizing the first wholly artificial gene. In the 1980s Khorana synthesized the gene for rhodopsin, a protein involved in vision.

Har Gobind Khorana, youngest of the five children of Shri Ganput Rai Khorana and Shrimat Krishna Devi Khorana, was born in Raipur, in the Punjab region of India (now part of West Pakistan). His birthdate was recorded as January 9, 1922, but the exact date of his birth is uncertain. Although his family was poor, his parents believed strongly in the importance of education. His father was a village agricultural taxation clerk in the British colonial government. Khorana attended D.A.V. High School in Multan (now West Punjab). After receiving his Bachelor of Science (1943, with honors) and Master's degree (1945, with honors) from Punjab University in Lahore, India, Khorana was awarded a Government of India Fellowship, which enabled him to study at Liverpool University, England, where he earned his Ph.D. in 1948. From 1948 to 1949, he worked as a postdoctoral fellow at the Federal Institute of Technology, Zurich, Switzerland, with Professor Vladimir Prelog, who had a major influence on his life-long approach to science.

After briefly returning to India, Khorana accepted a position in the laboratory of (Lord) Alexander Todd at Cambridge University (1950–52), where he studied proteins and nucleic acids. From 1952 to 1960, Khorana worked in the organic chemistry section of the British Columbia Research Council, Vancouver, Canada. The next year Khorana moved to the University of Wisconsin, Madison, Wisconsin, where he served as Co-director of the Institute for Enzyme Research and Professor of Biochemistry. In 1964, he became the Conrad A. Elvehjem Professor of the Life Sciences. In 1970, Khorana accepted the position of Alfred P. Sloan Professor, Departments of Biology and Chemistry, at the Massachusetts Institute of Technology, Cambridge, Massachusetts. From 1974 to 1980, he was Andrew D. White Professor-at-large, Cornell University, Ithaca, New York. During his long and distinguished career, Khorana has been the author or co-author of over 500 scientific publications.

In 1953, Khorana and Todd published their only co-authored paper; it described the use of a novel phosphorylating reagent. Khorana found that this reagent was very useful in overcoming problems in the synthesis of polynucleotides. Between 1956 and 1958, Khorana and his coworkers established the fundamental techniques of nucleotide chemistry.

Their goal was to develop purely chemical methods of synthesizing oligonucleotides (long chains of nucleotides). In 1961, Khorana synthesized Coenzyme A, a factor needed for the activity of certain key metabolic **enzymes**.

In 1955, Khorana learned about Severo Ochoa's discovery of the enzyme polynucleotide phosphorylase and met **Arthur Kornberg**, who described pioneering research on the enzymatic synthesis of **DNA**. These discoveries revolutionized nucleic acid research and made it possible to elucidate the genetic code. Khorana and his coworkers synthesized each of the 64 possible triplets (codons) by synthesizing polynucleotides of known composition. Khorana also devised the methods that led to the synthesis of large, well-defined nucleic acids.

By combining synthetic and enzymatic methods, Khorana was able to overcome many obstacles to the chemical synthesis of polyribonucleotides. Khorana's work provided unequivocal proof of codon assignments and defined some codons that had not been determined by other methods. Some triplets, which did not seem to code for any particular amino acid, were shown to serve as "punctuation marks" for beginning and ending the synthesis of polypeptide chains (long chains of **amino acids**). Khorana's investigations also provided direct evidence concerning other characteristics of the genetic code. For example, Khorana's work proved that three nucleotides specify an amino acid, provided proof of the direction in which the information in messenger RNA is read, demonstrated that punctuation between codons is unnecessary, and that the codons did not overlap. Moreover, construction of specific polyribonucleotides proved that an RNA intermediary is involved in translating the sequence of nucleotides in DNA into the sequence of amino acids in a protein. Summarizing the remarkable progress that had been made up to 1968 in polynucleotide synthesis and understanding the genetic code, Khorana remarked that the nature of the genetic code was fairly well established, at least for *Escherichia coli*.

Once the genetic code had been elucidated, Khorana focused on gene structure-gene function relationships and studies of DNA-protein interactions. In order to understand gene expression, Khorana turned to DNA synthesis and sequencing. Recognizing the importance of the class of ribonucleotides known as transfer RNAs (tRNAs), Khorana decided to synthesize the DNA sequence that coded for alanine tRNA. The nucleotide sequence of this tRNA had been determined in Robert Holley's laboratory. In 1970, when Khorana announced the total synthesis of the first wholly artificial gene, his achievement was honored as a major landmark in **molecular biology**. Six years later, Khorana and his associates synthesized another gene. In the 1980s, Khorana carried out studies of the chemistry and molecular biology of the gene for rhodopsin, a protein involved in vision.

In 1966, Khorana was elected to the National Academy of Sciences. His many honors and awards include the Merck Award from the Chemical Institute of Canada, the Dannie-Heinneman Preiz, the American Chemical Society Award for Creative Work in Synthetic Organic Chemistry, the Lasker Foundation Award for Basic Medical Research, the Padma Vibhushan Presidential Award, the Ellis Island Medal of Honor, the National Medal of Science, and the Paul Kayser

International Award of Merit in Retina Research. He holds Honorary Degrees for numerous universities, including Simon Fraser University, Vancouver, Canada; University of Liverpool, England; University of Punjab, India; University of Delhi, India; Calcutta University, India; University of Chicago; and University of British Columbia, Vancouver, Canada.

KIDNEYS

Human kidneys derive their name from their unique shape. The bi-lateral organ structures are kidney bean-shaped, and are located at the back of the abdominal cavity (retroperitoneal). In most adults they are small—about the size of the average adult fist—but this size is not indicative of their importance. The kidney's role in the body's chemistry and operation is absolutely vital to life.

An estimated 1,000 milliliters (ml) of fluid flows through the kidneys each minute. The water that is reconditioned by the kidneys is responsible for maintaining the proper balance of acid and base and of ions (such as sodium and potassium) in cellular and **tissue** fluids and in **blood plasma**. Kidneys also keep the volume of water in the body constant, help regulate blood pressure, and stimulate the production of red blood cells via hormone signals. Abnormalities in the ionic balances can cause functional disruption to **organs**, specialized physiological functions, and illness.

Blood enters the kidney through the renal artery, is processed, and exits via the renal vein. The removed waste flow down channels called the urethers to the bladder for storage. When the bladder is full, the collected urine is excreted from the body through the urethra (which, in males, empties via the penis).

The kidney consists of four regions; the renal capsule (a thin, protective, outer membrane), cortex (a light colored outer region), medulla (a darker, inner region), and the renal pelvis (a flat, funnel-shaped cavity that collect the urine into the urethers). Within the cortex and medulla are millions of tiny structures called nephrons. These are the basic units of operation of the kidney. A nephron is constructed of a long thin tube (collecting duct), closed at one end (Bowman's capsule), that has two twisted regions (proximal and distal tubules) at either end of a hairpin loop (Loop of Henle). The nephron is surrounded by **capillaries**.

The filtration of fluids takes place in the nephrons. The fluids are passed under pressure though he walls of the capillaries and Bowman's capsule. Once insider the nephron, small molecules such as ions, glucose and **amino acids** are selectively reabsorbed by specialized proteins, called transporters, in an active energy-demanding process. Water reabsorption occurs passively. The molecules and water are routed back into circulation.

The kidney can adjust the pace of the filtering activity in relation to the volume of ingested water based on the level of a hormone called anti-diuretic hormone. The greater this hormone's concentration, the more water is able to pass through the kidney.

Malfunction of the kidney is serious. Each year in the United States more than three million people have some type of kidney condition or renal insufficiency caused by an **infection**, kidney stones, or **cancer**. Over 300,000 people suffer from kidney failure each year, necessitating kidney dialysis (where fluid is mechanically removed from the body, filtered to remove waste, and put back into the body) or removal and replacement of the defective kidney.

See also Electrolytes and electrolytes balance; Elimination of waste; Glomerular filtration

KINESTHETICS AND KINESTHETIC SENSATIONS

Kinesthetics and kinesthetic sensations refer to the awareness of body position, movement, and **equilibrium** by receptors located in muscles, **tendons**, **joints**, and the internal **ear**. The receptors for kinesthetic sensations are special sensory receptors called proprioceptors. The ability to perceive kinesthetic sensations occurs because proprioceptors carry signals to the **central nervous system** where they are processed in the somatosensory region of the **cerebral cortex**.

Examples of proprioceptors include neuromuscular spindles, golgi tendon **organs**, joint kinesthetic receptors, and **hair** cells of the vestibular apparatus in the internal ear. Neuromuscular spindles are found in all voluntary **skeletal muscle**. However, a higher density of neuromuscular spindles are present in those muscles performing fine motor skills, such as the hands, compared to muscles that perform crude movements like those of the back. Neuromuscular spindles are located within normal muscle fibers called extrafusal muscle fibers. The neuromuscular spindles comprise 2–10 specialized intrafusal muscle fibers surrounded by a capsule of connective **tissue**. Intrafusal muscle fibers are so named because they are located within a fusiform-shaped capsule, while the extrafusal muscle fibers are located outside of the capsule. When the length of skeletal muscles change during contraction or stretch, the neuromuscular spindles fire sensory impulses to the **brain** where they are interpreted. Additionally, neuromuscular spindles monitor the rate of change in muscle length. Not only do neuromuscular spindles consist of sensory receptors, they also contain the motor **neurons** necessary for **muscle contraction**. This enables us to detect the orientation of our body in relation to other body parts as well as the environment around us and adjust our movements if we so choose.

Golgi tendon organs are located where the tendon intersects the muscle. Tendon organs consist of branching myelinated sensory nerve fibers surrounded by connective tissue. Whether the muscle is stretched or contracted, golgi tendon organs are able to perceive the force and tension produced in the muscle. Not only does this mechanism assist in kinesthetic sensation, it also acts as a protective mechanism by causing

muscles to relax so that they are not damaged from excessive tension.

Joint kinesthetic receptors respond to the pressure induced by movement of **synovial joints**, enabling us to determine the change in position of a joint. Synovial joints are surrounded by articular capsules where the joint kinesthetic receptors can be found. The joint kinesthetic receptors consist of several types of sensory receptors including Pacinian corpuscles, Ruffini corpuscles, free nerve endings, and golgi tendon organs.

The vestibular apparatus of the internal ear is comprised of receptor organs that function in equilibrium by providing awareness of orientation of the body in space. The receptor organs include the semicircular canals and the otolith organs; both function by hair cell receptors to assist in kinesthetic sensation.

See also Muscular innervation; Joints and synovial membranes

KOCH, ROBERT (1843-1910)

German physician

Robert Koch is considered to be one of the founders of the field of bacteriology. He pioneered principles and techniques in studying **bacteria** and discovered the specific agents that cause tuberculosis, cholera, and anthrax. For this he is also regarded as a founder of public health, aiding legislation and changing prevailing attitudes about hygiene to prevent the spread of various infectious diseases. For his work on tuberculosis, he was awarded the Nobel Prize in 1905.

Robert Heinrich Hermann Koch was born in a small town near Klausthal, Hanover, Germany, to Hermann Koch, an administrator in the local mines, and Mathilde Julie Henriette Biewend, a daughter of a mine inspector. The Kochs had a total of thirteen children, two of whom died in infancy. Robert was the third son. Both parents were industrious and ambitious. Robert's father rose in the ranks of the mining industry, becoming the overseer of all the local mines. His mother passed her love of nature on to Robert who, at an early age, collected various plants and insects.

Before starting primary school in 1848, Robert taught himself to read and write. At the top of his class during his early school years, he had to repeat his final year. Nevertheless, he graduated in 1862 with good marks in the sciences and mathematics. A university education became available to Robert when his father was once again promoted and the family's finances improved. Robert decided to study natural sciences at Gottingen University, close to his home.

After two semesters, Koch transferred his field of study to medicine. He had dreams of becoming a physician on a ship. His father had traveled widely in Europe and passed a desire for travel on to his son. Although bacteriology was not taught then at the University, Koch would later credit his interest in that field to Jacob Henle, an anatomist who had published a theory of contagion in 1840. Many ideas about contagious diseases, particularly those of chemist and microbiologist Louis Pasteur, who was challenging the prevailing myth of spontaneous generation, were still being debated in universities in the 1860s.

During Koch's fifth semester at medical school, Henle recruited him to participate in a research project on the structure of uterine nerves. The resulting essay won first prize. It was dedicated to his father and bore the Latin motto, *Nunquam Otiosus*, meaning never idle. During his sixth semester, he assisted Georg Meissner at the Physiological Institute. There he studied the secretion of succinic acid in animals fed only on **fat**. Koch decided to experiment on himself, eating a half pound of butter each day. After five days, however, he was so sick that he limited his study to animals. The findings of this study eventually became Koch's dissertation. In January 1866, he finished the final exams for medical school and graduated with highest distinction.

After finishing medical school, Koch held various positions; he worked as an assistant at a hospital in Hamburg, where he became familiar with cholera, and also as an assistant at a hospital for developmentally delayed children. In addition, he made several attempts to establish a private practice. In July, 1867, he married Emmy Adolfine Josephine Fraatz, a daughter of an official in his hometown. Their only child, a daughter, was born in 1868. Koch finally succeeded in establishing a practice in the small town of Rakwitz where he settled with his family.

Shortly after moving to Rakwitz, the Franco-Prussian War broke out and Koch volunteered as a field hospital physician. In 1871, the citizens of Rakwitz petitioned Koch to return to their town. He responded, leaving the army to resume his practice, but he didn't stay long. He soon took the exams to qualify for district medical officer and in August 1872 was appointed to a vacant position at Wollstein, a small town near the Polish border.

It was here that Koch's ambitions were finally able to flourish. Though he continued to see patients, Koch converted part of his office into a laboratory. He obtained a microscope and observed, at close range, the diseases his patients confronted him with.

One such disease was anthrax, which is spread from animals to humans through contaminated wool, by eating uncooked meat, or by **breathing** in airborne spores emanating from contaminated products. Koch examined under the microscope the **blood** of infected sheep and saw specific microorganisms that confirmed a thesis put forth ten years earlier by biologist C. J. Davaine that anthrax was caused by a bacillus. But Koch was not content to simply verify the work of another. He attempted to culture, or grow, these bacilli in cattle blood so he could observe their life cycle, including their formation into spores and their germination. Koch performed scrupulous research both in vitro and in animals before showing his work to Ferdinand Cohn, a botanist at the University of Breslau. Cohn was impressed with the work and replicated the findings in his own laboratory. He published Koch's paper in 1876.

In 1877, Koch published another paper that elucidated the techniques he had used to isolate *Bacillus anthracis*. He

had dry-fixed bacterial cultures onto glass slides, then stained the cultures with dyes to better observe them, and photographed them through the microscope.

It was only a matter of time that Koch's research eclipsed his practice. In 1880, he accepted an appointment as a government advisor with the Imperial Department of Health in Berlin. His task was to develop methods of isolating and cultivating disease-producing bacteria and to formulate strategies for preventing their spread. In 1881 he published a report advocating the importance of pure cultures in isolating disease-causing organisms and describing in detail how to obtain them. The methods and theory espoused in this paper are still considered fundamental to the field of modern bacteriology. Four basic criteria, now known as Koch's postulates, are essential for an organism to be identified as pathogenic, or disease-causing. First, the organism must be found in the tissues of animals with the disease and not in disease-free animals. Second, the organism must be isolated from the diseased animal and grown in a pure culture outside the body, or in vitro. Third, the cultured organism must be able to be transferred to a healthy animal, who will subsequently show signs of **infection**. And fourth, the organisms must be able to be isolated from the infected animal.

While in Berlin, Koch became interested in tuberculosis, which he was convinced was infectious, and, therefore, caused by a bacterium. Several scientists had made similar claims but none had been verified. Many other scientists persisted in believing that tuberculosis was an inherited disease. In six months, Koch succeeded in isolating a bacillus from tissues of humans and animals infected with tuberculosis. In 1882, he published a paper declaring that this bacillus met his four conditions—that is, it was isolated from diseased animals, it was grown in a pure culture, it was transferred to a healthy animal who then developed the disease, and it was isolated from the animal infected by the cultured organism. When he presented his findings before the Physiological Society in Berlin on March 24, he held the audience spellbound, so logical and thorough was his delivery of this important finding. This day has come to be known as the day modern bacteriology was born.

In 1883, Koch's work on tuberculosis was interrupted by the Hygiene Exhibition in Berlin, which, as part of his duties with the health department, he helped organize. Later that year, he finally realized his dreams of travel when he was invited to head a delegation to Egypt where an outbreak of cholera had occurred. Louis Pasteur had hypothesized that cholera was caused by a microorganism; within three weeks, Koch had identified a comma-shaped organism in the intestines of people who had died of cholera. However, when testing this organism against his four postulates, he found that the disease did not spread when injected into other animals. Undeterred, Koch proceeded to India where cholera was also a growing problem. There, he succeeded in finding the same organism in the intestines of the victims of cholera, and although he was still unable to induce the disease in experimental animals, he did identify the bacillus when he examined, under the microscope, water from the ponds used for drinking water. He remained convinced that this bacillus was the cause of cholera and that the key to prevention lay in improving hygiene and sanitation.

Koch returned to Germany and from 1885–1890 was administrator and professor at Berlin University. He was highly praised for his work, though some high-ranking scientists and doctors continued to disagree with his conclusions. But Koch was an adept researcher, able to support each claim with his exacting methodology. In 1890, however, Koch faltered from his usual perfectionism and announced at the International Medical Congress in Berlin that he had found an inoculum that could prevent tuberculosis. He called this agent tuberculin. People flocked to Berlin in hopes of a cure and Koch was persuaded to keep the exact formulation of tuberculin a secret, in order to discourage imitations. Although optimistic reports had come out of the clinical trials Koch had set up, it soon became clear from autopsies that tuberculin was causing severe inflammation in many patients. In January 1891, under pressure from other scientists, Koch finally published the nature of the substance, but it was an uncharacteristically vague and misleading report which came under immediate criticism from his peers.

Koch left Berlin for a time after this incident to recover from the professional setback, although the German government continued to support him throughout this time. An Institute for Infectious Diseases was established and Koch was named director. With a team of researchers, he continued his work with tuberculin, attempting to determine the ideal dose at which the agent could be the safest and most effective. The discovery that tuberculin was a valuable diagnostic tool (causing a reaction in those infected but none in those not infected), rather than a cure, helped restore Koch's reputation. In 1892, there was a cholera outbreak in Hamburg. Thousands of people died. Koch advocated strict sanitary conditions and isolation of those found to be infected with the bacillus. Germany's senior hygienist, Max von Pettenkofer, was unconvinced that the bacillus alone could cause cholera. He sneered at Koch's ideas, going so far as to drink a freshly isolated culture. Several of his colleagues joined him in this demonstration. Two developed symptoms of cholera, Pettenkofer suffered from **diarrhea**, but no one died; Pettenkofer felt vindicated in his opposition to Koch. Nevertheless, Koch focused much of his energy on testing the water supply of Hamburg and Berlin and perfecting techniques for filtering drinking water to prevent the spread of the bacillus.

In the following years, he gave the directorship of the Institute over to one of his students so he could travel again. He went to India, New Guinea, Africa, and Italy, where he studied diseases such as the plague, **malaria**, rabies, and various unexplained fevers. In 1905, after returning to Berlin from Africa, he was awarded the Nobel Prize in physiology or medicine for his work on tuberculosis. Subsequently, many other honors were awarded him recognizing not only his work on tuberculosis, but his more recent research on tropical diseases, including the Prussian Order Pour le Merits in 1906 and the Robert Koch medal in 1908. The Robert Koch Medal was established to honor the greatest living physicians, and the Robert Koch Foundation, established with generous grants from the German government and from the American philan-

thropist, Andrew Carnegie, was founded to work toward the eradication of tuberculosis.

Meanwhile, Koch settled back into the Institute where he supervised clinical trials and production of new tuberculins. He attempted to answer, once and for all, the question of whether tuberculosis in cattle was the same disease as it was in humans. Between 1882 and 1901, he had changed his mind on this question, coming to believe that bovine tuberculosis was not a danger to humans, as he had previously thought. He espoused his beliefs at conferences in the United States and Britain during a time when many governments were attempting large-scale efforts to minimize the transmission of tuberculosis through meat and milk.

Koch did not live to see this question answered. On April 9, 1910, three days after lecturing on tuberculosis at the Berlin Academy of Sciences, he suffered a **heart** attack from which he never fully recovered. He died at Baeden Baeden the next month at the age of 67. He was honored after death by the naming of the Institute after him. In the first paper he wrote on tuberculosis, he stated his lifelong goal, which he clearly achieved: "I have undertaken my investigations in the interests of public health and I hope the greatest benefits will accrue therefrom."

See also Bacteria and responses to bacterial infection

KOCHER, EMIL THEODOR (1841-1917)
Swiss surgeon

In 1870s Switzerland, goiter was a common ailment, usually marked by a glandular swelling on the front of the neck. In later years it would be understood that a simple iodine supplement to the diet could significantly reduce the disorder. But in the nineteenth century, surgical removal of the thyroid gland was the only known cure. However, in the absence of effective anesthetics and antisepsis, surgical attempts to remove a goiter meant almost certain **death** for the patient. This was the challenge faced by Swiss surgeon Theodor Kocher, who devoted his medical career to making thyroidectomy, or the removal of a thyroid gland, a relatively safe procedure by applying new notions of antisepsis. Kocher performed thousands of thyroidectomies in his career, and the post-operative research and data he collected helped amass new knowledge about the **physiology** of the thyroid gland and its related disorders. For his many contributions to medicine, and especially the treatment of goiter, Kocher received the Nobel Prize in medicine in 1909.

Emil Theodor Kocher was born the son of Jacob Alexander and Maria (Wermuth) Kocher, in Bern, Switzerland. His father was an engineer and his mother a descendant of the Moravian Brethren. She passed on to her son a deeply religious philosophy that would help him gain an empathetic understanding of his patients in years to come. Schooled in Berlin, Germany; London, England; Paris, France; and Vienna, Austria, Kocher received his M.D. from the University of Bern in 1869. That same year he married

Marie Witschi-Courant—the couple would have three sons. Newly married and newly graduated from medical school, Kocher visited various European clinics, including one in Vienna, where he studied under the most famous European surgeon of the day, Theodor Billroth. In 1872, Kocher, who was only thirty-one years old at the time, was named professor of clinical surgery at Bern University, a post he would hold for the next forty-five years.

Kocher first gained recognition for developing a method for treating a dislocated shoulder, a technique now known by his name. Subsequently, he also created new methods or improvements in existing methods for operations upon the **lungs**, stomach, gall bladder, intestine, **cranial nerves**, and hernia. He also developed a special pair of surgical forceps, now known as "Kocher's forceps," instruments that were used for many years after his death. Despite his many successes and contributions that improved surgical procedures, Kocher was open to other suggestions and ideas. "It is an indication of his scientific objectivity that he was always ready to abandon any of his own techniques or gadgets in favor of improvements introduced by other surgeons," Theodore L. Sourkes has written in *Nobel Prize Winners in Medicine and Physiology*. The example Sourkes provides is Kocher's ready abandonment of his own style of surgically correcting hernia's in favor of another approach.

Kocher further contributed to medicine with his *Textbook of Operative Surgery* (the book was translated into several languages, including an English edition in 1895), his pioneering of ovariotomy and, especially, his application of the antiseptic techniques of the English researcher and doctor Joseph Lister.

Kocher himself credited his success with thyroidectomy operations in part to Lister's method of antisepsis. He said while receiving his Nobel Prize that it was because of Lister that one of the "most dangerous operations, the removal of the thyroid gland, so often appearing urgently necessary because of severe respiratory disturbances, could be performed without substantial danger." However, despite his mastery over the operation, Kocher himself considered the increased knowledge about the *physiological* function of the thyroid gland an even greater advancement in medical science. In 1883, at the congress of the German Surgical Society, Kocher reported that out of his first 100 thyroidectomies, 30 had resulted in a serious disorder. This ailment was apparently a result of the whole, rather than partial, removal of the goiter. The symptoms Kocher described were called operative myxedema, and were akin to naturally occurring myxedema. Patients suffering from myxedema usually reported weight gain, slowing of intellect and speech, **hair** loss, tongue thickening, and abnormal **heart** rates, as well as developing blood-related problems of anemia and altered white blood-cell counts. Kocher further related that myxedema symptoms were similar to problems experienced by patients suffering from sporadic cretinism and cachexia strumipriva, diseases that resulted in mental retardation and dwarfism. Because of Kocher's postulations, it was discovered that a lack of thyroid secretions was the cause of all these diseases. Kocher further pointed out that hypothyroidism can be traced not only to absence of the gland,

whether congenital or surgical, but also to a goiter which has caused the gland to stop working. His descriptions of the thyroid disorder have clarified and brought together a series of medical observations on this subject over the years.

Kocher's observations also opened the way for future treatment of thyroid disorders. Although initial attempts to rectify the condition by administering thyroid hormone were not particularly successful, researchers recognized the importance of iodine, and in 1914 the effective part of the hormone, thyroxin, was isolated for effective treatment. Meanwhile, Kocher helped perfect surgical technique for thyroidectomy, and his surgical mortality rates dropped by a great margin over the years.

During his long surgical career Kocher performed more than 2,000 thyroidectomies. In time the need for the operation declined as iodine-deprived regions, like the "goiter belt" of the Great Lakes area in the United States and certain parts of Switzerland, incorporated supplements into their diets. Nevertheless, Kocher's contributions to combating endemic goiter continue to be recognized in a world where nearly five percent of the population still continues to suffer this disorder.

Kocher died in Bern eight years after winning the Nobel. While placing a wreath on his tomb, American neurosurgeon Harvey Cushing said in a speech at the First International Neurological Congress in 1931, "From hard work and responsibility surgeons are prone to burn themselves out comparatively young, but Kocher had been blessed with an imperturbility of spirit or had cultivated these habits of self-control which enabled him to bear his professional labors, his years, and his honors with equal composure to the very end."

See also Thyroid histophysiology and hormones

KÖHLER, GEORGES (1946-1995)
German immunologist

For decades, antibodies—substances manufactured by **plasma** cells to help fight disease—were produced artificially by injecting animals with foreign macromolecules, then extracted by bleeding the animals and separating the antiserum in their **blood**. The technique was arduous and far from foolproof. But the discovery of the hybridoma technique by German immunologist Georges Köhler changed revolutionize the procedure. Köhler's work made antibodies relatively easy to produce and dramatically facilitated research on many serious medical disorders, such as the acquired immunodeficiency syndrome (**AIDS**) and **cancer**. For his work on what would come to be known as monoclonal antibodies, Köhler shared the 1984 Nobel Prize in medicine.

Born in Munich, in what was then occupied Germany, on 17 April 1946, Georges Jean Franz Köhler attended the University of Freiburg, where he obtained his Ph.D. in biology in 1974. From there he set off to Cambridge University in England, to work as a postdoctoral fellow for two years at the British Medical Research Council's laboratories. At Cambridge, Köhler worked under Dr. **César Milstein**, an

Argentinean-born researcher with whom Köhler would eventually share the Nobel Prize. At the time, Milstein, who was Köhler's senior by nineteen years, was a distinguished immunologist, and he actively encouraged Köhler in his research interests. Eventually, it was while working in the Cambridge laboratory that Köhler discovered the hybridoma technique.

Dubbed by the *New York Times* as the "guided missiles of biology," antibodies are produced by human plasma cells in response to any threatening and harmful bacterium, virus, or tumor cell. The body forms a specific antibody against each antigen; and César Milstein once told the *New York Times* that the potential number of different **antigens** may reach "well over a million." Therefore, for researchers working to combat diseases like cancer, an understanding of how antibodies could be harnessed for a possible cure is of great interest. And although scientists knew the benefits of producing antibodies, until Köhler and Milstein published their findings, there was no known technique for maintaining the long-term culture of antibody-forming plasma cells.

Köhler's interest in the subject had been aroused years earlier, when he had become intrigued by the work of Dr. Michael Potter of the National Cancer Institute in Bethesda, Maryland. In 1962 Potter had induced myelomas, or plasma-cell **tumors** in mice, and others had discovered how to keep those tumors growing indefinitely in culture. Potter showed that plasma tumor cells were both seemingly immortal and able to create an unlimited number of identical antibodies. The only drawback was that there seemed no way to make the cells produce a certain type of antibody. Because of this, Köhler wanted to initiate a cloning experiment that would fuse plasma cells able to produce the desired antibodies with the "immortal" myeloma cells. With Milstein's blessing, Köhler began his experiment.

"For seven weeks after he had made the hybrid cells," the *New York Times* reported in October, 1984, "Dr. Köhler refrained from testing the outcome of the experiment for fear of likely disappointment. At last, around Christmas 1974, he persuaded his wife," Claudia Köhler, "to come to the windowless basement where he worked to share his anticipated disappointment after the critical test." But disappointment faded when Köhler discovered his test had been a success: Astoundingly, his hybrid cells were making pure antibodies against the test antigen. The result was dubbed monoclonal antibodies. For his contribution to medical science, Köhler—who in 1977 had relocated to Switzerland to do research at the Basel Institute for Immunology—was awarded the Nobel in 1984.

The implications of Köhler's discovery opened new avenues of basic research. In the early 1980s, Köhler's discovery led scientists to identify various lymphocytes, or white blood cells. Among the kinds discovered were the T-4 lymphocytes, the cells destroyed by AIDS. Monoclonal antibodies have also improved tests for hepatitis B and streptococcal infections by providing guidance in selecting appropriate antibiotics, and they have aided in the research on thyroid disorders, lupus, rheumatoid arthritis, and inherited **brain** disorders. More significantly, Köhler's work has led to advances in research that can harness monoclonal antibodies into certain

drugs and toxins that fight cancer, but could also cause damage in their own right. Researchers are also using monoclonal antibodies to identify antigens specific to the surface of cancer cells so as to develop tests to detect the spread of cancerous cells in the body.

Despite the significance of the discovery, which has also resulted in vast amounts of research funds for many research laboratories, for Köhler and Milstein—who never patented their discovery—there was little financial remuneration. Following the award, however, he and Milstein, together with Michael Potter, were named winners of the Lasker Medical Research Award.

In 1985, Köhler moved back to his hometown of Freiburg, Germany, to assume the directorship of the Max Planck Institute for Immune Biology. He died in 1995.

See also Immune system; Immunity

KÖLLIKER, ALBERT VON (1817-1905)
Swiss zoologist

Born as Rudolf Albert von Kölliker, the son of a banker in Zurich, Switzerland, Kölliker attended the Zurich Gymnasium and showed an early interest in botany. As a medical student at the University of Zurich from 1836 to 1839, he studied zoology under Lorenz Oken (1779–1851), **anatomy** under Friedrich Arnold (1803–1890), and botany under Oswald Heer (1809–1883). Both at the gymnasium and in Oken's classes, the future botanist Karl Wilhelm von Nägeli (1817–1891) was his schoolmate.

After one semester at the University of Bonn, Germany, Kölliker spent three semesters at the University of Berlin, studying **comparative anatomy** and **physiology** under **Johannes Müller**, microscopic anatomy under Friedrich Gustav Jakob Henle (1809–1885), and **embryology** under Robert Remak (1815–1865). While studying marine invertebrates on the North Sea coast during the winter of 1840–1841, he discovered that spermatozoa arise from cells specifically equipped to produce them, and that each spermatozoon is a single cell. These findings, published as a book in Berlin and submitted as his dissertation in Zurich, earned him his Ph.D. in 1841. The following year he received his M.D. from the University of Heidelberg with a dissertation on fly larvae.

Shortly after Henle moved to Zurich in 1840, Kölliker became his assistant, and when Henle left for Heidelberg in 1844, Kölliker became associate professor of physiology and comparative anatomy at Zurich. In 1847, he became full professor of physiology and comparative anatomy at the University of Würzburg, Germany. With Carl Theodor Ernst von Siebold (1804–1885), he co-founded *Zeitschrift für wissenschaftliche Zoologie* (Journal of Scientific Zoology) in 1848.

In the 1840s and 1850s, Kölliker contributed significantly to anatomy, **histology**, and physiology, but gradually became more interested in embryology and developmental biology. His research laid much of the groundwork for the science of cytology. In 1844, he published his studies of **cell division** in the eggs of the cephalopod sepia. Martin Barry (1802–1855) and Remak were also investigating cell division at this time, and the results of the three scientists were sometimes at odds. In the 1850s, wary of the doctrine of **Theodor Schwann** (1810–1882) that daughter cells arise freely in the zygote, Kölliker showed that the fertilized ovum is one cell and that its subsequent development is by cell division. In 1861, he published *Entwicklungsgeschichte des Menschen und der höheren Tiere* (History of the Development of Humans and Higher Animals), the first important book on comparative embryology since Aristotle's *On the Generation of Animals*.

Kölliker accepted **evolution** but opposed Charles Darwin (1809–1882) on the issue of whether evolutionary changes appear gradually or abruptly. In *Über die Darwin'sche Schöpfungstheorie* (On the Darwinian Theory of Creation), published in 1864, Kölliker argued that teleology, or Aristotelian final causality, the idea that nature moves toward a pre-established goal, is not supported by empirical science. **Thomas Henry Huxley** defended Darwin against Kölliker, who probably saw more teleology in Darwin than was really there. Kölliker's insistence on sudden evolution prefigured the theories of Hugo de Vries (1848–1935) that mutations occur so unexpectedly that a single generation could bring forth a new species.

In 1841, Kölliker suggested that the cell nucleus might be the locus of inheritance. A meticulous researcher who wanted to see everything first hand, he spent much of the next 40 years investigating this hypothesis, Finally, in 1885, following the work of Wilhelm Roux (1850–1924), he published an article in *Zeitschrift für wissenschaftliche Zoologie* claiming that inherited traits are transmitted within the cell nucleus.

Kölliker retired from teaching in 1897, published his autobiography, *Erinnerungen aus meinem Leben* (Memories from my Life) in 1899, and remained moderately active in research until his death in Würzburg from lung disease.

See also Cell cycle and cell division; Comparative anatomy; Embryology

KORNBERG, ARTHUR (1918-   )

American biochemist

Arthur Kornberg discovered **deoxyribonucleic acid (DNA)** polymerase, a natural chemical tool that scientists could use to make copies of DNA, the giant molecule that carries the genetic information of every living organism. The achievement won Kornberg the 1959 Nobel Prize in physiology or medicine (which he shared with **Severo Ochoa**). Since his discovery, laboratories around the world have used the enzyme to build and study DNA. This has led to a clearer understanding of the biochemical basis of **genetics**, as well as new strategies for treating **cancer** and hereditary diseases.

Kornberg was born in Brooklyn, New York, to Joseph Kornberg and Lena Katz. An exceptional student, he gradu-

ated at age fifteen from Abraham Lincoln High School. Supported by a scholarship, he enrolled in the premedical program at City College of New York, majoring in biology and chemistry. He received his B.S. in 1937 and entered the University of Rochester School of Medicine. It was here that his interest in medical research blossomed and he became intrigued with the study of enzymes—the protein catalysts of chemical reactions. During his medical studies, Kornberg contracted hepatitis, a disease of the **liver** that commonly causes jaundice, a yellowing of the skin. The incident prompted him to write his first scientific paper, "The Occurrence of Jaundice in an Otherwise Normal Medical Student."

Kornberg graduated from Rochester in 1941, and began his internship in the university's affiliated institution, Strong Memorial Hospital. At the outbreak of World War II in 1942, he was briefly commissioned a lieutenant junior grade in the United States Coast Guard and then transferred to the United States Public Health Service. From 1942 to 1945, Kornberg served in the **nutrition** section of the division of physiology at the National Institutes of Health (NIH) in Bethesda, Maryland. He then served as chief of the division's **enzymes** and **metabolism** section from 1947 to 1952.

During his years at NIH, Kornberg was able to take several leaves of absence. He honed his knowledge of enzyme production, as well as isolation and purification techniques, in the laboratories of Severo Ochoa at New York University School of Medicine in 1946, of **Carl Cori** and **Gerty Cori** at the Washington University School of Medicine in St. Louis in 1947, and of H. A. Barker at the University of California at Berkeley in 1951. Kornberg became an authority on the **biochemistry** of enzymes, including the production of coenzymes—the proteins that assist enzymes by transferring chemicals from one group of enzymes to another. While at NIH, he perfected techniques for synthesizing the coenzymes diphosphopyridine nucleotide (DPN) and flavin adenine dinucleotide—two enzymes involved in the production of the energy-rich molecules used by the body.

To synthesize coenzymes, Kornberg used a chemical reaction called a condensation reaction, in which phosphate is eliminated from the molecule used to form the enzymes. He later postulated that this reaction was similar to that by which the body synthesizes DNA. The topic of DNA synthesis was of intense interest among researchers at the time, and it closely paralleled his work with enzymes, since DNA controls the biosynthesis of enzymes in cells.

In 1953, Kornberg became professor of microbiology and chief of the department of microbiology at Washington University School of Medicine in St. Louis. That year was a time of great excitement among researchers studying DNA; **Francis Crick** and **James Watson** at Cambridge University had just discovered the chemical structure of the DNA molecule. At Washington University, Kornberg's group built on the work of Watson and Crick, as well as techniques Ochoa had developed for synthesizing RNA—the decoded form of DNA that directs the production of proteins in cells. Their goal was to produce a giant molecule of artificial DNA.

The first major discovery they made was the chemical catalyst responsible for the synthesis of DNA. They discovered the enzyme in the common intestinal bacterium *Escherichia coli,* and Kornberg called it DNA polymerase. In 1957, Kornberg's group used this enzyme to synthesize DNA molecules. Although the molecules were biologically inactive, this was an important achievement; it proved that this enzyme does catalyze the production of new strands of DNA, and it explained how a single strand of DNA acts as a pattern for the formation of a new strand of nucleotides—the building blocks of DNA.

In 1959, Kornberg and Ochoa shared the Nobel Prize for their "discovery of the mechanisms in the biological synthesis of **ribonucleic acid** and deoxyribonucleic acid." The *New York Times* quoted Nobel Prize recipient Hugo Theorell as saying that Kornberg's research had "clarified many of the problems of regeneration and the continuity of life."

In the same year he received the prize, Kornberg accepted an appointment as professor of biochemistry and chairman of the department of biochemistry at Stanford University. He continued his research on DNA biosynthesis there with Mehran Goulian. The two researchers were determined to synthesize an artificial DNA that was biologicaly active, and they were convinced they could overcome the problems which had obstructed previous efforts.

The major problems Kornberg had encountered in his original attempt to synthesize DNA were twofold: the complexity of the DNA template he was working with, and the presence of contaminating enzymes called nucleases which damaged the growing strand of DNA. At Stanford, Kornberg's group succeeded in purifying DNA polymerase of contaminating enzymes, but the complexity of their DNA template remained an obstacle, until Robert L. Sinsheimer of the California Institute of Technology was able to direct them to a simpler one. He had been working with the genetic core of Phi X174, a virus that infects *Escherichia coli.* The DNA of Phi X174 is a single strand of nitrogenous bases in the form of a ring which, when broken, leaves the DNA without the ability to infect its host.

But if the dilemma of DNA complexity was solved, the solution raised yet another problem. The DNA ring in Phi X174 had to be broken in order to serve as a template. But when the artificial copy of the DNA was synthesized in the test tube, it had to be reformed into a ring in order to acquire infectivity. This next hurdle was overcome by Kornberg's laboratory and other researchers in 1966; they discovered an enzyme called ligase, which closes the ring of DNA. With their new knowledge, Kornberg's group added together the Phi X174 template, four nucleotide subunits of DNA, DNA polymerase, and ligase. The DNA polymerase used the template to build a strand of viral DNA consisting of 6,000 building blocks, and the ligase closed the ring of DNA. The Stanford team then isolated the artificial viral DNA, which represented the infectious, inner core of the virus, and added it to a culture of *Escherichia coli* cells. The DNA infected the cells, commandeering the cellular machinery that uses genes to make proteins. In only minutes, the infected cells had ceased their normal synthetic activity and begun making copies of Phi X174 DNA.

Kornberg and Goulian announced their success during a press conference on December 14, 1967, pointing out that the achievement would help in future studies of genetics, as well as in the search for cures to hereditary diseases and the control of viral infections. In addition, Kornberg noted that the work might help disclose the most basic processes of life itself. The Stanford researcher has continued to study DNA polymerase to further understanding of the structure of that enzyme and how it works.

Kornberg has used his status as a Nobel Laureate on behalf of various causes. On April 21, 1975, he joined eleven speakers before the Health and Environment Subcommittee of the House Commerce Committee to testify against proposed budget cuts at NIH, including ceilings on salaries and the numbers of personnel. The witnesses also spoke out against the tendency of the federal government to direct NIH to pursue short-term projects at the expense of long-term, fundamental research. During his own testimony, Kornberg argued that NIH scientists and scientists trained or supported by NIH funding "had dominated the medical literature for twenty-five years." His efforts helped prevent NIH from being ravaged by budget cuts and overly influenced by politics.

Later that year, Kornberg also joined other Nobel Prize winners in support of Andrei Sakharov, the Soviet advocate of democratization and human rights who had been denied permission to travel to Sweden to accept the Nobel Prize in physics. Kornberg was among thirty-three Nobel Prize winners to send a cable to Soviet President Nikolai V. Podgorny, asking him to permit Sakharov to receive the prize.

Kornberg received the Paul-Lewis Laboratories Award in Enzyme Chemistry from the American Chemical Society, 1951, the Scientific Achievement Award of the American Medical Association, 1968, the Lucy Wortham James Award of the Society of Medical Oncology, 1968, the Borden Award in the Medical Sciences of the Association of American Medical Colleges, 1968, and the National Medal of Science, 1980. He is a member of the National Academy of Sciences, the American Academy of Arts and Sciences, the American Society of Biological Scientists, and a foreign member of the Royal Society of London. In addition, he is a member of the American Philosophical Society and, from 1965 to 1966, served as president of the American Society of Biological Chemists. Kornberg has been married to Sylvy Ruth Levy Kornberg since 1943. His wife, who is also a biochemist, has collaborated on much of his work. They have three sons.

See also Genetic code; Genetic regulation of eukaryotic cells; Genetics and developmental genetics; Molecular biology

Krebs, Edwin G. (1918-)
American biochemist

In the 1950s, Edwin G. Krebs and his longtime associate **Edmond Fischer** discovered reversible protein phosphorylation, a fundamental biological mechanism. Together Krebs and Fisher's work illuminates the basic processes that regulate many vital aspects of cell activity, such as **protein synthesis**, cell **metabolism**, and hormonal responses to stress. Medical application of their discoveries has helped in research on

Alzheimer's disease, organ transplants, and certain kinds of **cancer**, and in 1992 the two scientists shared the Nobel Prize in **physiology** or medicine. In addition to his contributions in the field of biochemical research, Krebs has also been recognized for his teaching and administrative abilities.

Edwin Gerhard Krebs was born to William Carl Krebs and Louisa Helena Stegeman Krebs in Lansing, Iowa. He was the third of four children. His father, a Presbyterian minister, died while Krebs was in his first year of high school. In order to keep Krebs's two older brothers enrolled at the University of Illinois in Urbana, Louisa Krebs moved the family from Greenville, where Edwin Krebs grew up, to the university town. There she rented a house big enough for her family, with extra rooms to rent out to students, keeping the family together and helping the children continue their education. She had been a teacher herself before her marriage.

In 1940, after completing his high-school and undergraduate work in Urbana, Krebs entered medical school at Washington University School of Medicine in St. Louis, Missouri. To Krebs's way of thinking, medicine had the advantage of being directly related to people. His general interest in science he attributed to concerns about economic security. At Washington University he received classical medical training and was also introduced to medical research. He had the opportunity to work under Arda A. Green, who was associated with **Carl Ferdinand Cori** and **Gerty T. Cori**. The Coris were a husband-and-wife team who had won the Nobel Prize in 1947 for research on carbohydrate metabolism and the enzyme phosphorylase. Krebs's later collaboration with Fischer at the University of Washington in Seattle had its beginning in the research conducted by the Coris.

After receiving his medical degree in 1943 and completing an eight-month residency in internal medicine at Barnes Hospital in St. Louis, Krebs became a medical officer in the navy, serving in that capacity until 1946. This was the only period in his career during which Krebs was a practicing physician. Due to the unavailability of a resident position, and on the advice of one of his professors, Krebs now began studying science. Because of his background in chemistry, Krebs chose to work in **biochemistry** and was accepted by the Coris as a postdoctoral fellow in their laboratory. For two years, while working for the Coris, Krebs studied the interaction of protamine (a basic protein) with rabbit muscle phosphorylase. This work seemed so rewarding to him that he decided to continue his efforts in the field of research, and when in 1948 he was invited by Hans Neurath to join the faculty in the department of biochemistry at the University of Washington, he jumped at the opportunity to become assistant professor.

At this time Neurath's department greatly emphasized protein chemistry and enzymology (**enzymes** are proteins that act as catalysts in biochemical reactions). Work in the Coris' laboratory had established that the enzyme phosphorylase existed in active and inactive forms, but what controlled its activity was unknown. Combining his experience on mammalian **skeletal muscle** phosphorylase with Edmond Fischer's experience with potato phosphorylase after Fischer joined the department, Krebs and Fischer teamed up to uncover the molecular mechanism by which phosphorylase makes energy available to a contracting muscle. What they discovered was reversible protein phosphorylation. An enzyme called protein kinase takes phosphate from **adenosine triphosphate (ATP)**, the supplier of energy to cells, and adds it to inactive phosphorylase, changing the shape of the phosphorylase and consequently switching it on. Another enzyme, called protein phosphatase, reverses this process by removing the phosphate from phosphorylase, thus deactivating it. Protein kinases are present in all cells.

Once it became evident that reversible protein phosphorylation was a general process, the impact of Krebs and Fischer's work was immeasurable. Their collaboration opened the field of biochemical research and paved the way to much of the work done in the area of biotechnology and genetic engineering. Protein phosphorylation has even been posited as the basis of learning and memory. Medical applications have included development of the drug cyclosporin, which blocks the body's immune response by interfering with phosphorylation to prevent rejection of transplants. As important as what happens when the process functions normally is what happens when it goes awry: protein kinases are involved in almost 50 percent of cancer-causing oncogenes.

Recognition for Krebs's work came through various awards besides the Nobel Prize. In 1988 Krebs and Fischer shared the Passano Award for their research, and Krebs was one of four scientists to share the Lasker Award for Basic Medical Research in 1989. He was co-recipient of the Robert A. Welch Award in Chemistry in 1991, followed by the Nobel Prize a year later. Besides concentrating his research on protein phosphorylation, Krebs has investigated signal transduction and carbohydrate metabolism.

In 1968, Krebs had left the University of Washington to accept the position of founding chairman of the department of biological chemistry at the University of California in Davis. When he returned to Washington in 1977, he became chairman of the department of **pharmacology**. In both positions he was able to assist in the recruitment of talented faculty, which Krebs considers critical to the continued development of the field of biochemistry. From 1977 until 1983, Krebs was associated with the Howard Hughes Medical Institute as well.

Krebs was married on March 10, 1945, to Virginia Deedy French, and the couple three children. As a young boy, Krebs loved to read historical novels about the Civil War and the settling of the West. He credits his wife for keeping him aware of the other aspects of living besides his work.

See also Cell cycle and cell division; Cell differentiation; Cell membrane transport; Cell structure

KREBS, HANS ADOLF (1900-1981)

German biochemist

Few students complete an introductory biology course without learning about the Krebs cycle, an indispensable step in the process the body performs to convert food into energy. Also known as the citric acid cycle or tricarboxylic acid cycle, the

Krebs cycle derives its name from one of the most influential biochemists of our time. Born in the same year as the twentieth century, Hans Adolf Krebs spent the greater part of his eighty-one years engaged in research on intermediary **metabolism**. First rising to scientific prominence for his work on the ornithine cycle of urea synthesis, Krebs shared the Nobel Prize in physiology and medicine in 1953 for his discovery of the citric acid cycle. Over the course of his career, the German-born scientist published, oversaw, or supervised a total of more than 350 scientific publications. But the story of Krebs's life is more than a tally of scientific achievements; his biography can be seen as emblematic of biochemistry's path to recognition as its own discipline.

In 1900, Alma Davidson Krebs gave birth to her second child, a boy named Hans Adolf. The Krebs family—Hans, his parents, sister Elisabeth and brother Wolfgang—lived in Hildesheim, in Hanover, Germany. There his father Georg practiced medicine, specializing in surgery and diseases of the ear, **nose**, and throat. Hans developed a reputation as a loner at an early age. He enjoyed swimming, boating, and bicycling, but never excelled at athletic competitions. He also studied piano diligently, remaining close to his teacher throughout his university years. At the age of fifteen, the young Krebs decided he wanted to follow in his father's footsteps and become a physician. But World War I had broken out, and before he could begin his medical studies, he was drafted into the army upon turning eighteen in August of 1918. The following month he reported for service in a signal corps regiment in Hanover. He expected to serve for at least a year, but shortly after he started basic training the war ended. Krebs received a discharge from the army to commence his studies as soon as possible.

Krebs chose the University of Göttingen, located near his parents' home. There, he enrolled in the basic science curriculum necessary for a student planning a medical career and studied **anatomy, histology, embryology** and botanical science. After a year at Göttingen, Krebs transferred to the University of Freiburg. At Freiburg, Krebs encountered two faculty members who enticed him further into the world of academic research: Franz Knoop, who lectured on physiological chemistry, and Wilhelm von Möllendorff, who worked on histological staining. Möllendorff gave Krebs his first research project, a comparative study of the staining effects of different dyes on muscle tissues. Impressed with Krebs's insight that the efficacy of the different dyes stemmed from how dispersed and dense they were rather than from their chemical properties, Möllendorff helped Krebs write and publish his first scientific paper. In 1921, Krebs switched universities again, transferring to the University of Munich, where he started clinical work under the tutelage of two renowned surgeons. In 1923, he completed his medical examinations with an overall mark of "very good," the best score possible. Inspired by his university studies, Krebs decided against joining his father's practice as he had once planned; instead, he planned to balance a clinical career in medicine with experimental work. But before he could turn his attention to research, he had one more hurdle to complete, a required clinical year, which he served at the Third Medical Clinic of the University of Berlin.

Krebs spent his free time at the Third Medical Clinic engaged in scientific investigations connected to his clinical duties. At the hospital, Krebs met Annelise Wittgenstein, a more experienced clinician. The two began investigating physical and chemical factors that played substantial roles in the distribution of substances between **blood, tissue,** and **cerebrospinal fluid**, research that they hoped might shed some light on how pharmaceuticals such as those used in the treatment of syphilis penetrate the nervous system. Although Krebs published three articles on this work, later in life he belittled these early, independent efforts. His year in Berlin convinced Krebs that better knowledge of research chemistry was essential to medical practice.

Accordingly, the twenty-five-year-old Krebs enrolled in a course offered by Berlin's Charité Hospital for doctors who wanted additional training in laboratory chemistry. One year later, through a mutual acquaintance, he was offered a paid research assistantship by **Otto Warburg**, one of the leading biochemists of the time. Although many others who worked with Warburg called him autocratic, under his tutelage Krebs developed many habits that would stand him in good stead as his own research progressed. Six days a week work began at Warburg's laboratory at eight in the morning and concluded at six in the evening, with only a brief break for lunch. Warburg worked as hard as the students. Describing his mentor in his autobiography, *Hans Krebs: Reminiscences and Reflections,* Krebs noted that Warburg worked in his laboratory until eight days before he died from a pulmonary **embolism**. At the end of his career, Krebs wrote a biography of his teacher, the subtitle of which described his perception of Warburg: "cell physiologist, biochemist, and eccentric."

Krebs's first job in Warburg's laboratory entailed familiarizing himself with the tissue slice and manometric (pressure measurement) techniques the older scientist had developed. Until that time biochemists had attempted to track chemical processes in whole **organs**, invariably experiencing difficulties controlling experimental conditions. Warburg's new technique, affording greater control, employed single layers of tissue suspended in solution and manometers (pressure gauges) to measure chemical reactions. In Warburg's lab, the tissue slice/manometric method was primarily used to measure rates of respiration and glycolysis, processes by which an organism delivers oxygen to tissue and converts **carbohydrates** to energy. Just as he did with all his assistants, Warburg assigned Krebs a problem related to his own research—the role of heavy metals in the oxidation of sugar. Once Krebs completed that project, he began researching the metabolism of human **cancer** tissue, again at Warburg's suggestion. While Warburg was jealous of his researchers' laboratory time, he was not stingy with bylines; during Krebs's four years in Warburg's lab, he amassed sixteen published papers. But Warburg had no room in his lab for a scientist interested in pursuing his own research. When Krebs proposed undertaking studies of intermediary metabolism that had little relevance for Warburg's work, the supervisor suggested Krebs switch jobs.

Unfortunately for Krebs, the year was 1930. Times were hard in Germany, and research opportunities were few. He accepted a mainly clinical position at the Altona Municipal Hospital, which supported him while he searched for a more research-oriented post. Within the year he moved back to Freiburg, where he worked as an assistant to an expert on metabolic diseases with both clinical and research duties. In the well-equipped Freiburg laboratory, Krebs began to test whether the tissue slice technique and manometry he had mastered in Warburg's lab could shed light on complex synthetic metabolic processes. Improving on the master's methods, he began using saline solutions in which the concentrations of various ions matched their concentrations within the body, a technique which eventually was adopted in almost all biochemical, physiological, and pharmacological studies.

Working with a medical student named Kurt Henseleit, Krebs systematically investigated which substances most influenced the rate at which urea—the main solid component of mammalian urine—forms in **liver** slices. Krebs noticed that the rate of urea synthesis increased dramatically in the presence of ornithine, an amino acid present during urine production. Inverting the reaction, he speculated that the same ornithine produced in this synthesis underwent a cycle of conversion and synthesis, eventually to yield more ornithine and urea. Scientific recognition of his work followed almost immediately, and at the end of 1932—less than a year and a half after he began his research—Krebs found himself appointed as a *Privatdozent* at the University of Freiburg. He immediately embarked on the more ambitious project of identifying the intermediate steps in the metabolic breakdown of carbohydrates and fatty acids.

But Krebs was not to enjoy his new position in Germany for long. In the spring of 1933, along with many other German scientists, he found himself dismissed from his job as a result of Nazi purging. Although Krebs had officially and legally renounced the Jewish faith twelve years earlier at the urging of his patriotic father, who believed wholeheartedly in the assimilation of all German Jews, this legal declaration proved insufficiently strong for the Nazis. In June of 1933, he sailed for England to work in the **biochemistry** lab of Sir Frederick Gowland Hopkins of the Cambridge School of Biochemistry. Supported by a fellowship from the Rockefeller Foundation, Krebs resumed his research in the British laboratory. The following year, he augmented his research duties with the position of demonstrator in biochemistry. Laboratory space in Cambridge was cramped, however, and in 1935 Krebs was lured to the post of lecturer in the University of Sheffield's Department of Pharmacology by the prospect of more lab space, a semi-permanent appointment, and a salary almost double the one Cambridge was paying him.

His Sheffield laboratory established, Krebs returned to a problem that had long preoccupied him: how the body produced the essential **amino acids** that play such an important role in the metabolic process. By 1936, Krebs had begun to suspect that citric acid played an essential role in the oxidative metabolism by which the carbohydrate pyruvic acid is broken down so as to release energy. Together with his first Sheffield graduate student, William Arthur Johnson, Krebs observed a process akin to that in urea formation. The two researchers showed that even a small amount of citric acid could increase the oxygen absorption rate of living tissue. Because the amount of oxygen absorbed was greater than that needed to completely oxidize the citric acid, Krebs concluded that citric acid has a catalytic effect on the process of pyruvic acid conversion. He was also able to establish that the process is cyclical, citric acid being regenerated and replenished in a subsequent step. Although Krebs spent many more years refining the understanding of intermediary metabolism, these early results provided the key to the chemistry that sustains life processes. In June of 1937, he sent a letter to *Nature* reporting these preliminary findings. Within a week, the editor notified him that his paper could not be published without a delay. Undaunted, Krebs revised and expanded the paper and sent it to the new Dutch journal *Enzymologia,* which he knew would rapidly publicize this significant finding.

In 1938, Krebs married Margaret Fieldhouse, a teacher of domestic science in Sheffield. The couple eventually had three children. In the winter of 1939, the university named him lecturer in biochemistry and asked him to head their new department in the field. Married to an Englishwoman, Krebs became a naturalized English citizen in September, 1939, three days after World War II began.

The war affected Krebs's work minimally. He conducted experiments on vitamin deficiencies in conscientious objectors, while maintaining his own research on metabolic cycles. In 1944, the Medical Research Council asked him to head a new department of biological chemistry. Krebs refined his earlier discoveries throughout the war, particularly trying to determine how universal the Krebs cycle is among living organisms. He was ultimately able to establish that all organisms, even microorganisms, are sustained by the same chemical processes. These findings later prompted Krebs to speculate on the role of the metabolic cycle in **evolution**.

In 1953, Krebs received one of the ultimate recognitions for the scientific significance of his work—the Nobel Prize in physiology and medicine, which he shared with **Fritz Lipmann**, the discoverer of co-enzyme A. The following year, Oxford University offered him the Whitley professorship in biochemistry and the chair of its substantial department in that field. Once Krebs had ascertained that he could transfer his metabolic research unit to Oxford, he consented to the appointment. Throughout the next two decades Krebs continued research into intermediary metabolism. He established how fatty acids are drawn into the metabolic cycle and studied the regulatory mechanism of intermediary metabolism. Research at the end of his life was focused on establishing that the metabolic cycle is the most efficient mechanism by which an organism can convert food to energy. When Krebs reached Oxford's mandatory retirement age of sixty-seven, he refused to stop researching and made arrangements to move his research team to a laboratory established for him at the Radcliffe Hospital. Krebs died at the age of eighty-one.

See also Cell structure; Metabolism

KROGH, AUGUST (1874-1949)
Danish physiologist

August Krogh (pronounced "Krawg") won the 1920 Nobel Prize in physiology or medicine for the discoveries he made concerning human respiration. Krogh showed that most **capillaries** (the smallest **blood** vessels) of the body's **organs** and tissues are closed when they are at rest, but when there is activity and the need for oxygen increases, more capillaries will open. Krogh's explanations of respiration and capillary action were of major significance for the understanding of the physiology of the human pulmonary and circulatory systems. His research into respiration and circulation also had practical applications for the development of modern medical science.

Schack August Steenberg Krogh was born in Grenaa, Jutland. His father was a brewer, shipbuilder, and a newspaper editor. His mother, Marie Drechmann Krogh, was of gypsy ancestry. Throughout his life Krogh was active in both zoology and human physiology, accomplishing his major discoveries in the physiology of respiration. In 1910, he was able to settle an important biological controversy by establishing that, in human **lungs**, the absorption of oxygen and the elimination of carbon dioxide takes place by diffusion rather than secretion. From his work on human respiration, Krogh went on to describe the operation of the capillary system and the mechanisms that regulate it.

Krogh's interest in various areas of science can be traced back to his childhood. His father, who had been trained as a naval architect, influenced Krogh's interest in ships and the sea; the young Krogh also spent time in his youth studying the insects he found in the fields near his home. When he was fourteen he left school to serve on a Danish patrol boat assigned to protect fisheries in Iceland. Instead of pursuing a naval career, however, Krogh decided to complete his education. His love for the sea, nonetheless, remained with him for the rest of his life.

Krogh returned to school at the Gymnasium at Aarhus, and then went on to the University of Copenhagen in 1893. He first entered Copenhagen with the intention of studying physics and medicine, but under the influence of zoologist William Sørensen, he changed to the study of zoology and physiology. So slash rensen had advised Krogh to attend the lectures of Christian Bohr, an expert in circulatory and respiratory physiology. After attending Bohr's lectures at Copenhagen, Krogh began to work in Bohr's laboratory in 1897, and after receiving his master of science degree in 1899, he became Bohr's laboratory assistant.

One of Krogh's earliest achievements was his invention of a microtonometer—an instrument that measures gas pressure in fluids—which he developed to help in his research with a marine organism named *Corethra*. As a student, Krogh had done research on the larvae of *Corethra* to determine how its air bladders operated (he found that they worked like the diving tanks of submarines). Traveling in 1902 to Greenland, Krogh studied the amounts of oxygen and carbon dioxide dissolved in fresh and seawater. His research cast a new understanding on the role of the oceans in carbon dioxide regulation

and at the same time he was able to improve his techniques for measuring gas pressures in fluids.

In 1903, Krogh received a Ph.D. in zoology from the University of Copenhagen, where, in his doctoral dissertation, he demonstrated the difference between the skin and lung respiration of the frog. Whereas the frog's skin respiration was constant and regular, Krogh found that the frog's lung respiration varied and was controlled by the autonomic system through the mechanism of the vagus nerve. Oxygen passed from the air sacs (**alveoli**) of the lung through a membrane to the capillaries and then to the blood stream where it formed carbon dioxide after it was used by the different tissues in the body. The process then reversed when the blood carried carbon dioxide to the alveoli of the lungs, where it was exhaled. Respiration would then vary according to the organism's need for oxygen.

Krogh was married in 1905 to Marie Jørgensen, a physiologist who also worked in Bohr's laboratory. (The couple would eventually have three daughters and one son, with the son and the youngest daughter becoming physiologists.) In 1906, the first of Krogh's papers to receive international recognition, a work that showed that nitrogen is not involved in animal **metabolism**, was awarded the Seegen Prize from the Vienna Academy of Sciences. In 1907, Krogh received further international attention at Heidelberg, Germany, when he discussed his findings on the diffusion of pulmonary gases at the International Congress of Physiology.

In 1908, Krogh made another trip to Greenland with his wife to study the Eskimo's meat-eating dietary habits and the effects it had on their respiration and metabolism. He was also given an associate professorship of zoo physiology at the University of Copenhagen that year. Two years later Krogh and his wife were given a laboratory at Ny Vestergade for physiological research; the couple worked there together for a number of years. Krogh then became a full professor at Copenhagen in 1916.

From 1908 to 1912, Krogh was engaged in research to resolve the question of how oxygen was transferred in the lungs to the blood. Bohr and John Burdon Sanderson Haldane (who was well-known for his own research into the mechanics of respiration), along with other scientists, believed that the lung acted as a gland in the alveolar transfer of oxygen to the blood; in other words, the lung secreted the oxygen. Krogh, in 1912, convincingly delivered the fatal blow to the secretion theory by first showing that in fishes there is no secretion of oxygen into the air sacs, and then by demonstrating that the amount of oxygen in the blood always equaled the amount that should be provided by his diffusion theory. The development of Krogh's microtonometer proved to be critical for verifying the results of these demonstrations.

It was not until 1916, however, that Krogh accomplished the work that would, in 1920, earn him the Nobel Prize in physiology or medicine. He showed that muscle tension was always slightly lower than the tensions in the capillaries, even when the muscle was at work. Noting that there were few open capillaries when a muscle was at rest, Krogh demonstrated that as soon as the muscle became active many capillaries began to open up. He was also able to show that blood did not enter the capillaries through the pressure of the blood vessels but from the relaxed tonus (partial contraction) of the active muscle. The relaxation of the muscle allowed the field of capillaries to open and the blood to flow in, thus providing more oxygen to the muscle, organ, or **tissue**.

Krogh's discoveries relating to gas exchanges in the lung and to the operation of the capillary system helped to develop medical techniques for **breathing** through the trachea. His work also improved surgical methods for open **heart** surgery, such as the procedure for reducing body temperature to below normal levels to slow down the rate of **gaseous exchange**.

In 1922 Krogh became interested in insulin (which had been discovered by Frederick G. Banting and John James Rickard Macleod the year before), partly because his own wife

had diabetes. Besides being active in insulin research, Krogh helped to promote manufacturing facilities in Denmark for its production. Krogh also maintained his interest in zoology, writing about insects and becoming particularly attentive to theories about the way honey bees communicate.

During World War II, Krogh lived in Sweden, having been forced to flee Nazi-occupied Denmark because of his open opposition to Nazism. Krogh returned to Denmark in 1945, and died in Copenhagen in 1949. His research in his last years was performed at the laboratories of the Carlsberg and Scandinavian Insulin Foundations.

See also Angiology; Circulatory system; Muscle contraction; Muscular system overview; Respiration control mechanisms; Respiratory system

L

LABIA • *see* GENITALIA (FEMALE EXTERNAL)

LACTATION • *see* MAMMARY GLANDS AND LACTATION

LACTIC ACID

Lactic acid is a naturally occurring hydroxy-carboxylic acid. It occurs in many acidic fermentation products, e.g., silage, sour milk, sauerkraut, and pickles, and also has numerous commercial uses such as in drug preparation, acidification of wine for vinegar production, preparation of foods, plastics, and lacquers.

In **physiology**, lactic acid is produced as the end product of anaerobic respiration or glycolysis in muscle. Anaerobic respiration is generally defined as respiration in the complete absence of oxygen, but this is not the case for muscles. They are able to carry out anaerobic respiration even when oxygen is present. In **skeletal muscle**, respiration is normally aerobic and oxygen is used to metabolise glucose. As the skeletal muscle contracts during exercise, the rate of aerobic respiration increases so as to provide more energy for **muscle contraction**. This is facilitated by a concomitant increase in the supply of oxygen through the **blood** due to a greater ventilation and cardiac output. Glucose is taken up more rapidly and converted to pyruvate by **enzymes** of aerobic glycolysis and then to carbon dioxide and water by the enzymes of the Krebs cycle in **mitochondria**. The glycolytic breakdown of glucose to pyruvate yields a small amount of **ATP** while mitochondrial oxidation yields a great deal more. As aerobic respiration reaches its maximum, as determined by the rate of oxygen supply to the skeletal muscle, anaerobic respiration is also mobilized simultaneously, to provide an additional energy source for maximum muscular exertion. Thus more ATP is provided for muscular contraction than is possible by aerobic respiration alone. The mobilization of anaerobic respiration in this way allows a rapid production of ATP within a short period for maximum muscular exertion.

Under this situation, the muscular **tissue** is actually consuming a much greater amount of oxygen than at rest, and a large amount of glucose is oxidized aerobically. Thus anaerobic respiration takes place in the skeletal muscle even when there is an increased supply of oxygen. The products of anaerobic respiration in muscle are water and lactic acid. If allowed to accumulate in the body over long periods, lactic acid becomes toxic so anaerobic respiration can only be carried out for brief periods, after which the lactic acid has to be converted back to glucose or **glycogen**, or changed to carbon dioxide through aerobic respiration. Nevertheless, the maximum rate of muscular exertion made possible by the simultaneous occurrence of aerobic and anaerobic respiration is most essential for the survival of an animal at critical moments when an extra energy is needed for activity.

Lactic acid is utilized for the reformation of glucose in the process of gluconeogenesis and this occurs in the **liver**. There, lactic acid is converted first back to pyruvate and then through the reverse steps of glycolysis back to glucose. The events encompassing the production of lactic acid in the muscle, its conversion to glucose in the liver and the return of the glucose to the muscle with the eventual reformation of lactic acid constitutes the Cori cycle.

See also Metabolism; Respiration control mechanisms

LAËNNEC, RENÉ-THÉOPHILE-HYACINTHE (1781-1826)

French physician and inventor

Laënnec's 1816 invention of the stethoscope enabled physicians to examine the sounds of the living **heart** and **lungs** in detail. His 1819 book about his use of this instrument inaugurated the modern era of diagnosing chest, heart, and lung diseases accurately and scientifically.

Born in Quimper, a seacoast town in West Brittany, France, Laënnec was baptized "Théophile-René-Marie-Hyacinthe Laënnec," but was known professionally as either "René-Théophile-Hyacinthe Laënnec," which appears on his tombstone, or just "R.-T.-H. Laënnec." Sometimes his surname is spelled without the diaeresis over the first "e." His mother died of tuberculosis when he was five. His father had little interest in his children and in 1788, sent him to live with his Uncle Guillaume, a prominent physician in Nantes. Laënnec studied hard to become an engineer, but the French Revolution was especially violent in Nantes, and interrupted his education. In 1795, he became apprenticed to his uncle as a surgeon in several hospitals in Nantes. He toured northwestern France for a few years and served briefly in Napoleon's army before moving to Paris in 1801 to begin formal medical study.

Laënnec learned dissection from Guillaume Dupuytren (1777–1835) at the École Pratique, but soon became the star pupil of Napoleon's personal physician, Baron Jean Nicolas Corvisart des Marets (1755–1821) at the Hôpital de la Charité. He received his M.D. in 1803 with the top prize in medicine and the only prize in surgery. He was soon regarded as among the best clinicians and medical scientists in France. With Corvisart's help, he gained prestigious appointments in Paris. For the rest of his life he divided his time between working in Paris and vacationing for his health in Brittany. Always weak and sickly, suffering from asthma, tuberculosis, and migraine, Laënnec died young in Kerbouarnec, France.

As diagnosticians, both Corvisart and Laënnec understood the importance of listening carefully to the patient's chest. They practiced a technique called "auscultation," or listening directly to chest sounds with the physician's ear on the patient's body. The problem with auscultation was mechanical: how to get the ear close enough.

One day in 1816, Laënnec was frustrated trying to examine an obese woman with symptoms of heart disease. Almost instinctively, he rolled up a few sheets of paper into a tight tube, put one end of this cylinder to his ear and the other on her chest. What he heard amazed him. The chest sounds were louder and clearer than he had ever heard them before, and the background noise was mostly eliminated when he blocked his other ear. This was the first stethoscope. He named it from two Greek words, *stethos* ("chest") and *skopos* ("one who watches"), because he and Corvisart believed that with it they could "see" inside the chest. Laënnec called stethoscopic technique "mediate auscultation" to distinguish it from "immediate auscultation," putting the ear directly on the patient. He promptly replaced his earliest paper stethoscopes with wooden ones, about the same size and shape as a modern flashlight.

After three years of correlating the many different chest sounds and with his findings at autopsy, he introduced his discovery in a massive two-volume work: *De l'auscultation médiate, ou traité du diagnostic des maladies des poumons et du coeur* [On Mediate Auscultation, or: A Treatise on the Diagnostics of the Diseases of the Lungs and Heart] (1819), translated into English by Sir John Forbes (1788–1861) in 1821. The second, even more thorough, edition of Laënnec's book (1826) instantly became a classic in the English-speaking world when Forbes published his translation in 1827.

Besides the stethoscope, Laënnec's greatest contribution to medicine was redefining the concept of disease itself. Influenced by **Xavier Bichat**, Laënnec and many of his colleagues shifted the focus of **pathology** from the whole disease to the particular **organs** or tissues that were affected by it. They centered their attention on the "lesion," that is, on any change from a normal anatomical structure. Several lesions, including "Laënnec's thrombus," a **blood** clot in the heart, and "Laënnec's cirrhosis," a certain kind of progressive **liver** destruction, are named after him. His pathological research contributed most to the knowledge of respiratory tract diseases, especially tuberculosis. Ironically, it was this disease that killed him.

See also Anatomy; Cardiac disease; Pathology; Respiratory system

LANDSTEINER, KARL (1868-1943)

American immunologist

Karl Landsteiner was one of the first scientists to study the physical processes of immunity. He is best known for his identification and characterization of the human **blood** groups, A, B, and O, but his contributions spanned many areas of **immunology**, bacteriology, and **pathology** over a prolific forty-year career. Landsteiner identified the agents responsible for immune reactions, examined the interaction of **antigens and antibodies**, and studied allergic reactions in experimental animals. He determined the viral cause of poliomyelitis with research that laid the foundation for the eventual development of a polio vaccine. He also discovered that some simple chemicals, when linked to proteins, produced an immune response. Near the end of his career in 1940, Landsteiner and immunologist Philip Levine discovered the Rh factor that helped save the lives of many unborn babies whose Rh factor did not match their mothers. For his work identifying the human blood groups, Landsteiner was awarded the Nobel Prize in physiology or medicine in 1930.

Karl Landsteiner was born on June 14, 1868 in Vienna, Austria. In 1885, at the age of 17, Landsteiner passed the entrance examination for medical school at the University of Vienna. He graduated from medical school at the age of 23 and immediately began advanced studies in the field of organic chemistry, working in the research laboratory of his mentor, Ernst Ludwig. In Ludwig's laboratory Landsteiner's interest in chemistry blossomed into a passion for approaching medical problems through a chemist's eye.

For the next ten years, Landsteiner worked in a number of laboratories in Europe, studying under some of the most celebrated chemists of the day: Emil Fischer, a protein chemist who subsequently won the Nobel Prize in chemistry in 1902, in Wurzburg; Eugen von Bamberger in Munich; and Arthur Hantzsch and Roland Scholl in Zurich. Landsteiner published many journal articles with these famous scientists. The knowledge he gained about organic chemistry during these formative years guided him throughout his career. The nature of antibodies began to interest him while he was serving as an assistant to Max von Gruber in the Department of Hygiene at the University of Vienna from 1896 to 1897. During this time Landsteiner published his first article on the subject of bacteriology and **serology**, the study of blood.

Landsteiner moved to Vienna's Institute of Pathology in 1897, where he was hired to perform autopsies. He continued to study immunology and the mysteries of blood on his own time. In 1900, Landsteiner wrote a paper in which he described the agglutination of blood that occurs when one person's blood is brought into contact with that of another. He suggested that the phenomenon was not due to pathology, as was the prevalent thought at the time, but was due to the unique nature of the individual's blood. In 1901, Landsteiner demonstrated that the blood serum of some people could clump the blood of others. From his observations he devised the idea of mutually incompatible blood groups. He placed blood types into three groups: A, B, and C (later referred to as O). Two of his colleagues subsequently added a fourth group, AB.

In 1907, the first successful **transfusions** were achieved by Dr. Reuben Ottenberg of Mt. Sinai Hospital, New York, guided by Landsteiner's work. Landsteiner's accomplishment saved many lives on the battlefields of World War I, where transfusion of compatible blood was first performed on a large scale. In 1902, Landsteiner was appointed as a full member of the Imperial Society of Physicians in Vienna. That same year he presented a lecture, together with Max Richter of the Vienna University Institute of Forensic Medicine, in which the two reported a new method of typing dried blood stains to help solve crimes in which blood stains are left at the scene.

In 1908, Landsteiner took charge of the department of pathology at the Wilhelmina Hospital in Vienna. His tenure at the hospital lasted twelve years, until March of 1920. During this time, Landsteiner was at the height of his career and produced 52 papers on serological immunity, 33 on bacteriology and six on pathological **anatomy**. He was among the first to dissociate **antigens** that stimulate the production of immune responses known as antibodies, from the antibodies themselves. Landsteiner was also among the first to purify antibodies, and his purification techniques are still used today for some applications in immunology.

Landsteiner also collaborated with Ernest Finger, the head of Vienna's Clinic for Venereal Diseases and Dermatology. In 1905, Landsteiner and Finger successfully transferred the venereal disease syphilis from humans to apes. The result was that researchers had an animal model in which to study the disease. In 1906, Landsteiner and Viktor Mucha, a scientist from the Chemical Institute at Finger's clinic, developed the technique of dark-field microscopy to identify and study the microorganisms that cause syphilis.

One day in 1908, the body of a young polio victim was brought in for autopsy. Landsteiner took a portion of the boy's spinal column and injected it into the spinal canal of several species of experimental animals, including rabbits, guinea pigs, mice, and monkeys. Only the monkeys contracted the disease. Landsteiner reported the results of the experiment, conducted with Erwin Popper, an assistant at the Wilhelmina Hospital.

Scientists had accepted that polio was caused by a microorganism, but previous experiments by other researchers had failed to isolate a causative agent, which was presumed to be a bacterium. Because monkeys were hard to come by in Vienna, Landsteiner went to Paris to collaborate with a Romanian bacteriologist, Constantin Levaditi of the Pasteur Institute. Working together, the two were able to trace poliomyelitis to a virus, describe the manner of its transmission, time its incubation phase, and show how it could be neutralized in the laboratory when mixed with the serum of a convalescing patient. In 1912, Landsteiner proposed that the development of a vaccine against poliomyelitis might prove difficult but was certainly possible. The first successful polio vaccine, developed by Jonas Salk, wasn't administered until 1955.

Landsteiner accepted a position as chief dissector in a small Catholic hospital in The Hague, Netherlands, where he performed routine laboratory tests on urine and blood from 1919 to 1922. During this time he began working on the concept of haptens, small molecular weight chemicals such as fats or sugars that determine the specificity of antigen-antibody reactions when combined with a protein carrier. He combined haptens of known structure with well-characterized proteins such as albumin, and showed that small changes in the hapten could affect antibody production. He developed methods to show that it is possible to sensitize animals to chemicals that cause contact dermatitis (inflammation of the skin) in humans, demonstrating that contact dermatitis is caused by an antigen-antibody reaction. This work launched Landsteiner into a study of the phenomenon of allergic reactions.

In 1922, Landsteiner accepted a position at the Rockefeller Institute in New York. Throughout the 1920s Landsteiner worked on the problems of immunity and allergy. He discovered new blood groups: M, N, and P, refining the work he had begun 20 years before. Soon after Landsteiner and his collaborator, Philip Levine, published the work in 1927, the types began to be used in paternity suits.

In 1929, Landsteiner became a United States citizen. He won the Nobel Prize in physiology or medicine in 1930 for identifying the human blood types. In his Nobel lecture, Landsteiner gave an account of his work on individual differences in human blood, describing the differences in blood between different species and among individuals of the same species. This theory is accepted as fact today but was at odds with prevailing thought when Landsteiner began his work. In 1936, Landsteiner summed up his life's work in what was to become a medical classic: *Die Spezifität der serologischen Reaktionen*, which was later revised and published in English, under the title *The Specificity of Serological Reactions*.

Landsteiner retired in 1939, at the age of seventy-one, but continued working in immunology. With Levine and Alexander Wiener he discovered another blood factor, labeled the Rh factor, for Rhesus monkeys, in which the factor was first discovered. The Rh factor was shown to be responsible for the infant disease erythroblastosis fetalis that occurs when mother and fetus have incompatible blood types and the fetus is injured by the mother's antibodies. Landsteiner died in 1943, at the age of 75.

See also Immune system; Immunity

LARYNX AND VOCAL CORDS

The larynx is an organ in the neck associated with **swallowing** and sound (voice) production. The larynx is located at the point where incoming air (inspirational air) is directed to the trachea leading to the **lungs**, and where incoming food is routed into the esophagus for passage to the stomach.

The location and design of the larynx allows it to protect the airway from inadvertently routed food. The larynx is also commonly called the "voice box," in recognition of its crucial role in the formation of speech. The design of the larynx allows it to produce the sounds that comprise speech. As well, the design and location of the larynx play a role in the control of air movement during **breathing**.

The larynx is a tubular chamber about two inches high. The walls of the larynx are made of a tough material called **cartilage** bound by ligaments and membranes. The cartilaginous larynx rests on another piece of cartilage called the cricoid. Immediately below the cricoid begin the rings of the trachea. Immediately above the larynx lies a U-shaped structure called the hyoid. The larynx extends out of the trachea, which is easily detected in males as the "Adam's apple."

A pair of vocal cords (also known as vocal folds) lies at the center of the larynx. The vocal cords are essentially bands of muscle with a structure called the arytenoid at the center of each cord. The arytenoid has many muscles attached to it. They supply the muscular power that moves the larynx. During silent breathing the muscles are relatively slack, allowing the vocal cords to lie along the walls of the larynx. This orientation leaves the air passage open. During speech, contraction of the arytenoid muscles pulls the vocal cords are brought close together and are folded into a closed position across the larynx. Air passing by the vocal cords vibrates each cord, which produces a sound. An analogy is the vibration of the reed on a wind instrument like the oboe. Various muscles can adjust the tension of the vocal cords and the space between, to produce sounds of various pitches. The tauter the cords, the higher the pitch. Modification of the sound by the throat, tongue and lips produces recognizable speech.

The larynx is rich in nerves and so is very sensitive to touch. This permits the larynx to respond very quickly to any unintended material that touches it, such as food. A reflexive cough expels the material.

See also Cough reflex; Neck muscles and fasciae; Swallowing and dysphagia

LAVERAN, ALPHONSE (1845-1922)

French physician

Alphonse Laveran was a French army physician who took advantage of his period of service in Algeria to study **malaria**, a disease known since ancient times and common in tropical and subtropical areas. Using primitive technology, he discovered and ultimately proved that malaria was caused by a minute animal parasite; he also suggested, though he did not

himself prove, that the parasite was transmitted to human beings by some species of mosquito. He later went on to study other diseases caused by parasites. For the work he did in this field throughout his career, he was awarded the Nobel Prize in medicine in 1907.

Charles Louis Alphonse Laveran was born into a military family in Paris. He was the second child and only son of Louis-Theodore Laveran, a career military physician, and Marie-Louise Anselme Guénard de la Tour Laveran. When Laveran was five years old, he went with his parents to Algeria, where his father was stationed. His father was his first teacher, and after the family returned to Paris in 1856, Laveran received his secondary education at the College Sainte-Barbe and the Lycée Louis-le-Grand. In 1863, he entered the military medical school at Strasbourg, which his father had also attended; Laveran graduated in 1867. He joined the military medical service following graduation and saw active duty during the Franco-Prussian War of 1870–1871. It was at this time that he first witnessed the ravages that diseases can cause in an army at war. In 1874, he won by competitive examination an appointment to a professorship earlier held by his father at the École du Val-de-Grace, a military medical school in Paris. This was a temporary appointment, and at its conclusion in 1878, he was sent to the military hospital at Bône (now Annaba) in Algeria.

It was while at Bône that Laveran began a careful study of malaria, common in many parts of Algeria, in an effort to learn its cause. He set up a small laboratory and with the primitive, low-powered microscope available to him, he spent much time examining **blood** samples from malaria patients both living and deceased. His studies were briefly interrupted when he was transferred to Biskra, Algeria, where malaria was rare, but they were resumed when he moved on to Constantine, also in Algeria. There, on November 6, 1880, he first observed under the microscope circular and cylindrical bodies that had moving filaments, or flagella. This confirmed his earlier suspicion that malaria was caused by living animal cells, minute single-celled creatures called protozoa, which acted as parasites in the human body. The particular protozoan which Laveran had discovered to be the cause of malaria later came to be called plasmodium.

Laveran's discovery was presented to the Academy of Medicine in Paris on November 23, 1880. A second paper, based upon further research, was published by the Société Médicale des Hopitaux on December 24 of that year. In 1881, Laveran published a brief monograph, *Nature parasitaire des accidents de l'impaludisme,* which provided more details of his findings. Laveran's conclusions were not immediately accepted by other scientists studying malaria. His microscopic research proved difficult to replicate. Moreover, in the wake of the discoveries of the German scientist **Robert Koch** and others, it was widely assumed at the time that **bacteria** were the causes of most diseases, including malaria.

Laveran, however, continued his research, examining the blood of hundreds of malaria patients, both in Algeria and in Italy. By 1884, in a personal microscopic demonstration, he was able to persuade Louis Pasteur that his theory was correct. Other noted scientists such as William Osler were convinced during the course of the 1880s. Also in 1884, Laveran published a book, *Traité des fièvres palustres avec la description des microbes du paludisme,* which summarized all of his research on malaria. In this work, he revealed his suspicion that the malaria protozoa were nurtured and transmitted to human beings by some species of mosquito. It remained for the British physician, **Ronald Ross**, working in India in the late 1890s, to prove that the malaria parasite was indeed transmitted by the Anopheles mosquito.

Laveran returned to Paris from Algeria in 1883 and became professor of military hygiene at the École du Val-de-Grace in 1884. He married Sophie Marie Pidancet in 1885. When his professorship came to an end in 1894, he was offered only temporary administrative positions at the military hospitals at Lille and at Nantes. Angry because he was not offered a post at a military laboratory where he could continue his research, he resigned from the military medical service in December of 1896, and accepted a position at the Pasteur Institute in Paris. There he pursued his research for the rest of his life.

Laveran's demonstration that protozoa, as well as bacteria, could be the causes of disease in both human beings and animals led many other researchers into the field. Laveran himself did much significant work on disease-causing parasites. He was especially concerned with the trypanosome family of protozoa, one of which is the cause of the disease trypanosomiasis, or African sleeping sickness, transmitted by the tsetse fly. He also studied the trypanosome responsible for another tropical disease, kala azar, or dumdum fever. He was awarded the Nobel Prize in 1907 for his work on all disease-causing protozoa. He used half of the prize money to establish a laboratory for research on tropical diseases at the Pasteur Institute.

Laveran was honored with membership in the French Academy of Sciences in 1901. The French government made him a Commander of the Legion of Honor in 1912. During World War I he served on several committees concerned with preserving the health of French soldiers, and he served as president of the Academy of Medicine in 1920. He died after a short illness in 1922.

Leeuwenhoek, Anton van (1632-1723)
Dutch biologist

Anton van Leeuwenhoek is best remembered as the first person to study **bacteria** and "animalcules," or one-celled animals, now known as protozoa. Unlike his contemporaries **Robert Hooke** and **Marcello Malpighi**, Leeuwenhoek did not use the more advanced compound microscope; instead, he strove to manufacture magnifying lenses of unsurpassed power and clarity that would allow him to study the microcosm in far greater detail than any other scientist of his time.

Though Leeuwenhoek's family was relatively prosperous, he received little formal education. After completing grammar school in Delft, Netherlands, he moved to Amsterdam to work as a draper's apprentice. In 1654, he returned to Delft to establish his own shop and worked as a

Anton van Leeuwenhoek. *The Library of Congress.*

draper for the rest of his life. In addition to his business, Leeuwenhoek was appointed to several positions within Delft's city government, which afforded him the financial security to spend a great deal of time and money in pursuit of his hobby—lens grinding. Lenses were an important tool in Leeuwenhoek's profession because cloth merchants often used small lenses to inspect their products. His hobby soon turned to fascination, however, as he searched for more and more powerful lenses.

In 1671, Leeuwenhoek constructed his first simple microscope. It consisted of a tiny lens that he had ground by hand from a globule of glass and was placed within a brass holder. To this he had attached a series of pins designed to hold the specimen. It was the first of nearly six hundred lenses ranging from 50 to 500 times magnifications that he would grind during his lifetime. Through his microscope, Leeuwenhoek examined such substances as skin, **hair**, and his own **blood**. He studied the structure of ivory as well as the physical composition of the flea, discovering that fleas, too, harbored parasites.

Leeuwenhoek began writing to the British Royal Society in 1673. At first, the Society gave his letters little notice, thinking that such magnification from a single lens microscope could only be a hoax. However, in 1676, when he sent the Society the news that he had discovered tiny one-celled animals in rainwater, the interest of member scientists was piqued. Following Leeuwenhoek's specifications, they built microscopes of comparable magnitude and confirmed his findings. In 1680, the Society unanimously elected Leeuwenhoek as a member.

Until this time, Leeuwenhoek had been operating in an informational vacuum; he read only Dutch and, consequently, was unable to learn from the published works of Hooke and Malpighi (though he often gleaned what he could from the illustrations within their texts). As a member of the Society, he was finally able to interact with other scientists. In fact, the news of his discoveries spread worldwide, and he was often visited by royalty from England, Prussia, and Russia. The traffic through his laboratory was so persistent that he eventually allowed visitors by appointment only. Near the end of his life he had reached near-legendary status and was often referred to by the local townsfolk as a magician.

Amid all the attention, Leeuwenhoek remained focused upon his scientific research. Specifically, he was interested in disproving the common belief in spontaneous generation, a theory proposing that certain inanimate objects could generate life. For example, it was believed that mold and maggots were created spontaneously from decaying food. He succeeded in disproving spontaneous generation in 1683 when he discovered bacteria cells. These tiny organisms were nearly beyond the resolving power of even Leeuwenhoek's remarkable equipment and would not be seen again for more than a century.

Leeuwenhoek created and improved upon new lenses for most of his long life. For the forty-three years that he was a member of the Royal Society, he wrote nearly 200 letters that described his progress. However, he never divulged the method by which he illuminated his specimens for viewing, and the nature of that illumination is still a mystery. Upon his death, Leeuwenhoek willed twenty-six of his microscopes, a few of which survive in museums, to the British Royal Society.

See also History of anatomy and physiology: The Renaissance and Age of Enlightenment

LEG • *see* LOWER LIMB STRUCTURE

LEUKEMIA

The abnormal proliferation of white **blood** cells causes a type of **cancer** known as leukemia. Leukemia is primarily caused by chromosomal instability, which leads to the transfer of genes from one chromosome to another, with or without fusion or juxtaposition of two different genes. This phenomenon is known as translocation and causes a change in the rate expression of the translocated genes. Translocations can therefore lead either to gene fusions or to the juxtaposition of oncogenes to other genes. At presently, it is known at least ten different types of gene fusion and nine types of juxtapositions associated with leukemia.

Leukemia itself is divided into two different types: lymphocytic leukemias and myelogenous leukemias. Lymphocytic leukemias occur when a given translocation affects lymphoid cells, which induces its overproduction and accumulation in

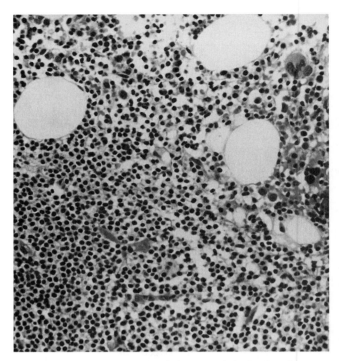

Close up of bone marrow in chronic lymphocytic leukemia cells. © Biophoto Associates, Science Source/Photo Researchers. Reproduced by permission.

lymphocytic tissues and lymph nodes, with the consequent spread to other areas of the body. When the translocation affects the rate of production of young myelogenous cells in the bone marrow, it also leads to an extramedullar production of these cells in the lymph nodes, spleen, and the **liver**. Myelogenous leukemia may occur under the form of neutrophilic, basophilic, eosinophilic, or monocytic subtypes, when the mutated cells are partially differentiated, i.e., have some degree of maturation and functional specialization. However, the majority of leukemia cases do present white cells without specialization, which are very different from their normal analogues.

Leukemia may also occur in two different forms: acute or chronic. Chronic leukemia is characterized by more differentiated cells and evolves slowly, over a period of 10–20 years, whereas the acute form is due to the production of more undifferentiated cells. Leukemic cells are usually ineffective in providing immune protection, as do their normal counterparts. Therefore, patients with leukemia are prone to infections, severe anemia, as well as hemorrhages (due to a decrease in blood **platelets**). Moreover, leukemic cells migrate to other tissues of the body (metastasis), such as bones, spleen, liver, and lymph nodes, where they rapidly overgrow and destroy the surrounding tissues. Because the metabolic rate of abnormally rapid-proliferating cells dramatically increases the demand for the nutrient reserves of the body (i.e., glucose, **amino acids, vitamins**), leukemia can induce a metabolic starvation of other tissues and **organs**.

New molecular treatments against most types of leukemia have being recently developed and have proved highly effective in controlling the disease, such as a drug that inhibits c-kit, an oncogene essential for the induction of leukemic cell-proliferation, as well as several redifferentiation arsenic-like compounds. However, these treatments do not imply cure. In the case of c-kit inhibitor, for instance, the drug has to be orally taken throughout the patient's life.

See also B lymphocytes; Cell differentiation; Erythrocytes; Hemopoiesis; Immune system; Lymphatic system

LEUKOCYTES

Leukocytes are white **blood** cells made in the bone marrow that defend the body from infective organisms and foreign substances. When **infection** occurs, the numbers of leukocytes increase.

The five types of leukocytes are: neutrophils, eosinophils, basophils, monocytes, and lymphocytes. Neutrophils, eosinophils, and basophils are in a class called polymorphonucelar granulocytes. Their nucleus appears to have segments within it and the remaining volume of the cell, which is called the cytoplasm, is granular in appearance. Each type of leukocyte has granules of distinctive appearance. Monocytes and lymphocytes are grouped together as mononuclear agranulocytes. Their nucleus does not have segments and the cytoplasm is free of granules.

Neutrophils comprise up to 70% of the leukocyte population. They are chemically attracted to the site of injury or infection, where they adhere to blood vessel wall and migrate into the surrounding **tissue**. Infecting **bacteria** are engulfed in a process known as phagocytosis. Flexible "arms" of the leukocyte surround the target and then fuse together, drawing the target into the neutrophil in a vacuole (a structure that can be envisioned as a bag within the cytoplasm). The bacteria are subsequently destroyed by the release of degradative **enzymes**.

Eosinophils make up two to five percent of the total leukocytes. They can survive for weeks and function in allergic responses and against parasites. Eosinophils are also chemically attracted to the trouble spot in the body. Basophils make up less than one percent of the total leukocyte population. They function in the immediate type of immune response that occurs in conditions such as asthma, hay fever, and anaphylaxis.

Monocytes are the largest type of leukocyte. They make up five to eight percent of the total leukocyte population and function as macrophages, which are essentially roving garbage collectors in areas such as the **liver**, lymph nodes, and the **lungs**. Like neutrophils, they engulf debris and microorganisms by the process of phagocytosis and digest them in lysosomal granules. Monocytes are also precursors for several other types of phagocytic cells, such as the Kupffer cells of the liver and bone osteoclast cells.

Lymphocytes are important as B cells and T cells. B cells become activated by the presence of an antigen and produce antibody specific to the antigenic target. T cells are vital components of cell-mediated immunity (which does not depend on the presence of antibodies circulating in body fluids and in tissues).

Abnormally low numbers of leukocytes (called leukopenia) leaves an individual susceptible to infections.

See also Immune system; Infection and resistance

LEVI-MONTALCINI, RITA (1909-)
Italian-born American neurobiologist

Rita Levi-Montalcini is recognized for her pioneering research on nerve cell growth. During the 1950s she discovered a protein in the nervous system, which she named the nerve growth factor (NGF). Her subsequent collaboration with biochemist **Stanley Cohen** at Washington University in St. Louis, Missouri, led to the isolation of that substance. Later applications of their work have proven useful in the study of several disorders, including **Alzheimer's disease, cancer,** and **birth defects**. Levi-Montalcini's and Cohen's work was recognized in 1986 when they were jointly awarded the Nobel Prize in physiology or medicine. Levi-Montalcini became the fourth woman to receive the Nobel in that field.

Levi-Montalcini, the third of four children of Adamo Levi and Adele Montalcini, was born into an upper-middle-class Jewish family in Turin, Italy, in 1909. She grew up in a traditional family and was steered by her father to pursue an education at an all-girls' high school that prepared young women for marriage. She graduated from high school when she was eighteen, having demonstrated exceptional intellectual ability, but was unable to enter a university because of the limited education that had been offered to her. Levi-Montalcini was uncertain what she wanted to do with her life (though she was certain she did not want to marry and have children), and it wasn't until three years later, when her beloved governess was stricken with cancer, that she decided to become a doctor.

After having convinced her father she wanted to enter medical school, Levi-Montalcini passed the entrance exams with distinction. She enrolled in the Turin School of Medicine in 1930, where she studied under Dr. Giuseppe Levi, a well-known histologist and embryologist who introduced Levi-Montalcini to research on the nervous system. She graduated from medical school in 1936 and became Levi's research assistant. With the rise of Fascism in the late 1930s, Jews were restricted from academic positions as well as the medical profession, and Levi-Montalcini was forced to resign from her academic and clinical posts in 1938. The following year, she accepted a position at the Neurological Institute in Brussels, where she worked until the Nazi invasion in 1939 precipitated her return to Italy.

Upon returning to Italy, she took up residence in Turin with her family. Restrictions imposed upon Jews had increased during her absence, and Levi-Montalcini was forced to set up a private laboratory in her bedroom. Again working with Levi, who had also been banned from his academic post, Levi-Montalcini began researching the nervous system of chicken embryos. In a memoir published in *Women Scientists: The Road to Liberation,* Levi-Montalcini recalls, "Looking back to that period I wonder how I could have found so much interest in, and devoted myself with such enthusiasm to, the study of a small neuroembryological problem, when all the values I cherished were being crushed, and the triumphant advance of the Germans all over Europe seemed to herald the end of Western civilization. The answer may be found in the well-known refusal of human beings to accept reality at its face value, whether this be the fate of the individual, of a country, or of the human race." Her research at the time, in fact, laid the groundwork for her discovery of NGF.

By 1942, the Allied bombing of Turin forced Levi-Montalcini and her family to move to the countryside, where she continued experimentation on chicken embryos to study the mechanisms of nerve **cell differentiation**, or the specialization of nerve cells. Contrary to previous studies conducted by the respected neuroembryologist Viktor Hamburger, who theorized that nerve cells reached their destinations because they were directed by the **organs** to which they grew, Levi-Montalcini hypothesized that a specific nutrient was essential for nerve growth. When Nazi troops invaded northern Italy in 1943, Levi-Montalcini was again forced to relocate, this time to Florence, where she remained for the duration of the war under an assumed name. Following the liberation of Florence in 1944, Levi-Montalcini worked as a doctor in a refugee camp, and, when northern Italy was liberated the following year, she resumed her post as research assistant to Levi in Turin. Hamburger, who was interested in a paper Levi-Montalcini had published on her wartime experiments, contacted her in 1946, inviting her to fill a visiting research position at Washington University in St. Louis. This temporary position ultimately lasted over three decades.

Levi-Montalcini's early work at Washington University concerned further experimentation on the growth processes of chicken embryos in which she observed a migratory sequence of nerve cells. Her observations validated her theory on the existence of a "trophic factor," which provided the essential nutrients for nerve cell differentiation. In 1950 she began studying mouse **tumors** that had been grafted onto chicken embryos, and which Elmer Bueker had earlier demonstrated were capable of eliciting a proliferation of nerve cells. After repeating Bueker's results, Levi-Montalcini reached a different conclusion. Instead of maintaining that the nerve cells proliferated in response to the presence of the tumor, she deduced that the nerve cells grew out of the tumor and that, thus, the tumor released a substance that elicited the growth. Traveling to Rio de Janeiro in 1952, Levi-Montalcini further tested her hypothesis using **tissue** cell cultures. Her tissue culture experiments regarding the presence of a substance in the tumor proved highly successful. However, there remained the important step of isolating this substance, which she called "the nerve-growth promoting agent" and later labeled nerve growth factor. Upon returning to Washington University, Levi-Montalcini began working with American biochemist Stanley Cohen between 1953 and 1959. During that time, they extracted NGF from snake venom and the salivary glands of male mice. Through these experiments, Cohen was able to

determine the chemical structure of NGF, as well as produce NGF antibodies. Levi-Montalcini maintained her interest in the research of NGF; and, when she returned to Italy in 1961, she established a laboratory at the Higher Institute of Health in Rome to perform joint NGF research with colleagues at Washington University.

By 1969 Levi-Montalcini established and served as director of the Institute of Cell Biology of the Italian National Research Council in Rome. Working six months out of the year at the Institute of Cell Biology and the other six months at Washington University, Levi-Montalcini maintained labs in Rome and St. Louis until 1977, at which time she resumed full-time residence in Italy. During this time she received numerous awards for her work, including becoming the tenth woman to be elected to the National Academy of Sciences in 1968. Despite her success, Levi-Montalcini was the only director of a laboratory conducting NGF research for many years. Later researchers, realizing the significance of understanding the growth of nerve fibers in treating degenerative diseases, have continued the work that Levi-Montalcini began in the late 1930s.

Levi-Montalcini remained active in the scientific community in her later years, upholding status as professor emeritus at Washington University from 1977 until her retirement in 1989, as well as contributing greatly to scientific studies and programs in her native country. After winning the Nobel Prize in 1986, she was appointed president of the Italian Multiple Sclerosis Association and also became the first woman to attain membership to the Pontifical Academy of Sciences in Rome. In 1987, she was awarded the National Medal of Science, the highest honor among American scientists.

Levi-Montalcini kept abreast with current scientific trends into her last years, conducting further research at the Institute of Cell Biology in Rome that focused on the importance of NGF in the immune and endocrine systems. Additionally, with her twin sister, who is an artist, Levi-Montalcini established educational youth programs that provide counseling and grants for teenagers interested in the arts or sciences. Recognized not only for her astute intuitive mind and her dedication to fully understanding the mechanisms of NGF, Levi-Montalcini, frequently described by her congenial manner and wit, influenced three generations of scientists during her own lifetime.

See also Biochemistry; Nerve impulses and conduction of impulses; Nervous system overview; Nervous system: embryological development; Neural damage and repair

LEWIS, EDWARD B. (1918-)
American developmental geneticist

Edward B. Lewis has dedicated a lifetime of research to the study of gene clusters responsible for early embryonic development. His tenacity resulted in important discoveries and led to formal recognition of his work. In 1995, Lewis was awarded the Nobel Prize in physiology or medicine for his groundbreaking genetic research. He shared the prize with two other scientists, Eric Wieschaus of Princeton University and **Christiane Nüsslein-Volhard** of the Max Planck Institute for Developmental Biology in Germany. Working independently of his co-recipients, Lewis studied "master control" gene clusters in fruit flies and subsequently discovered their corresponding human counterparts. Such a discovery promises to explain and eventually prevent congenital human malformations (about 40% of all human **birth defects**). It may also lead to improved in-vitro **fertilization** techniques, as well as a better understanding of substances harmful to early **pregnancy**.

Edward B. Lewis was born in Wilkes-Barre, Pennsylvania, to Edward B. Lewis and Laura (Histed) Lewis. His early years were spent trying to satiate his thirst for scientific knowledge in an environment that did not lend itself to learning. Books were not commonplace at home and as he remembered, "the high school library had nothing at all on genetics." Lewis found solace in playing the flute. He practiced daily, and during high school played with the local symphony orchestra. His musical abilities led to a scholarship at Bucknell University; however, Lewis transferred to the University of Minnesota, which offered course work in **genetics**. In 1939, Lewis received a B.A. degree in biostatistics from the University of Minnesota. He went on to earn a Ph.D. in genetics at the California Institute of Technology (Caltech) in 1942 and a M.S. in meteorology the following year. After serving as a weatherman in the Army during World War II, Lewis returned to Caltech to reestablish his affiliation with his alma mater.

Since the 1940s, Lewis has been a pioneer in the field of developmental genetics. The direction of his research was already set as a sophomore in high school: with the encouragement of a biology teacher, Lewis and a friend, Edward Novitski, purchased 100 fruit flies from Purdue University for one dollar. Lewis and Novitski let the flies breed, checking each day for any unusual new hatchlings. Their eagerness to learn something from a living specimen sparked careers in biology for both boys. In Lewis it created a lifelong obsession with the genetic workings of the fruit fly. In fact, it was a mutated fruit fly discovered by Novitski that led to Lewis's first postulations about the genetic factors causing mutations in the flies. Like Lewis, Novitski spent his professional life immersed in genetics research. Now retired, he resides in Eugene, Oregon.

Continuing his work with fruit fly specimens, Lewis was able to collect, crossbreed, and ultimately study an enormous amount of mutant flies. By mutating fly embryos so that the flies developed extra pairs of wings, Lewis was able to discern that it was not only the wings that were duplicated, but the whole body segment that contained the wings. Because the fruit fly has only eight **chromosomes** (humans have 23 sets), Lewis was able to pinpoint the gene sequence responsible for the development and order of each fly-body segment. His findings were published in a 1978 *Nature* paper entitled "A Gene Complex Controlling Segmentation in *Drosophila*." Since then, geneticists have discovered that the gene sequences are almost identical for all other animal species as well.

Lewis has often received recognition for his contributions to developmental genetics. In 1981, he was honored with a Ph.D. from the University of Umeå in Sweden. He received the Thomas Hunt Morgan Medal from the Genetics Society of America in 1983. He was awarded the Canadian Gairdner Foundation International Award in 1987 and Israel's Wolf Prize in Medicine in 1989. In 1990, he received three separate awards: the Lewis S. Rosenstiel Award in basic medical research, the National Medal of Science, and an honorary membership in the Genetics Society in Great Britain. Lewis won the prestigious Albert Lasker Basic Medical Research Award in 1991, the Louisa Gross Horwitz Prize in 1992, and was given an honorary Doctor of Science degree from the University of Minnesota in 1993.

See also Embryology; Embryonic development: early development, formation, and differentiation

LI, CHOH HAO (1913-1987)
Chinese-born American biochemist

Among other important contributions to endocrinology and neurochemistry, Choh Hao Li identified and isolated seven of the eight peptide **hormones** secreted by the pituitary gland.

Li was born in Canton, China, to an affluent family, and graduated in science at the University of Nanking in 1933. After teaching chemistry for two years at that institution, he applied for a Ph.D. program of the department of organic chemistry at the University of California at Berkeley. In 1950, he became professor of **biochemistry** at University of California at Berkeley. At Berkley in 1940, Li isolated and identified the **luteinizing hormone**, ACTH (adrenocorticotrophic hormone) that activates adrenocortical cells to produce steroid hormones. He was also able to partially synthesize it after establishing the ACTH molecular structure, in 1956. Li also identified the **human growth hormone** (HGH), also known as somatotropin in 1956, and fourteen years later, he succeeded in synthesizing this molecule.

Li found that HGH is a complex molecule made of 256 **amino acids**, which greatly differed from the growth hormone of other animals. This accomplishment was particularly important for the treatment of human dwarfism, a condition caused by underproduction of HGH. However, because HGH was then extracted from the pituitary of human corpses, it offered a certain degree of risk by contamination that could cause **brain** infections. With the advance of biotechnology, a much safer recombinant (genetically engineered) HGH was developed in 1985.

Choh Hao Li was appointed director of the Hormone Research Laboratory at Berkeley in 1950, and later served in the same position at University of California at San Francisco (UCSF) from 1967 until his death in 1987. In the 1960s while still at Berkeley, he investigated pituitary hormones involved in the **metabolism** of fat. Because it was difficult to obtain enough amounts from human sources, he acquired 500 dried camel pituitaries, only to discover that camels did not have the kind of fat metabolizer he was looking for. However, Li found and isolated another substance from these samples, a molecule termed beta-endorphin that he stored for future investigations. Years later, while working at UCSF, he resumed the study of beta-endorphin and found out that it contained encephalin. He isolated and purified the encephalin, in order to discover whether this molecule had a role in **pain** perception. His further experiments with encephalin showed that this endorphin was 48 times more powerful than morphine in relieving pain when injected in the brain, and three times more powerful when injected intravenously. He also observed that encephalin was as addictive as morphine, although the former was an endogenous (originating within the body) substance. His last scientific achievement was the identification and purification of insulin-like growth factor I (IGF-I), also called somatomedin C. Low levels of IGF-I are thought to account for the low stature of African pygmies, who have a congenital inability to synthesize significant amounts of this hormone. Low endogenous levels of IGF-I are also accountable for at least one type of dwarfism, the Levi-Lorain dwarfism.

Choh Hao Li published 1,100 scientific articles during his carrier, with about 300 collaborators. He was twice nominated for the Nobel Prize, and was awarded the Lasker Award for Basic Medical Research.

See also Endocrine system and glands; Hormones and hormone action; Pituitary gland hormones

LIFE · *see* FUNCTIONAL CHARACTERISTICS OF LIVING THINGS.

LIMB BUDS AND LIMB EMBRYOLOGICAL DEVELOPMENT

In **human development**, each limb results from a developmental field. The developmental fields are determined during gastrulation. These limb fields are established by the expression of HOX genes. The expression of the factor Tbx-5 causes the limb to develop into a hind limb, and expression of Tbx-4 causes the limb to develop into a forelimb. Beginning from the fourth week from **fertilization**, over a period of 25 days, a complex of genetic signals control the intricate pathways that result in a limb with the correct orientation, size, and number of digits. Limb development is a continuous process divided into four stages: the bud stage (initial outgrowth), the paddle stage (dorsoventral flattening), the plate stage (relative expansion of the distal end), and rotation stage (rotation around the proximodistal axis).

The limbs of the embryo develop from buds that protrude from the side of the main body axis. Limb buds arise on the lateral body at the level of sclerotomes as **ectoderm** and **mesoderm** (somite) proliferations. In particular, each limb bud consists of a mesenchymal core of mesoderm covered by an ectodermic cap. Limb buds will become the early arms and legs of the embryo. The upper limbs appear before the lower limbs that are delayed about two days in respect to the upper limbs. At the early stages of embryonic development, the forelimb and hind limb buds look like paddles on either side of the embryo and are indistinguishable from one another. The limb

buds continue their formation by the migration and proliferation of the differentiating mesenchymal tissues. The ectoderm at the tip of the bud thickens to form a specialized structure, called the apical ectodermal ridge. This structure is the signaling center that allows proper growth along the proximodistal axis (shoulders to digits). Concomitantly, the limb becomes flattened along the dorsoventral axis (back of hand to palm) and asymmetric along the anteroposterior axis (thumb to little finger). Proximodistal, dorsoventral, and anteroposterior axes represent the routes of the normal limb growth. The most proximal structure (stylopod) begins to differentiate first, followed by the progressive differentiation of more distal structures (zeugopod and autopod). This outgrowth and patterning depends on the establishment and maintenance of other signaling centers within the limb bud, named the zone of polarizing activity located in the mesenchyme at the posterior margin of the bud, and the non ridge ectoderm of the bud. These developmental components are interdependent and, through a series of reciprocal signals and feedback systems, yield the correct **tissue** patterning and growth. Each bud develops to form a complex of interconnected limb elements comprised of bone, muscle, and tendon characteristic of either the fore limb or hind limb. The actual trigger for limb bud initiation is still unknown, although likely candidates have been identified as Fibroblast Growth Factor 8 and 10.

The plate stage is characterized by the formation of flattened plate-like areas on the distal ends of the limbs called the hand plates and foot plates They are flattened along the dorsoventral axis. Within these distal plates, some structure is noticeable. There are radially arranged thickenings called digital rays (precursors of the digits). Between the digital rays are thin areas where cells begin to undergo apoptosis (programmed cell **death**) that allow the separating of the digits. The thin areas are called interdigital grooves. This arrangement gives rise to free digits. A constriction on the limb just proximal to the hand and footplates, called primary constrictions of the limb, is evident in this stage. These constrictions will develop into wrists and ankles. At approximately seven weeks, the longitudinal axes of the upper and lower limb buds are parallel.

In the rotation stage, the position of the limb buds relative to the trunk change is a predetermined manner not related to muscle activity or inherent osseous torsion. During this stage, the rotation of the limbs creates a three dimensional structure. Because of the differentiate growth of the **cartilage** model that continue to elongate the limb, different parts grow at different rates. This causes a twisting or rotation of each limb around its proximodistal axis. Upper limbs rotate one way (laterally or externally), while lower limbs rotate the other way (medially or internally) bringing the great toe to the midline from its initial postaxial position. This creates the characteristic positions of the limbs with the point of the elbow facing caudally and dorsally, and the knee facing cranially and ventrally. Consequently, the equivalent bones and muscles of the upper and lower limbs are oriented 180 degrees apart. This means that in the structural organization of the upper and lower limbs, their flexors and extensors are positioned on opposite sides and the movements at equivalent **joints** are in opposite directions.

The skeleton of the limb arises from somatic mesoderm by means of endochondral **ossification**. Formation of the intermediate segment (forearm) involves programmed apoptosis to separate a single mesenchymal condensation into two cartilage models (one for the radius and one for the ulna). In addition, separation of the digits, as previously mentioned, depends on apoptosis within the interdigital grooves. Cartilage breaks down to form the joints in specific points. Periosteum, ligaments, **tendons**, and intramuscular **connective tissues** form from the non-condensed mesenchyme. In addition to somatic mesoderm, there are cells that migrate into the limb bud from the body wall. These cells are identified into three groups: (1) somitic components (somitic myotomes in particular) that migrate into the limb buds and give rise to all of the musculature of the limb, (2) spinal nerves from the **brachial plexus** that go to the upper limb and from the lumbosacral plexus that go to the lower limb, and (3) **blood** vessel precursors going into the limb to provide the vasculature.

By the end of the eighth week, the limb is perfectly formed. From there on out, the only remaining development is growth that is synchronized with that of the fetal body.

See also Anatomical nomenclature; Embryology; Embryonic development, early development, formation, and differentiation; Lower limb structure; Upper limb structure

LIMBIC LOBE AND LIMBIC SYSTEM

The inferior prefrontal areas and perihilar regions of the **brain** cortex of both **cerebral hemispheres** are termed the limbic lobe, or limbic system. The limbic cortex is the oldest part of the **cerebral cortex** and is involved in the neurophysiological processing of emotions, instinctive behavior, and the regulation of **vegetative functions**, such as body temperature, hunger, **thirst**, volume of body fluids, cardiovascular regulation, and the contractibility of smooth muscles. Similar to other cerebral areas, the limbic lobe also presents a modular functional division of subcortical structures such as the portions of the **basal ganglia**, the amygdala, the septum, the anterior nucleus of the **thalamus**, and the paraolfactory, which surround the **hypothalamus**. The limbic cortex involves these subcortical limbic structures in each brain hemisphere and connects this older cortical portion (paleocortex) with the more recent cortex, (neocortex). The hypothalamus and other limbic centers send and receive signals from areas of the brain stem, such as reticular formation and the medial forebrain bundle. The hypothalamus is the main control-center of the limbic system, and is involved in the control of both endocrine and specific vegetative functions of the body, such as cardiovascular regulation, gastrointestinal and feeding regulation, temperature, water (thirst and urine excretion), uterine contractions of labor and birth, and milk ejection from the **breasts**. The hypothalamus also regulates hormonal secretion by the anterior portion of the pituitary gland.

The limbic system structures control behavior-related signals as well, such as satiety and tranquility (ventromedial nucleus), fear, punishment (thin zone of the periventricular nuclei and central gray area of the mesencephalon), and sexual drive (hypothalamus). Some other limbic areas also control reward and punishment sensations. The medial forebrain bundle in the lateral and ventromedial nuclei of the hypothalamus is associated with reward, although over-stimulation of the lateral nucleus can cause rage. The central gray area around the aqueduct of Sylvius and the periventricular zones of the hypothalamus and the **hippocampus** area associated with punishment sensations.

See also Brain stem function and reflexes; Brain: intellectual functions; Central nervous system

LIPIDS AND LIPID METABOLISM

A lipid is a molecule that is at least partially composed of a chain of hydrocarbons—carbons in association with hydrogen molecules. Hydrocarbon chains tend to exclude water, because there are no vacant sites along the chain for the water to bind to, and are thus described as being hydrophobic (from Latin meaning, "water hating"). Lipids are important in biological systems for a number of reasons, including the synthesis of important compounds and their use in membranes.

In biological systems, there are four classes of lipids. One class consists of fatty acids, compounds that contain a hydrogen chain that ends in a carboxylate group (COOH). Fatty acids can be saturated (there are no double bonds linking the carbons in the chain) or unsaturated (there are one or more double bonds present). Unsaturated fatty acids, such as those that make up margarine, tend to be softer at lower temperatures than saturated fatty acids (contrast the consistency of margarine and butter at room temperature). Some fatty acids are essential, that is, they cannot be made by the body and must be supplied in the diet. Linoleic and linolenic acids are two examples. Fatty acids also make up biological molecules such as lipopolysaccharide of **bacteria** and the phospholipids that form biological membranes

Another group of lipids is called glycerides. These are fatty acids that contain a phosphate group. The phosphate group is charged, and so ends to associate with water (hydrophilic, from Latin meaning, "water loving"). The presence of hydrophilic and hydrophobic regions on the same molecule allows phosholipids to arrange themselves in so-called bilayers, with the phosphate portions oriented outward toward the water and the fatty acid tails buried inside away from the water. This allows membranes to form. The presence of proteins in the phospholipid bilayer allows membranes in different **organs** or areas of the body to function in different ways.

A third group of lipids is the so-called complex lipids, that consist of a lipid portion and another portion made of protein or other constituent. Lipoprotein and cholesterol are two complex lipids. There are several different kinds of lipoproteins, having different physiological functions, and which are beneficial or detrimental to health. The fourth group of lipids is called nonglycerides. This group includes steroids, which can function in keeping membranes pliable (cholesterol, which can also be a health hazard if it accumulates and clogs **arteries**) and stimulating changes in females (**progesterone**) and males (**testosterone**). Another nonglyceride is cortisone, which is useful in the treatment of various maladies including rheumatoid arthritis and asthma.

Lipids such as fatty acids are an important source of energy for humans. They are stored in **fat** droplets in specialized storage cells called adipose cells. If needed, the lipids are broken down into simpler compounds, converted to small spherical structures called chylomicrons, and routed to the bloodstream. There, the lipids can be carried throughout the body for use as fuel. Conversion of the chylomicrons to energy takes place in a specialized compartment of cells known as the **mitochondria**.

See also Adipose tissue; Fat, body fat measurements

LIPMANN, FRITZ (1899-1986)

American biochemist

Fritz Lipmann's landmark paper, "Metabolic Generation and Utilization of Phosphate Bond Energy," published in 1941, laid the foundation for research into concepts involving group potential and the role of group transfer in biosynthesis. Lipmann had revealed the basis for the relationship between metabolic energy production and its use, providing the first coherent picture of how living organisms operate. His discovery in 1945 of coenzyme A (CoA), which occurs in all living cells and is a key element in the **metabolism** of **carbohydrates**, fats, and some **amino acids**, earned him the 1953 Nobel Prize in physiology or medicine. Lipmann also conducted groundbreaking research in **protein synthesis**. Lipmann was an instinctual researcher with a knack for seeing the broader picture. Lacking the talent or inclination for self-promotion, he struggled early in his career before establishing himself in the world of **biochemistry**.

Fritz Albert Lipmann was born on June 12, 1899, in Königsberg, the capital of East Prussia (now Kaliningrad, Russia). After graduating from the gymnasium, Lipmann decided to pursue a career in medicine, largely due to the influence of an uncle who was a pediatrician and one of his boyhood heroes. In 1917 he enrolled in the University of Königsberg but had his medical studies interrupted in 1918 as he was called to the medical service during World War I. In 1919, he was discharged from the army and went to study medicine in Munich and Berlin. He eventually returned to Königsberg to complete his studies and obtained his medical degree from Berlin in 1922. Lipmann's interest in biochemistry and physiology was further cultivated when during his practical year of medical studies he worked in the **pathology** department in a Berlin hospital and took a three-month course in modern biochemistry taught by Peter Rona. Returning to Königsberg, Lipmann, chose to do his thesis work with bio-

chemist **Otto Meyerhof**, whose physiological investigations focused on the muscle.

For the most part, Lipmann worked on inhibition of glycolysis (the breakdown of glucose by **enzymes**) by fluoride in **muscle contraction** and did his doctoral dissertation on metabolic fluoride effects. During this time in Berlin, Lipmann met many of the era's great biochemists, including Karl Lohmann, who discovered **adenosine triphosphate** (ATP—a compound that provides the chemical energy necessary for a host of chemical reactions in the cell) and taught Lipmann about phosphate ester chemistry, which was to play an important role in Lipmann's later research. During this time Lipmann also met his future wife.

Over the next ten years, Lipmann continued with a varied research career. Lipmann spent a short time in Heidelberg when Meyerhof moved his laboratory there but then returned to Berlin and worked with Albert Fischer on **tissue** culturing and the study of metabolism as a method to measure cell growth. Soon, uniformed followers of Hitler began to appear in the streets of Berlin; both Lipmann and his wife had unpleasant encounters—and Lipmann was physically assaulted. Realizing that they would soon have to leave Germany, Lipmann, through Fischer's intervention, received an offer to work at the Rockefeller Foundation (now Rockefeller University). At the Rockefeller Foundation, Lipmann worked in the laboratory of chemist Phoebus Aaron Theodor Levene, who had conducted research on the egg yolk protein, which he called vitellinic acid, and found that it contained 10% bound phosphate (that is, phosphate strongly attached to other substances). Lipmann's interest in this protein, which served as food for growing animal tissues, led him to isolate serine phosphate from an egg protein.

At the end of the summer of 1932, Lipmann and his wife returned to Europe to work with Fischer, who was now in the Biological Institute of the Carlsberg Foundation in Copenhagen, Denmark. Free to pursue his own scientific interests, Lipmann delved into the mechanism of fermentation and glycolysis and eventually cell energy transformation. In the course of these studies, Lipmann found that pyruvate oxidation (a reaction that involves the loss of electrons) yielded ATP. Lohmann, who first discovered ATP, had also found that creatine phosphate provides the muscle with energy through ATP. Further work led Lipmann to the discovery of acetyl phosphate and the recognition that this phosphate was the intermediate of pyruvate oxidation. A discovery that Lipmann said was his most impressive work and had motivated all his subsequent research.

Before Lipmann could piece together his formula for the foundation of how organisms produce energy, once again the rise of the Nazis forced him to flee to the United States. Lipmann acquired a research fellowship in the biochemistry department of Cornell University Medical College. His work with pyruvate oxidation and ATP had germinated and set him on a series of investigations that led to his theories of phosphate bond energy and energy-rich phosphate bond energy. During a vacation on Lake Iroquois in Vermont, Lipmann began his essay "Metabolic Generation and Utilization of Phosphate Bond Energy," in which Lipmann first proposed the notion of group potential and the role of group transfer in biosynthesis.

Although his essay covered a wide range of topics, including carbamyl phosphate and the synthesis of sulfate esters, Lipmann's explanation of the role of ATP in group activation (such as amino acids in the synthesis of proteins), foretold the use of ATP in the biosynthesis of macromolecules (large molecules). In more general terms, he identified a link between generation of metabolic energy and its utilization. A prime example of ATP's role in energy transmission was the transfer of phosphor potential from ATP to provide the energy needed for muscles to contract.

Subsequently, Lipmann gained an unusual appointment in the Department of Surgery at the Massachusetts General Hospital through the support of a Ciba Foundation fellowship. Building upon his group transfer concept, Lipmann delved into the nature of the metabolically active acetate, which had been postulated as an "active" intermediary in group activation. In 1945, working with a potent enzyme from pigeon **liver** extract as an assay system for acetyl transfer in animal tissue, Lipmann and colleagues at Massachusetts General Hospital discovered Coenzyme A (CoA), the "A" standing for the activation of acetate. (Coenzymes are organic substances that can attach themselves to and supplement specified proteins to form active enzyme systems.) Eventually, CoA would be shown to occur in all living cells as an essential component in the metabolism of carbohydrates, fats, and certain amino acids. In 1953, Lipmann received the Nobel Prize in physiology or medicine for his discovery specifically of the acetyl-carrying CoA, which is formed as an intermediate in metabolism and active as a coenzyme in biological acetylations. (Lipmann shared the prize with his old colleague and friend, **Hans Krebs**, from Berlin.) Although proud of the Nobel Prize, Lipmann often stated that he believed his earlier work on the theory of group transfer was more deserving.

In 1957 Lipmann once again found himself at the Rockefeller Institute, twenty-five years after his first appointment there. Lipmann was to spend the next thirty years at the institute, primarily working on the analysis of protein biosynthesis. He and his colleagues contributed greatly to our understanding of the mechanisms of the elongation step of protein synthesis (stepwise addition of single amino acids to the primary protein structure).

In addition to the Nobel Prize, Lipmann received the National Medal of Science in 1966 and was elected a foreign member of the Royal Society in London.

See also Glucose utilization, transport and insulin; Glycogen and glycolysis; Lipids and lipid metabolism; Protein metabolism

LIVER

The liver is the largest internal organ located in the upper right quadrant of the **abdomen** just below the **diaphragm**. It is a highly vascular organ containing 10–20% of total **blood** volume.

Liver tissue containing fatty, cirrhotic lesions, a complication of cystic fibrosis. *Photograph by Joseph Siebert. Custom Medical Stock Photo. Reproduced by permission.*

The basic functional units of the liver are the hexagonal lobules, which are surrounded by connective **tissue**. In the corner of each lobule is a portal triad composed of the portal vein, hepatic artery, and hepatic duct. Through the middle of the lobule runs a central vein surrounded by the hepatic cords composed of hepatocytes (liver epithelial cells). The spaces between the cords are filled by liver sinusoids connecting the portal triads with the central vein and containing endothelial cells as well as Kupffer cells (hepatic phagocytic cells). The lobules are organized into four lobes, two major (right and left) and two minor (caudate and quadrate).

Oxygenated blood comes into the liver through the hepatic artery and deoxygenated blood flows through the portal vein (coming from the **gastrointestinal tract**). An exchange of substances occurs in the liver sinusoids and the blood is collected in the central vein and then the hepatic vein. Due to the large number of sinusoids (**capillaries**) that can be filled up with blood, the liver can become a blood reservoir if required.

The liver plays an important role in **metabolism, digestion**, detoxification and removal of **bacteria**, debris, and waste products through phagocytosis and **bile** synthesis. The most important metabolic activities of the liver are the regulation of blood sugar levels through **glycogen** formation (high sugar) or gluconeogenesis (low sugar), **protein synthesis** (albumin, fibrinogen, heparin, clotting factors, and globulins) and recycling of **amino acids**, **fat** metabolism (synthesis of phospholipids, lipoproteins, and breakdown of fatty acids). Another important liver function is storage of copper, iron, **vitamins** A, B_{12}, D, E, and K.

The detoxification role of the liver is centered on the conversion of toxic ammonia to urea, which is then secreted into the circulation and finally excreted by the **kidneys**. Drugs present in the circulation are inactivated in the liver at various rates lowering their availability in blood. Removal of particulate matter, such as dead blood cells, bacteria, and general debris is carried out by Kupffer cells.

An important role is played by the liver in the digestion process through bile production by hepatocytes. Bile is mainly an excretory compound but is also essential for the dilution of stomach acids and emulsification of fats.

Liver diseases are often fatal and can occur through different disease processes. One non-fatal disease is hemochromatosis, an iron storage defect leading to accumulation of iron in the liver. Problems with liver function can be manifested through appearance of jaundice that is yellowing of the tissues due to a high concentration of **bilirubin** in the blood. Jaundice is commonly associated with hepatitis (inflammation of the liver) caused mainly by **viral infection**. Some hepatitis **viruses**, (e.g., B and C) can cause chronic **infection** leading to the formation of scar tissue, loss of liver function, and often to liver **cancer**.

Cirrhosis is a process that can occur as a result of infection, prolonged inflammation, or as a result of alcohol abuse. It involves the death of hepatocytes and their replacement with the fibrous tissue. As a result, liver tissue becomes paler due to lower blood flow and has a nodular appearance due to accumulation of fibrous material.

The liver has a remarkable ability to regenerate tissue, provided that the circulation is not fully destroyed. However, in cases of severe damage, a liver transplant may be the only option for treatment of liver disease.

See also Bile, bile ducts, and the biliary system; Metabolic waste removal; Organs and organ systems

LOCOMOTION · *see* POSTURE & LOCOMOTION

LOEWI, OTTO (1873-1961)

German-born American pharmacologist and physiologist

Otto Loewi (pronounced *lōee*) made important early discoveries about how nerve impulses are transmitted. He found that nerve transmissions depend both on electrical stimulation and certain chemical substances (neurotransmitters) produced by nerve cells. For this discovery he shared the 1936 Nobel Prize in physiology or medicine with **Henry Hallett Dale**, a British scientist who showed that chemical transmissions of nerve impulses take place in the voluntary, as well as the involuntary, nervous system.

Loewi was born in Frankfurt am Main, Germany, on June 3, 1873, the son of Jakob Loewi, a wealthy Jewish wine merchant, and Anna Willstadter Loewi. As a young man, Loewi wanted to be an art historian, but he was dissuaded from pursuing this career by his parents, who urged him to study medicine instead. Although he had done poorly in mathematics and physics at the Frankfurt Gymnasium, he took their advice and, in 1891, entered the University of Strasbourg to study medicine. He was fortunate at Strasbourg; there were many excellent teachers and researchers there who developed his interest in biology and physiology. Loewi received his medical degree in 1896.

After graduation, Loewi briefly visited Italy, and then returned to Germany for more training in chemistry and experimental methods. During this period, he also worked in the tuberculosis and pneumonia wards at the City Hospital of Frankfurt, where he was discouraged from continuing with clinical medicine because of high **death** rates. Instead, he turned his attention to an academic career in scientific research, and in 1898 he joined the department of **pharmacology** at the University of Marburg, first with an assistantship and then as a lecturer.

By 1902, Loewi had published the results of his scientific research at Marburg. His work dealt with the functioning of the **kidneys** and the effects on these **organs** of substances that increase the production of urine, known as **diuretics**. In 1903, along with other researchers including Henry Hallett Dale, Loewi began to consider the chemical transmission of nerve impulses. The hormone adrenaline and the chemical muscarine had already been identified as possible nerve transmitters by several English physiologists. In 1905, Loewi followed Hans Meyer, under whom he had worked at Marburg, to the University of Vienna. That same year he met Gulda Goldschmiedt, the daughter of a chemistry professor, in Switzerland, and he married her the following year. They would have four children.

At the University of Vienna, Loewi concentrated on the effects of adrenaline and noradrenaline on diabetes and **blood** pressure. He also studied the response of the **heart** to the stimulation of the vagus nerve, one of the main **cranial nerves** in the autonomic system. In 1909, he was appointed to the University of Graz as a professor of pharmacology, where he remained until the German occupation of Austria in 1938.

By 1921, fifteen years after the idea of chemical transmission of nerve impulses had first been proposed by the English physiologists, scientists had still not discovered definite evidence of the existence of a chemical transmitter within the nervous system. At this time, Loewi was fully engaged in the search for such evidence, and one night he had a dream for the design of an experiment that would determine the existence of a chemical transmitter. He jotted down some notes from the dream, still half asleep, but when he awoke the next morning, he could not read his scrawl. The next night at three o'clock the idea returned to him; this time he immediately went to his laboratory.

For this path breaking experiment, Loewi used two hearts from frogs. He removed the vagus nerve from the first heart, and he stimulated the same nerve in the second one. After stimulating the nerve in the second heart, he removed some fluid and injected it into the heart without the vagus nerve. He observed that the rate of this heart slowed as if the vagus nerve had been stimulated. Then he stimulated the heart with the vagus nerve so it would beat faster. He again removed fluid and injected it into the heart without the vagus nerve. Its rate increased as if it had been stimulated directly by the missing nerve.

Loewi had established the role chemicals play in the transmission of nerve impulses, but he was not sure at first what these chemicals were. He called one "vagus substance" and the other "accelerator substance." Over the next fifteen years, Loewi, along with his colleagues, published a number of papers on the results of his initial experiment. What he had called vagus substance was identified as acetylcholine in 1926; other transmitters were later identified. In 1936, Loewi identified adrenaline as one of the **sympathetic nervous system** transmitters and noradrenaline as the most important one. Henry Hallett Dale shared the 1936 Nobel Prize with Loewi for his discovery of chemical transmitters in the voluntary nervous system.

After the German occupation of Austria in 1938, Loewi was only allowed to leave because he turned over his Nobel Prize money to the Nazis. His family was also able to escape, and they joined him in New York City in 1940. He became a United States citizen in 1946. He spent the rest of his life writing articles, delivering lectures, and writing his memoirs.

See also Action potential; Nerve impulses and conduction of impulses; Nervous system overview; Nervous system: embryological development; Neurology; Neurons; Neurotransmitters

LOWER LIMB STRUCTURE

The lower limbs of humans consist of the paired appendages called the legs. Each leg is made up of an upper leg, lower leg, and the foot.

The upper portion of the leg begins at the junction with the pelvic girdle. From the girdle region a bone called the femur inclines downward to the knee joint. The inclination of the femur brings the knees in line with the center of gravity of the body. The femur is the longest and strongest bone in the body. It is a nearly perfect cylinder. At its top end the femur is ball-shaped, so as to fit into the socket of the pelvic girdle. Near the ball-shaped region, the femur flares upward before the main part of the bone descends. The two protrusions that are present are known as the trocanters. They provide leverage for the muscles that manipulate the thigh. The body of the femur (also known as the shaft or the corpus femoris) is strengthened along its length by a ridge (the linea aspera) running down the bone. Testing has demonstrated that the femur is designed to withstand a great deal of stress and load, consistent with its location in a weight bearing part of the body.

Both the front (calf) and the back of the upper leg are well muscled, and supply much of the strength and power for the leg. In athletes, the gastrocnemius and soleus muscles of the calf are often very pronounced. Prominent **tendons** are also found here. For example, the Tendo Calcaneus is the thickest and strongest tendon in the body.

The knee forms the articulated (capable of movement) joint between the femur and the next lower bone of the leg, the tibia. The knee in protected and stabilized by the patella (also known as the kneecap), a flat, triangular bone. Ligaments and other **tissue** also help stabilize the knee against excessive side-to-side movement, and cartilaginous material called the meniscus acts to cushion the junction of the femur and the tibia.

The tibia is the second longest bone in the body, after the femur. It extends the length of the lower leg, from the knee to the ankle. It is closely associated with a long and slender bone called the fibula (or calf bone)

The back of the lower leg is also well muscled. The muscles assist in leg and foot motion.

The lower leg connects to the foot. The foot is a complex structure. There are over 20 bones in each foot alone. The foot is also well muscled and is supported by ligaments and tissue known as **fascia**. Support is of prime importance in the foot, as it bears the weight of the body and must adopt different configurations to permit **locomotion**.

See also Skeletal muscle

LOWER, RICHARD (1631-1691)
English physician

Richard Lower made significant contributions to cardiology, was among the earliest physicians to perform **blood transfusions**, and assisted with the famous neuroanatomical investigations of **Thomas Willis**.

Lower was born the son of a country gentleman on the family estate in Tremeer, Cornwall, England. His exact birthdate is unknown but he was baptized as an infant on 29 January 1631. From 1643 to 1649 he attended the most celebrated British preparatory school of the time, Westminster School of St. Peter's College, London. There he won the praise of the headmaster, Richard Busby, who sent him to Christ Church College, Oxford University, in 1649. He received his B.A. in 1653, his M.A. in 1655, and both his B.Med. and D.Med. in 1665, all from Oxford.

Lower studied chemistry at Oxford under Peter Stahl (d. 1675) and became the protege of Willis, who was then renowned as the greatest medical scientist in England. Lower was part of an informal group of researchers known as the "Oxford physiologists." Among his scientific collaborators at Oxford, besides Willis, were Ralph Bathurst (1620–1704), Robert Boyle (1627–1691), **Robert Hooke**, John Locke (1632–1704), John Mayow (1641–1679), Thomas Millington (1628–1704), Walter Needham (1631–1691), William Petty (1623–1687), Henry Stubbe (1632–1676), John Wallis (1616–1703), John Ward (1617–1689), and Christopher Wren (1632–1723). Lower, Millington, and Wren all helped Willis to produce the monumental *Cerebri anatome* [Anatomy of the Brain] in 1664.

In 1666, Lower married Elizabeth Billing, with whom he had two daughters. When Willis moved to London in 1666, Lower soon followed him, set up his own medical practice, but continued working with Willis on several research projects. Lower was elected a fellow of the Royal Society in 1667 and a fellow of the Royal College of Physicians in 1675. After Willis died in 1675, Lower became London's leading physician for a short while, but his outspoken Whig politics drove most of his highborn patients away after 1678. He died in London.

In the early seventeenth century, mainly on the European continent, several medical and surgical researchers considered the possibility of using blood as a rejuvenating or healing agent, either by transfusing human blood from the old into the young or by injecting blood from healthy animals into diseased humans. Giovanni Colle (1558–1631) performed many transfusion experiments, Johannes Scultetus (1595–1645) reported his successful transfusion of dog blood into humans, but Lower is generally credited with inventing the first reliably safe transfusion process among dogs, humans, and sheep. He reported these experiments in the *Philosophical Transactions of the Royal Society* from 1665 to 1667. In connection with this work, he significantly improved the design of the syringe.

Tensions ran high in the mid-seventeenth century between adherents of the ancient medical theories of **Galen** and followers of the new **physiology** of **William Harvey**. The Oxford physiologists were all Harveians. Harvey himself was at Oxford from 1642 to 1646. One staunch Galenist, Edmund Meara (d. 1680), denounced Willis and the entire Harveian interpretation of physiology in *Examen diatribae Thomae Willisii de febribus* [Examination of the Discourse of Thomas Willis on Fevers] (1665). As Willis's disciple, Lower rushed to his master's defense. His first publication, *Diatribae Thomae Willisii de febribus vindicatio* [Vindication of the Discourse of Thomas Willis on Fevers], appeared only four months later. It was a vigorous polemic, which not only counterattacked Meara, but also defended Boyle and Harvey and laid out a Harveian agenda for further research into blood physiology. It contained some of the earliest correct observations about lung function, heart-lung interaction, the differences between arterial and venous blood, and the relation of respiration to blood color.

Lower's *Tractatus de corde* [Treatise on the Heart] (1669) contained many important advances in cardiology, including the first accurate anatomical description of the structure of **heart** muscle. He improved Harvey's theory of blood circulation and speculated about why dark blood from a vein turns bright red when exposed to air. The best known section of this book, "Dissertatio de origine catarrhi" ["Dissertation on the Origin of Catarrh"], disproved the traditional belief that nasal congestion was caused by **mucus** dripping down from the **brain** or the pituitary.

See also Cardiac cycle; Circulatory system; History of anatomy and physiology: The Renaissance and Age of Enlightenment; History of anatomy and physiology: The science of medicine; Lungs; Mucus; Pulmonary circulation

LUMBAR SPINE · *see* VERTEBRAL COLUMN

LUNGS

The lungs are vital **organs** located in the chest (thoracic) cavity of the human body. They function in gas exchange by

obtaining oxygen from the air to provide energy for cells throughout the body and by eliminating carbon dioxide. Air enters through the nasal cavities, travels into the main airway (called the trachea) and then into the lungs. There are two lungs. The left lung is smaller (to accommodate the **heart**), while the right lung is slightly larger. The trachea branches into the **bronchi**, which are tree-like tubes that connect to each lung. The bronchi branch further into smaller passageways called the bronchioles, which lead to the **alveoli**. The alveoli are sac-like structures (that look like clusters of grapes on a stem) where gas exchange occurs. Unlike the trachea, the bronchioles can be recognized by an absence of **cartilage**. Pulmonary arterioles are **blood** vessels that travel juxtaposition to the bronchi and branch more and more extensively, turning into pulmonary **capillaries** that cover the alveoli. Oxygen diffuses from the alveoli into the pulmonary capillaries that surround the alveolar sacs. This is how oxygen gets transported from the lungs to tissues throughout the body.

The airways are categorized into two major sections, the conducting portion and the respiratory portion. The conducting portion of the airways spans from the trachea to the terminal bronchioles, while the respiratory portion includes from the respiratory bronchioles to the alveoli. The conducting portion serves not only as a passageway for air to travel to the alveoli, but also to warm and humidify the air. It is also important in protecting the body from aerosolized pathogens (disease-causing microorganisms) and contaminants in the air. This is achieved this by tiny finger-like structures called **cilia** that beat synchronously to remove particulate material trapped in mucous secreted by airway cells, a process called mucociliary clearance. The surface area increases exponentially from the conducting to the respiratory portions of the airways. With over 300 million alveoli, there is a large enough surface area (enough to be the size of a tennis court) for massive amounts of gas exchange and coupled to a very short diffusion path, the body can obtain sufficient amounts of oxygen, especially helpful during strenuous cardiovascular exercise. Additionally, the lung volume has a large reserve capacity beyond normal **breathing** to account for increases in ventilation, or the movement of oxygen into the lungs by bulk flow.

The bronchus, bronchiolus, and alveolus can be distinguished morphologically in terms of differing cellular and structural organization. The airway epithelium (or the outermost cells in contact with the airway surface) is composed of a continuous network of cells that are anchored by a thin membrane called the basement membrane. The basement membrane is found throughout the airways. Below the basement membrane is a type of **tissue** called connective tissue that tapers off in the bronchiolar epithelia and is absent in the alveoli. The epithelium of the bronchi consists of Goblet cells, which are the primarily mucous secreting cells, as well as cells that have cilia. Morphologically, the cells in the bronchi are columnar meaning that they appear rectangular or column-shaped. There are also glands called submucosal glands in the epithelia that secrete various substances such as mucous. As the airways branch into the bronchioles, the cells become cuboidal, or cube-shaped, and there are no submucosal glands. The alveoli do not contain ciliated or secretory epithelial cells

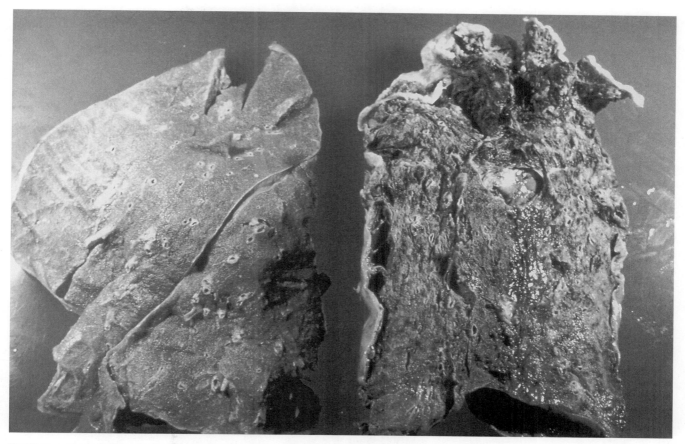

Healthy lung tissue (left) contrasts with diseased lung tissue (right) damaged by cigarette smoking. *Photograph by A. Glauberman. National Audubon Society Collection/Photo Researchers, Inc. Reproduced by permission.*

but are composed of Type I and Type II alveolar epithelial cells. These cells are squamous (or flattened) in shape and are widely spaced to allow for the diffusion of oxygen into the pulmonary capillaries.

The lungs can expand and contract without friction during breathing due to the **pleura**, a thin membranous structure. The visceral pleura surrounds the lungs, while the parietal pleura lines the wall of the thoracic cavity. The visceral and parietal pleura are separated from each other by a small fluid-filled space called the pleural cavity. Ventilation requires work and before the lungs can become inflated, a pressure change must take place. The elastic properties of the lung allow ventilation to take place more efficiently and the fluid in the pleural cavity serves as a lubricant that allows the lungs to slide against the chest wall.

See also Gaseous exchange; Respiration control mechanisms; Respiratory system embryological development; Respiratory system

LUPUS • *see* AUTOIMMUNE DISORDERS

LUTEINIZING HORMONE (LH) • *see* FOLLICLE STIMULATING HORMONE (FSH) AND LUTEINIZING HORMONE (LH)

LUZZI, MONDINO DE (1270-1326)
Italian anatomist

Born in medieval Florence, Mondino de Luzzi was the first and most important anatomist of the Middle Ages, a time when the medical knowledge of the human body was still generally based upon both Galen's **anatomy** and Aristotle's natural philosophy, due to the Catholic Church partial prohibition (and in earlier periods total prohibition) against human cadaver dissection. (Pope Boniface VIII prohibited in 1299 the dismembering of human cadavers, the removal of internal **organs**, and the separation of flesh from bones.) However, the Greek physician Claudius **Galen** had only dissected animals, and most of his **physiology** was based on ancient assumptions derived from the theory of the four elements as governing the different organs, taught in Ancient Greece. Mondino took the risks of clerical punishment (excommunication by the Church) and became the first medieval physician to dissect human cadavers at the University of Bologna, in his search for a better understanding of the constitution of the human body.

Mondino studied medicine at the University of Bologna, where he later became a professor of anatomy and surgery. He was the first to re-introduce the systematic teaching of anatomy in the medical curriculum, and performed dissections in front of the students during anatomy classes. Mondino wrote in 1316 his *Anathomia Mundini (Mondino's Anatomy),* a didactic treatise based on his direct observation during dissections, which was first published in 1478. His book remained as the standard medical textbook of anatomy for almost two centuries, and was reprinted 39 times until the publication of the anatomical treatise in 1543, *De Humani Corporis Fabrica Libri Septem,* by **Andreas Vesalius** (1514–1564).

Mondino's Anatomy also showed the influence of Galen's work as well as that of **Avicenna's,** but he aimed to demonstrate the importance of anatomy for the evolution of surgical techniques. He also established a teaching methodology, recommending that the dissection should be performed in four days, each of the three first days dedicated to a different part of the body: **abdomen,** thorax and head respectively, and the fourth day to the dissection of body extremities. Therefore, the more perishable parts were studied first, such as the organs of the digestive tract, followed by the **lungs** and the **heart,** then the head organs, and finally the limbs. He also described in his work the dissection techniques for each part of the body. For instance, Mondino advocated the vertical incision of the abdomen as the better way of opening it. Mondino's emphasis in teaching was that medical knowledge should be based on reason and direct observation, but should also develop in the student's mind an admiration for the creative work of God, the ultimate goal of the natural philosophy in the Middle Ages. *Mondino's Anatomy* also contains a rudimentary and imprecise attempt to describe the nervous system. However, in the Middle Ages his work shed a new light on the knowledge of the **brain** and the **spinal cord** anatomy.

See also History of anatomy and physiology: The Classical and Medieval periods

LYMPHATIC SYSTEM

The lymphatic system is a complex network of thin vessels, **capillaries,** valves, ducts, nodes, and **organs** that runs throughout the body, helping protect and maintain the internal fluids system of the entire body by both producing and filtering lymph, and by producing various **blood** cells. The three main purposes of the lymphatic system are to drain fluid back into the bloodstream from the tissues, to filter lymph, and to fight infections.

The lymphatic system includes the spleen, the thymus, the lymph nodes, and the lymph ducts. It has a major component in immunity.

The lymphatic system branches through all parts of the body carrying lymph, a clear, watery, blood **plasma** fluid containing proteins, leucocytes, and glucose. Lymph contains red

blood cells and many white blood cells, especially lymphocytes. Lymphocytes are the cells that attack **bacteria** in the blood. Lymph is produced by small bean-shaped lymph nodes, soft nodules that are not usually externally visible or easily felt. They are located in clusters in various parts of the body, such as the neck, armpit, and groin.

Lymph seeps outside the blood vessels in spaces in body tissues and is stored in the lymphatic system to flow back into the bloodstream. Lymph nodes help prevent **infection** by filtering out foreign material, such as bacteria and **cancer** cells, and destroying toxins and germs. They also make lymphocytes that produce antibodies and attack cells infected with virus. Antibodies are proteins that detect invading foreign cells and help to kill them. This process is sometimes called the immune response.

When bacteria are detected in the lymph fluid, the lymph nodes swell as they produce and supply additional white blood cells to help fight off whatever pathogen has entered the body. At this stage, lymph nodes may be felt by infection sufferers. An extreme example is that of bubonic plague victims, who could be identified by the large bumps on their body called "buboes." These were swollen lymph nodes filled with infected material.

There are two main types of lymphocytes: **T lymphocytes** and B lymphocytes. The B lymphocytes make antibodies. One kind of T lymphocyte assists in the production of antibodies, and the other fights infection directly. These killer T cells are being studied for their potential to fight cancer.

The lymphatic system runs parallel to the bloodstream. As the blood circulates, fluid seeps into the body tissues. This fluid carries **nutrition** to the cells and wastes back to the bloodstream. The lymphatic system also returns water, fats, proteins and other substances to the blood. The spleen filters the lymph to remove old red blood cells. These are destroyed and replaced by new red blood cells made in the bone marrow.

The lymphatic system may be considered as a slow-flowing drainage network that collects **interstitial fluid** throughout the body, filters it, destroys foreign matter and germs, and returns it to the bloodstream.

See also Immune system; Immunity, cell mediated; Immunity, humoral regulation; Immunology; Infection and resistance

LYNEN, FEODOR FELIX KONRAD
(1911-1979)
German biochemist

Feodor Felix Konrad Lynen's work led to a better understanding of how cells make and use cholesterol and other materials necessary for life. His discovery of the structure of acetyl-coenzyme A led to a detailed description of the steps of several important life processes, including the **metabolism** of both cholesterol and fatty acids. Aside from influencing **biochemistry,** his work was also important to medicine because cholesterol was known to contribute to **heart** attacks, strokes, and

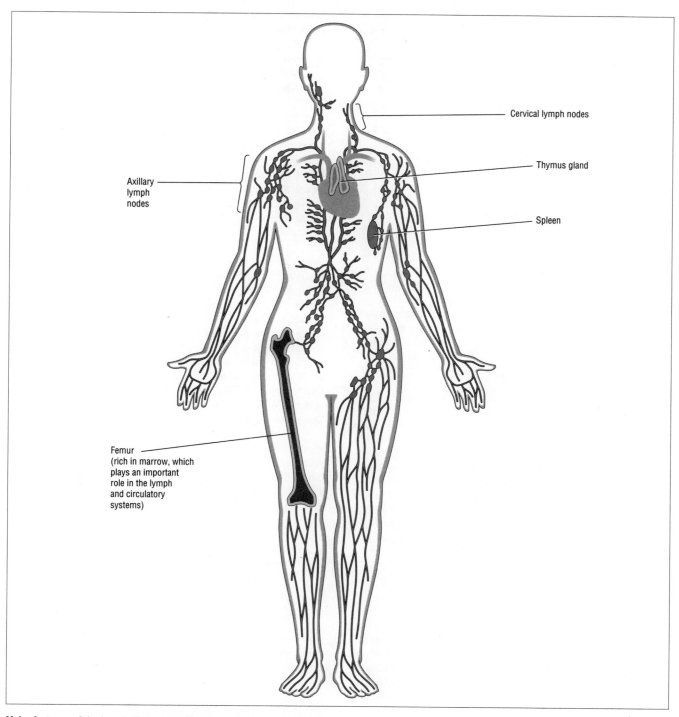

Cervical lymph nodes

Thymus gland

Spleen

Axillary lymph nodes

Femur
(rich in marrow, which plays an important role in the lymph and circulatory systems)

Major features of the lymphatic system. *Illustration by Argosy Publishing.*

other circulatory diseases. For his work on cholesterol and the fatty acid cycle, Lynen shared the 1964 Nobel Prize in physiology or medicine with German-American biochemist **Konrad Emil Bloch**.

Feodor Lynen was born in Munich, Germany, on April 6, 1911, the seventh of eight children. His father, Wilhelm L.

Lynen, was a professor of engineering at the Munich Technical University. His mother, Frieda (Prym) Lynen, cared for the family. Lynen showed an early interest in his older brother's chemistry experiments but remained undecided about his career throughout secondary school. He considered medicine and even thought of becoming a ski instructor. Ultimately, he enrolled in the Department of Chemistry at the University of Munich in 1930, where he studied with German chemist and Nobel laureate Heinrich Wieland. Wieland was Lynen's principal teacher both as an undergraduate and graduate student. In 1937, Lynen received his doctorate degree. Three months later, he married Wieland's daughter, Eva, with whom he would have five children.

Upon graduation, Lynen stayed at the University of Munich in a postdoctoral research position. In 1942, he was appointed a lecturer, and he became an associate professor in 1947. Throughout his years with the University, where he stayed until his death, Lynen supervised the research of nearly ninety students, many of whom reached leading positions in academia or industry.

In Nazi Germany, Lynen remained exempt from both military service and work in Nazi paramilitary organizations because of a permanently damaged knee that resulted from a ski accident in 1932. The onset of World War II, however, made it difficult to continue working in Munich and, in an effort to maintain his research, Lynen moved his lab to a small village, Schondorf on the Ammersee, about eighteen miles from Munich. In 1945, the University's Department of Chemistry in Munich was destroyed. In the chaotic aftermath of Germany's surrender, scientific research halted altogether. Lynen eventually continued his work at various lab facilities, but did not return to the rebuilt Department of Chemistry until 1949.

In the first years after the war, German scientists were spurned by their European and American colleagues. Only four German biochemists were invited to attend the First International Congress of Biochemistry held in Cambridge, England, in July of 1949. Lynen, one of the four, made an ideal good-will ambassador for Germany. His cheery nature and solid research drew many foreign scientists to Munich. His magnetic personality was formally recognized years later when, in 1975, he was chosen to serve as president of the Alexander von Humboldt Foundation, an institution devoted to fostering relations between Germany and the international scientific community.

During the 1940s, Lynen began studying how the living cell changes simple chemical compounds into sterols and **lipids**, complex molecules that the body needs to sustain life. The long sequence of steps and the roles various **enzymes** and **vitamins** played in this complicated metabolic process were not well understood. After World War II, Lynen began to publish his early findings. At the same time, he became aware of similar work being conducted in the United States by Bloch. Eventually, Lynen and Bloch began to correspond, sharing their preliminary discoveries with each other. By working in this manner, the scientists determined the sequence of thirty-six steps by which animal cells produce cholesterol.

One of the breakthroughs in the cholesterol synthesis work came in 1951 when Lynen published a paper describing the first step in the chain of reactions that resulted in the production of cholesterol. He had discovered that a compound known as acetyl-coenzyme A, which is formed when an acetate radical reacts with coenzyme A, was needed to begin the chemical chain reaction. For the first time, the chemical

structure of acetyl-coenzyme A was described in accurate detail. By solving this complex biochemical problem, Lynen established his international reputation and created a new set of challenging biochemical problems. Determining the structure of acetyl-coenzyme A supplied Lynen with the discovery he needed to advance his research.

Lynen, who had remained an enthusiastic skier even after his 1932 accident, suffered a second serious ski injury at the end of 1951. (Although the second accident left him with a pronounced limp, Lynen continued to hike, swim, mountain climb, and ski.) During his rehabilitation, he contemplated how the structure and action of acetyl-coenzyme A made it a likely participant in other biochemical processes. Upon his return to the lab, Lynen began investigating the role of acetyl-coenzyme A in the biosynthesis of fatty acids and discovered that, as with cholesterol, this substance was the necessary first step. Lynen also investigated the catabolism of fatty acids, the chemical reactions that produce energy when fatty acids in foods are burned up to form carbon dioxide and water.

In 1953, Lynen was made full professor at the University of Munich. A year later, he was named director of the newly established Max Planck Institute for Cell Chemistry. At a time when other universities were attempting to coax Lynen away from Munich, this position ensured that he would stay.

In addition to elucidating the role of acetyl-coenzyme A, Lynen's research revealed the importance of many other chemicals in the body. One of the most significant of these was his work with the vitamin biotin. In the late 1950s, Lynen demonstrated that biotin was needed for the production of **fat**.

Lynen and Bloch shared the Nobel Prize in physiology or medicine in 1964, largely because the Nobel Committee recognized the medical importance of their work. Medical authorities knew that an accumulation of cholesterol in the walls of **arteries** and in **blood** contributed to diseases of the **circulatory system**, including **arteriosclerosis**, heart attacks, and strokes. In its tribute to Lynen and Bloch, the Nobel Committee noted that a more complete understanding of the metabolism of sterols and fatty acids promised to reveal the possible role of cholesterol in heart disease. Any future research into the link between cholesterol and heart disease, the Nobel committee observed, would have to be based on the findings of Lynen and Bloch.

In 1972, Lynen moved to the Max Planck Institute for Biochemistry, which had just recently been founded. Between 1974 and 1976, Lynen was acting director of the Institute. He continued to oversee a lab at the University of Munich.

In 1976, on the occasion of his sixty-fifth birthday, more than eighty of Lynen's friends, students, and colleagues contributed essays to a book, *Die aktivierte Essigsäure und ihre Folgen,* in which they described their relationship with Lynen. They celebrated Lynen as a renowned scientist and a proud Bavarian. The author of over three hundred scholarly pieces, Lynen was also praised as a hard-working man who expected much of himself and his students. Six weeks after an aneurism operation, Lynen died on August 6, 1979.

See also Embolism

M

MacLeod, Colin Munro (1909-1972)
Canadian-born American microbiologist

Colin Munro MacLeod is recognized as one of the founders of **molecular biology** for his research concerning the role of **deoxyribonucleic acid (DNA)** in **bacteria**. Along with his colleagues Oswald Avery and Maclyn McCarty, MacLeod conducted experiments on bacterial transformation which indicated that DNA was the active agent in the genetic transformation of bacterial cells. His earlier research focused on the causes of pneumonia and the development of serums to treat it. MacLeod later became chairman of the department of microbiology at New York University; he also worked with a number of government agencies and served as White House science advisor to President John F. Kennedy.

MacLeod, the fourth of eight children, was born on January 28, 1909, in Port Hastings, in the Canadian province of Nova Scotia. He was the son of John Charles MacLeod, a Scottish Presbyterian minister, and Lillian Munro MacLeod, a schoolteacher. During his childhood, MacLeod moved with his family first to Saskatchewan and then to Quebec. A very bright youth, he skipped several grades in elementary school and graduated from St. Francis College, a secondary school in Richmond, Quebec, at the age of fifteen. He was granted a scholarship to McGill University in Montreal but was required to wait a year for admission because of his age; during that time he taught elementary school. After two years of undergraduate work in McGill's premedical program, during which he became managing editor of the student newspaper and a member of the varsity ice hockey team, he entered the McGill University Medical School. He received his medical degree in 1932.

Following a two-year internship at the Montreal General Hospital, MacLeod moved to New York City and became a research assistant at the Rockefeller Institute for Medical Research. His research there, under the direction of Oswald Avery, focused on pneumonia and the pneumococcal infections that cause it. He examined the use of animal anti-serums (liquid substances that contain proteins that guard against **antigens**) in the treatment of the disease. MacLeod also studied the use of sulfa drugs, synthetic substances that counteract bacteria, in treating pneumonia, as well as how pneumococci develop a resistance to sulfa drugs. He also worked on a mysterious substance then known as "C-reactive protein," which appeared in the **blood** of patients with acute infections.

MacLeod's principal research interest at the Rockefeller Institute was the phenomenon known as bacterial transformation. First discovered by Frederick Griffith in 1928, this was a phenomenon in which live bacteria assumed some of the characteristics of dead bacteria. Avery had been fascinated with transformation for many years and believed that the phenomenon had broad implications for the science of biology. Thus, he and his associates, including MacLeod, conducted studies to determine how the bacterial transformation worked in pneumococcal cells.

The researchers' primary problem was determining the exact nature of the substance that would bring about a transformation. Previously, the transformation had been achieved only sporadically in the laboratory, and scientists were not able to collect enough of the transforming substance to determine its exact chemical nature. MacLeod made two essential contributions to this project: He isolated a strain of pneumococcus which could be consistently reproduced, and he developed an improved nutrient culture in which adequate quantities of the transforming substance could be collected for study.

By the time MacLeod left the Rockefeller Institute in 1941, he and Avery suspected that the vital substance in these transformations was DNA. A third scientist, Maclyn McCarty, confirmed their hypothesis. In 1944 MacLeod, Avery, and McCarty published "Studies of the Chemical Nature of the Substance Inducing Transformation of Pneumococcal Types: Induction of Transformation by a Deoxyribonucleic Acid Fraction Isolated from Pneumococcus Type III" in the *Journal of Experimental Medicine*. The article proposed that DNA was the material which brought about genetic transformation. Though the scientific community was slow to recognize the article's significance, it was later hailed as the beginning of a revolution that led to the formation of molecular biology as a scientific discipline.

MacLeod married in 1938. He and his wife eventually had one daughter. In 1941, he became a citizen of the United States and was appointed professor and chairman of the department of microbiology at the New York University School of Medicine, a position he held until 1956. At New York University he was instrumental in creating a combined program in which research-oriented students could acquire both an M.D. and a Ph.D. In 1956, he became professor of research medicine at the Medical School of the University of Pennsylvania. He returned to New York University in 1960 as professor of medicine and remained in that position until 1966.

From the time the United States entered World War II until the end of his life, MacLeod was a scientific advisor to the federal government. In 1941, he became director of the Commission on Pneumonia of the United States Army Epidemiological Board. Following the unification of the military services in 1949, he became president of the Armed Forces Epidemiological Board and served in that post until 1955. In the late 1950s, he helped establish the Health Research Council for the City of New York and served as its chairman from 1960 to 1970. In 1963 President John F. Kennedy appointed him deputy director of the Office of Science and Technology in the Executive Office of the President; from this position he was responsible for many program and policy initiatives, most notably the United States/Japan Cooperative Program in the Medical Sciences.

In 1966 MacLeod became vice-president for Medical Affairs of the Commonwealth Fund, a philanthropic organization. He was honored by election to the National Academy of Sciences, the American Philosophical Society, and the American Academy of Arts and Sciences. MacLeod was en route from the United States to Dacca, Bangladesh, to visit a cholera laboratory when he died in his sleep in a hotel at the London airport on February 11, 1972.

See also Bacteria and responses to bacterial infection; Careers in anatomy and physiology; Medical training in anatomy and physiology

MAGENDIE, FRANÇOIS (1783-1855)
French physiologist

François Magendie was the son of a surgeon who was known for his radical views and his active support of the French Revolution. The elder Magendie raised his two sons to be fiercely independent and to think for themselves—two traits that François never lost.

Apprenticed at sixteen to a surgeon friend of his father's, the young man began his formal medical studies a few years later, obtaining his medical degree in 1808 from the University of Paris. Although he was interested at first in

anatomy, Magendie later switched to **physiology**, and almost at once ran into trouble. At the time, most European scientists believed strongly in vitalism—the idea that biological processes were governed by "vital forces" that could not be explained in strictly scientific terms. Magendie disagreed. He was convinced that in the biologic sciences, just as in the physical sciences, facts were more important than theories—and even here, all facts had to be verified in the laboratory.

Magendie's strong opinions—and his desire to experiment in sometimes forbidden areas—won him numerous enemies and even, at times, the reputation of being a vivisectionist. Nevertheless, he established the idea of experimental physiology (an idea further popularized by his disciple, **Claude Bernard**) and made a number of important discoveries. Magendie, for instance, was interested in the action of various plant-derived drugs on the body. One of the drugs he studied was strychnine and, in 1809, he described in detail the effects of strychnine injections on animal subjects—and also proved that the **poison** reached the animal's **spinal cord** by the **blood** stream and not (as was then commonly believed) by the **lymphatic system**. Because of such experiments, Magendie was able to introduce into French medicine a variety of new drugs, including morphine, codeine, quinine and, strychnine.

In 1815, post-revolutionary France was short of food, and Magendie was asked to serve as chairman of a special commission set up to investigate the nutritional value of various food extracts. Intrigued by the problem, Magendie continued his nutritional investigations long after the commission disbanded. Among other things, he found that mammals could not be kept alive by diets that lacked any nitrogen-containing foods (in other words, proteins). He found, as well, that not all these substances were equally life-sustaining—a few, like gelatin, had very little nutritional value. Magendie's findings pointed the way for later nutritional researchers, like Frederick Gowland Hopkins, Thomas B. Osborne and William Rose, who were able to provide more definitive answers. Magendie devoted much of his time to studies on the nervous system and, in 1822, published a classic paper in which he distinguished the separate motor and sensory roots of the spinal nerves. He showed that the ventral roots, those entering the spine in front of the dorsal roots, carry impulses to the muscles; and the dorsal roots, those entering the spine slightly behind the ventral roots, carry impulses to the **brain** from receptor **neurons**

See also Circulatory system; History of anatomy and physiology: The Renaissance and Age of Enlightenment; Nutrition and nutrient transport to cells; Pharmacology; Poison and antidote actions; Protein metabolism

MALARIA AND THE PHYSIOLOGY OF PARASITIC INFECTIONS

Malaria is a disease caused by a unicellular parasite known as Plasmodium. Although more than 100 different species of Plasmodium exist, only four types are known to infect humans

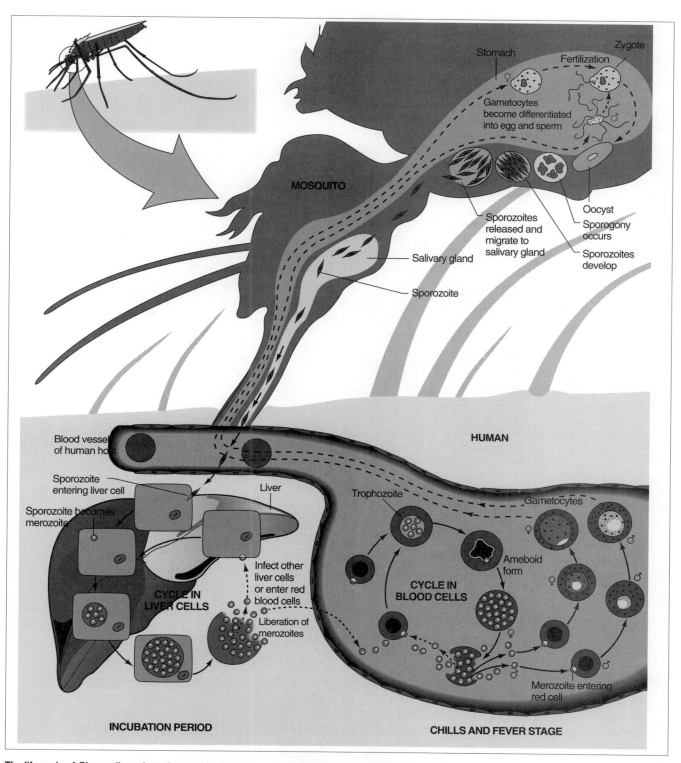

The life cycle of *Plasmodium vivax,* the parasite that causes malaria. *Illustration by Hans & Cassidy.*

including, *Plasmodium falciparum, vivax, malariae,* and *ovale.* While each type has a distinct appearance under the microscope, they each can cause a different pattern of symptoms. *Plasmodium falciparum* is the major cause of **death** in

Africa, while *Plasmodium vivax* is the most geographically widespread of the species and the cause of most malaria cases diagnosed in the United States. *Plasmodium malariae* infections produce typical malaria symptoms that persist in the **blood** for very long periods, sometimes without ever producing symptoms. *Plasmodium ovale* is rare, and is isolated to West Africa. Obtaining the complete sequence of the Plasmodium genome is currently under way.

The life cycle of Plasmodium relies on the insect host (for example, the Anopheles mosquito) and the carrier host (humans) for its propagation. In the insect host, the Plasmodium parasite undergoes **sexual reproduction** by uniting two sex cells producing what are called sporozoites. When an infected mosquito feeds on human blood, the sporozoites enter into the bloodstream. During a mosquito bite, the **saliva** containing the infectious sporozoite from the insect is injected into the bloodstream of the human host and the blood that the insect removes provides nourishment for her eggs. The parasite immediately is targeted for a human **liver** cell, where it can escape from being destroyed by the **immune system**. Unlike in the insect host, when the sporozoite infects a single liver cell from the human host, it can undergo asexual reproduction (multiple rounds consisting of replication of the nucleus followed by budding to form copies of itself).

During the next 72 hours, a sporozoite develops into a schizont, a structure containing thousands of tiny rounded merozoites. Schizont comes from the Greek word 'schizo,' meaning to tear apart. One infectious sporozoite can develop into 20,000 merozoites. Once the schizont matures, it ruptures the liver cells and leaks the merozoites into the bloodstream where they attack neighboring **erythrocytes** (red blood cells, RBC). It is this stage of the life cycle that can cause disease and death if not treated. Once inside the cytoplasm of an erythrocyte, the parasite can break down **hemoglobin** (the primary oxygen transporter in the body) into **amino acids** (the building blocks that makeup protein). A by-product of the degraded hemoglobin is hemozoin, or a pigment produced by the breakdown of hemoglobin. Golden-brown to black granules are produced from hemozoin and are considered to be a distinctive feature of a blood-stage parasitic **infection**. The blood-stage parasites produce schizonts, which rupture the infected erythrocytes, releasing many waste products, explaining the intermittent fever attacks that are associated with malaria.

The parasites' own propagation is ensured by a certain type of merozoite, that invades erythrocytes but does not asexually reproduce into schizonts. Instead, they develop into gametocytes (two different forms or sex cells that require the union of each other in order to reproduce itself). These gametocytes circulate in the human's blood stream and remain quiescent (dormant) until another mosquito bite, where the gametocytes are fertilized in the mosquito's stomach to become sporozoites. Gametocytes are not responsible for causing disease in the human host and will disappear from the circulation if not taken up by a mosquito. Likewise, the salivary sporozoites are not capable of re-infecting the salivary gland of another mosquito. The cycle is renewed upon the next feeding of human blood. In some types of Plasmodium, the sporozoites turn into hypnozoites, a stage in the life cycle that allows the

parasite to survive but in a dormant phase. A relapse occurs when the hypnozoites are reverted back into sporozoites.

An infected erythrocyte has knobs on the surface of the cells that are formed by proteins that the parasite is producing during the schizont stage. These knobs are only found in the schizont stage of *Plasmodium falciparum* and are thought to be contacted points between the infected RBC and the lining of the blood vessels. The parasite also modifies the erythrocyte membrane itself with these knob-like structures protruding at the cell surface. These parasitic-derived proteins that provide contact points thereby avoid clearance from the blood stream by the spleen. Sequestration of schizont-infected erythrocytes to blood vessels that line vital organ such as the **brain**, lung, **heart**, and gut can cause many health-related problems.

A malaria-infected erythrocyte results in physiological alterations that involve the function and structure of the erythrocyte membrane. Novel parasite-induced permeation pathways (NPP) are produced along with an increase, in some cases, in the activity of specific transporters within the RBC. The NPP are thought to have evolved to provide the parasite with the appropriate nutrients, explaining the increased permeability of many solutes. However, the true nature of the NPP remains an enigma. Possible causes for the NPP include: 1) the parasite activates native transporters; 2) proteins produced by the parasite cause structural defects; 3) plasmodium inserts itself into the channel thus affecting it's function, and; 4) the parasite makes the membrane more leaky. The properties of the transporters and channels on a normal RBC differ dramatically from that of a malaria-infected RBC. Additionally, the lipid composition in terms of its fatty acid pattern is significantly altered, possibly due to the nature in which the parasite interacts with the membrane of the RBC. The dynamics of the membranes, including how the fats that makeup the membrane are deposited, are also altered. The increase in transport of solutes is bi-directional and is a function of the developmental stage of the parasite. In other words, the alterations in erythrocyte membrane are proportional to the maturation of the parasite.

See also Cell membrane transport; Hemopoiesis; Immunology; Infection and resistance

MALPIGHI, MARCELLO (1628-1694)
Italian physician

In the second half of the seventeenth century, Marcello Malpighi used the newly invented microscope to make a number of important discoveries about living tissues and structures, earning himself enduring recognition as a founder of scientific microscopy, **histology** (the study of tissues), **embryology**, and the science of plant **anatomy**.

Malpighi was born at Crevalcore, just outside Bologna, Italy. The son of small landowners, Malpighi studied medicine and philosophy at the University of Bologna. While at Bologna, Malpighi was part of a small anatomical society headed by the teacher Bartolomeo Massari, in whose home the

Marcello Malpighi. *The Library of Congress.*

group met to conduct dissections and vivisections. Malpighi later married Massari's sister.

In 1655, Malpighi became a lecturer in logic at the University of Bologna; in 1656, he assumed the chair of theoretical medicine at the University of Pisa; in 1659, he returned to Bologna as lecturer in theoretical, then practical, medicine; from 1662 to 1666, he held the principal chair in medicine at the University of Messina; finally in 1666, he returned again to Bologna, where he remained for the rest of his teaching and research career. In 1691, at the age of sixty-three, Malpighi was called by his friend Pope Innocent XII to serve as the pontiff's personal physician. Reluctantly, Malpighi agreed and moved to Rome, where he died on November 29, 1694, in his room in the Quirinal Palace.

Early in his medical career, Malpighi became absorbed in using the microscope to study a wide range of living tissue—animal, insect, and plant. At the time, this was an entirely new field of scientific investigation. Malpighi soon made a profoundly important discovery. Microscopically examining a frog's **lungs**, he was able for the first time to describe the lung's structure accurately—thin air sacs surrounded by a network of tiny **blood** vessels. This explained how air (oxygen) is able to diffuse into the blood vessels, a key to understanding the process of respiration. It also provided the one missing piece of evidence to confirm William Harvey's revolutionary theory of the blood circulation: Malpighi had discovered the **capillaries**, the microscopic con-

necting link between the **veins** and **arteries** that Harvey—with no microscope available—had only been able to postulate. Malpighi published his findings about the lungs in 1661.

Malpighi used the microscope to make an impressive number of other important observations, all "firsts." He observed a "host of red atoms" in the blood—the red blood corpuscles. He described the papillae of the tongue and skin—the receptors of the senses of **taste** and touch. He identified the rete mucosum, the Malpighian layer, of the skin. He found that the nerves and spinal column both consisted of bundles of fibers. He clearly described the structure of the kidney and suggested its function as a urine producer. He identified the spleen as an organ, not a gland; structures in both the kidney and spleen are named after him. He demonstrated that **bile** is secreted in the **liver**, not the gall bladder. In showing bile to be a uniform color, he disproved a 2,000-year-old idea that the bile was yellow and black. He described glandular adenopathy, a syndrome rediscovered by Thomas Hodgkin (1798–1866) and given that man's name 200 years later.

Malpighi also conducted groundbreaking research in plant and insect microscopy. His extensive studies of the silkworm were the first full examination of insect structure. His detailed observations of chick embryos laid the foundation for microscopic embryology. His botanical investigations established the science of plant anatomy. The variety of Malpighi's microscopic discoveries piqued the interest of countless other researchers and firmly established microscopy as a science.

See also History of anatomy and physiology: The Renaissance and Age of Enlightenment; Respiratory system

MAMMARY GLANDS AND LACTATION

Mammary glands are the accessory reproductive glands within the **breasts** that function in milk production. Fifteen to twenty-five lobes of glandular **tissue** containing mammary **alveoli** synthesize and secrete the milk in the mammary glands, a process known as lactation. The milk is transferred via a system of lactiferous ducts that converge to the nipple. Ligamentous and fatty tissue surround the ducts and give support and shape to the breasts.

Lactation begins after delivery of a baby. The initial secretion of the mammary glands before true lactation begins is termed colostrum. Colostrum is a thin yellowish fluid that contains proteins and immunoglobins, which initiates temporary passive immunity in the newborn. Delivery of the **placenta** decreases the levels of **estrogen** and **progesterone** that, in conjunction with prolactin, initiate the production of milk. In the days following delivery, milk production ceases and will only occur if the breast is suckled. Suckling causes a surge of prolactin that is responsible for milk production. **Oxytocin** is responsible for the release of milk by causing contraction of smooth muscles in the breasts.

Breast-feeding provides many benefits to the newborn as well as the mother. In addition to providing antibodies to

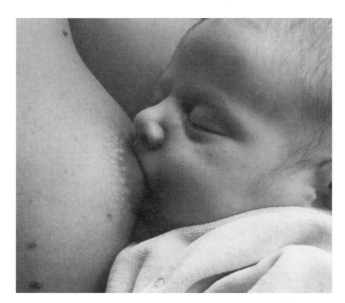

The suckling action of a baby triggers hormones that cause the mammary glands within the breasts to secrete and continue producing breast milk. *Photo Researchers, Inc. Reproduced by permission.*

MANGOLD, HILDE PROESCHOLDT (1898-1924)

German biologist

German biologist Hilde Mangold studied the process by which part of an embryo—sometimes called an organizer—causes other parts of the embryo to form specific types of tissues and **organs**. This process is called embryonic induction. Working under the direction of German biologist **Hans Spemann**, Mangold helped discover the location of the organizer in amphibians.

Mangold had attended the University of Frankfurt, where she had once been to a lecture by Hans Spemann. Spemann was studying the processes that control the development of amphibian embryos, and Mangold was intrigued by his experiments. After she graduated, she moved to Freiburg, Germany, where Spemann was head of the Zoological Institute. Mangold began working on an advanced degree, and Spemann suggested a series of experiments for her doctoral thesis.

Spemann proposed that Mangold should work with early embryos of two species of newts. Her experiments would involve transplanting a portion of one embryo to a different part of an embryo of the other species. Because one newt species had light cells, and one had dark cells, she would be able to determine exactly what happened to the transplant as the embryo developed.

Earlier experiments (using a single species of newt) had shown that the fate of most transplanted cells was determined by the recipient embryo rather than by the donor. For instance, if cells that would normally become belly skin were transplanted to an area that would normally become back skin, the transplanted cells would become back skin. This result indicated that the fate of the cells was not yet determined when they were transplanted.

This was not the case for one region of the early embryo—the region that would eventually form the neural tube. (The neural tube, in turn, eventually forms the newt's **brain** and spinal cord.) When these cells were transplanted, they continued to form a neural tube regardless of their location. In other words, the fate of these cells was already determined in the early embryo.

It was these cells that would eventually form a neural tube that Mangold transferred between the two species. To her and Spemann's surprise, Mangold found that in such a transplant, most of the cells making up the resulting neural tube came from the recipient of the transplant rather than the donor. Therefore, the transplant cells were somehow able to control not only their own fate, but also the fate of the cells surrounding them. They induced these cells to develop into tissues and organs that they would not normally form. In fact, not only would a second **spinal cord** and brain be produced, but other internal organs as well. The eventual result was a secondary embryo attached to the main embryo (similar to a conjoined twin).

Spemann named this transplanted region the organizer because it was capable of reorganizing the cells in the recipient embryo. Mangold and Spemann wrote a paper on their

the newborn, the mother's milk also contains antibodies that provide important protection from disease. **Digestion** is made easier for the baby because of the many digestive **enzymes** found in the mother's milk. Studies indicate that breast milk also reduce the occurrence of **allergies**, **diarrhea**, and **ear** infections in babies.

Studies also indicate that lactation reduces the risk of breast **cancer** for the mother. Lactation inhibits ovulation, and the lack of ovulation helps to maintain low levels of estrogen. Decreased exposure to estrogen over a lifetime has been correlated to a decreased risk of breast cancer. Additionally, lactation can only occur in cells that have undergone differentiation and maturation. This decreases the possibility that cells will proliferate uncontrollably, which would result in breast cancer.

Other studies have concluded that lactation does not decrease the risk of breast cancer. Results of studies may differ because of the varying factors involved. For example, the total duration of lactation differs from the number of years a mother breast-feeds each child. Some mothers may intermittently breast-feed their children, causing a rise and fall in estrogen levels while other mothers may only breast-feed. The age at which a woman first becomes pregnant and the age she first begins to breast-feed, as well as the number of children she has, may also account for the varying results of the studies. Additionally, other factors such as the age at which the woman began **menstruation**, the age at which the woman went through **menopause**, and **genetics** all influence the risk of developing breast cancer.

See also Hormones and hormone action; Reproductive system and organs (female) results. (The paper was also Mangold's thesis for her doctorate degree.) Just as the paper was being published, Mangold was killed (at the age of 26) when a heater in her kitchen exploded. Spemann, however, went on to win a Nobel Prize in 1935 for the discovery of the organizer. (Mangold's death made her ineligible.) This Nobel Prize was one of the few that have ever been awarded for work based on a doctoral thesis, in this case, the thesis of Hilde Mangold.

See also Cell differentiation; Embryonic development, early development, formation, and differentiation; Organizer experiment

MASTICATION

Mastication is also referred to as chewing. It is the first step in the breakdown of complex blends of nutrients into their energy-producing basic components.

When chunks of food are masticated, they are broken into smaller pieces. This increases the surface area that is available for digestive **enzymes** to attack. The act of chewing also includes the release of **saliva**, which softens the food and, because of salivary **mucus**, makes the food slippery enabling the pieces to be swallowed. Larger pieces of dry food are difficult to swallow and can even become lodged in the esophagus (the digestive tract tube that leads from the back of the throat to the stomach).

Saliva also functions to begin the **digestion** of food. Saliva contains the enzymes ptyalin and amylase, which digests starches into sugars, and lipase, which digests fats. Furthermore, saliva contains a compound called bicarbonate of soda that can neutralize acidic or alkaline compounds present in the food or drink.

The motions of mastication are stimulated by the presence of food in the mouth. A reflexive action is the relaxation of the muscles of the lower jaw, which causes some muscles that control the mouth to contract. This in turn causes the mouth to close, helping to contain food in the mouth. Mastication then becomes an orchestrated effort involving the tongue, lips, cheeks and the lower jaw to maintain the food between the grinding teeth. For example, the tongue continuously re-positions food by squeezing food against the hard upper palate of the mouth, sweeping the food forward and, by lowering the back portion of the tongue to create negative pressure, causing the food to drop down onto the lower teeth. These motions can be occurring within seconds and occur unconsciously.

The reflexive nature of mastication has been known for a long time. Over 150 years ago the rhythmic jaw movements of chewing were observed in anencephalic human infants. Anencephaly is an always lethal condition (minutes to days after birth) where the top of the **skull** fails to develop, leaving the **brain** exposed and sometimes malformed in those infants who survive to birth.

See also Gustatory structures; Taste, physiology of gustatory structures

MECHANORECEPTORS

Sensory receptors are transducers that convert some form of energy into a change in electrical potential across a neuron's membrane. Mechanoreceptors specifically respond to mechanical energy, or movement. Mechanoreceptors are found in many bodily locations, including the **hair** cells of the inner ear, the stretch receptors within the muscles, and the various types of receptors in the skin. Mechanoreceptors located in the bladder and parts of the alimentary canal alert the body to pressure within these **organs**.

Despite their varied locations and structures, all mechanoreceptors work in the same general way. Stimulation of these receptors changes the shape of the membranes in which they are embedded, opening mechanosensitive ion channels that either initiate an **action potential**, or directly cause the release of neurotransmitter molecules that affect the neighboring neuron.

The touch receptors of the skin vary in size and in the extent of their receptive fields. The touch receptors differ also in how long they continue to respond during a long-lasting stimulus and in the frequency of vibrations to which they are sensitive. Two types of skin mechanoreceptors are Pacinian corpuscles located deep in the **dermis** layer and Meissner's corpuscles located on the ridges of the fingerprints.

Vibrissae and other hairs on the body surface serve as mechanoreceptors. When a hair is disturbed, its bending changes the shape of its follicle and the skin **tissue** surrounding it. This in turn activates nerve endings in the area, initiating an action potential. Another group of mechanoreceptors are the proprioceptors found in the muscles and between the muscles and **tendons**. These receptors inform the **brain** of the position of various body parts in space; and if they are moving, in which direction, and how fast. There are two classes of the proprioceptors, Golgi tendon organs, and muscle spindles. Golgi tendon organs, located between a muscle and a tendon, monitor the force of a **muscle contraction**. As the force of a muscle contraction increases, the Golgi tendon organ indirectly causes a slowing of the muscle contraction. This adjustment is necessary for fine-motor coordination.

The muscle spindles monitor the degree to which a muscle is stretched, as in having a weight put on the muscle, for example. As the muscles are stretched, the muscle spindles cause the opening of mechanosensitive ion channels in the afferent neuron, depolarizing it and causing synaptic depolarization of the corresponding motor neuron, which shortens the spindle by causing the muscle to contract.

In the inner ear, the stereocilia, or hairs, of the hair cells are bent by vibrations. This bending causes one hair to pull on its immediate neighbor via tiny filaments called tip links. This briefly opens **calcium** ion channels in the neighboring hair, causing the release of the neurotransmitter glutamate from the base of the hair cell, stimulating the afferent neuron carrying its information.

See also Hearing (physiology of sound transmission); Nerve impulses and conduction of impulses; Nervous system overview

MEDAWAR, PETER BRIAN (1915-1987)

English biologist

Peter Brian Medawar made major contributions to the study of **immunology** and was awarded the Nobel Prize in physiology or medicine in 1960. Working extensively with skin grafts, he and his collaborators proved that the **immune system** learns to distinguish between "self" and "non-self." During his career, Medawar also became a prolific author, penning books such as *The Uniqueness of the Individual* and *Advice to a Young Scientist.*

Medawar was born on February 28, 1915, in Rio de Janeiro, Brazil, to Nicholas Medawar and the former Edith Muriel Dowling. When he was a young boy, his family moved to England, which he thereafter called home. Medawar attended secondary school at Marlborough College, where he first became interested in biology. The biology master encouraged Medawar to pursue the science under the tutelage of one of his former students, John Young, at Magdalen College. Medawar followed this advice and enrolled at Magdalen in 1932 as a zoology student.

Medawar earned his bachelor's degree from Magdalen in 1935, the same year he accepted an appointment as Christopher Welch Scholar and Senior Demonstrator at Magdalen College. He followed Young's recommendation that he work with pathologist **Howard Florey**, who was undertaking a study of penicillin, work for which he would later become well-known. Medawar leaned toward experimental **embryology** and **tissue** cultures. While at Magdalen, he met and married a fellow zoology student. Madawar and his wife had four children.

In 1938, Medawar, by examination, became a fellow of Magdalen College and received the Edward Chapman Research Prize. A year later, he received his master's from Oxford. When World War II broke out in Europe, the Medical Research Council asked Medawar to concentrate his research on tissue transplants, primarily skin grafts. While this took him away from his initial research studies into embryology, his work with the military would come to drive his future research and eventually lead to a Nobel Prize.

During the war, Medawar developed a concentrated form of fibrinogen, a component of the **blood**. This substance acted as a glue to reattach severed nerves, and found a place in the treatment of skin grafts and in other operations. More importantly to Medawar's future research, however, were his studies at the Burns Unit of the Glasgow Royal Infirmary in Scotland. His task was to determine why patients rejected donor skin grafts. He observed that the rejection time for donor grafts was noticeably longer for initial grafts, compared to those grafts that were transplanted for a second time. Medawar noted the similarity between this reaction and the body's reaction to an invading virus or **bacteria**. He formed the opinion that the body's rejection of skin grafts was immunological in nature; the body built up an immunity to the first graft and then called on that already-built-up immunity to quickly reject a second graft.

Upon his return from the Burns Unit to Oxford, he began his studies of immunology in the laboratory. In 1944, he became a senior research fellow of St. John's College, Oxford, and university demonstrator in zoology and **comparative anatomy**. Although he qualified for and passed his examinations for a doctorate in philosophy while at Oxford, Medawar opted against accepting it because it would cost more than he could afford. In his autobiography, *Memoir of a Thinking Radish,* he wrote, "The degree served no useful purpose and cost, I learned, as much as it cost in those days to have an appendectomy. Having just had the latter as a matter of urgency, I thought that to have both would border on self-indulgence, so I remained a plain mister until I became a prof." He continued as researcher at Oxford University through 1947.

During that year Medawar accepted an appointment as Mason professor of zoology at the University of Birmingham. He brought with him one of his best graduate students at Oxford, Rupert Everett "Bill" Billingham. Another graduate student, Leslie Brent, soon joined them and the three began what was to become a very productive collaboration that spanned several years. Their research progressed through Medawar's appointment as dean of science, through his several-month-long trip to the Rockefeller Institute in New York in 1949—the same year he received the prestigious title of fellow from the Royal Society—and even a relocation to another college. In 1951, Medawar accepted a position as Jodrell Professor of Zoology and Comparative Anatomy at University College, London. Billingham and Brent followed him.

Their most important discovery had its experimental root in a promise Medawar made at the International Congress of Genetics at Stockholm in 1948. He told another investigator, Hugh Donald, that he could formulate a foolproof method for distinguishing identical from fraternal twin calves. He and Billingham felt they could easily tell the twins apart by transplanting a skin graft from one twin to the other. They reasoned that a calf of an identical pair would accept a skin graft from its twin because the two originated from the same egg, whereas a calf would reject a graft from its fraternal twin because they came from two separate eggs. The results did not bear this out, however. The calves accepted skin grafts from their twins regardless of their status as identical or fraternal. Puzzled, they repeated the experiment, but received the same results.

They found their error when they became aware of work done by Dr. **Frank Macfarlane Burnett** of the University of Melbourne, and Ray D. Owen of the California Institute of Technology. Owen found that blood transfuses between twin calves, both fraternal and identical. Burnet believed that an individual's immunological framework developed before birth, and felt Owen's finding demonstrated this by showing that the immune system tolerates those tissues that are made known to it before a certain age. In other words, the body does not recognize donated tissue as alien if it has had some exposure to it at an early age. Burnet predicted that this immunological tolerance for non-native tissue could be reproduced in a lab. Medawar, Billingham, and Brent set out to test Burnet's hypothesis.

The three-scientist team worked closely together, inoculating embryos from mice of one strain with tissue cells from donor mice of another strain. When the mice had matured, the trio grafted skin from the donor mice to the inoculated mice. Normally, mice reject skin grafts from other mice, but the inoculated mice in their experiment accepted the donor skin grafts. They did not develop an immunological reaction. The prenatal encounter had given the inoculated mice an acquired immunological tolerance. They had proven Burnet's hypothesis. They published their findings in a 1953 article in *Nature*. Although their research had no applications to transplants among humans, it showed that transplants were possible.

In the years following publication of the research, Medawar accepted several honors, including the Royal Medal from the Royal Society in 1959. A year later, he and Burnet accepted the Nobel Prize physiology or medicine for their discovery of acquired immunological tolerance: Burnet developed the theory and Medawar proved it. Medawar shared the prize money with Billingham and Brent.

Medawar's scientific concerns extended beyond immunology, even during the years of his work toward acquired immunological tolerance. While at Birmingham, he and Billingham also investigated pigment spread, a phenomenon seen in some guinea pigs and cattle where the dark spots spread into the light areas of the skin. "Thus if a dark skin graft were transplanted into the middle of a pale area of skin it would soon come to be surrounded by a progressively widening ring of dark skin," Medawar asserted in his autobiography. The team conducted a variety of experiments, hoping to show that the dark pigment cells were somehow "infecting" the pale pigment cells. The tests never panned out.

Medawar also delved into animal behavior at Birmingham. He edited a book on the subject by noted scientist Nikolaas Tinbergen, who ultimately netted a Nobel Prize in 1973. In 1957, Medawar also became a book author with his first offering, *The Uniqueness of the Individual,* which was actually a collection of essays. In 1959, his second book, *The Future of Man,* was issued, containing a compilation of a series of broadcasts he read for British Broadcasting Corporation (BBC) radio. The series examined the impacts of **evolution** on man.

Medawar remained at University College until 1962 when he took the post of director of the National Institute for Medical Research in London, where he continued his study of transplants and immunology. While there, he continued writing with mainly philosophical themes. *The Art of the Soluble,* published in 1967, is an assembly of essays, while his 1969 book, *Induction and Intuition in Scientific Thought,* is a sequence of lectures examining the thought processes of scientists. In 1969 Medawar, then president of the British Association for the Advancement of Science, experienced the first of a series of strokes while speaking at the group's annual meeting. He finally retired from his position as director of the National Institute for Medical Research in 1971. In spite of his physical limitations, he went ahead with scientific research in his lab at the clinical research center of the Medical Research Council. There he began studying **cancer.**

Through the 1970s and 1980s, Medawar produced several other books—some with his wife as co-author—in addition to his many essays on growth, aging, immunity, and cellular transformations. In one of his most well-known books, *Advice to a Young Scientist,* Medwar asserted that for scientists, curiosity was more important that genius.

See also Bacteria and responses to bacterial infection; Embryonic development: early development, formation, and differentiation; Genetics and developmental genetics; Integument; Transfusions; Transplantation of organs

MEDIASTINUM

The mediastinum is the central part of the thoracic (chest) cavity. It is not a distinct organ, but a volume of space that has been arbitrarily defined to make it easier to discuss the location of structures in the chest.

The mediastinum is bounded by the front by the **sternum** or breastbone, to the back by the spine, and to the left and right by the pleurae of the **lungs.** The **heart,** the thymus, and a number of lymph nodes are wholly contained inside the mediastinum; passing through it are a number of major **blood** vessels, lymph vessels, and nerves, along with portions of the esophagus, trachea, and **bronchi.**

The mediastinum is divided for convenience into several sub-volumes that are named using the conventional medical terminology for top (superior), bottom (inferior), front (anterior), and back (posterior). The superior mediastinum is that part of the chest cavity which is above the highest point of the heart. The great vessels of the heart (aorta, superior vena cava, pulmonary vessels) extend up from the heart into the superior mediastinum. The superior mediastinum also contains the thymus, portions of the esophagus and trachea, and other structures. The inferior mediastinum includes the whole chest cavity below the superior mediastinum, and is subdivided into the middle mediastinum (the heart and pericardial sac), the anterior mediastinum (the small space between heart and sternum, mostly filled with fatty **tissue**), and the posterior mediastinum (the space between the heart and the spine, containing much of the esophagus and a variety of **arteries, veins,** and nerves).

See also Anatomical nomenclature; Pericardium; Pleurae; Thymus histophysiology

MEDICAL TRAINING IN ANATOMY AND PHYSIOLOGY

Medical training in **anatomy** and **physiology** begins early for physicians, nurses, and other health care professionals. Because an understanding of general structure and the physical relationship of systems is critical to an eventual appreciation and understanding of the interplay of **structure and function** in physiology and pathological processes (**pathol-**

ogy), a course in gross anatomy is the traditionally one of the first courses in a medical school curriculum. In addition, many medical scientists also undertake formal study in anatomy and physiology.

Following a course in gross anatomy (the large scale morphology of the human body), students usually move on to study **tissue** level anatomy (**histology**) and specialized anatomical systems (e.g., **neurology** is a course dedicated to the study of the **brain** and neural system) where additional emphasis is placed in the relationship between structure and function.

A formal course in physiology is also a required course of study in medical and nursing programs. Although there are many different systems and schemes for the timing of courses in anatomy and physiology (e.g., Physiology is usually studied after Gross Anatomy, but in some cases the two courses may overlap).

The study of gross anatomy is usually accomplished through the regional dissection of cadavers. Cadavers (dead human bodies) are most often procured through programs that allow people to will or dedicate their bodies to medical science after **death**. For this reason, medical studies and dissections are usually carried out under circumstances designed to assure the dignity of the donor. Dissections are accompanied by lectures and readings that help students understand and relate the observation of structure in the dissection room to the overall body plan. Specialized lectures or courses in **embryology** allow students to understand often confusing anatomical relationships based upon the fetal development of structures. The coursings and branches of the **aortic arch** are, for example, only fully understandable when related the development of the fetal **heart** and arterial system.

Dissections proceed regionally, and may take four to eight months. Usually starting with the limbs or thoracic cavity, students explore the structures and relationships of **organs** within each region. Typical courses of study allow regional exploration of the upper limbs, lower, limbs, thorax, **abdomen**, urogenital structures, and the head and neck. Examinations usually require students not only to be able to properly identify structures, but to also be able to integrate knowledge regarding structure, function, and embryology gained from lecture.

In the physiology laboratory, students undertake a series of experiments that demonstrate various physiological principles dealing with topics that range from osmotic pressure changes to the electrical induction of **muscle contraction**. Many physiology experiments require students to draw of knowledge obtained during their studies of **biochemistry**. Physiology courses also attempt to integrate lecture and practical laboratory exercises.

Microscopic anatomy, involving study of tissues involves substantial use of the microscope to identify and characterize different tissue types. For example, students must differentiate between the histology of smooth, striated (skeletal), and cardiac muscle. As with gross anatomy, one major goal of histological studies is to provide a basis for the continued appreciation of the interplay of structure and function. Also critical in histological studies, is the study of normal tissue types upon which subsequent studies of the pathology (changes brought about by disease processes or injury) can be based.

In many medical schools, courses in anatomy are now conducted within departments of dedicated to the study of cell biology. This change reflects an increasing emphasis on the role of **molecular biology** and **genetics** in modern medicine.

See also Ethical issues in embryological research; Ethical issues in genetics research; Histology and microanatomy; History of anatomy and physiology: The Classical and Medieval periods; History of anatomy and physiology: The Renaissance and Age of Enlightenment; History of anatomy and physiology: The science of medicine

MEDULLA OBLONGATA

Along with the **cerebellum** and the **pons**, the medulla oblongata makes up that portion of the **brain** called the hindbrain. So necessary are the functions of the medulla oblongata that with its loss comes instant death. Lying crossways between the higher brain and the body, it controls several basic autonomic functions including respiration. Located on the lowest portion of the brain stem it looks like the swollen tip of the spinal chord. Not only does it serve as the main conduit for nerve impulses that enter and leave the higher neural systems, it also functions as the pathway for communication between the right and left hemispheres.

It is within the medulla that both the sensory and motor **neurons** from each hemisphere cross over. This is why the right hemisphere controls the left side of the body and visa versa. As one of the most primitive areas of the brain, it regulates some of the more basic functions required for life. These include the involuntary processes of **swallowing** and **digestion** as well as **breathing**. It also regulates the heartbeat and the diameter of certain **blood** vessels-thus controlling blood flow. One could consider it the master control center for the **autonomic nervous system**. Partnered with the cerebellum it controls movement and along with the **thalamus**, it regulates states of arousal and **sleep**.

See also Brain stem function and reflexes; Brain, intellectual functions

MEIOSIS · *see* CELL CYCLE AND CELL DIVISION

MELANOCYTES

Melanocytes are pigment (melanin)-producing cells found in the deepest or basal layer of the **epidermis**. They are dendritic cells with cytoplasmic extensions that stretch out to contact other cells (keratocytes) in the epidermis.

Melanocytes originate as melanoblasts in the neural crest cells of the embryo at about seven weeks gestation. The melanoblasts migrate into the developing **dermis** and differentiate into melanocytes.

Melanocytes contain submicroscopic organelles called melanosomes that synthesize melanin. Melanin is transferred into keratocytes and **hair** follicles via the dendritic processes of melanocytes. Skin and hair color is primarily determined by the presence and distribution of melanin and can be brown-black (eumelanin) and yellow-red (pheomelanin) in coloration. This gives skin and hair its distinctive color and hue.

The various races of people all have approximately the same number of melanocytes in their skin. Darker-skinned people have appreciable amounts of melanin in all layer of the epidermis. Light-skinned people have relatively little melanin in the epidermis except in deeply pigmented areas such as the nipples. Albino individuals do not have fewer melanocytes in their epidermis than non-albinos, but rather they have a mutant gene that prevents the synthesis of melanin due to a missing enzyme.

Within the melanosome of the melanocyte, the enzyme tyrosinase oxidizes tyrosine to dihydroxyphenylalanine (DOPA) and then to DOPA quinone. Further non-enzymatic oxidation and polymerization results in the production of melanin.

The primary function of melanocytes is to protect the skin from solar UV damage through the production of melanin. The polymer melanin has the unique capability to absorb light of wavelengths ranging from 200–2400 nm. It has also been shown to be a free oxygen-radical scavenger, protecting the metabolically active keratinocytes of the epidermis from any free radicals that are generated by UV irradiation. In so doing, melanin reduces UV damage effects such as aging, wrinkling, and cutaneous neoplasms (skin cancers). The protective function of melanin is illustrated by the high incidence of skin cancers in sun-exposed epidermis of light-skinned, blue-eyed, easily sunburned individuals and albinos. Dark skin is much less likely to form skin cancers and ages more slowly than light skin.

Levels of **hormones** such as **estrogen, progesterone,** melanocyte stimulating hormone (MSH), and adrenocorticotropic hormone (ACTH), can affect distribution of melanin in the epidermis. In pregnant women, this can be seen as "the mask of pregnancy." Chronic illnesses such as Addison's disease, **liver** disease and **hyperthyroidism** can also present with unusual distribution of melanin.

See also Integument; Integumentary system, embryonic development; Sweat glands

MEMBRANE POTENTIAL

Every neuron has a separation of electrical charge across its cell membrane, and the membrane potential results from a separation of positive and negative. The relative excess of positive charges outside and negative charges inside the membrane of a nerve cell at rest is maintained because the lipid bilayer acts as a barrier to the diffusion of ions. It gives rise to an electrical potential difference, which ranges from about 60 to 70 mV. The potential across the membrane when the cell is at rest (i.e., when there is no signalling activity) is known as the resting potential. Because, by convention, the potential outside the cell is arbitrarily defined as zero, and given the relative excess of negative charges inside the membrane; the potential difference across the membrane is expressed as a negative value: $V_r = -60$ to -70 mV, where V_r, is the resting potential voltage.

The charge separation across the membrane is disturbed whenever there is a net flux of ions into or out of the cell. A reduction of the charge separation is called depolarisation, and an increase in charge separation is called hyperpolarization. Transient current flow and, therefore, rapid changes in potential are made possible by ion channels, a class of integral proteins that traverse the cell membrane. Ionic species are not distributed equally on the two sides of a nerve membrane. Na^+ and Cl^- are more concentrated outside the cell, while K^+ and organic anions (organic acids and proteins) are more concentrated inside. The overall effect of this ionic distribution is the resting potential. There are essentially two forces acting on a given ionic species. The driving force of the chemical concentration gradient tends to move ions down this gradient (chemical potential). On the other hand, the electrostatic force due to the charge separation across the membrane tends to move ions in a direction determined by its particular charge. Thus, for instance, Cl^- ions, which are concentrated outside the cell, tend to move inward down the concentration gradient through non-gated chloride channels. However, the relative excess of negative charge inside the membrane tends to push Cl^- ions back out of the cell. Eventually, an **equilibrium** can be reached so that the actual ratio of intracellular and extracellular concentration ultimately depends on the existing membrane potential.

The same argument applies to the K^+ ions. However these two forces act together on each Na^+ ion to drive it into the cell. First, Na^+ is more concentrated outside than inside and, therefore, tends to flow into the cell, down its concentration gradient. Second, Na^+ is driven into the cell by the electrical potential difference across the membrane. Therefore, if the cell is to have a steady resting membrane potential, the movement of Na^+ ions into the cell must be balanced by the efflux of K^+ ions. Although these steady ionic interchanges prevent irreversible depolarization, the process cannot be allowed to continue unopposed. Otherwise, the K^+ pool would be depleted, intracellular Na^+ would increase, and the ionic gradients would gradually run down, reducing the resting membrane potential. Dissipation of ionic gradients is ultimately prevented by **ATP** dependent Na^+K^+ pumps, which extrude Na^+ from the cell while taking in K^+.

See also Cell membrane transport; Cell structure; Nerve impulses and conduction of impulses

MEMORY AND LEARNING (PHYSIOLOGICAL BASIS) • *see* BRAIN, INTELLECTUAL FUNCTIONS

MENDEL, GREGOR (1822-1884)

Austrian biologist

The science of **genetics** can trace its origins to biologist Gregor Mendel. In meticulous studies with pea plants, Mendel acquired the experimental data necessary to formulate the laws of heredity.

Born in Heinzendorf, Austria (now the Czech Republic), Mendel was the son of a peasant farmer and the grandson of a gardener. As a child, Mendel benefited from the progressive education provided by the local vicar, and he eventually enrolled at the Philosophical Institute in Olmutz (now Olomouc). However, Mendel's worsening financial condition repeatedly forced him to suspend his studies, and in 1843, he entered the Augustinian monastery at Brünn (now Brno).

Although Mendel felt no personal vocation at the time, he believed that the monastery would provide him the best opportunity to pursue his education without the financial worries. He took the name Gregor and eventually was placed in charge of the monastery's experimental garden. In 1847, he was ordained as a priest. Four years later, he was sent to the University of Vienna to study zoology, botany, chemistry, and physics. Following his studies, he returned in 1854 to the monastery and also began teaching the natural sciences at the Brno Technical School.

From then until 1868, in his limited spare time, Mendel performed most of his now-famous heredity experiments. No one had yet been able to make any statistical analysis in breeding experiments, but Mendel's strong background in the natural sciences and his coursework in principles of combinatorial operations prompted him to try. Mendel worked mostly with pea plants, carefully selecting pure varieties that had been cultivated for several years under strictly controlled conditions. He crossed different plants until he produced seven easily distinguishable seed and plant variations (yellow vs. green seeds, wrinkled vs. smooth seeds, tall vs. short plant stems, etc.).

Mendel discovered that, while short plants produced only short offspring, tall plants produced both tall and short offspring. Since only about one-third of the tall plants produced other tall plants, Mendel concluded that there must be two types of tall plants: those that bred true and those that did not.

Mendel continued experimenting, attempting to find intermediate varieties of the offspring by crossing these different plants. In other words, if a tall plant was crossed with a short plant, Mendel expected the result would be a medium-sized plant. Mendel soon found that this was not the case. Mendel crossed short plants with tall plants, planted the seeds from that union, and then self-pollinated the plants from this second generation. He followed the results by counting and recording each generation. All of the offspring that sprouted from the short-tall cross were tall, but the offspring from the self-pollination of those tall plants gave him half tall plants (non-pure), one-quarter pure tall, and one-quarter pure short. Tallness, the more powerful characteristic (the one that shows up the most), was dubbed the dominant trait. Shortness, the weaker characteristic (the one that is frequently masked), was called the recessive trait. It did not seem to matter whether

Gregor Mendel. *Corbis-Bettmann. Reproduced by permission.*

Mendel used male or female plants, the results were always the same. Mendel's quiet, methodical investigation took over eight years to complete and involved more than 30,000 plants.

The results of Mendel's initial plant breeding experiments formed the basis of his first law of heredity: the law of segregation. This law states that hereditary units (genes) are always in pairs, that genes in a pair separate during division of a cell (the **sperm** and egg each receive one member of the pair), and that each gene in a pair will be present in half the sperm or eggs.

Mendel's further experiments established a second law: the law of independent assortment. This law states that each pair of genes is inherited independently of all other pairs. However, it holds true only if the characteristics are located on different **chromosomes**. By sheer coincidence, Mendel had indeed selected such characteristics. But genes located on the same chromosome, as was Hunt Morgan later discovered, are usually inherited together.

In all, Mendel uncovered the following basic laws of heredity: 1) heredity factors must exist; 2) two factors exist for each characteristic; 3) at the time of sex cell formation, heredity factors of a pair separate equally into the gametes (the sperm or the egg); 4) gametes bear only one factor for each characteristic; 5) gametes join randomly no matter what factors they carry; 6) different hereditary factors sort independently when gametes are formed.

Mendel, however, never received acknowledgment during his lifetime for the important contribution he had made to the study of heredity. Although Mendel carefully documented his experiments, presented his findings to the Brünn Society for the Study of Natural Science in 1865, and published *Experiments with Plant Hybrids* the following year, the scientific community was indifferent. Botanists, including Karl Wilhelm von Nägeli, to whom Mendel sent his work, were unaccustomed to statistical analysis. Also, scientists as a whole were hesitant to give credence to such novel theories regarding heredity from such an obscure man.

Mendel died in Brünn on January 6, 1884. Ironically, because of Mendel's refutation of the intermediacy theory that he himself had once posited, Charles Darwin's evolutionary theory was greatly bolstered, for prior to Mendel, natural selection was believed to be counteracted or compromised by repeated blending of gene characteristics throughout the hereditary cycle. Not until 1900, when Mendel's pioneering work was rediscovered by Hugo de Vries and others, did Mendel begin to receive scientific recognition.

See also Genetics and developmental genetics; Evolution and evolutionary mechanisms

MENINGES

The meninges are a series of three membranes covering the **brain** and **spinal cord** that act to protect and partition the **central nervous system** (CNS). The membranes comprising the meninges are the dura mater, arachnoid layer, and the pia mater.

In addition to its protective function, the meninges protect the vascular supply to the CNS and provide a protective barrier for **cerebrospinal fluid.**

The dura mater consists of a periosteal layer, a meningeal layer, and a dural sinus. The dura mater is tough and fibrous and offers a great deal of protective and cohesive structural support to the CNS. The web-like arachnoid mater—also known as the arachnoid layer—is divided into the subdural space, subarachnoid space, and arachnoid villi. The arachnoid layer is much less fibrous and more fragile than the overlying dura mater. The pia mater is the innermost layer of the meninges.

The dural layer acts as a septum to partition the brain. The dural septa consists of a folded layer of dura matter (a falx) that in the form of a falx cerebri, for example, creates a longitudinal fissure separating two **cerebral hemispheres.** The falx cerebelli is another fold of the dura mater that acts to separate the cerebellar hemispheres. The tentorium cerebelli covers, supports, and protects the occipital lobes and **cerebellum.**

Inflammation of the meninges (meningitis) can be a serious, life-threatening, condition. Meningitis may result from a number of causes and is associated with a number of disease processes. For example, infectious meningococcal meningitis is caused by a meningococci bacterial infection. Viral infections may result in viral meningitis. With meningitis, the neck often becomes rigid and is accompanied by a severe **headache** caused by increased pressure on the meninges resulting from inflammation. The increased intracranial pressure may also cause severe vomiting and disorientation. Diagnosis of the cause of meningitis often requires the examination of cerebrospinal fluid enclosed within the membrane. The cerebrospinal fluid, usually obtained by puncturing the meninges of the spinal cord at the level of the lumbar spine (lumbar puncture), can be directly examined and tested for specific infective agents.

See also Cerebral morphology; Connective tissues; Inflammation of tissues; Nervous system: embryological development; Neurology

MENOPAUSE

Menopause is defined as the cessation of menstrual periods in women that occurs at about age 50 years. It is associated with a marked decline in oocyte number that is attributable to progressive atresia (absence) of the original complement of oocytes, secondary to a genetically programmed loss of ovarian follicles. Unlike the age of menarche (the beginning of the first menstrual period), which is affected by nutritional status and general health, the age of menopause has not changed since ancient times. Genetic variation in the **estrogen** receptor gene appears to be one determinant of the age of menopause. Among 900 postmenopausal women in the Netherlands, the mean age at menopause varied by 1.1 year among women with different alleles at the same locus of the estrogen receptor gene. Despite the prolongation of human life through the centuries, the outcome of menopause is almost unchanged.

The climacterium is the whole transitional period from normal ovarian activity (and so normal reproductive ability) to the definitive ovarian failure (and the end of reproduction ability). The term "climacterium" is a clinical term because describes a period in which, with or without the presence of menstrual bleeding, the symptoms of estrogen deficiency arise. The climacterium is divided into three stages, premenopause, perimenopause, and menopause.

Premenopause is the last period of reproductive age in which some alterations in menses may occur in an otherwise normal menstrual pattern. The menstrual cycle may become prolonged or too short, or bleeding may occur between menstrual cycles. Some women may begin to experience typical menopausal symptoms intermittently, such as hot flushes.

Perimenopause is the transitional period two to eight years preceding menopause. It is characterized by a normal ovulatory cycle interspersed with anovulatory (estrogen-only, without ovulation) cycles of varying length. As a result, menses become irregular, and heavy breakthrough bleeding, termed dysfunctional uterine bleeding, can occur during longer periods of anovulation. Thus, vaginal bleeding becomes unpredictable in both timing and amount. In addition, some women complain of hot flushes and vaginal dryness, more typical of the postmenopausal period. In perimenopause, there is a marked decline in the number of oocytes due to progressive atresia of the original complement of oocytes in ovary.

Menopause is the presence of amenorrhea (no menstrual periods) for twelve months, together with the occurrence of symptoms of menopause such as hot flushes. The **blood** contains a higher serum concentration of the **follicle stimulating hormone** (**FSH**). A serum FSH value more than 50 mIU/mL associated to 12 months amenorrhea are classical findings of menopause.

High FSH levels cause recruitment of many follicles at the same time, causing a faster consumption of the remaining oocytes. Eventually, the follicles that remain in the ovaries have very poor sensitivity to FSH stimulation. The follicles able to develop will be less in number each monthly cycle. Inhibin B acts also on the circulating leutenizing hormone (LH) levels; the normal LH peak in the middle of menstrual cycle becomes rare. In this endocrine condition, ovulation cannot be present. When ovulation fails, the corpus luteum is absent and **progesterone** synthesis does not start. Every step of this complex endocrine transition has an impact on menstrual pattern. Irregular bleeding and menopausal symptoms during this perimenopausal transition, because of the waxing and waning of ovarian function and intermittent ovulation, are frequently experienced.

The most common acute change during menopause is the hot flash, which occurs in 75–85% of women. Hot flashes typically begin as the sudden sensation of heat centered on the face and upper chest that rapidly becomes generalized. The sensation of heat lasts from two to four minutes, is often associated with profuse perspiration and occasionally palpitations, and is often followed by chills and shivering. Hot flashes may occur several times per day, although the range may be from only one or two each day to as many as one per hour during the day and night. Hot flashes represent thermoregulatory dysfunction; there is inappropriate peripheral vasodilatation with increased digital and cutaneous blood flow and perspiration. These changes result in rapid heat loss and a decrease in core body temperature below normal. Shivering then occurs as a normal mechanism to restore the core temperature. Hormone replacement therapy is the most often prescribed treatment for hot flashes.

Hot flashes are invariably associated with arousal from **sleep**. Because they can occur as often as once per hour, the result is marked sleep disruption in some women. A continuing sleep disturbance may lead to fatigue, irritability, depression, difficulty concentrating, and other emotional and psychological symptoms that have been attributed to the menopause. Although the relationship between depression and menopause continues to be studied, psychological distress may be related to nonhormonal stressful events occurring around the time of menopause, such as aging, children leaving home, and career or physical limitations. The efficacy of estrogen in menopausal women with mood disturbances is due in part to the improvement in sleep patterns from the prevention of hot flushes.

The incidence of myocardial infarction (heart attack) in women, although lower than in men, increases dramatically after the menopause so that the overall incidence of coronary heart disease in postmenopausal women approaches that of men. Surgical menopause or premature ovarian failure results in an excess risk of cardiovascular disease above that attributed to natural menopause. Early natural menopause also has been postulated to increase the risk of cardiovascular disease, but this effect may be explained by the observation that many women who have an early menopause are smokers. The reason for the increased susceptibility to coronary heart disease in women with estrogen deficiency is not entirely clear. One possibility is the change in **metabolism** of **lipids**, which occurs in the postmenopausal period. Postmenopausal women have an increase in serum total cholesterol, a decrease in serum high-density lipoprotein (HDL) cholesterol, and an increase in serum low-density lipoprotein (LDL) cholesterol concentrations; all three changes may increase the risk of coronary heart disease. These changes can often be reversed to a premenopausal pattern by estrogen replacement therapy. Furthermore, estrogens have direct effects on blood vessels to increase vasodilatation and to prevent or retard atherogenesis.

Estrogen replacement therapy in postmenopausal women may prevent or delay the onset of dementia. Studies to date have demonstrated up to a 30% decrease in the risk of **Alzheimer's disease** with estrogen therapy or a neutral effect. One potential mechanism by which estrogen may prevent cognitive decline is via an increase in cerebral blood flow. Although there are plausible biological mechanisms by which estrogen may lead to a reduced risk or improvement in the severity of dementia, more research is necessary to confirm this relationship.

See also Oogenesis; Ovarian cycle and hormonal regulation

MENSTRUATION

Menstruation is the periodic discharge of **blood**, **mucus**, and tissues from the uterus associated with the cyclical phenomenon of ovulation. Menstruation is the epiphenomenum (an event caused by another and accompaning it) of the ovarian uterine cycles.

The human menstrual cycle involves a complex and regular change in female **anatomy** and **physiology** over an approximate monthly time period. This cycle commences at pubery and ceases at **menopause**. The average adult menstrual cycle lasts 28 days, with approximately 14 days in the follicular phase and 14 days in the luteal phase. It is less variable among women between the ages of 20 and 40, but more variable for the first 5–7 years after menarche (onset) and for the last 10 years before cessation of menses.

The first day of menstruation (or menses) represents the first day of the cycle (day 1). The cycle is then divided into 4 phases: follicular, ovulatory, luteal and menstrual. The follicular phase begins with the onset of menses and ends on the day of the **luteinizing hormone** (LH) surge. The first day of menstruation is considered to be day 1 of both the menstrual cycle and the follicular phase. Menstruation marks the luteal-follicular transition and continues throughout the first few days of the follicular phase.

In follicular phase, with the beginning of menstruation, **plasma** concentration of estradiol, **progesterone**, and LH reach their lowest level. Only **follicle stimulating hormone (FSH)** produced by the hypophysis increases at this time. FSH is active early in the cycle and is responsible for maturation of the ovum. The rising of FSH levels begins about two days before bleeding and is involved in maturation of another group of ovarian follicles and the selection of a dominant follicle for ovulation in the next cycle. The dominant ovarian follicle emerges days 5–7 of the cycle. Twin gestation-prone women sometimes generate two dominant follicles. Follicular growth is a process described as a continuum. Until their numbers are exhausted, follicles begin to grow under almost all physiologic circumstances, including **pregnancy** and periods of anovulation (no ovulation).

At the beginning of the cycle, FSH is found in the follicular fluid of small follicles. FSH binds to the receptors located in granulosa cells of the ovary, and stimulates their differentiation from squamous cells into cuboidal cells. Furthermore, FSH stimulates mitosis of granulose cells and their secretion of estradiol. Estradiol begins to rise in plasma by the forth day of cycle and stimulates LH receptors on teca cells of the ovary to prepare them for progesterone production after ovulation. When estradiol rises, the pituary gland is inhibited in FSH secretion (negative feed-back) and stimulated in LH secretion (positive feed-back). As follicles enlarge in size, they secrete both androgen and **estrogen**. The dominant follicle is the one that has a follicular fluid estradiol: androgen ratio greater than 1.

The ovulatory phase immediately follows the follicular phase. As its name implies, this is when ovulation occurs. It is characterized by a hormone surge as estrogen rapidly peaks at days 10–15, then decreases. Luteinizing hormone (LH) has large surge at this time. LH is responsible for ovulation. By day 11–13 of the normal cycle, the LH rising triggers ovulation. The LH surge initiates substantial changes in the ovary. The egg can be fertilized for up to 36 hours but lives about 72 hours. Progesterone levels dramatically rise just after ovulation.

The oocyte in the dominant follicle completes its first meiotic division. In addition, the local secretion of plasminogen activator and other cytokines required for the process of ovulation is increased. The oocyte is released from the follicle at the surface of the ovary approximately 36 hours after the LH surge. It then travels down the fallopian tube to the uterine cavity. The follicle remnant is converted in corpus luteum to facilitate progesterone production.

The endometrium is relatively indistinct during menses and then becomes a thin line once menses is complete. It is normal to see small follicles of 0.12-0.31 in (3-8 mm) in diameter at this time. Within about seven days from the onset of menses, several 0.35-0.39 in. (9-10 mm) antral follicles are visible on ovarian ultrasonography. The rising serum estradiol concentrations result in proliferation of the uterine endometrium, which becomes thicker. By the late follicular phase, the dominant follicle increases in size by about 0.08 in. (2 mm) per day until a mature size of 0.79-1.02 in. (20-26 mm) is reached.

Rising serum estradiol concentrations result in gradual thickening of the uterine endometrium and an increase in the amount and character of the cervical mucus. Many women are able to detect this change in mucus character. Studies of cervical mucus samples during the menstrual cycle demonstrate a late follicular-phase peak in the mucin protein MUC5B that may be important for **sperm** transit to the uterus.

The gradually increasing serum progesterone concentrations have a profound impact on the endometrial lining, leading to cessation of mitoses and "organization" of the glands. In the luteal phase, progesterone has negative feedback on the secretion of both FSH and LH. As the corpus luteum fails (pregnancy absent) and the progesterone secretion diminishes FSH begins to rise to prepare a women for the next reproductive cycle. The decline in estradiol and progesterone release from the resolving corpus luteum results sequentially in the loss of endometrial blood supply, endometrial sloughing, and the onset of menstruation (menses) approximately 14 days after the LH surge.

The menstrual phase (bleeding) is a relatively imprecise marker of hormonal events in the menstrual cycle, since there is considerable inter-individual variability in the relationship between the onset of endometrial sloughing and the fall in serum hormone concentrations during the luteal phase. The menstrual phase is characterizzed by demise of corpus luteum, estrogen and progesterone withdrawal, and sloughing of the endometrium. Normal blood loss ranges from 20–80 ml, averaging 35 ml over the 3–7 day menstrual period. Women also lose an average of 13 mg of iron during menstruation.

See also Fertilization; Hormones and hormone action; Reproductive system and organs (female)

MESENCHYME • *see* EMBRYOLOGICAL DEVELOPMENT, EARLY DEVELOPMENT, FORMATION, AND DIFFERENTIATION

MESODERM

Mesoderm is one of three principal germinal layers of cells that are formed in early in embryonic development. Mesoderm comprises the middle germinal layer that lies between the ectodermal and endodermal layers. Mesodermal cells ultimately form tissues, **organs**, and systems including **blood** vessels, blood, **connective tissues**, muscles, **cartilage**, bone, lymph glands, **pericardium, peritoneum, pleura, gonads**, and **kidneys**.

In the embryonic disk **ectoderm** and **endoderm** sandwich mesoderm, the third primitive germinal layer. When the embryonic disk ultimately folds into a tube the basic "tube within a tube" plan of development becomes evident. A core endodermal tube establishes a primitive digestive pathway bounded by an oral orifice and an anal orifice. Around that innermost tube is an outer tube comprised of ectoderm. The ectoderm serves as a protective layer and the layer from which the nervous system and sense organs develop. Mesodermal cells fill the space between the inner (endodermal) and outer (ectodermal) tube.

About a week following **fertilization**, the human embryonic blastocyst is embedded in the endometrium of the uterus. The blastocyst is a proliferating ball of cells with a cavity termed a blastocoele. At one pole of the blastocyst there is a thickened mass of cells termed the inner cell mass. The inner cell mass contains communicating slit-like openings that form the amniotic cavity and the embryonic disk. Ectodermal cells lie on the dorsal side of the embryonic disk, the endodermal layer lies on the ventral side of the disk. Initially there are only two germ layers but by 18 days following fertilization a thickening occurs in the ectoderm to form a primitive streak. The walls of the primitive groove continue to thicken and, at the anterior (cephalic) end of the groove, expand into a primitive knot also known as Henson's node. Anterior to the primitive knot, primitive ectoderm and endoderm are in direct contact with each other to form a prochordal plate. The primitive streak also establishes the general head-to tail (cephalo-caudal) axis for subsequent development. Starting at the primitive knot, cells from the ectodermal layer migrate into the primitive groove and invaginate into the space between the ectoderm and endoderm to form mesodermal cells. Other cells, derived from other **fetal membranes** also contribute cells to the mesodermal layers.

As ectodermal cells stream invaginate to form a trilaminar embryonic disk, a head process (ultimately to become the **notochord**) forms in the middle mesodermal layer.

The mesodermal germ layer divides into an outer layer of somatic mesoderm and inner layer of splanchnic mesoderm. Intraembryonic mesoderm also becomes divided by a longitudinal groove into paraxial (along the axis) and lateral plate mesoderm. The paraxial mesoderm further divides by segmentation into a series of blocks termed **somites**

See also Embryonic development: Early development, formation, and differentiation; Fetal membranes and fluids; Implantation of the embryo; Limb buds and limb embryological development; Nervous system: embryological development; Organizer experiment; Sexual reproduction; Skeletal and muscular systems: embryonic development; Somite

METABOLIC RATE · *see* METABOLISM

METABOLIC WASTE REMOVAL

Cellular metabolic activity, i.e., the many vital biochemical processes executed by different cell types in the various tissues and **organs**, result in a variety of unnecessary metabolites or end products, which must be removed from the system. Excessive amounts of nutrients, such as water, **electrolytes**, and minerals, must also be eliminated in order to maintain the physiological **equilibrium** of many body systems, such as neural activity, cardiac function, and **blood** pressure. Through respiration, the **lungs** deliver oxygen into the blood circulation and collect carbon dioxide from the venous vessels, to be eliminated through exhalation. The process of **digestion** requires the elimination of non-absorbable substances and other compounds that are excreted from the body in the feces. One-fourth of the fecal matter that is eliminated from the intestines is constituted by solid matter and the remainder is water. The fecal solid matter usually contains between 10–20% of **fat**, 30% of non-digested substances, about three per cent of proteins, and over 30% of dead **bacteria**.

The **kidneys** excrete in the urine several ion solutes, and several metabolites such as urea, **uric acid**, creatinine, **bilirubin**, and a variety of toxins that result from either endogenous or exogenous metabolized products by the cellular enzymatic detoxification process. For instance, several genotoxic compounds, such as some therapeutic drug and food byproducts, environmental carcinogens, and endogenous toxic metabolites (i.e., toxins produced by cellular **metabolism** itself) do form chemical insertions in the genome, known as **DNA** adducts, what can lead to genetic mutation, cellular oxidative stress, triggering degenerative processes, and functional impairments. The enzymatic systems of DNA repair removes most of this chemical adducts from the DNA while complementary cellular enzymatic systems modify their molecular structures to facilitate their excretion from the cell into the circulation. These metabolites will be ultimately eliminated from the body mostly in the urine. The enzymatic excision and metabolization of endogenous DNA adducts, result in about 20 different metabolites, and some of those already measured in the urine are thymine-glycol, thymidine-glycol, hydroximethyl-uracil, and hydroxymethyl-deoxyuridine. A healthy human being eliminates in the urine approximately 100 nanno Moles (nM) of the first three compounds every day.

Another important enzymatic system involved in cellular metabolism and removal of toxic compounds (i.e., detoxification) is constituted by two types of **enzymes**: 1) those pertaining to the oxidative metabolism of Phase I; and 2) the conjugated enzymes of Phase II. The main Phase I enzymes are members of the super family of cytochrome P450 or CYPs; and the Phase II enzymes are glutathione-S-transferases (GSTs), UDP-glucoroniltransferases, and N-acetyltransferases (NATs). These enzymes are most active in the **liver** and to a minor degree in the intestine, lungs, kidneys, and other tissues, such as **mammary glands**. Most of the metabolites that result from this enzymatic metabolic pathway are removed through the kidneys as urine solutes. The liver for instance, degrades steroid **hormones** through the mediation of Phase II enzymes, forming glucoronides and sulfates, which are excreted part in the **bile** and part in the urine.

Enzymatic metabolic processing and detoxification as well as metabolic waste removal are important for several vital purposes, such as the maintenance of blood pressure and body fluid appropriate levels, protection of cells and organs against oxidative stress, electrolyte balance, DNA integrity, control of hormonal levels, and detoxification from environmental pollutants and solar radiation.

See also Antioxidants; Breathing; Circulatory system; Gaseous exchange; Glomerular filtration; Homeostatic mechanisms; Renal system

METABOLISM

Metabolism is the sum total of chemical changes that occur in living organisms and which are fundamental to life. Nutrients from the environment are used in two ways by living organisms. They can serve as constituents in the synthesis of components of the organism (assimilation) or they can be oxidatively degraded for the purposes of energy production (dissimilation). All constituents of living organisms are subject to continual processes of breakdown and re-synthesis, normally referred to as "turnover." Turnover is defined as the balance of synthesis and degradation of biomolecules in living organisms. Processes of breakdown are collectively called catabolism and all reactions concerned with synthesis are called anabolism. Amphibolic pathways may serve both functions of degradation and synthesis. Because metabolism is an open system (i.e., having constant inputs and outputs) its multistage processes approach a state of **equilibrium** that is never quite achieved, at least not while the organism is alive. Metabolism is, therefore, better described as a steady state, rather than an equilibrium. A chain of metabolic reactions is said to be in a "steady state" when the concentration of all intermediates remains constant, despite the net flow of material through the system. That means the concentration of intermediates remains constant, while a product is formed at the expense of the substrate.

Primary metabolism comprises those metabolic processes that are basically similar in all living cells and are necessary for cellular maintenance and survival. They include the fundamental processes of growth (e.g., the synthesis of biopolymers and the macromolecular structures of cells and organelles), energy production (glycolysis and the tricarboxylic acid cycle) and the turnover of body and cell constituents. Secondary metabolism refers to the production of substances, such as **bile** pigments from porphyrins in humans, which only occur in certain tissues and are distinct from the primary metabolic pathways.

Metabolism is regarded as a cybernetic network with self-regulatory properties and containing many cyclic processes. Processes of metabolic regulation are frequently presented as analogous to electronic and mechanical regulatory processes known in technology. Regulation is a fundamental principle in the organization of a living organism and, depending on the nature of the signal or method of information transfer, there are four broad types in **physiology**. In the nervous system, for example, the nerve impulse is an electronic signal and the regulatory response may also be electrical or chemical. In the humoral system, **hormones** act as chemical signals in a regulatory system that is superimposed on the more basic levels of metabolic control. Cyclic AMP acts as a second messenger for many hormones, which are synthesized at specific sites (endocrine glands) and are then transported to the target **tissue** or organ. Neural and humoral regulation processes represent intercellular (between cells) metabolic control.

Intracellular (within cells) metabolic control processes include regulation of gene expression and metabolic feedback or feed-forward processes. The triggers of differential gene expression may be chemical (hormones) or environmental (e.g., light). Differential gene expression is responsible for the regulation, at the molecular level, of differentiation and development, as well as the maintenance of numerous cellular "house-keeping" reactions, which are essential for the day-to-day functioning of a cell. In many metabolic pathways, the metabolites (substances produced or consumed by metabolism) themselves can act directly as signals in the control of their own breakdown and synthesis. Feedback control can be negative or positive. Negative feedback results in the inhibition by an end product, of the activity or synthesis of an enzyme or several **enzymes** in a reaction chain. The inhibition of the synthesis of enzymes is called enzyme repression. Inhibition of the activity of an enzyme by an end product is an allosteric effect and this type of feedback control is well known in many metabolic pathways (e.g., glycolysis). In positive feedback, an endproduct activates an enzyme responsible for its own production e.g. thrombin activates factors VIII and V during **blood** clotting, thus contributing to the speed of the cascade system and the rapid formation of a clot. An example of feed-forward enzyme activation may be found in the activation of **glycogen** synthetase by glucose-6-phosphate, that is a metabolite activates an enzyme concerned in its utilization.

Many reactions in metabolism are cyclic. A metabolic cycle is a catalytic series of reactions, in which the product of one bimolecular (involving two molecules) reaction is regenerated as follows: $A + B \rightarrow C + A$. Thus, A acts catalytically and is required only in small amounts and A can be regarded as carrier of B. The catalytic function of A and other members of the metabolic cycle ensure economic conversion of B to C. B is the substrate of the metabolic cycle and C is the product. If intermediates are withdrawn from the metabolic cycle, e.g., for biosynthesis, the stationary concentrations of the metabolic cycle intermediates must be maintained by synthesis. Replenishment of depleted metabolic cycle intermediates is called anaplerosis. Anaplerosis may be served by a single reaction, which converts a common metabolite into an intermediate of the metabolic cycle. An example of this is pyruvate to oxaloacetate reaction in the tricarboxylic acid cycle. Alternatively, it may involve a metabolic sequence of reactions, i.e., an anaplerotic sequence. An example of this is the glycerate pathway which provides phosphoenol pyruvate for anaplerosis of the tricarboxylic acid cycle.

Metabolic cycles are anabolic, catabolic or amphibolic (both catabolic and anabolic). The Calvin cycle is an anabolic (synthetic) cycle. A truly catabolic cycle, one that does not supply intermediates for biosynthesis, probably does not exist. The tricarboxylic cycle is an important central metabolic pathway, serving both the terminal oxidation of substrates and the provision of intermediates for biosynthesis e.g. the biosynthesis of porphyrins and certain **amino acids**, and is therefore an amphibolic cycle. Similarly the pentose phosphate cycle has a catabolic function and at the same time provides ribose phosphate for the synthesis of nucleic acids and certain coenzymes.

The first metabolic cycle to be recognized was the urea cycle described by **Hans Krebs** and Kurt Heseleit in 1932. This may be considered as an anabolic cycle since it results in the energy-dependent synthesis of urea. However, with respect to its metabolic role in the degradation of protein and detoxification of ammonia, it is catabolic. Under certain condition, it also provides arginine for **protein synthesis** and anaplerosis occurs by the synthesis of ornithine from glutamate.

See also Adenosine triphosphate; Biochemistry; Enzymes and coenzymes

METATHALAMUS

The metathalamus, which is also known as the corpora geniculata, is located in a region of the forebrain called the diencephalon. It is comprised of two structures, the medial geniculate body and the lateral geniculate body.

In cross-section the **brain** appears similar to a cauliflower. In this visual example the metathalamus would be located near the stem of a cauliflower cut lengthwise in half.

The diencephalon region of the forebrain is divided into three smaller regions. One of these is called the thalamencephalon. This is the area of the metathalamus, **thalamus**, and the **epithalamus**. These areas of the brain function in the receipt and processing of sensory information.

The metathalamus is involved with the processing of information received from the eyes and from the ears. Specifically, the lateral geniculate body contains the processing machinery for vision-related information, and the lateral geniculate body contains the machinery for hearing-related information. Nerves from the ears merge with the optic nerve that extends from the back of the **eye**. The optic nerve enters the brain and branches into the medial and lateral geniculate bodies of the metathalamus.

The medial geniculate body also goes by the names corpus geniculatum mediale, internal geniculate body, and postgeniculatum. It is located very near the thalamus. It is oval-shaped and fibers emerging from one part of the body pass to the optical tract.

The lateral geniculate body is also called the corpus geniculatum laterale, external geniculate body, or the pregeniculatum. It is also oval in shape but is darker in color and larger than the medial body. The gray and white matter (components of neuron cells) that make up the body alternate, giving a striped appearance. Many nerve fibers from the optic tract (specifically from the ears) merge with the lateral body.

See also Brain: intellectual functions; Cerebral hemispheres; Cerebral morphology; Nervous system overview

METCHNIKOFF, ÉLIE (1845-1916)
Russian immunologist

Élie Metchnikoff was a pioneer in the field of **immunology** and won the 1908 Nobel Prize in physiology or medicine for his discoveries of how the body protects itself from disease-causing organisms. Later in life he became interested in the effects

of **nutrition** on aging and health, which led him to advocate some controversial diet practices.

Metchnikoff, the youngest of five children, was born in the Ukrainian village of Ivanovka on May 16, 1845, to Emilia Nevahovna, daughter of a wealthy writer, and Ilya Ivanovich, an officer of the Imperial Guard in St. Petersburg. He enrolled at the Kharkov lycee in 1856, where he developed an especially strong interest in biology. At age 16, Metchnikoff published a paper in a Moscow journal criticizing a geology textbook. After graduating from secondary school in 1862, he entered the University of Kharkov, where he completed a four-year program in two years. He also became an advocate of the theory of **evolution** by natural selection after reading Charles Darwin's *On the Origin of Species by Means of Natural Selection.*

In 1864 Metchnikoff traveled to Germany to study, where his work with nematodes led to the surprising conclusion that the organism alternates between sexual and asexual generations. His studies at Kharkov, coupled with his interest in Darwin's theory, convinced him that highly evolved animals should show structural similarities to more primitive animals. He pursued his studies of invertebrates in Naples, Italy, where he collaborated with Russian zoologist Alexander Kovalevsky. They demonstrated the homology (similarity of structure) between the germ layers—embryonic sheets of cells that give rise to specific tissue—in different multicellular animals. For this work the scientists were awarded the Karl Ernst von Baer Prize.

Metchnikoff was only twenty-two when he received the prize and had a promising career ahead of himself. However, he soon developed severe eyestrain, a condition that hampered his work and prevented him from using the microscope for the next fifteen years. Nevertheless, in 1867 he completed his doctorate at the University of St. Petersburg with a thesis on the embryonic development of fish and crustaceans. He taught at the university for the next six years before moving to the University of Odessa on the Black Sea where he studied marine animals.

During the summer of 1880 he spent a vacation on a farm where a beetle infestation was destroying crops. In an attempt to curtail the devastation, Metchnikoff injected a fungus from a dead fly into a beetle to see if he could kill the pest. Following the assassination of Russian Czar Alexander II in 1884, Metchnikoff carried this interest in infectious processes with him when he left Odessa for Italy. Metchnikoff began to focus more on **pathology**, or the study of diseases.

This transformation was due primarily to his study of the larva of the *Bipinniara* starfish. While studying this larva, which is transparent and can be easily observed under the microscope, Metchnikoff saw special cells surrounding and engulfing foreign bodies, similar to the actions of white **blood** cells in humans that were present in areas of inflammation. During a similar study of the water flea *Daphniae*, he observed white blood cells attacking needle-shaped spores that had invaded the insect's body. He called these cells "phagocytes," from the Greek word *phagein*, meaning, "to eat."

While scientists thought that human phagocytes merely transported foreign material throughout the body, and therefore spread disease, Metchnikoff realized they performed a protective function. He recognized that the human white blood cells and the starfish phagocytes were embryologically homologous, both being derived from the **mesoderm** layer of cells. He concluded that the human cells cleared the body of disease-causing organisms. In 1884, he injected infected blood under the skin of a frog and demonstrated that white blood cells in higher animals served a similar function as those in starfish larvae. The scientific community, however, still did not accept his idea that phagocytic cells fought off infections.

Metchnikoff was twice widowed. Metchnikoff returned to Odessa in 1886 and became the director of the Bacteriological Institute. He continued his research on phagocytes in animals and pursued vaccines for chicken cholera and sheep anthrax. Hounded by scientists and the press because of his lack of medical training, Metchnikoff fled Russia a year later. A chance meeting with French scientist Louis Pasteur led to a position as the director of a new laboratory at the Pasteur Institute in Paris. There, he continued his study of phagocytosis for the next twenty-eight years.

But conflict with his fellow scientists continued to follow him. Many scientists asserted that antibodies triggered the body's immune response to infection. Metchnikoff accepted the existence of antibodies but insisted that phagocytic cells represented another important arm of the **immune system**. His work at the Pasteur Institute led to many fundamental discoveries about the immune response, and one of his students, **Jules Bordet**, contributed important insights into the nature of complement, a system of antimicrobial **enzymes** triggered by antibodies. Metchnikoff received the Nobel Prize in physiology and medicine in 1908 jointly with **Paul Ehrlich** for their pioneering work in immunology.

Metchnikoff's interest in immunity ultimately led to a series of unscientific writings on aging and death. On July 15, 1916, after a series of heart attacks, Metchnikoff died in Paris at the age of 71. He was a member of the French Academy of Medicine, the Swedish Medical Society, and the Royal Society of London, from which he received the Copley Medal. He also received an honorary doctorate from Cambridge University.

See also Embryology; Embryonic development: Early development, formation, and differentiation; Germ cells; Homologous structures

METENCEPHALON · *see* CEREBRAL MORPHOLOGY

MEYERHOF, OTTO (1884-1951)

German-born American biochemist

Otto Meyerhof helped lay the foundations for modern bioenergetics, the application of the principles of thermodynamics (the science of physics in relation to heat and mechanical action) to the analysis of chemical processes going on within the living cell. Meyerhof's research attempted to explain the function of a cell in terms of physics and chemistry; his research into the chemical processes of the muscle cell paved the way for the full understanding of the breakdown of glucose to provide body energy. For his discovery of the fixed relationship between the consumption of oxygen and the **metabolism** of **lactic acid** in the muscle, Meyerhof shared the 1922 Nobel Prize in physiology or medicine.

Born Otto Fritz Meyerhof on April 12, 1884, in Hannover, he was the second child and first son of Felix Meyerhof, a Jewish merchant, and Bettina May Meyerhof. Brought up in a comfortable middle class home, Meyerhof attended secondary school at the Wilhelms Gymnasium in Berlin where the family had moved. As a 16-year-old, he developed kidney trouble, necessitating a long period of bed rest. It was during this time that Meyerhof began reading extensively, especially the works of Goethe, which were to influence him deeply in his later life. A trip to Egypt in 1900 also provided him with a lasting love of archaeology. Once his health had improved, Meyerhof began studying medicine, receiving his medical degree from the University of Heidelberg in 1909. As a doctoral dissertation, Meyerhof wrote on the psychological theory of mental disturbances. His interests in psychology and psychiatry were ongoing, though he soon changed the direction of his professional passions.

From 1909 to 1912 Meyerhof was an assistant in internal medicine at the Heidelberg Clinic, and it was there he came under the influence of physiologist **Otto Warburg** who was researching the causes of **cancer**. Meyerhof soon joined this young man in a study of cell respiration, for Warburg was examining the changes that occur when a normal living cell becomes cancerous. It was this early work that set Meyerhof on a course of biochemical research. He also spent some time at the Stazione Zoologica in Naples, and by 1913 had joined the physiology department at the University of Kiel where, as a young lecturer, he first introduced his theory of applying the principles of thermodynamics to the analysis of cell processes. It was one thing to lecture about such an application; quite another to apply such principles to his own research.

Meyerhof decided to focus on the chemical changes occurring during voluntary **muscle contraction**, applying thermodynamics to the study of cellular function. The chemical and mechanical changes occurring during muscle contraction were on a large enough scale that he could measure them with the primitive instruments he had at Kiel, and there had been some earlier research on which to build. Specifically, physiologist Archibald Vivian Hill in England had observed the heat production of muscles, and others, including physiologist Claude Bernard of France and English biochemist Frederick Gowland Hopkins, had shown both that **glycogen**, a carbohydrate made up of chains of glucose molecules, is stored in **liver** and muscle cells, and that working muscles accumulated lactic acid. But Meyerhof's research had to be put on hold during the First World War.

In 1914, as Europe was becoming increasingly unsettled, Meyerhof married Hedwig Schallenberg, a painter. The couple had three children. During the war, Meyerhof served briefly as a German medical officer on the French front. After the war, he resumed his research into the physicochemical mechanics of cell function after being appointed assistant professor at the University of Kiel. In 1917, Meyerhof was able to show that the carbohydrate enzyme systems in animal cells and yeast are similar, thus bolstering his philosophical conviction in the biochemical unity of all life. It was while examining the contraction of frog muscle in 1919 that Meyerhof proved in a series of careful experiments that there was a quantitative relationship between depletion of glycogen in the muscle cell and the amount of lactic acid that was produced. He also demonstrated that this process could occur without oxygen, in what is termed anaerobic glycolysis. Meyerhof went on to demonstrate further that when the muscle relaxed after work, then molecular oxygen would be consumed, oxidizing part of the lactic acid—actually only about one-fifth of it. He concluded that the energy created by this oxidative process was used to convert the remaining lactic acid back to glycogen, and thus the cycle could begin again in the muscle cell.

Such an understanding, although not completely explaining all the steps in such a metabolic pathway, did pave the way for other research into the glucose cycle and energy production in the body. Meyerhof won the 1922 Nobel Prize for this pioneering work, sharing it with Hill whose work on heat in muscle contraction had inspired much of Meyerhof's own research.

The prize could not have come at a better time. Just prior to the announcement, Meyerhof had been passed over for the position of chair of the physiology department at the University of Kiel, largely because of anti-Semitism. The Nobel Prize secured for him laboratory space at the Kaiser Wilhelm Institute for Biology in Berlin, where he was appointed a professor in 1924. During his stay there, from 1924 to 1929, he trained many notable biochemists, Hans Adolf Krebs, **Fritz Lipmann**, and **Severo Ochoa** among them. He brought his colleague Karl Lohmann into the group, and Meyerhof and his team continued with further research at Berlin-Dahlem where the institute was located.

In 1925, Meyerhof managed to extract the glycolytic **enzymes** of muscle, thereby making it possible to isolate these enzymes and study the individual steps involved in the complex pathway in the muscle cell from glycogen to lactic acid. It took another two decades to fully delineate the pathway, and many other brilliant scientists worked on it, but it is today called the Embden-Meyerhof pathway in recognition of Meyerhof's groundbreaking work.

In 1926, researchers in the U.S. and in England discovered a new phosphorylated compound in muscle, phosphocreatine. Searching for another possible source of chemical energy in addition to lactic acid, Meyerhof tested this compound and found that its breakdown did produce heat. But more importantly, in 1929, Meyerhof's assistant, Lohmann, discovered **adenosine triphosphate** (**ATP**). Meyerhof and Lohmann went on to show that ATP is the most important molecule that powers the biochemical reactions of the cell, and to demonstrate its role in muscle contraction as well as in other energy-requiring processes of biological systems. That same year Meyerhof was appointed director of the new Kaiser Wilhelm Institute for Medical Research in Heidelberg, and for the first time had expansive and modern work space at his disposal.

Meyerhof remained at Heidelberg until 1938, continuing his work with ATP and providing striking evidence of the unity of biochemical processes amid the amazing diversity of life forms. Nevertheless, such philosophical proofs would not help him in the new Germany. Starting with Hitler's rise to power in 1933, Jewish scientists like Meyerhof steadily emigrating from Germany. Finally, in 1938, Meyerhof realized that he too could no longer stay. His daughter and oldest son were already out of the country. Together with his wife and youngest child, Meyerhof, on the pretext of taking a few weeks vacation, escaped to Paris where he continued his research at the Institute of Physicochemical Biology. In 1940, when the Germans invaded France, Meyerhof and his family were forced to flee once again, this time to Spain and eventually, via an arduous trip over the Pyrenees, to neutral Portugal where they caught a ship to the United States. There he joined his friend Hill—with whom he had shared the 1922 Nobel Prize—at the University of Pennsylvania where the Rockefeller Foundation had created a professorship in physiological chemistry for him.

Meyerhof continued research into the bioenergetics of cell function, summering at Woods Hole Marine Laboratory on Cape Cod and meeting with other refugees from the European conflagration. He suffered a severe heart attack in 1944, but recovered and became a United States citizen in 1946. In 1949, he was elected to the National Academy of Sciences in recognition of his life's work that included not only original research, but also the publication of over 400 scientific articles. He died of a second heart attack on October 6, 1951.

See also Biochemistry; Cell membrane transport; Cell structure; Glucose utilization, transport and insulin; Glycogen and glycolysis; Metabolism

MICROSCOPIC ANATOMY · *see* ANATOMY

MICTURITION · *see* ELIMINATION OF WASTE

MIDDLE EAR · *see* EAR (EXTERNAL, MIDDLE, AND INTERNAL)

MILLER, JACQUES (1931-　)
French-born Australian physician

Jacques Francis Albert Pierre Miller is mainly known by his seminal contributions to the understanding of the physiologic role of the thymus gland in the development and maturation of the **immune system**. He was born in Nice, France, and educated in Australia, where he attended Saint Aloysius' College in Sydney, before studying medicine at the University of Sydney School of Medicine. During his medical studies, Miller also achieved a B.S. in bacteriology, in 1953. After his medical residency at Royal Prince Alfred Hospital in Sydney, he was granted a Gaggin Fellowship to do research at the University College, London, United Kingdom, where he investigated mouse **leukemia** at the Chester Beatty Research Institute. Miller observed that many types of lymphocytic leukemia in mice, either induced by virus, radiation, chemical carcinogens, or spontaneous leukemia occurring in high-leukemic strain mice, always began in the thymus, and then spread. Other studies had previously reported that the surgical removal of the thymus in young mice would prevent both spontaneous and induced leukemia, although this procedure had not yet being used for virus-induced leukemia. Therefore, Miller infected mice with Gross virus at birth and removed their thymus when they were one month old. Leukemia was prevented. These studies gave Miller a strong indication that the thymus played an important role in the **physiology** of the immune system.

After completing a Ph.D. degree in 1960, Miller continued his research on mouse leukemia at the Institute until 1963. Miller was the first investigator to discover the relevant role of the thymus gland in body growth and in the immune system response. Until 1961, medical science had no clear idea of such a role, in spite of the standard experiments of surgically removing the gland from adult animals to observe the impact of the lack of the organ upon the organism or its behavior. No apparent change was reported apart from the fact that in adult mammals, the thymus seemed to have undergone atrophy (shrinkage). Miller decided to remove the thymus (thymectomy) of one-day-old mice in order to observe whether this early removal would have some consequence on the animal development. The mice could not develop properly, and died within two-three months later. Autopsies disclosed that the mice had lesions in the **liver**, similar to those found in viral hepatitis, and showed a deficiency of lymphocytes in the lymph nodes and spleen, which was never found in adult mice after thymus removal. At the time, another group had already shown that lymphocytes were the cells responsible for the initiation of the immune response, which led Miller to the suspicion that the deficiency he found in his animals was due to the early removal of the thymus. Contrary to the current belief of the time, Miller suspected that the thymus could be the site of production of lymphocytes, which would eventually mature into competence in other places other than the thymus. He published an article advancing this hypothesis in 1961 in the scientific journal *The Lancet*. Miller's suppositions were later proved correct by several independent research groups.

Miller spent most of 1963 at the National Institutes of Health in Maryland, working with germ-free mice, testing the ability of both immuno-incompetent (with an intact immune system) and thymectomised (thymus removed) animals to reject grafts, therefore assessing their immune system response. Because the mice showed tolerance to grafts from unrelated mice and even rats, it became clear that the thymus gland played some important role in the immune system. However, what this specific role could possibly be took another decade of research to be understood.

After returning to London where he stayed until 1966, Jacques Miller moved again to Australia to work as the head of experimental **pathology** at the Walter and Eliza Hall

Institute of Medical Research. Together with Graham Mitchell, he worked on the identification of the types of lymphocytes that could restore the immune response in immuno-incompetent mice. They showed for the first time the existence of two subsets of lymphocytes in mammalians, one originated from bone marrow and the other from the thymus, which have complementary functions in the immune response. Another accomplishment by Miller was the demonstration of the mechanisms of recognition and tolerance of tissues and proteins from the body by its own immune cells, as well as the causes of the delayed-type hypersensitivity reaction (i.e., **allergies**), using transgenic mice.

Jacques Millers's discoveries about the function of thymus was vital to current understanding of the immune system and had wide implications to the advancement of knowledge in areas such as **cancer**, autoimmune diseases, **AIDS**, allergy, viral immunity, and transplant rejection. He served on the World Health Organization in the field of eradicable diseases, as well as on the International Research Agency for Cancer. He became a Fellow of the Australian Academy of Science and of the Royal Society in 1970. He shared with Max Cooper the Sandoz Prize of Immunology in 1990, and received in the same year the Peter Medawar Prize of the Transplantation Society. He was awarded with the Florey Faulding Medal in 2000 for his pioneering contributions to immunology and the thymus physiology.

See also Thymus histophysiology

MILSTEIN, CÉSAR (1927-)
Argentinean-born English biochemist

César Milstein conducted one of the most important late twentieth century studies on antibodies. In 1984, Milstein received the Nobel Prize in physiology or medicine, shared with **Niels K. Jerne** and **Georges Kohler**, for his outstanding contributions to **immunology** and immunogenetics. Milstein's research on the structure of antibodies and their genes, through the investigation of **deoxyribonucleic acid** (**DNA**) and **ribonucleic acid** (**RNA**), has been fundamental for a better understanding of how the human **immune system** works.

Milstein was born on October 8, 1927, in the eastern Argentine city of Bahía Blanca, one of three sons of Lázaro and Máxima Milstein. He studied **biochemistry** at the National University of Buenos Aires from 1945 to 1952, graduating with a degree in chemistry. Heavily involved in opposing the policies of President Juan Peron and working part-time as a chemical analyst for a laboratory, Milstein barely managed to pass with poor grades. Nonetheless, he pursued graduate studies at the Instituto de Biología Química of the University of Buenos Aires and completed his doctoral dissertation on the chemistry of aldehyde dehydrogenase, an alcohol enzyme used as a catalyst, in 1957.

With a British Council scholarship, he continued his studies at Cambridge University from 1958 to 1961 under the guidance of Frederick Sanger, a distinguished researcher in the field of **enzymes**. Sanger had determined that an enzyme's functions depend on the arrangement of **amino acids** inside it. In 1960 Milstein obtained a Ph.D. and joined the Department of Biochemistry at Cambridge, but in 1961, he decided to return to his native country to continue his investigations as head of a newly-created Department of Molecular Biology at the National Institute of Microbiology in Buenos Aires.

A military coup in 1962 had a profound impact on the state of research and on academic life in Argentina. Milstein resigned his position in protest of the government's dismissal of the Institute's director, Ignacio Pirosky. In 1963, he returned to work with Sanger in Great Britain. During the 1960s and much of the 1970s, Milstein concentrated on the study of antibodies, the protein organisms generated by the immune system to combat and deactivate **antigens**. Milstein's efforts were aimed at analyzing myeloma proteins, and then DNA and RNA. Myeloma, which are **tumors** in cells that produce antibodies, had been the subject of previous studies by **Rodney R. Porter, Frank MacFarlane Burnett**, and **Gerald M. Edelman**, among others.

Milstein's investigations in this field were fundamental for understanding how antibodies work. He searched for mutations in laboratory cells of myeloma but faced innumerable difficulties trying to find antigens to combine with their antibodies. He and Köhler produced a hybrid myeloma called hybridoma in 1974. This cell had the capacity to produce antibodies but kept growing like the cancerous cell from which it had originated. The production of monoclonal antibodies from these cells was one of the most relevant conclusions from Milstein and his colleague's research. The Milstein-Köhler paper was first published in 1975 and indicated the possibility of using monoclonal antibodies for testing antigens. The two scientists predicted that since it was possible to hybridize antibody-producing cells from different origins, such cells could be produced in massive cultures. They were, and the technique consisted of a fusion of antibodies with cells of the myeloma to produce cells that could perpetuate themselves, generating uniform and pure antibodies.

In 1983 Milstein assumed leadership of the Protein and Nucleic Acid Chemistry Division at the Medical Research Council's laboratory. In 1984 he shared the Nobel Prize with Köhler and Jerne for developing the technique that had revolutionized many diagnostic procedures by producing exceptionally pure antibodies. Upon receiving the prize, Milstein heralded the beginning of what he called "a new era of immunobiochemistry," which included production of molecules based on antibodies. He stated that his method was a byproduct of basic research and a clear example of how an investment in research that was not initially considered commercially viable had "an enormous practical impact." By 1984, a thriving business was being done with monoclonal antibodies for diagnosis, and works on vaccines and **cancer** based on Milstein's breakthrough research were being rapidly developed.

In the early 1980s, Milstein received a number of other scientific awards, including the Wolf Prize in Medicine from the Karl Wolf Foundation of Israel in 1980, the Royal Medal from the Royal Society of London in 1982, and the Dale

Medal from the Society for Endocrinology in London in 1984. He is a member of numerous international scientific organizations, among them the U.S. National Academy of Sciences and the Royal College of Physicians in London.

See also Antigens and antibodies; Immune system; Immunity

MINOT, GEORGE RICHARDS (1885-1950)
American hematologist and physician

George Richards Minot was a pioneer in the medical field of hematology, the study of **blood** and blood-forming **organs**. His most important contribution was the discovery that pernicious anemia could be effectively treated by feeding patients large doses of **liver** or liver extract. For this discovery, he shared the 1934 Nobel Prize in medicine with his Boston colleague **William P. Murphy** and with **George Hoyt Whipple** of the University of Rochester medical school.

Minot was born on December 2, 1885, in Boston, Massachusetts, the eldest of three sons of James Jackson Minot and Elizabeth Whitney Minot. The Minots were an old and well-to-do family of Boston. James Jackson Minot was a physician, and several other men on both sides of the family had been distinguished medical practitioners as well. Minot was a sickly child who spent a good deal of his early life in bed, but he grew stronger as he got older, partly by spending much time out of doors. He became a keen observer of nature and developed a love of the sea and sailing. He entered Harvard College in 1904, did well in his studies, and graduated in 1908.

Like many upper-class young men of his time, Minot was casual about choosing a vocation and finally decided to enter Harvard medical school only five days before the opening of the fall term on October 1, 1908. Nevertheless, he performed well in his class work and gradually grew more serious about his career. It was in medical school that he became interested in the study of human blood, and after he received his M.D. in 1912 he immediately began his internship at the Massachusetts General Hospital in Boston. In 1914 he became an assistant at the Johns Hopkins University medical school in Baltimore. There he continued his studies of blood in the laboratory and did some of the research that led to Dr. William H. Howell's discovery of the anticoagulant drug heparin. Minot returned to Boston to join the staff of the Massachusetts General Hospital in January 1915, and on June 29 of that year, he married Marian Linzee Weld. The couple eventually had two daughters and a son.

In 1915 at the Massachusetts General Hospital, Minot began to focus his attention on various forms of anemia, but especially on pernicious anemia, a disease for which there was then no known cure and which was almost always fatal to the patient. In his prolonged study of blood smears under the microscope, Minot made the important discovery that the number of reticulocytes (young red blood cells) found in a sample provided a good index of the activity of the bone marrow, the part of the body that produces all red blood cells. He also began to suspect that the cause of pernicious anemia was some malfunction in the bone marrow, and that this in turn was somehow related to the diet of the patient. However, his study of this problem, while never entirely abandoned, was put partially aside for a number of years while he worked in several other areas of hematology.

Around 1917 he began to spend an increasing amount of time at the recently opened Collis P. Huntington Memorial Hospital, also in Boston, which specialized in **cancer** research and treatment. There he did significant research on several forms of **leukemia**, a cancerous disease of the blood. In 1918, he became an assistant professor at the Harvard Medical School.

In October 1921, Minot found himself in a serious medical crisis of his own. Always a man of delicate health, he now discovered that he was suffering from diabetes. His doctors placed him on a strict, semi-starvation diet. At the time, this was the only known way of alleviating the effects of the disease. Minot continued his work but he grew terribly weak. Then, in January 1922, the discovery of insulin was announced, and by January 1923, the hormone was available in sufficient quantity that Minot could begin taking it. His condition improved rapidly, and although required to remain on a strict diet (weighed out in grams at each meal) for the rest of his life, Minot was able to maintain a heavy schedule of activity until 1947. His return to relatively good health was signaled by his acceptance of the positions of chief of medical services at Huntington Hospital and associate in medicine at Peter Bent Brigham Hospital, both in 1923.

In 1925, Minot returned to the problem of the treatment of pernicious anemia. He was partly inspired by reading reports of experiments by Dr. George Hoyt Whipple of Rochester, who had bled dogs to make them anemic and then restored their health by feeding them a diet rich in red meat, especially liver. Minot speculated that feeding liver to human patients with pernicious anemia might have a beneficial effect. Minot enlisted a young colleague, Dr. William P. Murphy, to assist him in the experiment, and together they began feeding up to half a pound of liver per day to as many patients with pernicious anemia as they could persuade to eat it. This simple treatment produced dramatic results: nearly all the patients showed striking improvement, many within only two weeks or so. More important, they continued to improve with further liver feeding rather than suffering relapses following temporary remission of symptoms, as victims of pernicious anemia often did.

On May 4, 1926, at a meeting of the American Medical Association in Atlantic City, New Jersey, Minot and Murphy presented a report on the successful treatment of forty-five patients suffering from pernicious anemia. A year later, at a meeting in Washington, D.C., they reported favorable results in the treatment of 105 patients. The next step was to develop an extract of pure liver that would be less bulky and more palatable to the patient. Minot and Murphy persuaded Dr. Edward J. Cohn, a professor of physical chemistry at the Harvard medical school, to work on this problem, and Cohn soon isolated what was called Fraction G from pure liver. The Eli Lilly Company then began to manufacture the substance as

a commercial product. In 1929 Minot and others discovered that much smaller dosages of the extract, given intravenously, had the same effect as large doses taken by mouth.

The discovery of a cure for pernicious anemia marked the culmination of Minot's career as a scientific researcher. He was appointed professor of medicine at the Harvard medical school and director of the Thorndike Memorial Laboratory at Boston City Hospital in 1928. He remained very active in administration. Until April 16, 1947, when he suffered a severe stroke. He died on February 25, 1950, in Brookline, Massachusetts.

See also Hemopoiesis; Liver; Necrosis; Oxygen transport and exchange; Spleen histophysiology; Transfusions; Vascular exchange; Vitamins

MITOCHONDRIA AND CELLULAR ENERGY

Mitochondria are cellular organelles found in the cytoplasm in round and elongated shapes, that produce adenosine tri-phosphate (**ATP**) near intra-cellular sites where energy is needed. Shape, amount, and intra-cellular position of mitochondria are not fixed, and their movements inside cells are influenced by the cytoskeleton, usually being in close relationship with the energetic demands of each cell type. For instance, cells that have a high consumption of energy, such as muscular, neural, retinal, and gonadic cells do present much greater amounts of mitochondria than those with a lower energetic demand, such as fibroblasts and lymphocytes. Their position in cells also varies, with larger concentrations of mitochondria near the intra-cellular areas of higher energy consumption. In cells of the ciliated epithelium for instance, a greater number of mitochondria is found next to the **cilia**, whereas in spermatozoids they are found in greater amounts next to the initial portion of the flagellum, where the flagellar movement starts.

Mitochondria have their own **DNA** and also **RNA** (rRNA, mRNA and tRNA) and ribosomes, and are able to synthesize proteins independently from the cell nucleus and the cytoplasm. The internal mitochondrial membrane contains more than 60 proteins. Some of these are **enzymes** and other proteins that constitute the electron-transporting chain; others constitute the elementary corpuscle rich in ATP-synthetase, the enzyme that promotes the coupling of electron transport to the synthesis of ATP; and finally, the enzymes involved in the active transport of substances through the internal membrane.

The main ultimate result of respiration is the generation of cellular energy through oxidative phosphorilation, i.e., ATP formation through the transfer of electrons from nutrient molecules to molecular oxygen. Prokaryotes, such as **bacteria**, do not contain mitochondria, and the flow of electrons and the oxidative phosphorilation process are associated to the internal membrane of these unicellular organisms. In eukaryotic cells, the oxidative phosphorilation occurs in the mitochondria, through the chemiosmotic coupling, the process of transferring hydrogen protons (H+) from the space between the external and the internal membrane of mitochondria to the elementary corpuscles. H+ are produced in the mitochondrial matrix by the citric acid cycle and actively transported through the internal membrane to be stored in the inter-membrane space, thanks to the energy released by the electrons passing through the electron-transporting chain. The transport of H+ to the elementary corpuscles is mediated by enzymes of the ATPase family and causes two different effects. First, 50% of the transported H+ is dissipated as heat. Second, the remaining hydrogen cations are used to synthesize ATP from ADP (adenosine di-phosphate) and inorganic phosphate, which is the final step of the oxidative phosphorilation. ATP constitutes the main source of chemical energy used by the **metabolism** of eukaryotic cells in the activation of several multiple signal transduction pathways to the nucleus, intracellular enzymatic system activation, active transport of nutrients through the cell membrane, and nutrient metabolization.

See also Cell membrane transport; Cilia and ciliated epithelial cells; Hypoxia

Mitosis · *see* Cell cycle and cell division

Molecular biology

At its most fundamental level, molecular biology is the study of biological molecules and the molecular basis of **structure and function** in living organisms.

Molecular biology is an interdisciplinary approach to understanding biological functions and regulation at the level of molecules such as nucleic acids, proteins, and **carbohydrates**. Following the rapid advances in biological science brought about by the development and advancement of the Watson-Crick model of **DNA (deoxyribonucleic acid)** during the 1950s and 1960s, molecular biologists studied gene structure and function in increasing detail. In addition to advances in understanding genetic machinery and its regulation, molecular biologists continue to make fundamental and powerful discoveries regarding the structure and function of cells and of the mechanisms of genetic transmission. The continued study of these processes by molecular biologists and the advancement of molecular biological techniques requires integration of knowledge derived from physics, chemistry, mathematics, **genetics, biochemistry**, cell biology and other scientific fields.

Molecular biology also involves organic chemistry, physics, and biophysical chemistry as it deals with the physicochemical structure of macromolecules (nucleic acids, proteins, **lipids**, and carbohydrates) and their interactions. Genetic materials including DNA in most of the living forms or **RNA (ribonucleic acid)** in all plant **viruses** and in some animal viruses remain the subjects of intense study.

The complete set of genes containing the genetic instructions for making an organism is called its genome. It contains the master blueprint for all cellular structures and activities for the lifetime of the cell or organism. The human genome consists of tightly coiled threads of deoxyribonucleic acid (DNA) and associated protein molecules organized into

structures called **chromosomes**. In humans, as in other higher organisms, a DNA molecule consists of two strands that wrap around each other to resemble a twisted ladder whose sides, made of sugar and phosphate molecules are connected by rungs of nitrogen-containing chemicals called bases (nitrogenous bases). Each strand is a linear arrangement of repeating similar units called nucleotides, which are each composed of one sugar, one phosphate, and a nitrogenous base. Four different bases are present in DNA adenine (A), thymine (T), cytosine (C), and guanine (G). The particular order of the bases arranged along the sugar-phosphate backbone is called the DNA sequence; the sequence specifies the exact genetic instructions required to create a particular organism with its own unique traits.

Each time a cell divides into two daughter cells, its full genome is duplicated; for humans and other complex organisms, this duplication occurs in the nucleus. During **cell division** the DNA molecule unwinds and the weak bonds between the base pairs break, allowing the strands to separate. Each strand directs the synthesis of a complementary new strand, with free nucleotides matching up with their complementary bases on each of the separated strands. Nucleotides match up according to strict base-pairing rules. Adenine will pair only with thymine (an A-T pair) and cytosine with guanine (a C-G pair). Each daughter cell receives one old and one new DNA strand. The cell's adherence to these base-pairing rules ensures that the new strand is an exact copy of the old one. This minimizes the incidence of errors (mutations) that may greatly affect the resulting organism or its offspring.

Each DNA molecule contains many genes, the basic physical and functional units of heredity. A gene is a specific sequence of nucleotide bases, whose sequences carry the information required for constructing proteins, which provide the structural components of cells and tissues as well as **enzymes** for essential biochemical reactions.

The central dogma of molecular biology is that DNA is copied to make mRNA (messenger RNA), and mRNA is used as the template to make proteins. Formation of RNA is called transcription and formation of protein is called translation. Transcription and translation processes are regulated at various stages and the regulation steps are unique to prokaryotes and eukaryotes. DNA regulation determines what type and amount of mRNA should be transcribed, and this subsequently determines the type and amount of protein. This process is the fundamental control mechanism for **growth and development** (morphogenesis).

All living organisms are composed largely of proteins, the end product of genes. Proteins are large, complex molecules made up of long chains of subunits called **amino acids**. The protein-coding instructions from the genes are transmitted indirectly through messenger ribonucleic acid (mRNA), a transient intermediary molecule similar to a single strand of DNA. For the information within a gene to be expressed, a complementary RNA strand is produced (a process called transcription) from the DNA template in the nucleus. This messenger RNA (mRNA) moves from the nucleus to the cellular cytoplasm, where it serves as the template for **protein**

synthesis. Humans synthesize more than 30,000 different types of proteins.

Twenty different kinds of amino acids are usually found in proteins. Within the gene, sequences of three DNA bases serve as the template for the construction of mRNA with sequence complimentary codons that serve as the language to direct the cell's protein-synthesizing machinery. Cordons specify the insertion of specific amino acids during the synthesis of protein. For example, the base sequence ATG codes for the amino acid methionine. Because more than one codon sequence can specify the same amino acid, the **genetic code** is termed a degenerate code (i.e., there is not a unique codon sequence for every amino acid).

Areas of intense study by molecular biology include the processes of DNA replication, repair, and mutation (alterations in base sequence of DNA). Other areas of study include the identification of agents that cause mutations (e.g., ultra-violet rays, chemicals) and the mechanisms of rearrangement and exchange of genetic materials (e.g. the function and control of small segments of DNA such as plasmids, transposable elements, insertion sequences, and transposons to obtain recombinant DNA).

Recombinant DNA technologies and genetic engineering are an increasingly important part of molecular biology. Advances in biotechnology and molecular medicine also carry profound clinical and social significance. Advances in molecular biology have led to significant discoveries concerning the mechanisms of the embryonic development, disease, immunologic response, and **evolution.**

See also Embryonic development: Early development, formation, and differentiation; Gene therapy; Genetic regulation of eukaryotic cells; Genetics and developmental genetics

MONIZ, EGAS (1874-1955)
Portuguese neurologist

Egas Moniz was the professional name of Antonio Caetano de Abreu Freire, a scientist who made extensive contributions to the study of the human **brain.** A neurologist at the University of Lisbon, during the 1920s, Moniz developed cerebral angiography—an important breakthrough that is still used today to diagnose **tumors** and strokes. Moniz also pioneered surgical procedures to address psychiatric disorders; with the help of Almeida Lima, he developed the psychosurgical technique called frontal leucotomy, which severed the frontal lobes from the rest of the brain and reduced the patient's anxiety and other symptoms of neurosis. For his work on the frontal leucotomy, Moniz shared the 1949 Nobel Prize in physiology or medicine with **Walter Rudolf Hess.**

Moniz was born Antonio Caetano de Abreu Freire in Avança, Portugal, on November 29, 1874. A member of an aristocratic family, his father was Fernando de Pina Rezende Abreu and his mother was Maria do Rosario de Almeida e Sousa. Moniz received his early education from an uncle who was an abbot, and in 1891, he entered the University of Coimbra where he studied medicine. Graduating as an M.D. in

1899 with a thesis on diphtheria, Moniz chose **neurology** as his field of specialization. Although he began to suffer from gout, a disease that impaired the use of his hands, Moniz went on to study in France at the University of Paris and the University of Bordeaux. After writing a paper on the physiological **pathology** of sexual activity, he became a professor at the University of Coimbra in 1902, the same year he married. In 1911, Moniz was appointed professor of neurology at the University of Lisbon, where he would remain until his retirement in 1945.

Along with his medical career, Egas became immersed in Portuguese politics. In his years as a student at Coimbra, he had authored political literature promoting the cause of the liberal republicans that opposed Portugal's monarchical government. He first used the name Egas Moniz in writing these pamphlets, eventually adopting this moniker for all of his professional and political activities. Beginning in 1899, he served as a deputy in the Portuguese parliament. After the monarchy was overthrown in Portugal in 1910, Moniz became involved in rebuilding and reshaping his country's political system. In 1917, he was named Ambassador to Spain and Minister of Foreign Affairs, and he signed the Versailles Treaty as Portugal's delegate to the peace conference at the end of World War I. A political quarrel got him entangled in a duel in 1919, however, and he finally abandoned his political activities as a liberal republican when a conservative government took power in 1922.

Moniz studied head injuries during World War I, and in 1917 he published *A Neurología na Guerra,* in which he described some of his findings. One of the early obstacles for neurology was the absence of a reliable and safe technology for examining the living brain. Attempting to find an improved method of locating intracranial brain tumors, Moniz to begin experimenting on corpses by injecting radioactive solutions into the **arteries** and taking x rays of them. By 1927, he had developed this technique, known as cerebral angiography, to the point where the radioactive solution coursing through the brain's vessels and arteries made it possible to x-ray live brain tissues. Moniz mapped out the distribution of **blood** vessels in the head, and he was therefore able to detect and measure tumors that displaced the normal location of the arteries. To this day, angiography continues to be the most widely used method for diagnosing tumors, strokes, and other injuries.

The contribution Moniz made to psychosurgery was the result of his determination to find physical cures for mental illness. Attending the 1935 International Neurological Conference in London, Moniz was particularly impressed by the work of John F. Fulton and Carlyle G. Jacobsen, American scientists who had removed the frontal lobes from the brains of chimpanzees and observed certain behavioral changes. Known as frontal leucotomy, the procedure consists of severing the nerves connecting the frontal lobes to the rest of the brain. Moniz and his colleague Almeida Lima developed the technique so it could be applied to humans to alleviate certain psychiatric problems such as anxiety and neurosis. The procedure was considered successful; there were no fatalities in the original twenty patients, and the mental condition of most of them was declared improved or cured by the operation.

Widely hailed as the most important psychiatric procedure yet discovered, Moniz shared the Nobel Prize in 1949 for his frontal leucotomy research. This method was abandoned after World War II, however, when **psychopharmacology** made considerable inroads in treating nervous and mental disorders with drugs. Nonetheless, frontal leucotomy played an important role in educating neurologists about the human brain and the surgical procedures that can be applied to it.

See also Brain: Intellectual functions; Cerebral hemispheres; Nervous system overview; Nervous system: Embryological development; Neural damage and repair; Neurology; Psychopharmacology

MONOCYTES • *see* LEUKOCYTES

MONRO, *SECUNDUS*, ALEXANDER (1733-1817)

English physician

Three generations of Alexander Monros were professors of **anatomy** at the University of Edinburgh in unbroken succession from 1720 to 1846. All three contributed immeasurably toward making the medical school at Edinburgh among the world's best of its time. The subject of this article is called Monro *secundus* to distinguish him from his father, Alexander Monro *primus* (1697–1767), and his son, Alexander Monro *tertius* (1773–1859). Monro *secundus*, the greatest of the three, is chiefly remembered for discovering and accurately describing the foramen interventriculare, or "foramen of Monro," an important passage connecting the lateral and third ventricles of the **brain**. Knowledge of the communicative function of the foramen of Monro became crucial for the new science of neuroradiology early in the twentieth century because, for neurosurgeons like Walter Dandy (1886–1946), it paved the way for useful neurological diagnostic techniques such as pneumoventriculography, which injects air as an x-ray contrast medium directly into the ventricles.

Monro *secundus* was early groomed to succeed his already famous father. He began studying anatomy at the university under his father's tutelage when he was only eleven. By 1753, he was handling the student overflow from his father's popular classes. In 1755, he received his Edinburgh M.D. with a thesis on the testicles and **semen**. Unlike most medical doctoral dissertations of that time, which were only restatements of previous knowledge, his contained original research.

For the next two years, Monro traveled to the most prominent medical and surgical research centers of northern Europe to study under the best anatomists of the day, William Hunter (1718–1783) in London, Johann Friedrich Meckel (1714–1774) in Berlin, and Bernhard Siegfried Albinus (1697–1770) in Leiden. Back in Edinburgh in 1757, he assisted his father, substitute taught for him during his illnesses, and after 1758, co-taught the classes. He had already

assumed the main duties of the anatomy professorship long before he took it over completely upon his father's death in 1767. In 1798, Edinburgh named him and his son, Monro *tertius*, co-professors of anatomy. They shared the position until Monro *secundus* retired in 1808.

Monro was friendly and gregarious but argumentative, polemical, and quite critical of other anatomists. In 1770, he savagely attacked the ideas of William Hewson (1739–1774) on pneumothorax and lymphatics. In 1794, he fought publicly with Gilbert Blane (1749–1834) about the function of oblique muscle fibers.

Besides his polemics and several minor works on anatomy, Monro wrote three major anatomical treatises: *Observations on the Structure and Functions of the Nervous System* (1783), which introduced the foramen of Monro; *The Structure and Physiology of Fishes Explained and Compared with Those of Man and Other Animals* (1785), which became a standard textbook on **comparative anatomy**; and *A Description of All the Bursae Mucosae of the Human Body* (1788), his most original investigation.

See also Cerebral hemispheres; Genitalia (male external); Lymphatic system; Mammary glands and lactation; Nervous system overview

MORGAGNI, GIOVANNI BATTISTA (1682-1771)

Italian physician

Giavonni Morgagni founded pathological **anatomy**, the study of the structure and appearance of diseased or abnormal tissues. He laid the foundations for the pathological anatomical investigations of **Leopold Auenbrugger**, Matthew Baillie (1761–1823), **Xavier Bichat**, **René-Théophile-Hyacinthe Laënnec**, Jean Cruveilhier (1791–1874), and **Rudolf Virchow** (1821–1902), all of whom frankly acknowledged his influence.

Born in Forli, Italy, Morgagni attended schools there until entering the University of Bologna in 1698. The student and protégeacute; of Antonio Valsalva (1666–1723), he was elected to the Accademia degli Inquieti, a distinguished scholarly society, in 1699. He received in Bologna degrees in philosophy and medicine in 1701, became head of the Accademia in 1704, and practiced medicine and studied anatomy in three hospitals in Bologna until 1707, when he moved to Venice. Back in Forli in 1709, he practiced medicine until he accepted an invitation in 1711 to teach at the University of Padua. He taught medicine until 1716 and anatomy until his death, also in Padua.

Morgagni's masterpiece, *De sedibus et causis morborum per anatomen indagatis libri quinque* (The Seats and Causes of Diseases Investigated by Anatomy in Five Books) published in 1761, is a classic of **pathology**. Culminating a lifetime of meticulous study, it was the first medical treatise to correlate accurately a large variety of pathological findings in living patients with their anatomical conditions post mortem.

Its seventy chapters contain accounts of 700 cases. Among its findings are the first reliable descriptions of mitral stenosis, angina pectoris, various syphilitic manifestations, several kinds of aneurysm, atrophy of the **liver**, tuberculosis of the kidney, the **eye** disease called Morgagni's cataract, and the **heart** disease called Morgagni-Adams-Stokes syndrome.

Before *De sedibus*, Morgagni wrote a series of short anatomical tracts called *Adversaria anatomica*. In them he furthered Valsalva's work and that of Valsalva's teacher, **Marcello Malpighi**, especially as regards the **circulatory system**. Also in the *Adversaria Anatomica* Morgagni reported many of his minor anatomical discoveries, including structures in the trachea, urethra, and genitalia. About twenty-five anatomical structures are named after him. His early studies of normal anatomy prepared him well to recognize the pathological deviations from healthy **tissue** that he later reported in *De sedibus*.

See also Birth defects and abnormal development; Cardiac disease; History of anatomy and physiology: The Renaissance and Age of Enlightenment

MORGAN, THOMAS HUNT (1866-1945)
American geneticist

Thomas Hunt Morgan, along with William Bateson, was the cofounder of modern **genetics**. Morgan was the first to show that genetic variation occurs through numerous small mutations.

As child growing up in rural Kentucky, Morgan was surrounded by nature and wildlife. Perhaps that environment contributed to his intense interest in biology, for Morgan later majored in zoology at State College of Kentucky. After his graduation in 1886, Morgan investigated chemistry and morphology (the study of organism development to better understand evolutionary relationships) at Johns Hopkins University, completing his doctorate in 1890. From his graduate days on, Morgan believed that heredity was in some way central to understanding all biological phenomena—especially development and **evolution**. His persistence in trying to prove and develop heredity theories led to his winning the Nobel Prize in physiology or medicine in 1933.

In 1903, there were several attempts to explain variations in plants and animal species. One was Charles Darwin's theory of natural selection, a process by which organisms best adapted to local environments leave more offspring that survive to spread their favorable traits throughout a population. Morgan wondered how complex organisms such as humans could have evolved from such a process. To him, the theory seemed incomplete. Morgan viewed natural selection as a process that sorted out variations in an organism, not as one that created the variations. So what was it that determined whether a baby would be a boy or a girl, or whether it would have blue eyes or green eyes? The three widely known heredity theories of the time offered competing explanations: the Mendelian (or gene) theory, the chromosome theory, and the mutation theory.

Gregor Mendel, by cross-breeding pea plants, had first determined some of the rules of inheritable traits—those of **sex determination**, gene linkage (inheritance of characteristics together), and mimicry. Advocates of the chromosome theory maintained that genes located on **chromosomes** were responsible for specific inherited traits. Morgan was skeptical of the Mendelian and chromosome theories because the conclusions were speculative, based on nothing more than observation, inference, and analogy. Morgan wanted to be able to draw firm, rigorous, testable conclusions based on quantitative and analytical data. His strong belief in experimental analysis attracted Morgan to Hugo de Vries's mutation theory. De Vries, a Dutch botanist, had physical evidence that large-scale variations in one generation could produce offspring that were of a different species than their parent plants. Morgan set out to test de Vries's theory in animals and also to disprove the other heredity theories.

Morgan's first experiments using the fruit fly (*Drosophila melanogaster*) were unsuccessful. Morgan was not able to duplicate the magnitude of mutations that de Vries had claimed for plants. Then in 1910, Morgan noticed a natural mutation in one of the male fruit flies: it had white eyes instead of red. He began breeding the white-eyed male to its red-eyed sisters and found that all of the offspring had red eyes. When Morgan bred those offspring, he found that they produced a second generation of both red- and white-eyed fruit flies. Morgan was fascinated to find that all of the white-eyed flies were male. He traced the unusual finding to a difference between male and female chromosomes. The white-eye gene of the fruit fly was located on the male sex chromosome. By studying future generations of fruit flies, Morgan found that genes were linearly arranged on chromosomes. His work with the fruit fly strongly backed Mendel's gene concept and, moreover, established that chromosomes definitely carried genetic traits. For the first time, the association of one or more hereditary characteristics with specific chromosomes was clear, thereby unifying Mendelian "trait" theory and chromosome theory.

Morgan, working with students **Hermann Muller**, Alfred H. Sturtevant, and Calvin Bridges, later went on to develop and perfect these concepts of linkage by explaining why, for instance, occasionally white-eyed female flies were found in his studies. Morgan concluded that traits found on the same chromosome were not always inherited together. This genetic "mistake" was called crossing over, because one chromosome actually exchanged material with (or crossed over to) another chromosome. This process was an important source of genetic diversity. In 1915, Morgan, along with his students, published the culmination of his work, *The Mechanism of Mendelian Heredity*. These results provided the key to all further work in the area of genetics and laid the groundwork for all genetically-based research.

In 1904, Morgan married Lilian Vaughan Sampson, who assisted in his research. They had one son and three daughters. Morgan died in 1945, at age 79.

See also Genetics and developmental genetics; Human genetics

MOTOR FUNCTIONS AND CONTROLS

Each time a part of the body—no matter how small—moves, it is the result of a contraction of a muscle. Such a contraction is caused by binding of the neurotransmitter acetylcholine (ACh) to receptors on the muscle fibers. The ACh causes an **action potential** in the muscle fiber, admitting sodium and **calcium** ions and causing the actin filaments to slide across the myosin filaments, thus shortening the muscle. The **neurons** that deliver ACh to the muscle are called motor neurons, and the point at which a motor neuron and a muscle fiber meet is called a neuromuscular junction.

The axon of each motor neuron divides so that one neuron may innervate a large number of muscle fibers. The more fibers innervated, the stronger the contraction. Although the basic mechanism appears to be simple, there are many levels of control of movement arranged in a hierarchy.

Certain movements can occur even if motor neurons are disconnected from the **brain**. The neurons of the **spinal cord** are sufficient to cause these simple behaviors. More complex behaviors, however, are the result of the integration of information from several areas of the brain.

The right side of the brain directs movements in the left side of the body, and vice versa. Each side of the primary motor cortex contains the cell bodies of the neurons that eventually communicate with the motor neurons on the opposite side of the spinal cord, those that actually contact the muscle. If only this one **synapse** occurred, however, it would not be possible to "custom design" movements to fit the needs of the organism in its particular situation. Therefore, there are subcortical modulating systems that fine-tune commands coming from the primary motor cortex. When the motor neurons finally leave the brain and enter the spinal cord, crossing to the side opposite that from which they originated, their information has been adjusted by the modulating systems.

One of the modulating systems is called the **basal ganglia**, which comprises the caudate nucleus, putamen, globus pallidus, subthalamus, and substantia nigra. Many of the synapses in the basal ganglia are inhibitory, so certain commands from the primary motor cortex will be suppressed. The other modulating system is the **cerebellum**, which, also using inhibitory **neurotransmitters**, allows for the precise timing of muscle contractions in a sequence, resulting in coordinated rather than jerky movement. The lack of coordination associated with overindulgence in alcohol is the result of the depression of the cerebellar circuits by alcohol.

Both modulating systems receive sensory information from other parts of the brain, including input from proprioceptors. This information is used to tailor motions appropriately to fit the situation. Information from the modulating systems is then sent back to the cortex, where it affects the primary motor neuron's decision to fire, sending information to the spinal cord, or not.

The cortex is the site of still another level of control. The premotor cortex, which receives information from the cerebellum, and the supplementary motor cortex, which receives information from the basal ganglia, also influence neurons in the primary motor cortex. These areas are involved

Each time a part of the body—no matter how small—moves, it is the result of a contraction of a muscle. Such a contraction is caused by binding of the neurotransmitter acetylcholine (ACh) to receptors on the muscle fibers. *Photo by Mark Duncan. AP/Wide World Photos. Reproduced by permission.*

in planning of motions. PET scans reveal that thinking about raising the right arm causes increased **blood** flow to the left supplementary motor cortex.

See also Cerebral cortex; Mechanoreceptors; Muscle contraction; Muscular innervation

MRI • *see* IMAGING

MUCOCUTANEOUS JUNCTIONS

Mucocutaneous junctions are transition points between dry skin and the wet mucous membranes, the tissues lining body orifices such as the mouth, anus, nostrils, urethra, and vagina.

Mucous membranes consist of **epithelial tissue** over a basement membrane, attached to a layer of connective **tissue** called the lamina propria. Epithelium is a type of tissue that is generally moist, often containing mucus-secreting goblet cells, and regenerates very quickly.

The skin also contains epithelial tissue, but in the outer layers, the cells have lost their nuclei and flattened out. They contain a horny substance called keratin, forming a protective covering for the body. The skin has hairs, sebaceous glands, and **sweat glands**, which are absent from the mucous membranes. At the mucocutaneous junction, the epithelial tissue transitions between the dry keratinized skin and the wet epithelium of the mucous membrane.

At mucocutaneous junctions such as that of the lip, where the mucous membrane is exposed to the air, a type of tissue called parakeratinized epithelium is seen. Parakeratinized epithelium is tougher and drier than the typical mucous membrane, such as is found inside the mouth, but its surface cells still retain their nuclei. It exhibits vascular papillae, elevations containing many tiny **blood** vessels, or **capillaries**. Because there are no glands in the lips, they are subject to excessive drying in the air unless they are kept moist by **saliva**.

See also Integument

MUCUS

Mucus is a slippery secretion produced by a variety of slightly different types of mucous cells. It is protein based and has several molecular types of structures depending on the specific function. Mucus is found in the respiratory, digestive and reproductive systems. It is a viscous fluid composed of water, mucin, inorganic salts, epithelial cells, and **leukocytes**, all held in suspension.

In the **respiratory system**, mucus is found along the lining of the nasal cavities. Its function is to trap dust, **bacteria**, and other airborne contaminants. Once trapped, tiny **cilia** move the mucus down and into the **pharynx**. Deeper into the respiratory system, the **bronchi** and **alveoli** are covered in epithelial cells that constantly produce mucus. Leukocytes, cells of the **immune system**, are contained in the mucus and act as an antibacterial agent in the mucus. In this region the cilia move the mucus up and out of the respiratory system. This constant production of mucus is a first line of defense of the body. Many smokers irritate or destroy the mucus-producing goblet cells. After prolonged exposure the mucus and immune enhancing function of mucus is reduced or destroyed.

The digestive system uses mucus in quite different ways. Mucus is secreted in the mouth and assists in the **swallowing** and passage of food through the esophagus. In the stomach, mucus contains more bicarbonate and coats the gastric surface. This provides a barrier between the gastric acid and **enzymes** and the mucus neck cells of the epithelial surface. In addition to the stomach, mucus lines the entire intestinal tract and provides lubrication for digesting food as it passes along the intestines. It also serves as a type of binding agent for indigestible wastes.

The female reproductive system has several types of mucus. They are called G, L, and S types of mucus. Each type has a varied mix of the typical suspension molecules found throughout all mucus. Recent studies relate viscosity with function during reproduction. In other words, the texture of the mucus aids in **sperm** migration and exclusion into the uterus. Some researchers believe this function may prevent

unhealthy sperm from entering the uterus. Mucus also forms a plug at the opening of the cervix during **pregnancy** keeping the cervical canal sealed. This helps prevent contamination by bacteria and other agents.

See also Mastication; Reproductive system and organs (female); Sense organs: olfactory (sense of smell) structures

MULLER, HERMANN JOSEPH (1890-1967)
American geneticist

Hermann Joseph Muller was the first to show that genetic mutations can be induced by exposing **chromosomes** to x rays. For this demonstration, he was awarded the 1946 Nobel Prize in physiology or medicine. Muller also took up a crusade to improve the condition of the human gene pool by calling for a cessation of the unnecessary use of x rays in medicine, and a halt to nuclear bomb testing in order to prevent further damage to the genetic makeup of the human population.

Hermann Joseph Muller was born on December 21, 1890, in New York City. Hermann attended Morris High School in the borough of the Bronx in New York City. On a scholarship, Muller enrolled at Columbia University in 1907 and majored in **genetics**. He received his bachelors degree in 1910 and then continued at both Cornell Medical School and Columbia for his master's degree, studying the transmission of nerve impulses.

In 1912, Muller began working with A. H. Sturtevant, Calvin B. Bridges, and **Thomas Hunt Morgan**, a zoologist who was performing ground-breaking work in genetics. The group researched *Drosophila melanogaster*, a fruit fly with a brief three-week breeding cycle that makes them ideal for genetic study. *Drosophila* also have only four pairs of chromosomes, the dark-staining microscopic structures within the nucleus of each cell. Muller and other designed experiments to study mutations resulting in abnormal traits that seem to arise spontaneously in the fruit fly population. The mutations were tracked in order to infer which part of each chromosome contained the gene responsible for a particular trait, such as **eye** color or wing shape.

Muller's doctoral thesis in 1916 was on crossing over, a phenomenon discovered in 1909 by a Belgian scientist, F. A. Janssens, when he noticed that during the duplication and separation of like chromosomes, sometimes part of a chromosome would break off and reattach at a comparable place on the other chromosome. If two genes were far apart on a chromosome, then it would be more likely that a break could occur between them. Thus, a high frequency of crossing over observed between any two traits would mean a long distance between the genes, while a low frequency of crossing over between two traits would mean the genes were close together. The team used this information to map each chromosome in order to show how genes for each trait might be arranged along its length. The findings of the group were published in *The Mechanism of Mendelian Heredity*, written in 1915.

In 1916, Muller took a teaching position at Rice University in Texas, where he did further research in genetics, especially mapping modifier genes, which seem to control the expression of other genes. Upon his return to Columbia two years later, Muller did some of his most important theoretical work. Realizing that genes on the chromosomes are self-replicating and are responsible for synthesizing the other components of cells, he theorized that all life must have started out with molecules that were able to self-replicate, which he likened to naked genes. These molecules, he suggested, must have been something like **viruses**, a very astute hypothesis given the little that was known about viruses at the time.

In 1921, Muller began work as an associate professor teaching genetics and **evolution** and doing research on mutation at the University of Texas in Austin, where he remained until 1932. Muller had grown impatient with waiting for mutations to happen by chance, so he began seeking methods of hastening rates of mutation. In 1919, he had discovered that higher temperatures increase the number of mutations, but not always in both chromosomes in a chromosome pair. He deduced that mutations must involve changes at the molecular or sub-molecular level. He struck on the idea of using x rays instead of heat to induce mutations, and by 1926 he was able to confirm that x rays greatly increased the mutation rate in *Drosophila*. He also concluded that most mutations are harmful to the organism, but are not passed on to future generations since the individual affected is unlikely to reproduce; nonetheless, he suggested, if the rate of harmful mutations were to become too high, a species might die out.

Muller reported his success in inducing mutations in a 1927 article in *Science* entitled "Artificial Transmutation of the Gene." The article gained him international status as an innovator and introduced other scientists to a technique for studying a large number of mutations at once. This led to the realization that mutations are actually chemical changes that can be artificially induced with any number of other chemicals. It also helped spawn the infant study of radiation genetics.

Muller left Texas in 1932 and moved to Berlin to work at the Kaiser Wilhelm Institute. There he spent a year as a Guggenheim fellow doing research on mutations and exploring the structure of the gene. Muller's next stop was the Soviet Union, where he stayed from 1933 to 1937. At the Academy of Sciences in both Leningrad and Moscow he studied radiation genetics, cytogenetics, and gene structure.

The next year, Muller went to the Institute of Animal Genetics in Edinburgh, Scotland, again working on radiation genetics upon his return to the United Stated in 1940. He continued his research at Amherst College and, starting in 1945, at Indiana University, where he was appointed professor of zoology and where he stayed until his death on April 5, 1967.

In 1946, Muller was awarded the Nobel Prize in physiology or medicine for his important work on mutations. Muller was also a member of the National Academy of Sciences and a fellow of the Royal Society. He used the opportunity of his world fame to campaign for many social concerns sparked by his interest in the genetic health of the human population. Muller spoke out against needless x rays in medicine and for safety in protecting people regularly exposed to x rays.

In the 1950s, he campaigned to outlaw nuclear bomb tests because nuclear fallout would cause mutations in future generations. Toward the end of his life, Muller argued that the human race should take action in order to keep healthy genes in the population. His idea came out of the belief that modern culture and technology suspend the process of natural selection and thus increase the number of mutations in human genes. Further, Muller argued that there should be programs to promote eugenics, literally good genes. Muller the idea of establishing **sperm** banks in which the sperm of exceptionally healthy and gifted men would be frozen as an endowment to be used for future generations. The concept of such massive intervention in the human gene pool and in the private lives of individuals was and remains highly controversial.

See also Ethical issues in genetics research; Evolution and evolutionary mechanisms; Genetic code; Genetics and developmental genetics; Human genetics

MÜLLER, JOHANNES (1801-1858)
German physiologist

Johannes Müller was a notable physiologist of the nineteenth century who made important contributions in a wide range of subjects, including the selective nature of perception of different sensations through the electrical impulses transmitted by the nervous system. He also explored the relation of his studies to the development of a psychological system based upon the physiology of perception.

Johannes Peter Müller was born in Coblentz, Germany, and studied medicine at the University of Bonn where he graduated in 1822. After attending post graduation courses at the University of Berlin, he returned to Bonn in 1824 where he first worked at the university as private-docent, then as associate professor (1826), finally becoming professor in 1830. In the following year, he accepted the position of rector at the University of Berlin, where he worked, lectured, and researched until his death in 1858. At the times, natural philosophy was still a part of the rational system of the life sciences, and Müller at first adopted the vitalistic approach to explain the biologic events, which he tried to prove through experimental research. Vitalism was a theory that tried to explain life and biological phenomena as the result of specific vital energies that supposedly occurred exclusively in living beings, as opposed to the premise that all organic phenomena could be explained by the laws of physics and chemistry all alone. However, the experimental research that Müller introduced in medical research was, by itself, a new approach in the life sciences study, because the majority of the medical knowledge and theories of the times was based only on observation. Müller not only designed and carried out experimental physiological studies, but also trained other prominent physicians in this line of investigation.

Müller investigated the nervous system and physiology of sensation through experimentation, focusing on electrical nerve impulses. He then posed the following question: because all nerves transmit the same electrical impulses, why are these impulses perceived as different sensations? Further experiments led him to the conclusion that electrical impulses occurred in different nerve pathways, which he called channels. Therefore, different channels induced different sensations. His conclusions gave origin to the doctrine of specific nerve energies, also known as Müller's law, which he presented in his doctoral thesis. The Müller's law states that however excited, each sensory nerve gives rise to its own peculiar sensation.

Müller also researched embryo development and physiology, and described what was later known as Müllerian ducts, i.e., the embryonic **organs** formed during development through the invagination of the peritoneal epithelium. He also described that in female embryos, these two-fold and symmetric embryonic organs further evolved to form the Fallopian trumps, the uterus, and the vagina, whereas in male embryos they underwent atrophy, thus forming the Morgagni's caruncle and the prostatic utricle. His studies of the ocular tissues led to the identification of Müller's cells or fibers, which form the structure of **retina**, and among which the differentiated epithelial elements and neuronal cells of the retina are sustained. Müller's fibers traverse perpendicularly the layers of the retina and connect the internal and external limiting membranes. Another of his investigations was the comparison of cells and tissues extracted from **tumors** with those of the surrounding tissues. He advanced the hypothesis that every **tissue** that constitutes a tumor possesses its own analogue in the normal organism, either in its embryonic stage of development or in its differentiated and mature form. Scientists later proved this hypothesis correct. Müller was also the discoverer in 1837 of chondrin, the primary gelatin obtained from the collagen present in the **cartilage**, and of glutin, a protein present in gelatin.

See also Embryonic development: Early development, formation, and differentiation; Eye and ocular fluids; Nerve impulses and conduction of impulses

MÜLLER, PAUL HERMANN (1899-1965)
Swiss chemist

Paul Müller was an industrial chemist who discovered that dichlorodiphenyltrichloroethane (DDT) could be used as an insecticide. This was the first insecticide that could actually target insects; in small doses it was not toxic to humans and yet it was stable enough to remain effective over a period of months. When DDT was introduced in 1942, the effects it would have on the environment were not well understood. It was widely hailed, in particular for its ability to reduce the incidence of typhus, **malaria**, and a number of tropical diseases by reducing insect populations. Although the dangers of DDT use eventually prohibited its further general use, the World Health Organization estimates that approximately 25 million human lives were saved by the discovery and use of DDT. For his work with DDT and the role his discovery played in the fight against diseases such as typhus and malaria, Müller was awarded the 1948 Nobel Prize in physiology or medicine.

Paul Hermann Müller was born in Olten, Switzerland on January 12, 1899, to Gottlieb and Fanny Leypoldt Müller. His father was an official on the Swiss Federal Railway, and the family moved to Lenzburg and then to Basel, where Müller was educated until the age of seventeen. After finishing his secondary education, Müller worked for several years in a succession of jobs with local chemical companies. In 1919, he entered the University of Basel to study chemistry. He did his doctoral work under F. Fichter and H. Rupe, and his dissertation examined the chemical and electrochemical reactions of m-xylidine and some related compounds (xylidines are used in the manufacture of dyes).

After Müller received his Ph.D. in 1925, he went to work in the dye division of the J. R. Geigy Corporation, a prominent Swiss chemical company. Müller married in 1927 and they eventually had two sons and a daughter. Müller initially conducted research on the natural products that could be derived from green plants, and the compounds he synthesized were used as pigments and tanning agents for leather. In 1935, he was assigned to develop an insecticide. At that time, the only available insecticides were either expensive natural products or synthetics ineffective against insects; the only compounds that were both effective and inexpensive were the arsenic compounds, which were just as poisonous to human beings and other mammals. Müller noticed that insects absorbed and processed chemicals much differently than the higher animals, and he postulated that for this reason there must be some material that was toxic to insects alone. After testing the biological effects of hundreds of different chemicals, in 1939 he discovered that the compound DDT met most of his design criteria. First synthesized in 1873 by German chemist Othmar Zeidler, who had not known of its insecticide potential, DDT could be sprayed as an emulsion with water or could be mixed with talcum or chalk powder and dusted on target areas. It was first used against the Colorado potato beetle in Switzerland in 1939; it was patented in 1940 and went on the market in 1942.

Müller had set out to find a specific compound that would be cheap, odorless, long-lasting, fast in killing insects, and safe for plants and animals. He almost managed it. DDT in short term application is so non-toxic to human beings that it can be applied directly on the skin without ill effect. It is cheap and easy to make, and it usually needs to be applied only once during a growing season, unlike biodegradable pesticides which must often be applied several times, in larger amounts and at much higher cost. Typhus and malaria are very severe, often fatal illnesses, which are carried by body lice and mosquitoes respectively; in the 1940s, several potentially severe epidemics of these diseases were averted by dusting the area and the human population with DDT. The insecticide saved many lives during World War II and increased the effectiveness of Allied forces. Soldiers fighting in both the Mediterranean and the tropics were dusted with DDT to kill lice, and entire islands were sprayed by air before invasions.

Despite these successes, environmentalists were concerned from the time DDT was introduced about the dangers of its indiscriminate use. DDT was so effective that all the insects in a dusted area were killed, even beneficial ones, eradicating the food source from many birds and other small creatures. Müller and other scientists were actually aware of these concerns, and as early as 1945 they had attempted to find some way to reduce DDT's toxicity to beneficial insects, but they were unsuccessful. Müller also argued that insecticides must be biodegradable.

Hailed as a miracle compound, DDT came into wide use, and the impact on beneficial insects became an increasing problem. Because it was such a stable compound, DDT built up in the environment; this was a particular problem once DDT began to be used for agricultural purposes and applied over wide areas year after year. Higher animals, unharmed by individual small doses, began to accumulate large amounts of DDT in their tissues (called bio-accumulation). The high solubility of DDT in **fat** compounded its slow metabolic breakdown and retention in tissues. This had serious effects and several bird species, most notably the bald eagle, were almost wiped out because frequent exposure to the chemical caused the shells of their eggs to be thin and fragile. Many insects also developed resistances to DDT, and so larger and larger amounts of the compound needed to be applied yearly, increasing the rate of bio-accumulation. The substance was eventually banned in many countries; in 1972, DDT use was banned in the United States. Despite its documented dangers, DDT is still used in less developed nations.

In addition to the 1948 Nobel Prize in physiology or medicine, Müller received an honorary doctorate from the University of Thessalonica in Greece in recognition of DDT's impact on the Mediterranean region. He retired from Geigy in 1961, continuing his research in a home laboratory. He died on October 13, 1965.

See also Lipids and lipid metabolism; Metabolism; Poison and antidote actions

MURAD, FERID (1936-   )

American physician

Ferid Murad shared the 1998 Nobel Prize in physiology or medicine with **Robert Furchgott** and **Louis Ignarro**, for his seminal investigations in the 1970s on nitro compounds and nitric oxide's biological effects.

He is the son of Jabir Murat Ejupi, an Albanian immigrant who arrived at Ellis Island in 1913, and was registered by the immigration officer under the name John Murad. His mother, Henrietta Josephine Bowman, was American, from Alton, Illinois. Ferid decided to become a physician when he was 12 years old. He received a Rector Scholarship at DePauw University in Greencastle, Indiana in 1954. While attending college, he also worked as a waiter, taught the **anatomy** and **embryology** labs, and worked at one and sometimes two jobs during the summer to cover his expenses. When he had only one summer job, he took extra summer classes at Indiana University and at DePauw. During both medical school and graduate school, Murad achieved the best grades of his class every year.

In 1957, Murad applied for a M.D.-Ph.D. combined degree program at Western Reserve University, which had being recently initiated at Cleveland. After being evaluated by the faculty of the **pharmacology** department in February of 1958, Ferid Murad was accepted in the program under **Earl Sutherland Jr.** and Theodore Rall, who had discovered the previous year that cyclic AMP was the second messenger of **epinephrine**, as well as glucagon-mediated effects on glycogenolysis in **liver** preparations. Murad was assigned by his mentors with the task of demonstrating that the catecholamine effects on cyclic AMP formation resulted from events mediated by the beta-adrenergic receptors. The outcome of his research was the demonstration of beta-adrenergic-mediated effects on the activation of the enzyme adenylil cyclase in both **heart** and liver preparations. He also found that acetylcholine and other cholinergic agents inhibited adenylil cyclase preparations, which constituted the first demonstration that **hormones** inhibited cyclic AMP formation. These results made him interested in the further identification of other agents that could prevent cyclic AMP from activating the phosphorylase and phosphorylase kinase **enzymes**. His group of research also succeeded in demonstrating that several hormones, such as catecholamines, cholinergics, ACTH, vasopressin, among others, could modulate (i.e., increase or decrease) cyclic AMP synthesis, by interfering with the activity of adenylil cyclase.

Murad spent the period between 1965 and 1967 at Massachusetts General Hospital as an intern and medical resident, and returned to lab work in 1967 at the National Institutes of Health as clinical associate in Martha Vaughan's laboratory at the Heart Institute. Here, among other researches, he was able to continue his studies on cyclic AMP and hormonal regulation, until he was invited by the University of Virginia to develop the new clinical pharmacology division in the Department of Medicine, and was appointed associate professor in medicine and pharmacology. Murad remained at the University of Virginia until 1981, when he was appointed to the position of Director of the Clinical Research Center in 1971 and Director of Clinical Pharmacology in 1973, before being promoted to professor in 1975, at the age of 39. He designed a research program that included clinical and basic research, and formed a team of investigators comprised of talented students and fellows. Murad and his collaborators conducted the first experiments on the biological effects of nitric oxide gas, which were published in 1977.

In 1981, Murad went to Stanford University as chief of medicine of the Palo Alto Veterans Hospital, also accumulating the position of professor of medicine and pharmacology and associate chairman of medicine. He left Stanford in 1988 to become vice president at Abbot Laboratories where he directed the pharmaceutical discovery and development programs for four years. Aside from contributing to the discovery of many new drug targets, his team of about 1,500 scientists succeeded in carrying out clinical trials for 24 new compounds for several diseases. Meanwhile, Murad was also working with a group of 20 scientists on nitric oxide gas and cyclic AMP, under two National Institutes of Health extramural grants. In 1993, Murad left Abbot to found and head a new biotechnology company, Molecular Geriatrics Corporation; but in 1997, he decided to accept an invitation to become the first chairman of a newly combined basic science department of integrative biology, pharmacology and physiology at the University of Texas in Houston. Murad remains in Houston, where he continues his research on nitric oxide and molecular signaling.

See also Biochemistry

MURPHY, WILLIAM P. (1892-1987)

American physician and pathologist

William P. Murphy received the 1934 Nobel Prize in physiology or medicine for his role in the discovery of **liver** as the successful dietary treatment for pernicious anemia, a deadly disorder in which bone marrow ceases to produce the fully mature red **blood** cells needed to carry oxygen to all parts of the body. Murphy's professional persistence following the discovery led to the simple, effective, and inexpensive treatment of the disease by intramuscular injection of a highly-concentrated liver extract.

Murphy shared the Nobel Prize with **George Hoyt Whipple**, who had observed that a diet of liver, kidney, meat, and vegetables had a regenerative effect on the blood of dogs in which he had induced anemia; and **George Richards Minot**, who, building on Whipple's research, isolated liver as the effective dietary factor. Murphy and Minot collaborated on the highly successful study in which pernicious anemia patients were fed one-quarter to one-half pound of liver daily. Reputed for his diligence and dedication, Murphy assumed the painstaking, time-consuming responsibility of counting the microscopic reticulocytes (red blood cells) in the blood samples of pernicious anemia patients before and during the liver diet. The dramatic increase in reticulocytes in the samples following the patient's consumption of liver clearly identified the critical connection between liver ingestion and the production of mature red blood cells.

William Parry Murphy was born on February 6, 1892, in Stoughton, Wisconsin, to Congregational minister Thomas Francis Murphy and his wife, Rose Anna Parry. He attended public schools in Wisconsin and Oregon and received his B.A. in 1914 from the University of Oregon. Murphy taught high school math and physics for two years in Oregon before entering the University of Oregon Medical School in Portland, where he also worked in the **anatomy** department as a laboratory assistant. In 1918, he took a summer course at Rush Medical School in Chicago. He later received the William Stanislaus Murphy Fellowship award and entered Harvard Medical School in Boston, from which he graduated in 1922. He interned at Rhode Island Hospital, Providence, then returned to Boston to become an assistant resident physician at Peter Bent Brigham Hospital.

In 1925, Minot had put pernicious anemia patients at Boston's Huntington Memorial Hospital on a liver-rich diet and observed their improvement. Wanting more evidence, he told no one of his experiment, not even the resident from Boston's Peter Bent Brigham Hospital whose collaboration

he recruited. Minot was an attending physician at Brigham where Murphy, a hard-working resident with a keen interest in blood disorders, attracted his attention. Without saying why, Minot asked Murphy to feed liver to pernicious anemia patients at Brigham. Murphy followed Minot's instructions, and two independent surveys were underway at two different institutions.

Eventually, Murphy and Minot enlisted the expertise of Edwin J. Cohn, a physical chemistry professor at Harvard Medical School. Cohn chemically reduced large amounts of liver to a concentrated extract fifty to one hundred times more potent than the liver itself. Murray subsequently sought the help of Guy W. Clark of the Lederle Laboratories; soon they developed an extremely concentrated extract. Injected into the muscle only once a month, the extract provided the same therapeutic effect as the liver diet or the oral extract at a much lower cost to patients.

Murphy's lifesaving contribution to medicine was further advanced by Harvard physician William Castle, who, in 1948, isolated the active ingredient in liver which promoted the development of fully mature red blood cells in patients suffering from pernicious anemia. That factor, named cyanocobalamin for its high concentration of cobalt, is commonly called Vitamin B_{12}, which is now used universally via intramuscular injection for the lifesaving treatment of pernicious anemia.

In addition to working with Minot on the liver diet study, Murphy became Minot's partner in private practice in Boston. In 1924, he was appointed assistant in medicine at Harvard Medical School, promoted to associate in medicine at the Brigham Hospital in 1935, and became a senior associate in medicine and consultant in hematology there. Harvard and Brigham both granted him emeritus status in 1958, when he retired to a suburb of Boston. Murphy married in 1919, and he and his wife had two children. Murphy's honors include the Cameron Prize and Lectureship of the University of Edinburgh, the Bronze Medal of the American Medical Association, and the Gold Medal of the Massachusetts Humane Society. He died on October 9, 1987, in Brookline, Massachusetts.

See also Hemopoiesis; Necrosis; Oxygen transport and exchange; Spleen histophysiology; Transfusions; Vascular exchange

MURRAY, JOSEPH E. (1919-)

American surgeon

Joseph E. Murray, a basic scientific researcher and clinician, pioneered renal (kidney) transplantation. Since his first successful kidney transplant in 1954, over a quarter of a million have taken place worldwide. In 1990, the Nobel committee in Stockholm honored Murray for his work by awarding him the Nobel Prize in physiology or medicine.

Murray was born on April 1, 1919, in Milford, Massachusetts. The son of William Andrew Murray and Mary

DePasquale Murray, he grew up and went to high school in Milford, excelling in baseball as well as science. He entered Holy Cross College upon graduation from high school and earned an A.B. in 1940, and then went on to Harvard University to earn his medical degree in 1943. At Peter Bent Brigham Hospital, a Harvard-affiliated hospital, Murray interned in surgery for one year before the U.S. Army gave him a commission in the Medical Corps and assigned him to Valley Forge General Hospital. For the next three years, Murray worked in plastic surgery, his specialties being reconstructive surgery of the **eye** and hand. He performed over 1,800 operations in this period, working under and learning from Drs. J. Barrett Brown and Bradford Cannon. It was a training that would stand Murray in good stead with his later research, for one of the major problems plastic surgeons had to deal with was the rejection of skin grafts by the **immune system**. Murray and other plastic surgeons soon learned that grafts would take between identical twins.

Returning from the military in 1947, Murray assumed a residency at the Peter Bent Brigham Hospital (now Brigham and Women's Hospital). His first love continued to be plastic surgery, but this was a fledgling discipline in the late 1940s and Murray was encouraged to go into general surgery, keeping plastic surgery as a sideline. This he did, soon winning a reputation for his head and neck surgical reconstructions on **cancer** patients and gaining recognition for the discipline of plastic surgery in the process. Murray was also drawn to the work of a team of doctors at Brigham Hospital who were studying end-stage renal disease, and one of the directions their researches was taking was transplantation. At that time there had been kidney transplants in dogs, but there had never been a successful human transplant. Harvard researchers, led by John Merrill and David Hume, had been doing experiments transplanting **kidneys** from cadavers onto the thigh of patients with kidney failure, grafting the third kidney to the femoral vessel of the recipient. One such thigh transplant functioned for about six months, enough time to allow the patient's own kidneys to heal and resume functioning. Kidney dialysis was also being perfected at this time, but Murray felt that it was only a temporary solution. He developed a surgical technique to connect the **blood** vessels of the donor kidney with those in the **abdomen** of the recipient, implanting the ureter directly into the urinary bladder.

All was in readiness to test the new procedure, except for the right patient: he or she would have to be one of a pair of identical twins with the other twin willing and able to donate a kidney, thus avoiding rejection by the immune system of the recipient. Such an opportunity came in December, 1954, when the Herrick brothers turned up at Brigham Hospital. Richard Herrick had end-stage renal failure and his twin, Ronald, was prepared to donate a kidney. Murray reasoned there should be no problem with rejection as the introduced kidney would be genetically identical to the one being replaced. The subsequent operation, performed on December 23, lasted five and one-half hours and was an immediate success. Herrick lived another seven years on the transplanted kidney before dying of **heart** failure.

Murray continued to perform more successful operations on identical twins, including Edith Helm, who went on to have children and grandchildren, but the real problem now became how to suppress the immune reaction so that the operation would be more generally available. At first Murray and other researchers tried total body x rays and infusions of bone marrow from the donor to adapt the recipient's immune system. In most cases, the transplants functioned for several weeks, but there were many failures. Finally in 1959, after a course of total body x rays, a non-identical twin survived a kidney transplant from his brother and went on to lead a normal life. Later, in 1959, two Boston hematologists, William Dameshek and Robert Schwartz, demonstrated that the compound 6-mercaptopurine would prevent a host animal from rejecting a foreign protein. This was the opening Murray was looking for, and working with chemists and other researchers, Murray developed a drug regimen to suppress the immune system and thus allow an organ from a non-related donor to be accepted by the recipient's body. In 1962 Murray successfully completed the first organ transplant from a cadaver.

Murray's successes became known worldwide and inspired other surgeons to experiment with a variety of organ transplants. With the development of less toxic immune suppressants such as azathioprine, transplants became a growth industry with registries for **organs** documented worldwide. In the first three decades after the development of the surgical technique, there had been 8,890 kidney transplants, 2,160 **liver**, 1673 heart, 413 **pancreas**, and 67 heart-lung transplants in the United States alone. The success rate for kidney transplants is high: the new kidney thrives for at least ten years in 70% of the patients. A related medical benefit was the increase in research into the rejection phenomenon, and thus into the functioning of the human immune system, research that has proved invaluable with the onset of acquired immunodeficiency syndrome (**AIDS**).

After this work on renal transplants, Murray went back to his first love, plastic surgery, developing ways to repair inborn facial defects in children. He headed the plastic surgery divisions of Peter Bent Brigham Hospital from 1951–1986 and Children's Hospital Medical Center from 1972–1985, and he has been a professor of surgery at Harvard Medical School since 1970. Murray was the recipient of the Gold Medal from the International Society of Surgeons in 1963. Four years after retiring from surgery, but not from administrative duties at Brigham Hospital, Murray was awarded the Nobel Prize in physiology or medicine along with **E. Donnall Thomas**, whose work in bone marrow transplants was closely related to Murray's research.

Murray and his wife have six children. An avid mountaineer, at the age of 52, Murray climbed the Matterhorn, and he has trekked in high mountain areas in India, China, and Nepal.

See also Autoimmune disorders; Immunity; Immunology; Renal system; Transplantation of organs

MUSCLE CONTRACTION

The ability of muscles to contract makes movement possible. The association of muscle contraction with movement has been known for centuries. In the second century A.D., the physiologist **Galen** understood that muscles work by pulling rather than by pushing. However, the mechanism remained unresolved until the 1950s.

Muscle contraction does operate by a pulling force. The basic mechanism is widespread in nature, being found in humans and other animals, in smooth, skeletal and **heart** muscles as well as in movement events at the level of the single cell. Although there are differences in the mechanism of contraction of smooth and **skeletal muscle** there are also a number of similarities. Basically, two sets of filaments slide over one another, allowing the muscle to shorten. When the filaments slide back in the other direction the muscles attains its original length.

Skeletal muscle is composed of fibers that are in turn each made up of myofibrils. Each myofibril is composed of myofilaments. The filaments are called thick and thin filaments, and were named after their appearance under microscopic examination. Thick filaments are composed of a protein called myosin. Myosin is a fibrous protein. That is, the proteins tend to be arranged in a relatively linear fashion. Thin filaments are composed of two interwoven strands of a protein called actin. Each actin strand is arranged as a helix, somewhat like the arrangement of a spiral staircase. The two helical actin strands wind around each other to form the thin filament.

The myosin and actin filaments are organized in a muscle as hundreds of layers, with the thick myosin filaments sandwiched between the thin actin filaments. Viewed in cross section the filaments resemble a honeycomb. The length of each filament is only a few millionths of a meter. This short stretch is known as a sarcomere (also called an A band). At either end of the sarcomere is a wall structure called the Z disc. The Z disc provides an anchor for the myosin and actin filaments. In the middle of each sarcomere is a structure called the M line. The M line is composed of another protein and acts as an anchor for the thick filaments.

In this so-called sliding filament model, because the myosin and actin filaments can slide past one another, they themselves do not shorten in the process of muscle contraction. The pulling together of the Z discs is what shortens the muscle.

The overlapping sliding of filaments occurs for only a short distance between adjacent Z discs. But the millions of such events occurring along the many sarcomeres are sufficient to cause contraction. The energy of these many contraction is transmitted through a system of connective **tissue** wrapping around individual muscle fiber and blankets of connective tissue that wrap around the entire muscle. This allows the contraction energy to be translated into anatomical motion.

The energy for the process is provided by adenine triphosphate, which is commonly designated as **ATP**. The ATP is able to bind to power the binding of myosin to actin. Numerous such events act to ratchet the actin filament past a myosin filament. The dissolution of ATP to adenosine diphos- phate (ADP) releases the tension holding the filaments together, allowing relaxation of the filaments to occur. Also involved in this process in skeletal muscles are the so-called accessory proteins trophomyosin and troponin.

Smooth muscle lacks troponin, but does have a protein called caldesmon that acts in a similar fashion. **Calcium** is also necessary to cause a change in shape of the accessory proteins that exposes a site on the actin filament to which the myosin filament can bind, in the ATP-dependent process. Smooth muscle contraction also involves different molecules and unique myosin subunits, particularly the light chain or p-light chain.

See also Arthrology (joints and movement); Cardiac muscle; Muscle tissue damage, repair, and regeneration; Muscular innervation

MUSCLE TISSUE DAMAGE, REPAIR, AND REGENERATION

Muscle **tissue** damage can occur in all three types of muscle tissue: skeletal, cardiac, and **smooth muscle. Skeletal muscle** tissue is susceptible to damage from strenuous exercise. In particular, eccentric exercises (muscle lengthening) cause fatigue more rapidly than concentric exercises (muscle shortening), leading to muscle damage. Indicators of muscle damage include loss of strength and range of motion, accumulation of cellular **calcium**, delayed onset muscle soreness (DOMS), and increased **blood** levels of creatine kinase.

Almost immediately following damage to muscle, neutrophils travel to the injured area via the bloodstream, initiating the immune response. Neutrophils release free radicals and other toxins that help remove dead muscle fibers. Following the neutrophils, monocytes migrate to the area and differentiate into macrophages. These cells are responsible for phagocytosis of the tissue debris. Additionally, macrophages are responsible for activating the satellite cells essential for tissue repair. The greatest damage to the muscle occurs at least three days post exercise and is due to the production of free radicals and accumulation of calcium in the cells. These events are used to explain the delayed onset muscle soreness (DOMS) that accompanies damage of muscles. DOMS occurs 24–72 hours post exercise. A common misconception is that muscle soreness is a direct result of **lactic acid** build-up. However, this theory has been proven false and studies have shown that lactic acid is removed from skeletal muscles one hour after exercise. Macrophages release a hormone called prostaglandin that is theorized to stimulate the sensory nerves in muscles, thus creating the characteristic **pain**. Muscles are usually most sore two to three days following exercise and are not fully healed until 8–10 days post exercise. During repair, muscle tissue becomes resistant to damage from further strenuous exercise.

The extent of muscle damage can be correlated with levels of creatine kinase in the blood. In healthy muscle tissue, creatine kinase is an enzyme responsible for converting ADP to **ATP**. However, when muscle tissue is damaged, the cell membranes are destroyed and creatine kinase leaks out of the muscle cells. Blood levels of creatine kinase are highest five days following exercise. This evidence implies that muscle tissue is not damaged immediately following injury, but rather during the repair process.

Muscle precursor cells, or satellite cells, are responsible for muscle tissue regeneration that may take days to weeks to complete. Satellite cells remain dormant between the sarcolemma and basal lamina until injury occurs. Chemical signals stimulate these cells to migrate to the damaged muscle tissue. Satellite cells are then capable of forming myoblasts that fuse together to form immature muscle fibers, eventually developing into muscle fibers. Generally, there is a decrease in the number of muscle fibers present in regenerated muscle, resulting in decreased strength of the muscle. Blood vessels and nerves are important factors in the regeneration of muscle tissue. If vascularization is absent, macrophages will be unable to reach the damaged area and the healing process will be halted. Additionally, any living cells will die within hours if blood is absent from the injured region. In the case of severe skeletal muscle damage, fibrosis can occur, interrupting contraction of skeletal muscle. Fibrosis is the development of scar tissue in place of muscle fibers; it limits the ability of muscle tissue to regenerate.

Damage to cardiac muscle, once thought to be irreversible, is now shown to have the capability to regenerate somewhat after a **heart** attack. Evidence to support this claim is based on the findings of a protein, Ki67, expressed during the **cell cycle** in the nucleus of cardiac muscle cells. This protein appears to be the best indicator to date that **cell division** is occurring cardiac muscle. However, the degree to which regeneration can occur in cardiac muscle remains unclear. This is because damaged cardiac muscle has limited capacity to regenerate and is often replaced with collagenous connective tissue. Its inability to relax for a period of time is believed to inhibit the cell division necessary for regeneration. Cardiac muscle that does survive after injury is sometimes able to hypertrophy in order to make up for the lost muscle tissue. Of the three types of muscle tissue, smooth muscle tissue has the greatest capacity for regeneration. Pericytes are the **stem cells** of smooth muscle and are capable of regenerating small blood vessels that consist of smooth muscle. The bladder and the intestines are also capable of undergoing regeneration of smooth muscle.

See also Cell differentiation; Skeletal muscle

MUSCLES OF THE THORAX AND ABDOMEN

The thorax comprises the major greater portion of the chest, and contains the rib cage (costal cartilages). There are six principal muscles of the thorax. These are the intercostales externi, intercostales interni, subcostales, transversus thoracis, levatores costarum, serratus posterior superior, serratus posterior

inferior, and **diaphragm**. The **abdomen** lies below the thorax, overlying the stomach. The muscles of the abdomen are the obliquus externus, obliquus internus, transversus, rectus, and pyramidalis. The muscles work to assist respiration, **digestion**, and to protect internal **organs**.

The intercostal muscles are thin and plate-like segments made of muscle and fiber. They are found in the intercostal spaces; the spaces between the ribs. Their names are derived from their relation to one another. The intercostales externi are nearer to the surface than the intercostales interni. There are 11 of the external muscles on each side. Each one is attached from the lower part of a rib at one end and the upper part of the next lower rib, cross-meshing the ribs together. There are also 11 of the internal muscles on each side of the ribcage. They are also connected to adjacent ribs, but at the back of the ribs, and are meshed in the opposite pattern to the external muscles. The effect is to provide great structural support.

The subcostals connect a rib to the second or third rib below it. Each muscles runs in the same direction as the intercostales interni.

The transversus thoracis is also a thin plane of muscle and fiber. Found on the front wall of the chest, the muscle runs from the bottom of the thorax upward and laterally outward, and inserts into the **cartilage** of the second to the sixth ribs. The lowest of these inserted muscles is almost horizontal, while the topmost muscles is almost vertical.

There are 12 levatores costarum muscles on each side of the thorax. These muscles are small and tendon-like. They connect the ribs to the central **vertebral column**.

The serratus posterior superior is another thin muscle. Its shape differs from the other muscles, because it is shaped as a quadrilateral. It is located at the upper and back part of the thorax. From there, it is oriented downward (inferiorly) and outward, almost laterally, and inserts into the upper part of the second through fifth ribs. Thus, they provide support to these ribs from above. The serratus posterior inferior rises upward from the lower region of the thorax and inserts into the fourth pair of ribs.

The abdomen is also well muscled. The muscles of the abdomen are divided into two groups: the anterolateral muscles and the posterior muscles. As their names imply, the muscles in the first group are found at the front and side of the abdomen, while those of the second group are found in the lower back.

The anterolateral muscle group consists of: obliquus externus, obliquus internus, transversus, rectus, and pyramidalis. The obliquus externus muscles are the horizontal muscles, arranged in four pairs from the lower abdomen upward, which are prominent in bodybuilders. The obliquus internus muscles are smaller and more vertically oriented. Both muscle types help to anchor the ribs to the abdomen.

Underneath the obliquus muscles is the transversus. It is oriented more horizontally, fanning out to inserted in the pubic bone.

The rectus abdominus is a long flat muscle. It runs the entire length of the front of the abdomen, front the pubic area at its lower reaches to the fifth through seventh ribs at it upper reaches.

The pyramidalis is a small triangular-shaped muscle that is located at the lower part of the abdomen. It runs from the pubic area on either side of the abdomen and arches upward to near the navel.

The posterior muscle group is made of psoas major, psoas minor, iliacus, and the quadraturs lumborum. They all function to stabilize the lower portion of the rib cage from the back.

The muscles of the thorax and abdomen provide structural support for the rib cage and function in the rotational twisting, bending and flexing motions that these regions of the body exert. Injury to these muscle groups can severely compromise range of motion.

See also Anatomical nomenclature; Fascia; Muscular system overview

MUSCULAR DYSTROPHIES

Muscular dystrophies (dystrophinopathies) are progressive hereditary degenerative diseases of skeletal muscles due to an absence or deficiency of the protein dystrophin. Dystrophin and the associated proteins form a complex system that connects the intracellular cytoskeleton to the extracellular matrix. The normalcy of this system is critical for maintaining the integrity of the delicate, elastic muscle membrane (sarcolemma) and the muscle fiber. The responsible gene is located on the short arm of the X chromosome.

The most common mechanisms of mutation are deletions and duplications of base pairs largely clustered in the "hot spot," a **DNA** sequence associated with an abnormally high frequency of mutation or recombination. The type of deletion determines whether dystrophin is absent from the muscle or present in a reduced, altered form. This has an important clinical significance because the former is usually associated with the severe Duchenne's variety of the disease (DMD), whereas the latter situation may cause the milder Becker's variant (BMD).

In approximately two-thirds of cases, the dystrophin gene defect is transmitted to affected boys by carrier females following the pattern of Mendelian X-linked recessive inheritance. However, in one-third of cases, the defect arises as a result of a new mutation in the **germ cells** of parents or during very early embryogenesis.

DMD occurs at a frequency of 1 per 3,500 live births. The child who appears healthy at birth develops the initial symptoms around age five. Symptoms include clumsy gait, slow running, difficulty in getting up from the floor, difficulty in climbing stairs, and a waddling gait. Mental subnormality, if it occurs, is not progressive and is presumed to be due to the lack of **brain** dystrophin. The progression of muscle weakness and loss of function for the activities of daily living is relentless. In the typical situation, the child loses the ability to walk independently by age 9-12 years. Despite all therapeutic efforts, most patients die during the third decade. BMD is an allelic variant of DMD in which the mutation of the dystrophin gene produces a reduced amount of truncated dystrophin that is not capable of maintaining the integrity of the sarcolemma. However, the pace of muscle fiber loss is considerably slower

Jerry Lewis, talking with Sarah Schwegel, MDA National Goodwill Ambassador, during Muscular Dystrophy Association Labor Day Telethon. Each year Lewis's telethon raises millions of dollars for research.

than in DMD, which is reflected in a less severe clinical phenotype. The illness usually begins by the end of the first or at the beginning of the second decade. These boys, however, continue to walk independently past the age of 15 years and may not have to use a wheelchair until they are in their twenties or even later. In some people carrying a mutation in the dystrophin gene, no muscular symptoms are present at all.

Confirmation of clinical diagnosis of DMD/BMD is largely based on deletion analysis of DNA. In the 30% of patients in whom a deletion is not found, a muscle biopsy is necessary to establish the absence of dystrophin by immunohistochemistry, or Western blotting analysis (a form of protein analysis).

Treatment of dystrophinopathies is palliative (meaning aimed to relieve the symptoms not the cause), aimed at managing the symptoms in an effort to optimize the quality of life. **Gene therapy** which is oriented towards replacement of the defective dystrophin gene with a wild-type one (or a functionally adequate one) or upregulation of the expression of the surrogate molecules such as utrophin is still in the research phase.

Prevention of dystrophinopathies is based on genetic counseling and molecular genetic diagnosis. Female carriers can namely opt for prenatal diagnosis (evaluation in the first trimester of **pregnancy** of whether the fetus has inherited the mutation) or even preimplantation genetic analysis of the embryo. In the last case, **fertilization** *in vitro* is followed by isolation and genetic testing of the single cell from the few-cell-stage embryo. The embryo is implanted into the uterus if no dystrophin gene defect is found.

See also Birth defects and abnormal development; Ethical issues in embryological research; Ethical issues in genetics research; Genetics and developmental genetics; Human genetics; Muscular system overview; Skeletal and muscular systems: embryonic development

MUSCULAR INNERVATION

The contraction of muscles is controlled by signals from the nervous system. The signals reach the appropriate muscle via nerves that contact the muscle.

Typically, nerves contact muscles by means of what are known as motor end plates. The stimulus to contract passes from the end plate to the muscle fibers. The nerve cells (**neurons**) that function in **muscle contraction** are called motor neurons. The have a cell body at one end, a long filamentous region known as the axon and a branching region at the other end that is called the dendrites. The dendrites form the motor end plate

Each muscle fiber in vertebrate animals has a motor neuron associated with it. This is described as innervation. A single motor neuron can innervate other fibers as well. But each muscle fiber will have a motor neuron connection.

The connection between a muscle fiber and a motor neuron occurs at a specialized region called the neuromuscular junction. This is the area where the motor end plate is found. An endplate is analogous to the **synapse** that is between the axon terminus of one nerve cell and the dendrite branches of an adjacent nerve cell.

The motor end plate almost contacts an area of the muscle fiber called the sarcolemma. As with other nerve cells, a molecule called the neurotransmitter bridges the gap between the nerve cell and the muscle fibers, allowing a signal to flow. When a nerve signal, which is really an electrical **pulse**, reaches the end plate, the sarcolemma becomes depolarized. In other words, the difference in charge between the inside and the outside of the muscle membrane is changed. If the change is great enough, a wave of depolarization spreads outward over the surface of the fiber. This stimulates the fiber, and in fact all the fibers in the muscle, to contract.

The above scenario would be caused by an excitatory neuron. Neurons that inhibit muscular contract (inhibitory neurons) can also innervate muscles. In vertebrates, acetylcholine is an excitatory neurotransmitter and gamma-aminobutyric acid (GABA) is an inhibitory neurotransmitter.

The degree of muscle contraction depends on the number of motor units that are present and signaling. The more units there are, the finer the control can be exerted on muscle contraction. For example, in the arm there are many innervation points. Thus, contraction can range from slow to quick, and from gentle to powerful.

See also Autonomic nervous system; Membrane potential; Skeletal muscle; Smooth muscle

MUSCULAR SYSTEM, EMBRYONIC DEVELOPMENT • *see* SKELETAL AND MUSCULAR SYSTEMS, EMBRYONIC DEVELOPMENT

MUSCULAR SYSTEM OVERVIEW

The muscular system includes those tissues of the body, which by virtue of their composition of contractile **tissue**, are able to cause some type of movement. The term "contractile" refers to the fact that muscles are made of complex muscle fibers, composed of myofibrils. These myofibrils are composed of filaments that overlap each other in a way that allows them to slide past each other, changing the degree of overlap. This allows the muscle to grow shorter or longer, resulting in movement.

Some muscles are under voluntary control, meaning that an individual can decide to move those muscles. Voluntary muscles are also known as skeletal muscles, because most of these muscles are attached to the bones of the body, and are responsible for moving the skeletal system. Voluntary muscles not directly attached to bones include those muscles around the mouth and the anus. While skeletal muscles are under voluntary control, they also may receive some automatic input from areas of the **brain**. This is certainly true of many of the muscles of respiration (**breathing**), which are automatically programmed to continue contracting and relaxing, so that an individual does not have to decide to take each breath. There are more than 650 different voluntary muscles in the human body.

A second type of muscle, called **smooth muscle**, is considered to be under involuntary control. Smooth muscle makes up the muscles of the intestine, the uterus, the **blood** vessels, and the **eye**. The intestine contracts and relaxes without an individual even being aware of its actions, allowing food to be churned up and moved along its length. An individual cannot exert control over these muscles; one cannot, for example, decide to speed up the amount of time it takes for a meal to travel down the length of intestine. Similarly, when a woman goes into labor, she cannot prevent her uterus from contracting simply by willing it to stop. Once the correct hormone environment is in place, the uterus will continue contracting entirely without voluntary control. An individual's pupillary muscles contract or relax due to their response to the presence or absence of light, and not due to an individual's desire to have dilated or contracted pupils.

A third type of muscle is cardiac (**heart**) muscle, which is responsible for the forceful contraction of the heart beat. This type of muscle is primarily involuntary, although individuals have been taught to manipulate heart rate through a process known as **biofeedback**. Cardiac muscle is amazingly strong and resilient, given its ability to beat continuously over an individual's lifetime. In contrast, **skeletal muscle** needs

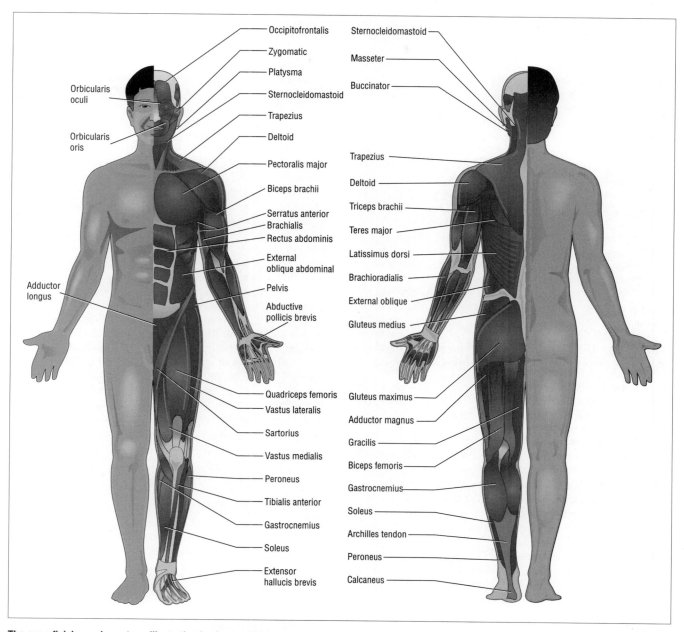

The superficial muscle system. *Illustration by Argosy Publishing.*

periods of rest, a luxury which cardiac muscles neither require nor can afford to take.

See also Muscle contraction; Muscle tissue damage, repair, and regeneration; Muscles of the thorax and abdomen; Muscular dystrophies; Muscular innervation

MYELENCEPHALON

The myelencephalon, or **medulla oblongata**, is one of the last portions of the brainstem to be contained in the **cranium**. It lies

just above the occipital condyles and connects to the **spinal cord**, which exits through the foramen magnum. It is a rather small swelling of the hindbrain that lies directly below the **pons**. Between each side of the anterior medulla there is a longitudinal fissure that separates each half. On the posterior side, a fissure extends upward to the fourth ventricle. The medulla itself is rather small. It measures about and inch in length, three quarters of an inch wide, and only about one half and inch in thickness.

At the superior (top) region of the medulla are two swellings of **tissue** called the pyramids. The two pyramids are bundles of white matter. Some tracts contain afferent nerve tracts while others contain efferent nerve tracts. This organization of incoming and outgoing signals is part of the neuronal coordination of the medulla

On each side of the pyramids is a small oval shaped structure called the olivary body or just plain olive. It contains a rippled layer of **gray matter**, the inferior olivary nucleus. This body acts as a relay station for signals on their way to the **cerebellum**. It is separated from main body of the medulla by a small depression from which the hypoglossal nerve emerges.

In addition to the origination of the hypoglossal nerve (XII), the glossopharangeal (XI), accessory (XI) **cranial nerves** originate in the medulla. Not only is the medulla the site of the nuclei for the formation of these cranial nerves it is also responsible for the control of coughing, sneezing, vomiting, and perspiring. This small structure is also responsible for three other important functions. The cardiac center makes adjustments to the rate and force of heartbeats. Another important location is the vasoconstrictor center, which regulates **blood** pressure by adjusting the vessel diameter. It may reroute blood from one place in the body to another depending on need. The respiratory center controls the rate and depth of **breathing**.

The medulla oblongata is believed to be one of the more primitive regions of the **brain**. It is formed early in development and regulates some of the basic functions of the body.

See also Brain stem function and reflexes; Brain, intellectual functions; Central nervous system (CNS); Cerebral cortex; Cerebral hemispheres; Cerebral morphology

MYOLOGY

Myology is defined as the science of growth, form, function, and diseases of muscles. While the term myology is fairly recent, the study of muscles dates back to the second century B.C. The Greek physician and anatomist **Galen** wrote extensively of his observations. During the Renaissance it was illegal to dissect cadavers, but artists continued this practice and brought highly accurate drawings of muscles to physicians. In the seventeenth century, the ways in which muscles contract and move the body was finally understood. In the eighteenth and nineteenth century, many researchers learned how muscles metabolize during **muscle contraction**, how much oxygen they use, and the process of **lactic acid** production. At the end of the nineteenth century, the role of **ATP** in muscle was discovered and the discipline has continued ever since.

Muscle **pathology** (disease) was first described by Duchenne in 1849. Many additional studies poured into the journals of muscle research immediately following. During the first half of the twentieth century, the **molecular biology** of all systems made tremendous advances, especially in myology. A major turning point was reached in 1987, when the gene for Duchenne muscular dystrophy was located. This has sped research and contributed greatly to the understanding of the inheritance of genetic muscle disorders.

Today, a considerable amount of research is benefiting from technological advances in molecular biology. Muscle proteins and genetic mutations are being studies as never before. It is hoped that someday scientists and medical researchers will be able to regenerate lost limbs. One of the important aspects of this type of research is to get new muscles to grow and function. Myology has never been as important as it is today. The future will reveal even more information in this interesting field of research.

See also Muscle contraction; Muscle tissue damage, repair, and regeneration; Muscles of the thorax and abdomen; Muscular dystrophies; Muscular innervation; Muscular system overview

N

NAILS

Each finger and toe on the human body has a flat plate commonly called a nail on its uppermost surface. Nails are epithelial cell structures that are composed of a fibrous protein called keratin. The same protein is also the constituent of wool, **hair**, horn, hoofs, and the quills of feathers.

The chemistry of keratin makes nail very resilient. The keratin protein is made up of large quantities of the amino acid cysteine. Adjacent cysteine molecules are able to link together by virtue of disulfide bonds between sulfur residues that are part of the cysteine structure. The formation of many of these disulfide bonds produces keratin chains that are tightly linked to each other, much like bundles of twigs lashed together. The result is a bundle of great rigidity that is resistant to breakage. When these linked polypeptides are on a flat plane, as in a nail, the plate they form acts like armor.

Each nail is formed from a nail root that is made up of normal layers of skin cells. As the nail grows, a layer of epidermal cells called the nail bed supports it. The nail grows from one end, with the older cells being pushed further along the nail bed. As the end of a nail extends beyond the nail bed, it is removed by filing or trimming, or is worn away.

Nails on the fingers grow at a rate of about 0.0039 in. (0.1 mm) per day. At that rate, a nail is completely replaced in about 4–6 months. The nail on the toes grows more slowly and 6–8 months is required to completely replace the toenail. The growth rate of nail can vary with age, the time of year, and with **nutrition**. A poorly nourished person will exhibit a slower nail growth rate, for example.

As resilient as they are, nails are still subject to damage and disease. Paronychia is a bacterial, viral, or fungal **infection** of the tissues surrounding the nail. The condition is painful and the nail can begin to form ridges because of underlying pressure. Onychomycosis is a fungal infection of the nail, which can be spread from a site of infection elsewhere in the body. Despite therapy with antifungal agents, it

Nails, as displayed by Olympic champion Florence Griffith-Joyner, are composed of the fibrous protein keratin. *Archive Photos, Inc. Reproduced by permission.*

can be difficult to eradicate. Skin conditions like dermatitis and psoriasis can also affect nail. Other ailments of the body's chemistry or function can be manifest as changes in the nail. For example, nail may grow thinner when one has a fever, is in **shock**, or experiences a myocardial infarction. An exaggerated curve of the nail plate can be a sign of congenital **heart** disease in children and pulmonary disease in adults.

See also Hair

NAUSEA

Nausea refers to the unpleasant sensation in the stomach that is usually accompanied by the urge to vomit.

The process of nausea is regulated in the **brain** by the chemoreceptor trigger zone, which is in an area of the medulla known as the vomiting center. The chemoreceptor trigger zone senses metabolic abnormalities that are contributing to stomach upset, such as toxins in the body, and stimulates the vomiting center. Once stimulated, the center directs the release of several neurotranmitters that drive the physical processes of vomiting; the forceful expulsion of the stomach contents.

Nausea itself is associated with several physiological processes. The contents of the stomach move less and contents of the intestine can flow back into the stomach. The intestine becomes more rigid, which restricts the passage of digested material down its length. The desire to eat disappears. A **headache** can develop and both perspiration and salivation increase. Finally, the windpipe (trachea) closes and the abdominal wall and the **diaphragm** muscles tighten quickly. The forceful squeezing of the stomach, combined with the upward pressure on the stomach from the intestinal tract, ejects the stomach contents up the esophagus and out of the mouth in the act of vomiting.

There are a variety of causes of nausea. For example, intense **pain**, such as often occurs from a migraine headache can be a trigger. Drugs, such as those used in **chemotherapy**, may also act as a triggering mechanism. Metabolic disorders also can be a factor. An example is the acidification of the **blood** that can occur in diabetes. The presence of toxins and other contaminants may also contribute to nausea symptoms. Nausea-inducing physiological changes can occur as a result of rapid hormonal shifts in the early weeks of **pregnancy**, from sea or other motion sickness, and from emotional stress.

In the case of food poisoning or the presence of other toxins, nausea is a beneficial response, allowing the body to purge itself of the offending agent. However, in other instances, control of nausea is sought (for example when nausea is a side effect of chemotherapy). Management of nausea via drugs can increase the well being of the patient during therapy and contribute to a patient's willingness to undergo therapy. Various chemicals can act as palliatives and alleviate the feelings of nausea, by blocking various receptors in the chemoreceptor trigger zone.

See also Medulla oblongata

NECK MUSCLES AND FASCIAE

The neck is a well-supported structure. A variety of muscles provide the firm attachment necessary for the safe support of the head, yet permit the flexibility needed for the neck to move from side-to-side and in an up-and-down motion. Additionally, the connective **tissue** fasciae protect and strengthen the neck. The supporting muscles and fasciae are compressed into a narrow area, so as to provide stability and flexibility while at the same time accommodating the various functional routes that pass through the neck, such as the air passage, food passage, **blood** vessels, and nerves.

At the front of the neck, strap muscles connect the neck to the **sternum** (the shoulder blade region). There are also muscular attachments to the tongue and the jaw, enabling the movements associated with speech and food processing in the mouth. A pair of muscles called the sternocleidomastoid is positioned on the right and left front lower regions of the neck, where they fan out to the shoulders.

The back of the neck is heavily layered with muscles. Two large, triangular-shaped trapezius muscles fan out over the shoulders from the upper neck. Another pair of muscles, called the splenii, point downward in a "V" shape. A pair of semispinalis capilis muscles are oriented similar to the splenii, but are narrower and more vertically arranged. The semispinalis cervicis pair, which has an inverted "V" shape, attaches the lower portion of the neck to the upper back. Along the long axis of the back of the neck are a series of short muscles. The multifidi radiate outward at a downward angle, while rotatores are at right angles to the long axis.

Fasciae is thickened connective tissue that is wrapped around muscles or groups of muscles. It provides protection and increased strength. The neck is also rich in so-called deep fascia and cervical fasciae, which are particularly dense layers of connective tissue fiber. The neck fascia extends some distance away from the neck, becoming continuous with fascia of the shoulders, chest, upper arms, and back.

The importance of the neck muscles becomes especially evident when these muscles, particularly those in the back of the neck, become fatigued or damaged. The resulting stiffness and **pain** can be pronounced and can require physiotherapeutic intervention to relieve the discomfort.

See also Connective tissues; Headache; Mastication

NECROSIS

Necrosis is the process of cell or **tissue death** resulting from the activities of disease organisms, toxins or physical factors (e.g., temperature), as well as inadequate **nutrition** or starvation (e.g., as a result of an interrupted **blood** supply). Necrosis contrasts with apoptosis, which is the genetically programmed death of cells.

Pathologically, necrotic tissues undergo characteristic changes over time. On the macroscopic (visible) level, all necrotic tissues except for the **brain** undergo a definite colour change and acquire a firm consistency within the first 24 hours. In two to three days, a border delineated by an inflammatory response is observed, which may have fibrinous exudate. After one week, a gray/white periphery zone, due to healing via fibrosis, is seen and after several months, a fibrous

Dr. Beck Weathers suffered frostbite while climbing Mount Everest, and displays the resulting necrosis on the face and hands. *AP/Wide World Photos. Reproduced by permission.*

scar has developed. In the brain and in infected tissues, macroscopic changes involve initially a softening and loss of tissue definition leading to extreme softening after two to three days and a rim of peripheral hyperemia (excess of blood in an area). After months, a cystic area traversed by fibrous strands may be observed.

The changes in cells undergoing necrosis arise from enzymatic degradation and protein denaturation of cellular components. Histologically, the process of necrosis can be categorized into several distinct types. Coagulation necrosis is characterized by the preservation of cell shape and tissue architecture and results from protein denaturation. It involves both nuclear and cytoplasmic changes and is the most common type of necrosis, occurring in all **organs** except the brain. Ultimately, it is repaired by fibrosis. Microscopically, an acute inflammatory infiltrate may be seen in these tissues, and after two to three days a vascular granulation may be observed which becomes fibrous granulation tissue after one week. Fibrosis may be seen after months.

Another type of necrosis is colliquative or liquifactive necrosis. This does not involve the preservation of tissue architecture and occurs as a result of enzymatic degradation. This occurs in the brain and in infected tissue. Brain cells undergoing this kind of necrosis ultimately become cystic because there are no fibroblasts in the brain to repair them. In the first 24 hours, neuronal degeneration and myelin pallor may be observed with perivascular neutrophil aggregation. After two to three days, peripheral vascular granulation tissue develops and after weeks, fibrillary gliosis and central cyst formation can be seen. In diseases such as tuberculosis, caseous necrosis, a mixture of both liquifactive and coagulative necrosis, may be observed.

A further category of necrosis is **fat** necrosis, which occurs in tissue with a high fat content, such as the **pancreas**,

and is due to the action of fat degrading **enzymes** known as lipases. Occasionally, tissues will receive an inadequate supply of oxygenated blood. An inadequate blood supply leads to cell **hypoxia**, and consequent cell injury and death. Tissues or organs that have been exposed to this kind of damage are said to undergo ischemic necrosis as a result of diminished blood flow.

See also Cell structure; Death and dying; Enzymes and coenzymes

NEHER, ERWIN (1944-   )
German biophysicist

Erwin Neher, along with **Bert Sakmann**, was awarded the 1991 Nobel Prize in physiology or medicine for the development of the patch clamp technique. The use of this technique enabled Neher and Sakmann to forge new paths in the study of membrane physiology and to understand the structure and functions of ion channels found in the **plasma** membranes of most body cells. The patch clamp technique has given physiologists a precise understanding of cellular microelectrical activity and has contributed significantly to the research and treatment of cystic fibrosis, diabetes, epilepsy, and other disorders of the cardiovascular and neuromuscular systems.

Neher was born in Landsberg, Germany, on March 20, 1944, the son of Franz Xavier Neher and Elisabeth Pfeiffer Neher. In 1965, he completed his undergraduate studies at the Institute of Technology in Munich with a major in physics. Two years later, he earned his master's degree from the University of Wisconsin under a Fulbright scholarship. He then went on to complete his doctorate at the Institute of Technology in Munich, Germany, in 1970.

While the existence of ion channels that transmit electrical charges was hypothesized as early as the 1950s, no one had been able to see these channels. As a doctoral student, Neher was drawn to the question of how electrically charged ions control such biological functions as the transmission of nerve impulses, the contraction of muscles, vision, and the process of conception. He realized that in order to get answers to these questions he would have to look for the ion channels.

It was in his doctoral thesis that Neher first developed the concept of the patch clamp technique as a way of discovering the ion channels. In 1974 he shared a laboratory space with Bert Sakmann at the Max Planck Institute in Göttingen. They both agreed that understanding the nature of ion channels was the most important problem in the biophysics of the cell membrane, and they set out to develop the techniques of patch clamping.

Neher briefly worked with Charles F. Stevens at the University of Washington. When Stevens moved to Yale, Neher followed him while maintaining his collaboration with Sakmann. From 1975 to 1976, Neher was a research associate in the department of physiology at Yale University, and much of the data for the paper on patch clamps came from the Yale studies.

In 1976 Neher and Sakmann published their landmark paper on the use of glass recording electrodes with microscopic tips, called micropipettes, pressed against a cell membrane. With these devices, which they called patch clamp electrodes, they were able to electrically isolate a tiny patch of the cell membrane and to study the protein s in that area. They could then see how the individual proteins acted as channels or gates for specific ions, allowing certain ions to pass through the cell membrane one at a time, while preventing others from entering. Their work with patch clamps allowed them to remove a patch of the membrane and to enter the interior of the cell. They then were able to conduct various experiments to observe the intricate mechanism of ion channels.

Several years passed after they presented their findings to an audience at the Biophysical Society meeting in 1976 in which Neher and Sakmann, along with their co-workers, refined the technique of patch clamping. Creating a better seal between the micropipette and the patch of cell membrane it pressed against was one of the refinements they sought. Without a tight seal there was interference by "noise" that overshadowed the smaller electrical currents.

The problem of outside noise interference was solved by Neher in 1980 when he was able to observe on his oscilloscope a marked drop in the noise level to almost zero. From this drop, he was able to infer that he had produced a seal that was one hundred times better than previously attained. While other researchers had noticed an abatement of noise at times, Neher was the first to realize the significance of the drop in noise level.

Neher found that by using a light suction with a super clean pipette, he could create a high-resistance seal of 10–100 gigohms (a gigohm is a measure of electrical resistance equal to one billion ohms). He called this seal a "gigaseal." With the gigaseal, background noise could be decreased, and a number of new ways could be used to control cells for patch clamp experimentation. Patches from the cell could now be torn away from the membrane to act as a membrane coating over the mouth of the pipette, thus allowing for more exact measurement of electrical ion movement. A strong suction could force the pipette into the cell while still maintaining a tight seal for the cell as a whole.

In 1976, Neher returned to the Max Planck Institute in Göttingen. On December 26, 1978, he married Eva-Maria Ruhr, a microbiologist. They have five children. Neher became director of the membrane biophysics department at the Max Planck Institute in 1983, and in 1987, he was made an honorary professor. In 1991, Neher and Sakmann won the 1991 Nobel Prize in physiology or medicine for proving the existence of ion channels.

Researchers using the patch clamp technique were able to discover a defective ion channel that was responsible for cystic fibrosis. Because of the use of patch clamps in research, there is now a better understanding of hormone regulation and the production of insulin as it relates to diabetes. The Nobel Committee also praised the work of Neher and Sakmann for helping in research on **heart** disease, epilepsy, and disorders affecting the nervous and muscle systems. Patch clamp research has helped in the development of new drugs for these conditions.

See also Action potential; Cell membrane transport; Cell structure; Electrolytes and electrolyte balance; Nerve impulses and conduction of impulses

NEONATAL GROWTH AND DEVELOPMENT

A neonate is an infant in the first 28 days of life, a critical period in a newborn's adaptation to life outside the womb. Most crucially, a neonate's health depends upon adequate **nutrition** (through mother's milk or formula) and successful respiratory functioning. At birth in the United States, a neonate typically weighs 7–8 lb. (3.1–3.6 kg), and measures 19–21 in. (48–53 cm).

In the United States, a neonate's condition and stability is first measured with the Apgar test, which is administered during the first five minutes after birth. The Apgar test measures an infant's color, respiratory affect, cardiac effort, body tone, and responsiveness to stimuli. Each health sign receives a 0, 1, or 2, with a total score of 7–10 being average–to–good, a score of 4–6 signaling potential developmental difficulties, and a score of 3 or below indicating doubtful survival. An Apgar score is not necessarily a reliable indicator of an infant's ability or failure to thrive.

The sensory and motor development of a normal neonate begins soon after birth. At birth, a neonate is equipped with the ability to suck, and to align his head with a mother's breast or a bottle. Neonates turn to milk over other substances, like water or sugar water, which suggests a functional sense of **smell**. Vision at birth is an average of 20/40, and is accurate at a distance of 1 ft. (0.3 m).

With hearing acuity at a near normal capability at birth, babies as young as one day old reflect auditory responses to human voices, as are measured by the relationship between their body movements and human speech. They do not register similar physical responses to non–human sounds, such as the ringing of a bell. Further, neonates in general respond more consistently and vigorously to female voices than male ones.

Neonates **sleep** between 10–21 hours a day, usually feeding every three to four hours. They rarely sleep through the night. The frequent feeding schedule causes them to wet their diapers up to ten times a day.

The relatively recent field of Neonatology, the study of the growth, care, and diseases of newborns, is a growing medical subspecialty that has radically improved the survival rate of neonates who are born pre–term or are underweight. A pre–term birth occurs anytime before the normal gestation period of 252–280 days from conception. Neonatal intensive care in the United States has saved babies born as early as 25 weeks of gestation (as opposed to the normal period of 38–40 weeks), largely through the use of incubators and ventilators.

Incubators operate through devices attached to an infant's skin that monitor and control body temperature. Ventilators can take over the respiratory function of a pre–term baby. The major cause of neonatal respiratory illness is hyaline membrane disease, a condition in which an infant's **lungs** are not sufficiently mature to manage ventilation and **blood** circulation. Diseases relating to the lungs, are the leading cause of **death** of neonates in the United States.

The physical immaturity of the preterm neonate also affects the **central nervous system**, and preterm babies often demonstrate an acute sensitivity to auditory and visual stimuli. Whereas a normal term baby turns towards a sound, like a rattle or bell, a preterm baby may register discomfort by turning away from it.

Neonatology indicates that conditioning traditionally considered emotional or affective can have a physiological effect on a neonate's development. In most cases, infants bonded successfully with parents will make greater gains in their sensory and motor developments than the infant who has not bonded.

See also Parturition; Embryology; Heart, embryonic development and changes at birth; Infant growth and development

Nerve impulses and conduction of impulses

The nervous system is responsible for short-term immediate communication and control between various body systems. In contrast to the **endocrine system** that achieves long-term control via chemical (hormonal) mechanisms, the nervous system relies on both chemical and electrical transmission of signals and commands.

Nerve cells (**neurons**) are specialized so that at one end there is a flared structure termed the dendrite. At the dendrite, the neuron is able to process chemical signals from other neurons and endocrine **hormones**. If the signals received at the dendrite end of the neuron are of a sufficient strength, and properly timed, they are transformed into action potentials that sweep down the neural cell body (axon) from the dendrite end to the other end of the neuron, the presynaptic portion of the axon that ends at the next **synapse** (the extra cellular gap between neurons)in the neural pathway. The arrival of the **action potential** at the presynaptic terminus causes the release of ions and chemicals (**neurotransmitters**) that travel across the synapse, the gap or intercellular space between neurons, to act as the stimulus to create another action potential in the next neuron, and thus perpetuate the neural impulse.

Nerve impulses are transmitted through the synaptic gap via chemical signals in the form of a specialized group of chemicals termed neurotransmitters. Neurotransmitters can also pass the neural impulse on to glands and muscles. Except where the neural synapses terminates on a muscle (neuromuscular synapse) or a gland (neuroglandular synapse), the synaptic gap is bordered by a presynaptic terminal portion of one neuron and the dendrite of the postsynaptic neuron.

As the action potential sweeps into presynaptic region, there is a rapid influx of **calcium** from the extra cellular fluid into a specialized area of the presynaptic terminus termed the synaptic knob. Via the process of exocytosis, specific neurotransmitters are then released from synaptic vesicles into the synaptic gap. The neurotransmitters diffuse across the synaptic gap and specifically bind to specialized receptor sites on the dendrite of the postsynaptic neuron.

Neurotransmitters are capable of exciting (creating an action potential) or inhibiting, the postsynaptic neuron. Excitation results from neurotransmitter driven shifts in ion balance that results in a depolarization. Inhibitory neurotransmitters generally work by inducing a state of hyperpolarization.

Excitatory neurotransmitters work by causing changes in sodium ion balance that, if the stimulus is strong enough (i.e., sufficient neurotransmitter binds to dendrite receptors) results in the postsynaptic neuron reaching threshold potential and the creation of an electrical action potential.

Excitation can also result from a summation of chemical neurotransmitters released from several presynaptic neurons that terminate on one postsynaptic neuron. In addition to such spatial control mechanisms, there are mechanisms that are time-dependent (temporal controls). Because neurotransmitters remain bound to their receptors for a time, excitation can also result from an increased rate of release of neurotransmitter from the presynaptic neuron.

Inhibitory neurotransmitters cause membrane changes that result in a movement of ions across the postsynaptic neural cell membrane that move the electrical potential away from the threshold potential.

Neural transmission across the synapse is, however, a result of a complex series of interactions that is far from the one-to-one presynaptic-postsynaptic neuron and many neurons can converge on a postsynaptic neuron.

Within the neuron, the mechanism of transmission is via the transmission of an electrical action potential that represents a change in electrical potential from the rest state the neuronal cell membrane. Electrical potentials are created by the separation of positive and negative ionic electrical charges that vary in distribution and strength on the inside and outside of the cell membrane. There are a greater number of negatively charged proteins on the inside of the cell, and an unequal distribution of positively charges cations both inside and outside the membrane.

The standing potential is maintained because, although there are both electrical and concentration gradients (a variation of high to low concentration) that induce the excess sodium cations to enter the cell and potassium cations to migrate out, the channels for such movements are normally closed so that the neural cell membrane remains impermeable or highly resistant to ion passage in the rest state.

The structure of the cell membrane and a physiological **sodium-potassium pump** maintain the neural cell resting **membrane potential** (RMP). Driven by an ATPase enzyme, a physiological sodium potassium pump moves three sodium cations from the inside of the cell for every two potassium cations that it moves back in. The ATPase is necessary because of this movement of cations against their respective resting concentration and electrical gradients.

Neural transmission, in the form of the creation of action potentials, results from sufficient electrical, chemical, or mechanical stimulus to the postsynaptic neuron that is greater than or equal to a threshold stimulus. The creation of an action potential is an "all or none" event and the level and form of stimulus must be sufficient and properly timed to create an action potential. When threshold stimulus is reached in the postsynaptic neuron, there is a rapid movement of ions and the resting membrane potential changes from −70mv to +30mv. This change of approximately 100mv is an action potential that travels down the neuron like a wave, altering the RMP in successively adjacent regions of neural cell membrane as it passes, until the action potential arrives at the presynaptic region of the axon to initiate the mechanisms of synaptic transmission.

Neural transmission is also subjected to refractory periods in which further excitation of the postsynaptic neuron is not possible. In addition, varying types of nerve fibers (e.g., myelinated or demyelinated) exhibit differences in how the action potential moved down the axon, or in the speed of transmission of the action potential.

See also Nerve impulses and conduction of impulses; Nervous system overview; Neural plexuses; Neurology; Reflexes

NERVOUS REFLEX • *see* REFLEXES

NERVOUS SYSTEM, EMBRYOLOGICAL DEVELOPMENT

The neural system develops from ectodermal cells, the outermost layer of embryonic germ cells. The covering of the neural system and the **blood** vessels that supply nervous **tissue** are mesodermal in origin.

Immediately after the formation of the embryonic disk of cells, there is a thickening in the **ectoderm** along the longitudinal axis to form the neural plate. The plate then folds along its long axis to form a groove. The sides of the groove are neural folds that fuse on the dorsal side to form a neural tube. A number of cells termed neural crest cells are pinched off the top of the primitive neural tube as it separates from the overlying ectodermal layer, The neural tube is divided into three basic layers, and the cell types within these layers take on the layer name. The outermost layer of cells in the marginal layer becomes marginal layer cells. The middle layer, known as the mantle layer gives rise to mantle cells. The innermost layer, the ependymal layer, forms ependymal cells.

The neural crest cells form the sensory ganglia of the **cranial nerves**. Neural crest cells also contribute to the formation of peripheral and **autonomic nervous system** ganglia.

The **brain** begins as a swelling at the cephalic end of the neural tube that ultimately will become the **spinal cord**. The neural tube is continuous and contains primitive cerebrospinal fluids. Enlargements of the central cavity (neural tube lumen) in the region of the brain become the two lateral, third, and forth ventricles of the fully developed brain. There are three regions that comprise the brain: the forebrain (also known as the prosencephalon), the midbrain (mesencephalon), and the hindbrain (rhombencephalon). The hindbrain consists of the **medulla oblongata**, **pons** and **cerebellum**.

The primitive brain is differentiated in several anatomical regions. The most cephalic region is the telencephalon. Ultimately, the telencephlon will develop the bilateral **cerebral hemispheres**, each containing a lateral ventricle, cortex (surface) layer of gray cells, a white matter layer, and basal nuclei. Caudal (later inferior) to the telecephalon is the diencephalon that will develop the **epithalamus**, **thalamus**, and **hypothalamus**. Caudal to the diencephalon is the mesencephalon, the midbrain region that includes the cerebellum and pons. Within the **myelencephalon** region is the medulla oblongata.

In contrast to the spinal cord, neural development inverts the **gray matter**, white matter relationship within the brain. The outer cortex is composed of gray matter, while the white matter (myelinated axons) lies on the interior of the developing brain. The forebrain contains the intellectual centers. Regions of the forebrain control attributes of personality, vision, speech, and hearing. Regions of the hindbrain become involved in autonomic functions including **breathing** and coordinated body movements. The midbrain is involved in body movement. The so-called pleasure center is located in the midbrain, which has been implicated in the development of addictive behaviors.

The **meninges** that protect and help nourish neural tissue form from **mesoderm** that surrounds the axis established by the primitive neural tube and **notochord**. The cells develop many fine **capillaries** that supply the highly oxygen demanding neural tissue.

Almost all the glial cells found within the **central nervous system** come from the mantle region of the neural tube (a special condensation of ectodermal cells). Schwann cells that form neural sheaths of some nerve fibers and of peripheral nerves form from neural crest cells and cells from the mantle. The cells migrate out to their ultimate location.

Within the autonomic nervous system, the sympathetic ganglionic chain forms from neural crest cells. Cells from the basal portion of the neural tube form the parasympathetic ganglia.

The **peripheral nervous system** that serves the limbs forms as the **limb buds** also form. Spinal nerves from the fifth cervical spinal nerve to the first thoracic spinal nerve fuse at the proximal end of the upper limb bud (the base end of the bud) to form a plexus. Ultimately, this combination of fusions and branches forms the **brachial plexus** that serves to innervate the upper limb, shoulder, and thorax.

Spinal nerves from the first lumbar to the third sacral nerve form two separate plexi for the **lower limb structure**. The lumbar plexi division into anterior and posterior segments allows for the separate innervation of flexor and extensor muscles in the thigh. The sacral plexus provides innervation to the remainder of the lower limbs.

See also Anatomical nomenclature; Basal nuclei; Brain stem function and reflexes; Brain, intellectual functions; Cerebral

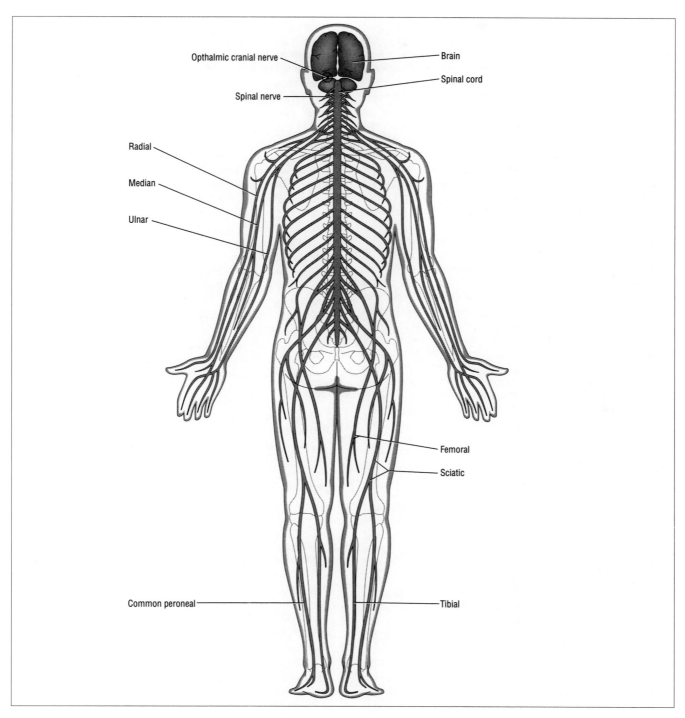

The nervous system. *Illustration by Argosy Publishing.*

morphology; Embryonic development, early development, formation, and differentiation; Human development (timetables and developmental horizons); Nervous system overview; Nervous system, embryological development; Organizer experiment; Sense organs: Embryonic development

NERVOUS SYSTEM OVERVIEW

The nervous system is responsible for short-term immediate control of the human body and for communication between various body systems. Although the **endocrine system** achieves long-term communication and control via chemical (hormonal) mechanisms, the nervous system relies on a faster method of alternating chemical and electrical transmission of signals, and commands through a network of specialized neural cells (**neurons**).

There are three differing types of neurons including sensory neurons, neurons associated with transmission of impulses, and effecter neurons such as motor neurons that transmit nerve impulses to specialized tissues (e.g., motor neurons to muscle **tissue**) and glands. In addition to neurons, there are a number of cell types that play a supportive role in the nervous system. Principal among these neuron-supporting cells are Schwann cells associated with an insulating myelin sheath that wraps around specific types of neural fibers or tracts.

Neurons contain key common components. At one end, the dendrite end, specialized cell processes and molecular receptor sites bind **neurotransmitters** released by other neurons and sensory **organs** across a gap known as the neural **synapse**. At the dendrite, the nerve impulse within a particular neuron is generated by a series of chemical and electrical events associated with the binding of specific neurotransmitters. The nerve impulse then travels down the neuron cell body, the axon, via an electrical **action potential** that results from rapid ion movements across the neuron's outer cell membrane. Ultimately, the action potential reaches the presynaptic terminus region where the electrical action potential causes the release of cell specific neurotransmitters that diffuse across the synapse (the gap between neurons) to start the impulse generation and conduction sequence in the next neuron in the neural pathway. The major chemical neurotransmitters include acetylcholine, norepinephrine, dopamine, and serotonin.

Neural transmission and the diffusion of neurotransmitters across the synapse do not always produce a subsequent action potential without the combined input of other neurons in a process termed summation. Depending on the specific neurotransmitters, receptor binding can produce either excitation or inhibition of action potential production. Subject to a refractory period during which a neuron returns to its normal state following the production of an earlier action potential, once the a neuron reaches a properly timed threshold stimulus, it will produce an action potential. The production of action potentials is an "all or none" process and once produced, the axon potential (nerve impulse) sweeps down the axon.

The nervous system is organized along morphological (structural) and functional lines. Structurally, the nervous system can be dived into the **central nervous system** (CNS) that includes the **brain** and **spinal cord**, and the **peripheral nervous system** (PNS) that contains all other nerves (e.g., sensory and motor neurons), ganglia, and associated cells.

The central nervous system is protected by a tri-fold layer of specialized membranes, termed the **meninges**. The brain and spine are organizationally reversed. The spinal cord contains **gray matter** tracts surrounded by white matter. In contrast, the brain contains centralized white matter.

Functionally, the nervous system can be divided into the somatic or voluntary nervous system (VNS) that coordinates voluntary muscles and **reflexes**, and the **autonomic nervous system** (ANS) associated with the regulation of viscera, **smooth muscle**, and cardiac muscle. The autonomic nervous system is further subdivided into sympathetic and parasympathetic systems.

The **sympathetic nervous system** (SNS), when related to the classic "fight or flight" response, heightens activity in bodily organs or systems (e.g., the **respiratory system**) and the metabolic rate (the rate at which energy is consumed by bodily processes such as respiration). In contrast, the **parasympathetic nervous system** (PNS) lowers response and decreases the metabolic rate. The sympathetic and parasympathetic systems work in opposition to control bodily systems.

The brain is divided into various areas or lobes. The large left and right anterior lobes represent the convoluted (wrinkled) **cerebral cortex** or cerebrum. Posterior lobes represent the **cerebellum**. At the top of the spinal cord lies the **pons** and medulla. The cerebellum, pons, and medulla together are referred to as the hindbrain and are associated with many basic process involved in body maintenance, **metabolism** (e.g. **breathing** and **heart** rate) and homeostasis. In general, the forebrain (the cerebrum and some related areas) is the area responsible for higher intellectual functions involved in sensory interpretations, memory, language, and learning. The midbrain tract act as switching systems that direct, coordinate, and integrate impulses among various regions of the brain.

Within the peripheral nervous system, **mechanoreceptors**, most of which are located in the skin (integumentary system) respond to physical stimuli such as pressure and motion. Thermoreceptors are specialized to respond to changes in temperature. Chemoreceptors associated with **taste** and **smell** senses respond to specific molecules. Highly developed complex sensory structure such as the eyes and ears respond to light (electromagnetic radiation) and sound.

In addition to a complex network of nerves throughout the body that act as a transmission system, the PNS contains specialized nerve cells to interface and transmit signals to muscles and glands.

Nerves usually contain neuron cell bodies that lie in tracts or fibers. Unmyelinated axons form gray matter. When Schwann cells wrap around the axon they create a myelin sheath around neurons (in the peripheral nervous system) that in tracts or fibers are termed white matter. Because the myelin sheath disrupts the normal transmission of the electrical action potential down the neuron, a specialized form of conduction of the nerve impulse or action potential occurs between spaces in the myelin sheath termed the nodes of Ranvier. Accordingly, diseases that disrupt or destroy the myelin sheath (demyelinating diseases) can impair or destroy normal nerve function.

Schwann cells are only one form of neuroglia or glial cells that are required to support normal neural function. Other glial cells include astrocytes, microglia, ependymal cells, oligodendrocytes, and satellite cells. Astrocytes are necessary for the proper vascularization of nerve cells and for the transport of nutrients and the removal of cellular waste products across the **blood** brain barrier. Microglia cells engage in phagcytosis and are capable of helping defend neural cells from attacks by a range of pathogenic agents. Ependymal cells line brain and spinal ventricles (fluid filled cavities in the brain and spine) and produce and maintain **cerebrospinal fluid**. Oligodendrocytes are responsible for the production of the myelin sheath in the CNS. Satellite cells protect neurons in ganglia.

See also Anatomical nomenclature; Anesthesia and anesthetic drug actions; Basal ganglia; Biofeedback; Brachial plexus; Brain stem function and reflexes; Brain: intellectual functions; Cerebral hemispheres; Cranial nerves; Depressants and stimulants of the central nervous system; Drug effects on the nervous system; Electroencephalograph (EEG); Gray matter and white matter; Handedness; Hearing (physiology of sound transmission); Hippocampus; Homeostatic mechanisms; Hypothalamus; Limbic lobe and limbic system; Metathalamus; Metencephalon; Muscular dystrophies; Muscular innervation; Myelencephalon; Nerve impulses and conduction of impulses; Nervous system: embryological development; Neural damage and repair; Neural plexuses; Neurology; Pain; Paraganglia; Reciprocal innervation; Referred pain; Reflexes; Reticular activating system (RAS); Sense organs: balance and orientation; Sense organs: embryonic development; Sense organs: ocular (visual) structures; Sense organs: olfactory (sense of smell) structures; Sense organs: otic (hearing) structures; Sneeze reflex; Sodium-potassium pump; Spinal nerves and rami; Touch, physiology of; Vegetative functions; Vision: histophysiology of the eye

NEURAL DAMAGE AND REPAIR

Nerve injuries may occur in the **central nervous system** (CNS) due to trauma or disease, but they are more frequent in peripheral nerves. Peripheral neural damage may be caused by perforations, cuts, stretch, traction, and **burns**. The nerve cell (neuron), does not proliferate through mitosis as do other somatic cell types in other tissues; when its axon is cut from the cell, it is swallowed by macrophages. However, the remaining segment of the axon that is connected to the cell starts to grow and is involved with the proliferating Schwann cells present in the nervous fiber membrane, thus forming a new axonal tube. The Schwann cells provide the newly formed axonal tube with layers of covering myelin that are necessary for synaptic transmission and nervous reflex. Therefore, nerves torn apart by cuts, perforation, or partially lacerated by excessive stretching or traction accompanied by simple bone fractures, may have a rate of spontaneous recovery ranging between 65–85%. Nerve damage from these types of injury heals mostly within four months. Conversely, more seriously damaged fibers must require surgical intervention, as in the case of extensive cuts or perforations that completely sever a peripheral nerve. Severe burns usually destroy nerves involved in motor and sensory functions at the affected area in a manner beyond repair. Accidentally severed fingers, toes, and even limbs, when quickly preserved in ice and surgically reattached in a few hours, may offer a satisfactorily level of

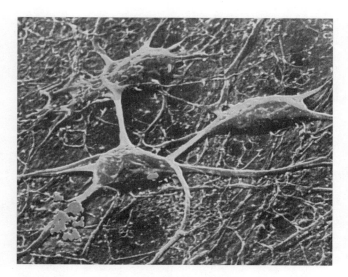

Scanning electron micrograph (SEM) showing three nerve cells of the human branching fibers (dendrites). *Photograph by Secchi-Lecague/Roussel-UCLAF/CNRI/Science Photo Library. National Audubon Society Collection/Photo Researchers, Inc. Reproduced by permission.*

neural repair, due to the advancements of modern neuro-microsurgery techniques.

Injuries in the central nervous system (CNS) structures, such as the **spinal cord** and the **brain**, may occur in consequence of either diseases or several other factors, such as accidents, radiotherapy, exposure to neurotoxins, inadequate body **posture**, occupational straining, or excessive sport exercise. Some initially minor spinal cord injuries may trigger a chronic inflammatory process that leads to a further gradual destruction of nerve **tissue** due to progressive death of **neurons**, hence amplifying the damage. CNS injuries represent a formidable challenge to medical sciences and no satisfactory treatment is available so far for severed spinal cord or brain tissue destruction, because central nerve cells do not regenerate within distances bigger than 1 mm. In the brain, myelin is provided by a different class of cells, the oligodendrocytes that do not proliferate as do the Schwann cells of the **peripheral nervous system**. Researchers around the world are experimenting to identify genes involved in nerve regeneration and other possible solutions for these so far irreversible lesions. Techniques under investigation include transplant of embryonic neurons, *in vitro* cultivation of central nervous cells with nerve growth factor, **gene therapy** with integrins, and injection of **stem cells**. One interesting finding is the role of a family of genes, the Bcl, in neuronal repair. Two members of this family, Bcl-2 and Bcl-3, are active in the only part of the brain where axon regrowth is observed, the dentate gyrus of the **hippocampus**. Bcl-2 and its product (a protein with the same name) are also highly expressed in embryonic neurons, but they become silent in mature brain neurons, with the only known exception of the hippocampal dentate gyrus. However, Bcl-2 remains active in the adult cells of the peripheral nervous system (PNS), what points to the possibility of developing gene therapies promoting the expression of this gene to induce axonal regrowth in the CNS.

See also Histology and microanatomy; Inflammation of tissues; Nerve impulses and conduction of impulses; Nervous system: embryological development; Radiation damage to tissue

NEURAL PLEXI

Central and peripheral **neurons** and neural fibers that are responsible for the innervation of visceral tissues, such as smooth muscles of internal **organs** and **blood** vessels, glands, **heart**, and limbs, belong to the autonomic nervous system. Nerve plexi (plural of nerve plexus) may also be constituted by one or more types of nerve components, such as somatic sensory and/or motor, with sympathetic, parasympathetic and vagal nerves taking part in the network. The sympathetic nerves are responsible for visceral **reflexes**. They originate in the ganglia that constitute the sympathetic chain located in the **spinal cord**, whose nerves radiate outward, forming many neural plexi or neural networks in many different tissues.

The carotid plexus is connected to the sympathetic cervical ganglia, and it is constituted by a network of nerves responsible for vasoconstriction, motor and secretion stimuli, innervating the internal and external **carotid arteries**, the intra cranial vessels, the lachrymal glands, the pituitary gland, and the sudoriparous glands of the head. Nerves of this plexus also control the dilation of pupils. The cardiac plexus contains vagal and sympathetic nerves, and is located at the base of the heart, with its superficial portion innervating the coronary **arteries**, and its deep portion innervating the cardiac muscles.

The pulmonary plexus innervate the **bronchi**, pulmonary arteries, and pulmonary tissues. The submucous plexus, also known as Meissner's plexus, innervate the extra hepatic bile ducts, the esophagus, and portions of the stomach, the colon, and the intestines where they control the peristaltic movements and gastric secretions. The celiac plexus contains mesenteric ganglia and controls the tonus of the sphincters in the digestive tubes and extra-hepatic biliary ducts, the vasoconstriction of abdominal visceral blood vessels, as well as the secretion of adrenal glands. The celiac plexus also contains the Auerbach's plexus, or myenteric plexus, located between the circular and longitudinal muscular layers of the digestive tube. The hypogastric plexi inhibit peristaltic movements and secretion of the descending colon, sigmoid, rectocolon, urethra, and innervate the bladder and increases the tonus of the anal and bladder internal sphincters. The hypogastric plexus also controls **smooth muscle** contraction in the female and male genital organs.

See also Bile, Bile ducts, and the biliary system; Brachial system; Cardiac muscle; Heart: rhythm control and impulse conduction; Lungs; Muscle contraction; Muscular innervation; Nerve impulses and conduction of impulses; Peripheral nervous system; Spinal nerves and rami; Sweat glands

NEUROLOGY

Neurology is the study of the nervous system and its components. Neuroscience may involve the nervous systems of humans and higher animals as well as simple multicellular nervous systems, or investigate nervous phenomenon at the cellular, organelle, or molecular level.

Neuroscience principally originated with three European scientists working at the end of the nineteenth and the beginning of the twentieth century. **Camillo Golgi**, an Italian physician, perfected a vital laboratory technique that first allowed scientists to trace the workings of the nervous system. Golgi completed a medical degree at the University of Padua in 1865, and then became a medical researcher at the University of Pavia. He was interested in cells and tissues, and experimented with ways to stain cells so they could be seen. Researchers before him had prepared cells with organic dyes, but Golgi found that staining with silver salts gave much clearer results. He became fascinated with nerve **tissue**, and using his staining process, he was the first to see in fine detail how this tissue was organized. Golgi proved that the fibers of nerve cells did not meet completely, but left a gap, now known as a **synapse**. He devoted his life to mapping the structure of the nervous system. Golgi's work was furthered by a Spanish medical researcher, **Santiago Ramón y Cajal**. Ramón y Cajal, working at the University of Zaragoza, first improved on Golgi's staining method, then used it to discover the connection between the **gray matter** in the **brain** and the **spinal cord**. He shared the Nobel Prize in physiology or medicine with Golgi in 1906.

Golgi and Ramón y Cajal established the **anatomy** of the nervous system. The English neurologist Charles Scott Sherrington is credited with founding modern neuroscience with his work on the functioning of the nervous system. In other words, he brought the science from describing what the nervous system was to showing how it worked. His research explored the brains ability to sense position and **equilibrium**, and the reflex actions of muscles.

Many other researchers continued to explore the workings of the nervous system. As laboratory **imaging** techniques progressed, neuroscientists were able to look at nerve cells at the molecular level. This allowed scientists to map the growth of nerve cells and nerve networks, and to study how individual cells process, store, and recall information. Working with living brains to explore nerve function was all but impossible until the late 1970s, when sophisticated brain imaging machines were first developed. Positron emission tomography (PET) revolutionized neuroscience by allowing scientists to produce pictures of a working brain. Since then, scientists and engineers have come up with even better brain imaging systems, such as functional magnetic resonance imaging (fMRI). Using fMRI, neuroscientists can detect increases in **blood** oxygenation during brain function, and this shows which areas of the brain are most active. Brain activity occurs very quickly—nerve cells can respond to stimulus within 10 milliseconds—and very sophisticated equipment is needed to capture such fleeting movements. So-called neuroimaging is one of the hottest fields in neuroscience, as neurologists and technicians work together to find new ways of recording nerve action. Researchers in the late 1990s explored ways to map the flux of sodium ions within the brain, giving a direct record of neural activity, or to measure the scattering of light by brain tissues with fiber-optics. Both these techniques hope to give a more precise picture of which areas of the brain become active when a person thinks.

See also Nerve impulses and conduction of impulses; Nervous system overview; Nervous system: embryological development; Neural damage and repair; Neurons; Neurotransmitters

NEUROMUSCULAR JUNCTION · *see* MUSCULAR INNERVATION

NEURONS

The basic cellular unit of the nervous system is the neuron. The unique morphological and intercellular structure of the neuron is dedicated to the efficient and rapid transmission of neural signals. Within the neuron, the neural signal travels electrically. At the **synapse**, the gap between neurons, neural signals are conveyed chemically by a limited number of chemicals termed **neurotransmitters**. Specialized parts of the neuron facilitate the production, release, binding, and uptake of these neurotransmitters.

Although there are variations related to function, a typical neuron consists of dendrites (also termed dendritic processes), a cell body, an axon, and an axon terminus.

Dendrites are the (filamentous) terminal portions of neuron that bind neurotransmitter chemicals migrating across the synaptic gaps separating neurons. Depending on the type and function of a particular neuron, neurotransmitters may cause or inhibit the transmission of neural impulses. The cell body contains the cell nucleus and a concentration of cellular organelles. The cell body is the site of the normal metabolic reactions that allow the cell to remain viable. Neurotransmitters synthesized within the cell body are transported to the axon terminus by microfilaments and microtubules.

The axon is a cytoplasmic continuation of the cell body specialized for the electrical conduction of neural signals. The axon may be long—up to a yard in length in humans—or short, depending upon the neuron's position and function. The cell membranes of the neural axon transmit neural signals via changes in action potentials that sweep down the membrane.

At the junction of the cell body and axon is a region termed the axon hillock. At the axon hillock, chemical signals received by the dendrites may reach a threshold level to cause a wave of electrical depolarization and hyperpolarization of the axon cell membrane. The net movements of ions across the cell membrane are responsible for these changes that move down the axon to the axon terminus as an **action potential**.

At the axon terminus, neurotransmitters are released into the synaptic gap. Through synaptic gaps, a typical neuron

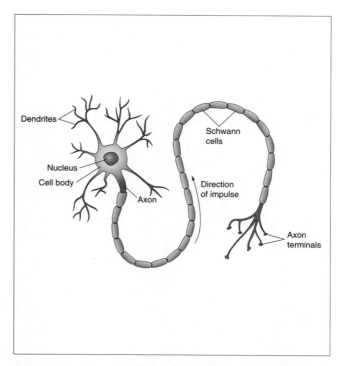

Anatomy of a neuron. *Illustration by Frank Forney. Reproduced by permission*

may interconnect with thousands and tens of thousands of other neurons.

Multipolar neurons have many processes and serve principally as motor neurons. Motor neurons, efferent because they conduct impulses away from the **central nervous system** (the **brain** and **spinal cord**) regulate the function of muscles and glands. Afferent neural pathways that send signals to the (CNS) are generally composed of unipolar neurons. Unipolar neurons also serve as sensory neurons—their filamentous dendritic processes exposed and elaborated into or connected to sensory receptor cells. Interneurons are neurons that connect neurons along a neural pathway.

See also Autonomic nervous system; Nervous system: embryological development; Neural damage and repair; Nerve impulses and conduction of impulses; Nervous system overview; Neurology; Parasympathetic nervous system; Reflexes

NEUROTRANSMITTERS

Neurotransmitters are chemicals released in minute amounts from the terminals of nerve cells in response to the arrival of an **action potential**. There are now more than 300 known neurotransmitters and they act either locally in point-to-point signal transmission (e.g., the motor nerve of a neuromuscular junction) or at a distal site (e.g., the hypothalamic releasing **hormones** acting on the anterior pituitary). Locally acting neurotransmitters relay the electrical signal travelling along a neuron as chemical information across the neuronal junction, or **synapse**, that separates one neuron from another neuron or a muscle. **Neurons** communicate with peripheral tissues, such as muscles, glands etc., or with each other largely by this chemical means rather than by direct electrical transmission.

Neurotransmitters are stored in the bulbous end of the nerve cell's axon. When an electrical impulse travelling along an axon reaches the junction, the neurotransmitter is released and diffuses across the synaptic gap, a distance of as little as 25 nm or as great as 100 μm. The interaction of the neurotransmitter with the postsynaptic receptor of the target cell generates either an excitatory postsynaptic potential (EPSP) or an inhibitory postsynaptic potential (IPSP). Transmitters that lead to EPSPs appear to open large, non-specific membrane channels, permitting the simultaneous movement of Na^+, K^+ and Cl^-. IPSPs are caused by Cl^- flux only.

Neurotransmitters include such diverse molecules as acetylcholine, noradrenalin, serotonin, dopamine, γ-aminobutyric acid, glutamate, glycine and numerous other small monoamines and **amino acids**. There are also small peptides, which appear to act as chemical messengers in the nervous system. They include substance P, vasopressin, **oxytocin**, **endorphins**, angiotensin and many others. A rather unusual but interesting neurotransmitter is the gas nitric oxide. This diverse range of chemical neurotransmitters may suggest that chemical coding could play as important a part in communication between neurons as do the strict point-to-point connections of neural circuitry.

Acetylcholine is one of the neurotransmitters functioning in the **peripheral nervous system**. It is released by all motor nerves to control skeletal muscles and also by autonomic nerves controlling the activity of **smooth muscle** and glandular functions in many parts of the body. Norepinephrine is released by sympathetic nerves controlling smooth muscle, cardiac muscle and glandular tissues. In these tissues acetylcholine and norepinephrine often exert diametrically opposed actions.

The neurotransmitters used by the majority of fast, point-to-point neural circuits in the **central nervous system** (CNS) are amino acids. Of these, the inhibitory substance γ-aminobutyric acid (GABA) is well characterized and it is present in all regions of the **brain** and **spinal cord**. GABA rapidly inhibits virtually all CNS neurons when applied locally by increasing cell permeability to chloride ions, thus stabilizing resting **membrane potential** near the chloride **equilibrium** level. Although GABAergic neurons also exist in the spinal cord, another inhibitory amino acid, glycine, predominates in this region of the CNS. Glycine is present in small inhibitory interneurons in the spinal cord **gray matter** and mediates the inhibition of most spinal neurons. The amino acids L-glutamate and L-asparagine depolarise neurons by activating membrane sodium channels and are ubiquitously distributed, appearing as the most common excitatory transmitters for interneurons in the CNS.

In contrast to the point-to-point signalling in which amino acids are involved, the monoamines are mainly associated with the more diffuse neural pathways in the CNS. The monoamines are present in small groups of neurons, primarily located in the brain stem, with elongated and highly branched axons. These diffuse ascending and descending monoaminergic innervations impinge on very large terminal fields and there is evidence that the monoamines may be released from many points along the varicose terminal networks of monoaminergic neurons. Most monoamines released in this way occur at nonsynaptic sites and a very large numbers of target cells may be affected by the diffuse release of these substances, which are therefore thought to perform modulatory functions of various types.

One of the most remarkable developments was the realization that most peptide hormones of the endocrine and neuroendocrine systems also exist in neurons. These are by far the largest group of potential chemical messengers. For example, the opioid peptides (endorphins) have attracted enormous interest because of their morphine-like properties. They are consequently of considerable interest in the understanding of **pain**. Endorphins represent a family of chemical messengers found in all regions of the CNS including the pituitary (e.g., beta-endorphin and dynorphin) and the peripheral enteric nervous system. Their presence in regions such as the **basal ganglia** and the **eye retina**, where it is unlikely that they have any connection with pain pathways, suggests that they may also have other diverse functions. There is still much to be learnt about the possible functions of neuropeptides in the CNS. In all cases so far examined the peptides seem to be capable of being released by a specialized secretory mechanism from stimulated CNS neurons. They can exert powerful effects on the CNS. For example, the direct administration of small amounts of peptide to the brain can elicit a variety of behavioural responses, including locomotor activity (substance P), analgesia (endorphins), drinking behavior (angiotensisn II), female sexual behavior (LHRH) and improved retention of learned tasks (vasopressin).

An interesting and novel neurotransmitter identified in the 1980s is nitric oxide (NO). This is a highly reactive naturally occurring gas generated in the body from arginine and has the alternative name "epithelium-derived-relaxing factor". Synthesis of NO in **blood** vessel epithelia occurs in response to the distortion of blood vessels by blood flow. The gas then rapidly diffuses into the surrounding muscle layers causing them to relax. It, therefore, has vasodilatory (dilation of blood vessels) properties and as a neurotransmitter occurs in a number of nerve networks. For example, it is known to be active in the dilation of **arteries** supporting the penis and in the relaxation of muscles of the corpora cavernosa (the two chambers filled with spongy **tissue** which run the length of the penis). NO released from stomach nerves causes the stomach to relax in order to accommodate food. Intestinal nerves also induce the relaxation of the intestinal muscle by releasing NO. In addition, nervous activity in the **cerebellum** is increased by NO and it appears that NO is an important neurotransmitter associated with memory. Despite its usefulness, nitric oxide can have a toxic effect on body cells and has been implicated in Huntington's disease and **Alzheimer's disease**.

See also Hormones and hormone action; Nervous system overview; Neurology; Pituitary gland and hormones

NICOLLE, CHARLES JULES HENRI (1866-1936)

French bacteriologist

Charles J. H. Nicolle, the recipient of the 1928 Nobel Prize in physiology or medicine, was recognized by the Swedish Academy for his research into the cause of typhus, a severe and widespread disease during the early twentieth century. Nicolle's discovery that typhus is transmitted by the human body louse—and therefore can be readily prevented—was of great benefit to both military and civilian medicine.

Born September 21, 1866, in Rouen, France, Charles Jules Henri Nicolle was the son of physician Eugène Nicolle. Charles's father was a medical doctor at the municipal hospital, as well as a professor of natural history at the École des Sciences et des Art. Encouraged by his brother, the noted bacteriologist Maurice Nicolle, Charles took a course in bacteriology at the Institute Pasteur in Paris, studying under the renowned bacteriologists, Émile Roux and Eacute;lie Metchnikoff. For his doctoral dissertation, Nicolle investigated the bacterium then called Ducrey's Bacillus (also known as Hemophilus ducreyi), the causative agent of soft chancre, a type of venereal disease.

Charles took his medical degree in 1893 in Paris, then returned to Rouen for a staff position in a hospital. Shortly thereafter, he married and the couple's two sons would eventually become physicians. Unable to develop a major biomedical research center in Rouen as he desired, Nicolle agreed in 1902 to assume the directorship of the Institute Pasteur in Tunis, Tunisia. For the remainder of his life, Nicolle lived and worked primarily in Tunis with occasional lecturing in Paris.

Affiliated with the original Institute Pasteur (which was founded in Paris in 1888), the institute in Tunis was basically an organization in name only. Over the years to come, however, Nicolle improved a run-down antirabies vaccination unit into a leading center for the study of North African and tropical diseases. It was in Tunis where Nicolle accomplished his groundbreaking work on typhus. He became intrigued by the observation that an outbreak of typhus did not seem to take hold in hospital wards as it did among the general populace of the city. Although the contagion infected workers who admitted patients into the hospital, it did not affect other patients or attendants in the actual wards. Those who collected or laundered the dirty clothes of newly admitted patients typically came down with the disease.

Realizing that the washing, shaving, and providing of clean clothes to the new patient was possibly the key to the pattern of **infection**, Nicolle initiated a series of experiments in 1909 to confirm his suspicion of the arthropod-borne nature of typhus. He theorized that lice, which attached themselves to the bodies and clothes of human beings, transmitted the disease, so he began his investigation by infusing a chimpanzee with human **blood** infected with typhus, then transferred the chimpanzee's blood to a healthy macaque monkey. When the fever and rash of typhus was seen on the monkey, Nicolle placed twenty-nine human body lice obtained from healthy humans on the skin of the macaque. These lice were later placed on the skin of a number of healthy monkeys, which all contracted the disease.

Once Nicolle isolated the relationship between typhus and the louse, preventative measures were established to counter unsanitary conditions. Nevertheless, the trenches of World War I remained major breeding places for the louse and typhus killed an enormous number of soldiers on all sides of the conflict. The development of the insecticide DDT by **Paul Müller** in 1939 was the most effective prophylactic against typhus, nearly eradicating the disease among soldiers during World War II.

Nicolle is also responsible for other important contributions to the science of bacteriology. Stemming from his research into typhus was his recognition of a phenomenon known as "inapparent infection," a state in which a carrier of a disease exhibits no symptoms. This theoretical discovery suggested how diseases survived from one epidemic to another.

Nicolle, along with a variety of other colleagues over time, also researched African infantile leishmaniasis, which affected humans, and a related disease in dogs. Another significant discovery concerned the role of flies in the transmission of the blinding disease trachoma. For these and other works, Nicolle received the French Commander of the Legion of Honor and was named to the French Academy of Medicine. In 1932 he became a professor in the College de France.

Besides his work in science, Nicolle was an accomplished literary figure, having published several novels. His scientific writings include five major books as well as numerous articles. Nicolle died on February 28, 1936 in Tunis.

See also Antigens and antibodies; Bacteria and responses to bacterial infection; Hemorrhagic fevers and diseases; Integument; Immune system; Malaria and the physiology of parasitic inflections

NICOTINE ACTIONS AND THE PHYSIOLOGY OF SMOKING

Nicotine is a naturally occurring compound that is classified as a liquid alkaloid. An alkaloid is an organic compound made up of carbon, hydrogen, nitrogen, and sometimes oxygen (nicotine lacks oxygen). Another example of an alkaloid is caffeine. Both caffeine and nicotine are similar in that they provide stimulation upon ingestion, a factor that encourages their habitual use.

Nicotine can gain entry to the bloodstream by several routes. Tobacco can also be applied as a wad to the inside of the mouth, enabling nicotine to diffuse across the mucous membranes of the gums to the bloodstream. As a component of a cigarette, nicotine can be inhaled into the **lungs**. There, it encounters the **alveoli**, diffuses across the alveolar membrane, and enters the bloodstream. Once in the bloodstream nicotine travels to the **brain** and then is delivered throughout the body.

The physiological effects of nicotine—both the good feelings and the irritability upon nicotine deprival—are due to

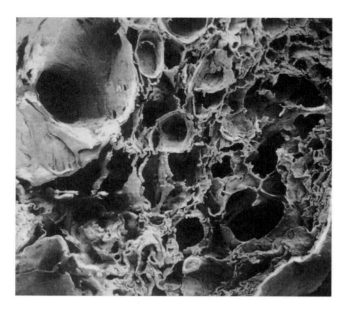

Autopsy photo depicts extent of damage to lung tissue in patient with smoking related emphysema. *Photograph by Hossler. Custom medical Stock Photo. Reproduced by permission.*

its effects on the brain. Nicotine reaches the brain within 10–15 seconds of entering the bloodstream. Initially, it causes the rapid release of adrenaline, the hormone that elicits a **"fight or flight" response**. Adrenaline causes the **heart** rate and **blood** pressure to increase and **breathing** to become rapid and shallow. Also glucose is released into the blood and, at the same time, the release of insulin can be blocked. Because insulin stimulates cells to take up excess glucose, the net effect is to increase the concentration of glucose in the blood. This increased sugar and the elevation of the metabolic rate—the rate at which nutrients are converted to energy—can act to produce a weight loss in smokers. However, this healthy effect is more than compensated for by the damage to the body from smoking. Also, evidence is conclusive now that long-term intake of nicotine increases the level of the so-called "bad" cholesterol, LDL, that damages **arteries**.

The target of nicotine in the brain are **neurons**. Neurons are the fundamental cells of the nervous system. Neural impulses move from neuron to neuron, even though the neurons may not physically contact one another. Chemicals called **neurotransmitters** provide the bridge. One such neurotransmitter is acetylcholine. Nicotine acts by occupying a space on a neuron that would otherwise be a binding site for acetylcholine. The binding of nicotine also causes the neuron to "fire" (i.e., to create an **action potential** that travels down the neuron). In contrast to the tightly regulated quantities of acetylcholine in the brain, nicotine levels are not regulated. Thus, upon exposure, brain activity increases in a haphazard fashion throughout the brain.

The global stimulation of brain activity produces an alert feeling. Also, the "reward center" of the brain is stimulated, producing a pleasant and happy feeling (for example, the hormone endorphin is released in greater quantity, produc-

ing the euphoric "runner's high") that encourages the repeating of the behavior that invoked the response. The need to take in nicotine can become addictive.

In response to the continued presence of nicotine, the neurons increase the number of neuroreceptors for nicotine. When nicotine use is abruptly stopped, these physiological adaptations remain. As a result, the body cannot function in the same way it did in the presence of the drug. For about a month, until the neuroreceptors readjust, feelings of irritability, anxiety, depression and craving for nicotine will be present. Indeed, of the millions of people who try and stop smoking each year, only 10% are successful.

See also Central nervous system (CNS); Drug effects on the nervous system; Drug interactions

NIRENBERG, MARSHALL WARREN (1927-)

American biochemist

Marshall Nirenberg is best known for deciphering the portion of **DNA (deoxyribonucleic acid)** that is responsible for the synthesis of the numerous protein molecules, which form the basis of living cells. His research has helped to unravel the DNA **genetic code**, aiding, for example, in the determination of which genes code for certain hereditary traits. For his contribution to the sciences of **genetics** and cell **biochemistry**, Nirenberg was awarded the 1968 Nobel Prize in physiology or medicine with **Robert W. Holley** and **Har Gobind Khorana**.

Nirenberg was born in New York City, and moved to Florida with his parents, Harry Edward and Minerva (Bykowsky) Nirenberg, when he was ten years old. He earned his B.S. in 1948 and his M.Sc. in biology in 1952 from the University of Florida. Nirenberg's interest in science extended beyond his formal studies. For two of his undergraduate years he worked as a teaching assistant in biology, and he also spent a brief period as a research assistant in the **nutrition** laboratory. In 1952, Nirenberg continued his graduate studies at the University of Michigan, this time in the field of biochemistry. Obtaining his Ph.D. in 1957, he wrote his dissertation on the uptake of hexose, a sugar molecule, by certain tumor cells.

Shortly after earning his Ph.D., Nirenberg began his investigation into the inner workings of the genetic code as an American Cancer Society (ACS) fellow at the National Institutes of Health (NIH) in Bethesda, Maryland. Nirenberg continued his research at the NIH after the ACS fellowship ended in 1959, under another fellowship from the Public Health Service (PHS). In 1960, when the PHS fellowship ended, he joined the NIH staff permanently as a research scientist in biochemistry.

After only a brief time conducting research at the NIH, Nirenberg made his mark in genetic research with the most important scientific breakthrough since **James D. Watson** and **Francis Crick** discovered the structure of DNA in 1953. Specifically, he discovered the process for unraveling the code of **deoxyribonucleic acid (DNA)**. This process allows scientists to determine the genetic basis of particular hereditary traits. In August of 1961, Nirenberg announced his discovery during a routine presentation of a research paper at a meeting of the International Congress of Biochemistry in Moscow.

Nirenberg's research involved the genetic code sequences for **amino acids**. Amino acids are the building blocks of protein. They link together to form the numerous protein molecules present in the human body. Nirenberg discovered how to determine which sequences patterns code for which amino acids (there are about 20 known amino acids).

Nirenberg's discovery has led to a better understanding of genetically determined diseases and, more controversially, to further research into the controlling of hereditary traits, or genetic engineering. For his research, Nirenberg was awarded the 1968 Nobel Prize in physiology or medicine. He shared the honor with scientists Har Gobind Khorana and Robert W. Holley. After receiving the Nobel Prize, Nirenberg switched his research focus to other areas of biochemistry, including cellular control mechanisms and the **cell differentiation** process.

Since first being hired by the NIH in 1960, Nirenberg has served in different capacities. From 1962 until 1966 he was Head of the Section for Biochemical Genetics, National Heart Institute. Since 1966 he has been serving as the Chief of the Laboratory of Biochemical Genetics, National Heart, Lung and **Blood** Institute. Other honors bestowed upon Nirenberg, in addition to the Nobel Prize, include honorary membership in the Harvey Society, the Molecular Biology Award from the National Academy of Sciences (1962), National Medal of Science presented by President Lyndon B. Johnson (1965), and the Louisa Gross Horwitz Prize for Biochemistry (1968). Nirenberg also received numerous honorary degrees from distinguished universities, including the University of Michigan (1965), University of Chicago (1965), Yale University (1965), University of Windsor (1966), George Washington University (1972), and the Weizmann Institute in Israel (1978). Nirenberg is a member of several professional societies, including the National Academy of Sciences, the Pontifical Academy of Sciences, the American Chemical Society, the Biophysical Society, and the Society for Developmental Biology.

Nirenberg married biochemist Perola Zaltzman in 1961. While described as being a reserved man who engages in little else besides scientific research, Nirenberg has been a strong advocate of government support for scientific research.

See also Genetic code; Human genetics

NMR · *see* IMAGING

NON-STRIATED MUSCLE · *see* SMOOTH MUSCLE

NOREPINEPHRINE · *see* NEUROTRANSMITTERS

NOSE

The nose is a prominent structural element of the face that contains cells that are a part of the olfactory sensory systems (the sense of **smell**). The nose is also a portal or orifice to **respiratory system**. The nose plays an important role in both warming and filtering air before it enters the trachea, **bronchi**, and **lungs**.

Bilateral nostrils (external nares) form the external opening of the nose. Composed chiefly of skin and **cartilage**, the nostrils open into bilateral nasal cavities that are separated by a partition termed the nasal septum that is composed of cartilage and bone. Each of the nasal cavities is partially partitioned into three subcavities or nasal passages termed the superior, middle, and inferior nasal conchae. The passages are covered with mucous membranes and at the internal nares (openings) are continuous with the **pharynx** as the back of the nose. Because the nasal septum runs from the external to internal nares, any deviation (deviated speptum) or breakage can occlude the free flow of air during respiration.

The nasal passages are connected with smaller openings to the paranasal sinuses. Fluids that accumulate in the nasal sinuses drain through the openings into the nasal passages, eventually to be discharges out through the nostrils. Blockage of the passages lead to headaches and feelings of congestion.

The nasal passages are also continuous with openings to the **Eustachian tubes** that lead to the ears and which play an important role in equalizing pressure in the ear. Nasolacrimal ducts allow drainage of fluids from the **eye** region.

The mucous membrane linings of the nose are hair-like projections termed **cilia** that act to move **mucus** and other fluids from the nose. The cilia also play an important role in the filtering out of dust, dirt, and particles such as pollen from inhaled air. The increased surface area of the nasal passages also serves to warm and humidify inhaled air before the air flows down the trachea and into the bronchi and lungs. In addition to keeping these passages free of particulate obstructions, the nose serves to keep the more sensitive respiratory structures moist, and at a temperature that permits efficient gas exchange.

Each nasal cavity also contains a small area of olfactory sensory receptors. These area are not covered with cilia.

The linings of the nose are highly vascularized. The increased **blood** supply is important to the oxygen demanding olfactory receptors and aids in the warming of inhaled air. Because the **capillaries** are small, numerous, and close to the surface, even mild trauma to the nose may result in a nosebleed.

Colds, hay fever, or other **allergies** may cause a generalized inflammation of the mucous membrane linings of the nasal passages (rhinitis) and act to block the nasal passages.

See also Cilia and ciliated epithelial cells; Face, nose, and palate embryonic development; Sense organs: Olfactory (sense of smell) structures

NOSE, EMBRYONIC DEVELOPMENT · *see*
FACE, NOSE, AND PALATE EMBRYONIC DEVELOPMENT

NOTOCHORD

The notochord is a cylindrical column of cells that appears early in embryonic development. The notochord forms on the dorsal side of the embryo and establishes a bilateral longitudinal axis of development before the formation of the neural tube and spinal vertebrae. The formation of the notochord is a common and uniting factor of all chordate animals. In humans, cells forming the notochord later become part of the vertebrae, nasopharynx, and **skull**.

In the early developmental stages of all vertebrates, a notochord forms within the embryonic disk. The notochord lies between the **ectoderm** and **endoderm**. The rod of cells forms between the primitive knot and the prechordal plate of cells.

As **mesoderm** forms, aggregations (clumps or blocks) of cells, derived from mesoderm, form **somites** along the sides of the notochord. Depending on their level these mesodermal somites ultimately develop into **vertebra**, ribs, muscles, and dermal structures.

Formation of the notochord is critical to the proper formation and organization of the neural plate, neural folds, and subsequent development of the **brain** and other neural structures. One of the classic experiments in developmental biology, the **organizer experiment**, showed that in amphibian embryos, cells taken from the notochordal region of one embryo could induce neural structure formation when grafted on to a second embryo.

See also Embryology; Embryonic development: Early development, formation, and differentiation; Nervous system, embryological development; Human development (timetables and developmental horizons)

NUCLEAR MAGNETIC RESONANCE (NMR) · *see* IMAGING

NURSE, PAUL MAXIME (1949-)
English cell biologist

Paul Nurse is a researcher in cell biology and **biochemistry**, and is best known as the discoverer of CDK (cyclin dependent kinase), one of the key regulators of the **cell cycle**. Nurse studied biological science at the University of Birmingham, England, gaining a B.Sc. in 1970. He went on to postgraduate study at the University of East Anglia, England, receiving his Ph.D. in 1973 in cell biology and biochemistry. After postdoctoral work at the Universities of Bern, Edinburgh and Sussex (1973–1984) he moved to the International Cancer Research Fund Laboratory (ICRF) in London, where he stayed for three

years. From 1987–1993, he held a position as a professor at the University of Oxford, and in 1993, he returned to the ICRF, and since 1996 has been its director-general.

Nurse pioneered research on the regulation of the division of eukaryotic cells. His work with the simple eukaryotic yeast cell *Schizosaccharomyces Pombe (S. Pombe)* led to the discovery of the cdc2-gene. This gene produces proteins that play an important part in some stages of the cell life cycle, specifically the G1 and G2/M stages of **cell division**. Nurse went on to discover many other genes, proteins, and **enzymes** associated with the regulation of cell division, and most importantly was able to extend his work from yeast to other organisms, including humans. The isolation of the human variant of the cdc2-gene, cyclin dependent kinase (CDK), showed that the process of cell division has been highly conserved during evolution, and all eukaryotic celled organisms appear to use the same basic process. CDK drives the cell through the cell cycle by modifying other proteins. These discoveries offer fundamental insights into the nature of all creatures, and also have practical applications in areas such as cancer research, where the role of CDK in the onset and spread of cancer cells is being studied.

His research work has bought Nurse many honors. He became a Fellow of the Royal Society in 1989, and a foreign associate member of the United States National Academy of Science in 1995. Nurse has won many international awards, including the Canadian Gairdner Foundation International Award (1992), and the 1997 General Motors Cancer Research Foundation Alfred P. Sloan, Jr. Prize and Medal. Nurse was also a co-recipient of the 2001 Nobel Prize in medicine along with **Leland H. Hartwell** and **R. Timothy Hunt**, for their discoveries of key regulators of the cell cycle. The ultimate goal of his research at the ICRF is to extend the knowledge of the eukaryotic cell cycle, how the process is controlled, and how cell division generates two spatially organized cells from the original single cell.

See also Cell cycle and cell division; Cell differentiation; Cell membrane transport; Cell structure; Cell theory; Enzymes and coenzymes; Evolution and evolutionary mechanisms

NÜSSLEIN-VOLHARD, CHRISTIANE (1942-)
German geneticist

Christiane Nüsslein-Volhard, along with two American molecular biologists made important discoveries about how genes control the early development of embryos.

Christiane Nüsslein-Volhard was born in Magdeburg, Germany, the daughter of Rolf Volhard, an architect, and Brigitte (Hass) Volhard, a musician and painter. While few women of her generation chose scientific careers, Nüsslein-Volhard found that being female in a male-dominated field presented little in the way of an obstacle to her studies. She received degrees in biology, physics, and chemistry from Johann Wolfgang Goethe University in 1964 and a diploma in biochemistry from Eberhard-Karls University in 1968. In 1973, she earned a Ph.D. in biology and **genetics** from the University of Tübingen. Nüsslein-Volhard was married for a short time and decided to keep her husband's last name, as it was associated with her developing scientific career.

In the late 1970s, Nüsslein-Volhard finished post-doctoral fellowships in Basel, Switzerland, and Freiburg, Germany, and accepted her first independent research position at the European Molecular Biology Laboratory (EMBL) in Heidelberg, Germany. She was joined there by **Eric F. Wieschaus** who was also finishing his training. Because of their common interest in *Drosophila*, or fruit flies, Nüsslein-Volhard and Wieschaus decided to work together to find out how a newly fertilized fruit fly egg develops into a fully segmented embryo.

Nüsslein-Volhard and Wieschaus chose the fruit fly because of its fast embryonic development. They began to pursue a strategy for isolating genes responsible for the embryos' initial growth. This was a bold decision by two scientists just beginning their scientific careers; it wasn't certain that they would be able to actually isolate specific genes.

Their experiments involved feeding male fruit flies sugar water laced with **deoxyribonucleic acid** (DNA)damaging chemicals. When the male fruit flies mated with females, the females often produced dead or mutated embryos. Nüsslein-Volhard and Wieschaus studied these embryos for over a year under a microscope which had two viewers, allowing them to examine an embryo at the same time. They were able to identify specific genes that basically told cells what they were going to be—part of the head or the tail, for example. Some of these genes, when mutated, resulted in damage to the formation of the embryo's body plan.

Nüsslein-Volhard and Wieschaus published the results of their research in the English scientific journal *Nature* in 1980. They received a great deal of attention because their studies showed that there were a limited number of genes that control development and that they could be identified. This was significant because similar genes existed in higher organisms and humans and, importantly, these genes performed similar functions during development. Nüsslein-Volhard and Wieschaus's breakthrough research could help other scientists find genes that could explain **birth defects** in humans. Their research could also help improve *in vitro* **fertilization** and lead to an understanding of what causes miscarriages.

In 1991, Nüsslein-Volhard and Wieschaus received the Albert Lasker Medical Research Award. During this time Nüsslein-Volhard had begun new research at the Max Planck Institute in Tübingen, Germany, similar to the work she did on the fruit flies. This time she wanted to understand the basic patterns of development of the zebra fish. She chose zebra fish as her subject because most of the developmental research on vertebrates in the past was on mice, frogs, or chickens, which have many technical difficulties, one of which was that one couldn't see the embryos developing. Zebra fish seemed like the perfect organism to study because they are small, they breed quickly, and the embryos develop outside of the mother's body. The most important consideration, however, was the fact that zebra fish embryos are transparent, which

would allow Nüsslein-Volhard a clear view of development as it was happening. Despite her prize-winning research on fruit flies, Nüsslein-Volhard received skeptical feedback on her zebra fish work.

On October 9, 1995, in the midst of criticism about her new research, Nüsslein-Volhard (the first German woman to win in this category), Wieschaus, and Edward B. Lewis of the California Institute of Technology won the Nobel Prize in Physiology or Medicine for their work on genetic development in *Drosophila*. Lewis had been analyzing genetic mutations in fruit flies since the forties and had published his results independently from Nüsslein-Volhard and Wieschaus.

See also Embryology; Embryonic development: Early development, formation, and differentiation; Human genetics

NUTRITION AND NUTRIENT TRANSPORT TO CELLS

Living organisms are dynamic systems always in need of energy and nutrients that are indispensable to multiple vital processes such as growth, maintenance and repair of body structures, protection and detoxification of cells, and reproduction. Nutrients are found in foods and are classified as **carbohydrates**, **lipids** (fats), and proteins. The main sources of carbohydrates in the human diet are sugars, such as sucrose (from sugar cane), lactose (from dairy products), and starches (from grains and its derivates). Lipids are found in butter, cheeses, margarines, whole milk, eggs, nuts, olive oil, meats, fish, and other sources. The main source of proteins is animal products (milk, eggs, and meats), although they are also present in soybeans, fruits, and nuts. Moreover, small quantities of micronutrients such as essential minerals and **vitamins** are also indispensable for both structural repair and crucial metabolic functions in cells, tissues and **organs**. The best sources of vitamins are whole milk, whole grains, raw vegetables, and fruits. Trace minerals are present in practically every kind of food in physiological amounts.

Nutrition starts with the intake of foods and the process of **digestion** that will reduce them to molecules suitable for absorption by the gastrointestinal mucosa. The basic process involved in digestion is hydrolysis, which reduces carbohydrates to monosaccharides and water. The dietary lipids are usually neutral fats (**triglycerides**) constituted by three molecules of fatty acids condensed with a single glycerol molecule. Hydrolysis adds three molecules of water to each triglyceride molecule, and separates the fatty acid from the glycerol.

From protein, the digestion process ultimately obtains the **amino acids** needed to a wide range of physiologic functions, such as the synthesis of body **enzymes** and proteins, nucleic acids, and regulation of metabolic processes in every **tissue**. Proteins are first digested through the action of pepsin, an enzyme present in the stomach, and then by several pancreatic enzymes delivered in the upper small intestine, duodenum, and jejunum. A final stage occurs in the intestinal lumen, mediated by peptidase enzymes. The resultant products are smaller proteins (dipeptides, tripeptides) and some free amino acids that can be absorbed by the intestinal membranes.

Absorption of nutrients by the gastrointestinal mucosa may take place by active transport or by diffusion. Active transport involves the spending of body energy, whereas diffusion occurs simply through random molecular movement and, therefore, without the use of body energy. Water for example, is transported through the intestinal mucosa by diffusion (isosmotic absorption); on the other hand, the absorption of sodium, involves active transport. Nutrients absorbed through the intestinal membranes are delivered in the **blood** circulation and transported to all tissues and organs, where they are absorbed by the cells.

The transport of most nutrients through the cell membrane depends on its solubility in lipids, because cell membranes are constituted by a double lipid layer, with the polar ends (hydrophilic ends) on the external side and the apolar chains (hydrophobic ends) in the interior face. Therefore, fatty acids and steroid **hormones** can easily penetrate through the membrane, whereas other substances will be transported through other mechanisms, such as active transport, or facilitated diffusion. Many molecules enter the cell through passive diffusion or osmotic transport, simply because they are dissolved in an extra cellular solution. When the concentration of a given solute inside the cell decreases, passive diffusion from the outside takes place through osmosis, without energy spent. However, the active transport through cell membranes is done with the release of energy from **ATP (adenosine triphosphate)**, which is converted into ADP (adenosine diphosphate) in the process. Another form of transport of nutrients to cells, termed facilitated diffusion, occurs in the delivery of several substances, such as glucose, galactose, and certain amino acids. In this process, permeases, or transporting molecules present in the plasmatic membrane, are associated with the molecules of the substance to be transported. Cells also have the ability to transfer great quantities of macromolecules (proteins, polysaccharides, polynucleotides) to their interior through endocytosis. Endocytosis may occur under the form of phagocytosis or pinocytosis. Phagocytosis is the process by which cells envelop a solid particle by modifying its membrane to form a sac or phagosome around the particle. Phagosomes are then pulled by the cytoskeleton motion to the cytosol, where they are incorporated into one or more lysosomes for digestion of the particles. Pinocytosis is used to transport liquid substances (solutes) into the cell. Several pockets occur in a given area of the cell membrane, capturing the liquid and forming vesicles, which are then pulled by the cytoskeleton motion to the cytoplasm. Pinocytosis may be selective or non-selective.

Selective pinocytosis occur in two stages: 1) the liquid substance first adheres to membrane receptors at the membrane surface; and 2), the membrane then sinks and the substance is transferred to vesicles that leave the surface and transport their content to the cytoplasm. In non-selective pinocytosis, the vesicles envelop all the solutes eventually present in the extra cellular fluid.

The ultimate goal of nutrition is the conversion of nutrients into bioenergy. Energy is indispensable for muscular and neural activity, synthesis of enzymes and proteins, repair of **DNA** and tissues, absorption of nutrients in the **gastrointestinal tract**, cell proliferation, excretion of toxic metabolites from the cells, body temperature maintenance, and many other physiological functions.

See also Breathing; Enzymes and coenzymes; Interstitial fluid; Lipids and lipid metabolism; Metabolic waste removal; Osmotic equilibria between intercellular and extracellular fluids

O

OCHOA, SEVERO (1905-1993)
Spanish-born American biochemist

Severo Ochoa is best known for being the first to synthesize **ribonucleic acid (RNA)** outside the cell. He has also discovered several important metabolic processes. For his work with RNA, he received the 1959 Nobel Prize in physiology or medicine, along with his colleague, American biochemist **Arthur Kornberg**.

Ochoa was born in Luarca, Spain, where his father was a lawyer, and graduated from the University of Malaga in 1921. He received a medical degree in 1928 from the University of Madrid. After further studies in experimental biology, in 1940, Ochoa joined the Medical School faculty of Washington University in St. Louis. In 1942, he moved to New York University's College of Medicine, becoming chairman of the **biochemistry** department in 1954. Ochoa became an American citizen in 1956.

Ochoa's synthesis in 1955 of RNA was pure serendipity—an unexpected byproduct of his study of the way cells use glucose that is stored as **ATP (adenosine triphosphate)**. Ochoa and a French associate, Marianne Grunberg-Manago, had purified an enzyme (now called polynucleotide phosphorylase) from the **bacteria** *Azotobacter vinelandii*. The two scientists were trying to study its reactions with ATP and other base-sugar combinations (called nucleosides) with one or three phosphate groups attached. No reaction occurred. However, when they added the enzyme and some magnesium to a nucleoside with two phosphate groups (diphosphate), over half of the nucleoside disappeared and some phosphorus was freed.

Ochoa traced the nucleoside to a new molecule that ultraviolet chromatography identified as a nucleotide. He then repeated the reaction with other nucleoside-diphosphates, in each case finding a nucleotide. Further analysis showed that the sugar was ribose, meaning that the reaction produced ribonucleic acid (RNA). Since the reaction was also reversible, Ochoa concluded that adding and removing phosphorus groups is a major mechanism in the synthesis and breakdown of nucleotide chains. Ochoa and other scientists used this method to decipher the **genetic code**. Later studies by others showed that RNA polymerase, instead of Ochoa's enzyme, is the main RNA synthesizing enzyme.

Ochoa is also known for his work on how the body uses carbon dioxide, and he helped identify a key compound in the **metabolism** of carbon dioxide. Ochoa also identified Krebs cycle reactions leading to energy storage in phosphate bonds.

See also Biochemistry; Enzymes and coenzymes

OOCYTE • *see* OOGENESIS

OOGENESIS

Oogenesis, or female gametogenesis, is the process by which diploid progenitor cells are reduced to haploid gametes in a female. This is a complicated pathway that begins in each female fetus and doesn't finish until the **fertilization** of the egg.

In early fetal development, the primordial **germ cells** migrate to the gonad. Mitotic divisions produce a population of oogonia with each oogonium in its own follicle. Development continues, and at about three months gestation, the oogonia have become primary oocytes. Shortly thereafter, the oocytes will randomly enter meiosis I but will proceed only to the dictyotene stage of prophase I where the **cell division** arrests. At the birth of a female infant, there will be approximately 2.5×10^6 primary oocytes that will remain in mid-meiosis I until they either degenerate, or ovulation occurs later in life.

At **puberty**, oogenesis begins again. Once a month from approximately twelve to fifty years of age, a single ovarian follicle will mature, and ovulation will occur releasing a single primary oocyte that will then complete meiosis I. This phase is the critical reduction division of meiosis that divides the total number of **chromosomes** per cell in half, from 46 to 23. The division is asymmetric and results in one secondary oocyte that contains the majority of the cytoplasm plus a nucleus with 23 chromosomes, and one polar body that has very little cytoplasm but includes a nucleus carrying the other half of the chromosome set. The polar body may degenerate or may undergo a second division. However, in terms of reproduction, it is a dead end and does not participate in zygote formation.

At this point, the secondary oocyte will have entered the fallopian tube, and meiosis II is initiated. This division proceeds to metaphase and stops. If there is no fertilization, the division will go no further. The cell will lodge in the uterine wall and will be discarded during the following **menstruation**. If fertilization occurs, a second asymmetric division takes place producing an egg with the cytoplasm and a haploid nucleus, known as the female pronucleus, and a secondary polar body that is the receptacle of the remaining haploid nucleus. Fusion of the male pronucleus from the **sperm** (23 chromosomes) and the female egg pronucleus (23 chromosomes) results in a zygote that has the reconstituted number of 46 human chromosomes, one half received from the mother and one half from the father.

Over the course of an average female life, only about 400 oocytes will be ovulated. There are a total of 20–30 cell divisions for each egg produced, so, the likelihood of any given egg acquiring a deleterious **DNA** based mutation is relatively low. However, a high degree of meiotic division error is seen in oogenesis. It is well documented that older mothers have an increased risk for conceptions that are either trisomic (one extra chromosome) or monosomic (one missing chromosome). These errors are most likely associated with the long period of stasis between birth and ovulation when oocytes are held suspended in mid cell division. It has been suggested that, over time, the cell division mechanism becomes less stable leading to the increased frequency of errors. Unfortunately, abnormal numbers of chromosomes are associated with physical and/or mental defects, so such conceptions often result in spontaneous termination. A few chromosome imbalances, including trisomy 13, trisomy 18, trisomy 21 (Down syndrome), and monosomy X (Turner syndrome), are sometimes tolerated and may give rise to a liveborn infant, but these children will have a host of problems ranging from **heart** abnormalities, structural malformations, and growth retardation, to mild to severe mental retardation.

Gametogenesis in both the male and female accomplish the same objective, i.e., generation of haploid gametes. However, there are a few important differences in the overall processes. In females, only one viable gamete will be produced from each primary cell as compared to male **spermatogenesis** that produces four gametes per primary cell. Oogenesis in females begins *in utero* and ceases around the fifth decade of life, whereas, in males, spermatogenesis does not start until puberty and then continues throughout the remainder of the man's lifetime. Furthermore, division errors in females give rise to gametes carrying abnormal numbers of chromosomes which contrasts to spermatogenesis in which accumulation of DNA based mutations may have a deleterious effect on the child. Although both types of gamete formation have the potential for mistakes, error-checking mechanisms significantly limit the likelihood that embryos with deleterious abnormalities will survive.

See also Cell cycle and cell dvision; Cell differentiation; Embryonic development, early development, formation, and differentiation; Gonads and gonadotropic hormone physiology; Ovarian cycle and hormonal regulation

OPTICS (PHYSIOLOGY OF VISION) • *see*
VISION: HISTOPHYSIOLOGY OF THE EYE

ORGANIZER EXPERIMENT

The organizer experiment was a classical experiment in developmental biology carried out by German anatomist, **Hans Spemann** and **Hilde Mangold** that demonstrated the principle of cell induction.

Spemann and Mangold transplanted a group of cells from one region of an amphibian blastopore (a hollowed out group of rapidly developing cells that occurs soon after **fertilization** and the onset of early embryonic **cell division**) to another region of a similar blastopore. The transplanted cells continued along their normal path of development but also induced cells in the new host to change their developmental path.

The transplanted cells developed as they would have on the dorsal side of the original blastopore. The cells invaginated and differentiated into **notochord** and **somites**. The transplanted cells induced cells in the new host blastopore to become associated mesodermal structures. The transplanted cells essentially caused the development of a second neural plate, neural tube, and associated axial and paraxial structures on the ventral side of the new blastopore that matched those found on the dorsal side. The capacity to cause this development inspired Spemann and Mangold to name the region of cells in the blastopore from which the transplanted cells were originally taken the organizer region.

The organizer experiment also demonstrated the important link between technological and laboratory procedural advances and the advancement of knowledge related to developmental processes. Especially in **embryology** and developmental biology, fundamental conceptual advances often result from technological advances that allow researchers to experiment with delicate and fragile embryonic cells and **tissue**.

Spemann and Mangold were able to mark the cells of the amphibian embryos in such a way that different cell groups remained discernable and traceable. Specifically, Spemann and Mangold were able to mark cells, and to then determine the fates and influences of those cells during embryological development. Cells from the organizer region, especially the notochord forming areas, were shown to induce the formation of a second neural tube, and rows of somites derived from cells within the new host would form around a transplanted notochord.

Although differing vastly in techniques and specimen, conceptually, the marking experiments are similar to the tagging of molecules performed in many modern **molecular biology** and **genetics** experiments.

Spemann's original inspiration to perform the organizer experiment came as a result of his experimental observations of amphibian embryos. In particular, Spemann was able to experimentally induce tadpole to develop one large cycloptic **eye** instead of two eyes. The single eye also contained a large single lens. In both amphibian and human embryos, regions of the eye develop from **ectoderm** and the optic vesicle. The lens of the eye forms in an area of contact between the ectoderm and optic vesicle. Spemann cut and removed cells (tissue ablation) from the region of an early amphibian embryo still in its blastopore stage Spemann then recombined those cells with native cells in a different region of another amphibian blastopore and observed developmental variations associated with the recombination of cells from different regions of the donor blastopore.

Spemann, joined by Mangold, subsequently extended his work into the formal organizer experiments for which Spemann was awarded the 1935 Nobel Prize in physiology or medicine. Mangold died the year after her results were published, and the Nobel Prize is not awarded posthumously.

Spemann and Mangold's organizer experiment demonstrated that **cell differentiation** could occur early in embryonic development—well prior to any gross or outward differentiation of cells. Moreover, the organizer experiment showed that cells were differentiated at least partially based upon their location within the embryo. Long before the structure of **DNA** was known, this argued for the influence of environment on development.

In some experimental variations, transplanted cells developed in accord with their location within the new host. In these cases, the cells were said to be undifferentiated at the time of transfer and thus, able to take on a new fate within the new host embryo. Cells that were already differentiated could not only continue to develop in the new host along the same path they would have taken in the original embryo; the organizer experiment showed that they could also induce and influence the fate of undifferentiated cells within their new host embryo.

See also Anatomical nomenclature; Embryology; Embryonic development, early development, formation, and differentiation; Endocrine system and glands; Endoderm; Ethical issues in embryological research; Ethical issues in genetics research; History of anatomy and physiology: The science of medicine

ORGANS AND ORGAN SYSTEMS

An organ is composed of two or more different types of **tissue** that work together to carry out a complex function. An organ system consists of a group of organs that perform intricate functions necessary for the survival of an organism. Sometimes an organism can survive with an impaired or non-functioning organ. However, when a whole system of organs shuts down, the life of the organism becomes compromised.

Thus, the organ systems work together to maintain a constant internal environment called homeostasis within the body to ensure survival of the organism.

There are 11 organ systems within the human body: integumentary, skeletal, muscular, nervous, endocrine, circulatory, lymphatic, respiratory, digestive, urinary, and reproductive.

The integumentary system acts as a protective barrier for the human body against microorganisms, dehydration, and injuries caused by the outside environment. Additionally, the integumentary system regulates body temperature. Organs of the integumentary system include **hair**, **nails**, sebaceous glands, sudoriferous glands, and the largest organ of the body, the skin.

The skeletal system is a structural framework providing support, shape, and protection to the human body. Additionally, the skeletal system provides attachment sites for organs. The skeletal system also stores minerals and **lipids** and forms **blood** cells. Bones, **cartilage**, **tendons**, and ligaments are all organs of the skeletal system.

The muscular system provides movement to the human body as a whole as well as movement of materials through organs and organ systems. Additionally, the muscular system functions to maintain **posture** and produce heat. The muscular system consists of **skeletal muscle**, **smooth muscle**, and cardiac muscle.

The nervous system conducts electrical impulses throughout the body to regulate and control physiological processes of the other organ systems. Organs of the nervous system include the **brain**, **spinal cord**, and nerves.

The **endocrine system** also functions to regulate and control physiological processes of the body. However, the endocrine system accomplishes its functions by sending out chemical signals called **hormones** into the blood. Glands, the organs of the endocrine system, secrete hormones and include: pituitary gland, pineal gland, **hypothalamus**, thyroid gland, parathyroid glands, thymus, adrenal glands, **pancreas**, ovaries, and the testes.

The **circulatory system** circulates blood throughout the body and in doing so transports gases, nutrients, and wastes to and from tissues. Organs of the circulatory system include the **heart**, blood vessels, and blood.

The lymphatic system, also known as the **immune system**, defends the body against microorganisms and other foreign bodies. Additionally, the lymphatic system transports fluids from the body's tissues to the blood, thus helping to control fluid balance in the body. This system also absorbs substances from the digestive system. The organs of the lymphatic system include the lymph, lymph nodes, lymph vessels, thymus, spleen, and **tonsils**.

The **respiratory system** exchanges gases between the body's tissues and the external environment. Oxygen is inhaled from the external environment and passes from the **lungs** into the blood, where it is exchanged for carbon dioxide that passes from the blood to the lungs and is expelled. The respiratory system consists of the **nose**, **pharynx**, **larynx**, trachea, **bronchi**, and lungs.

The digestive system functions to digest and absorb nutrients from the food ingested into the body. Additionally,

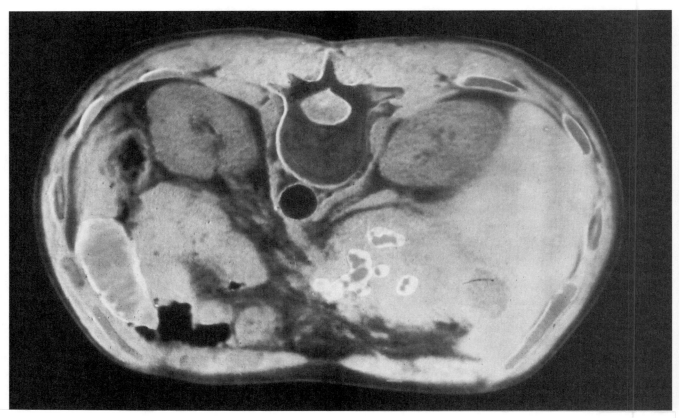

Computed tomography (CT) scan through the human abdomen. Visible structures include the liver (right), spleen (left), abdominal aorta (center), vertebral column, spinal cord, and kidneys (upper left and upper right). *Photo Researchers, Inc. Reproduced by permission.*

the digestive system transports foodstuff through the **gastrointestinal tract**. The primary organs of the digestive system include the mouth, pharynx, esophagus, stomach, small intestine, large intestine, rectum, and anal canal. Accessory organs that aid the primary organs include the teeth, salivary glands, tongue, **liver**, gallbladder, pancreas, and **appendix**.

The urinary system removes excess water and nutrients and filters wastes from the circulatory system. Additionally, the urinary system aids in red blood cell formation and metabolizes vitamin D. The urinary system's organs include the **kidneys**, ureters, urinary bladder, and urethra.

The reproductive system of the human body can be either male or female. The male reproductive system synthesizes gametes called spermatozoa that are responsible for fertilizing the female gametes, or oocytes, during reproduction. The female reproductive system is designed to undergo conception, gestation, and birth once a spermatozoon fertilizes an oocyte. The male reproductive system is composed of the testes, vas deferens, urethra, penis, scrotum, and prostate. The female reproductive system consists of the ovaries, uterus, fallopian tubes, vagina, vulva, and **mammary glands**.

OSMOSIS • *see* CELL MEMBRANE TRANSPORT

OSMOTIC EQUILIBRIA BETWEEN INTERCELLULAR AND EXTRACELLULAR FLUIDS

Intercellular and extracellular fluids are separated by the membranes of the body's cells. An example of such fluids is the fluid inside of **blood** cells (intercellular fluid) and that outside the blood cells (**extracellular fluid**). Blood cell membranes generally are not completely permeable. That is, they do not allow the unrestricted passage of any molecule back and forth across the membrane. Rather, the construction of the membranes makes them very permeable to the passage of water, but more selective in the other types of molecules and **electrolytes** (e.g. ions) that can move through the membrane. This difference in permeability results in a different chemical composition for intercellular and extracellular fluids.

The different ionic concentrations on one side of a membrane versus the other side can result in osmotic flow, the flow of water from a less concentrated region to a more concentrated region (e.g., against a particular molecular or ionic concentration gradient), in an effort to balance the ionic concentrations. In the body, even though the ionic concentrations of the intercellular and extracellular fluids can be quite different, the two fluid compartments are always in osmotic **equilibrium**. This is accomplished by the movement of water across the cell membrane.

Maintenance of osmotic equilibrium via water flow depends on the flexibility of the cell membrane. As water moves into a cell, for example, the membrane expands, allowing the cell to swell. Thus, more water is able to be present, which dilutes the concentration of the ion inside the cell. Conversely, as water moves out of a cell into the surrounding extracellular fluid, the membrane can accommodate the shrinkage of the cell. If a cell membrane were not flexible, osmotic equilibrium could not be achieved.

In the body, potassium ions often move back and forth across cell membranes. Because ions carry electrical charges, the differing potassium concentration across the membrane results in an electrical potential. Unless controlled, this potential could be damaging to the cell. But, because of osmotic equilibrium, the electrical potentials for potassium can be the same in the intercellular and extracellular fluids. An equation called the Nernst equation gives the potential difference across the membrane when ions are in equilibrium.

The ability of a cell to actively take up substances can be blocked. For example, if a membrane is permeable to some ions, such as potassium, sodium and chloride, but is not permeable to some other large negatively charged ion, then the concentration disparity of the large ion can become so great that movement of water into and out of the cell is stopped. This is also known as the Gibbs-Donnan equilibrium. A Gibbs-Donnan equilibrium exists between the **interstitial fluid** and blood **plasma**.

See also Acid-base balance; Cell membrane transport; Cell structure; Fluid transport; Membrane potential; Sodium-potassium pump

OSSIFICATION

Ossification is a term that refers to the formation of bone. There are two kinds of ossification: intramembranous ossification and endochondral ossification. In both types, pre-existing connective **tissue** is replaced by bone. The two processes are distinguished by their mechanics.

Intramembranous ossification is concerned with so-called mesenchyme cells—cells that have the ability to differentiate into any number of structurally and functionally different tissues. In intramembranous ossification, some mesenchyme cells form osteoblasts, which function to produce bone. This type of ossification transforms membrane into bone. The bone formation occurs at the periphery, with layers of bone being laid down, analogous to the ring-like diameter growth of a tree. This type of growth is called appositional growth. An example of intramembranous ossification is the formation of the topmost area of the **skull**.

Endochondral ossification involves the conversion of a type of **cartilage** called **hyaline cartilage** to bone. Hyaline cartilage is formed into a likeness of the future bone by cells called chondroblasts. This likeness is covered by a perichondrium. The perichondrium is a dynamic region, supplied with **blood** vessels and is thus capable of physiological activity. Cells in the perichondrium change to osteoblasts, which begin

to produce a bony region in the middle area of the hyaline cartilage. The region becomes mineralized with **calcium** carbonate to form what is called calcified cartilage. Further activity in the calcified cartilage yields what is termed the primary ossification center. Over time, this central region lengthens, as the entire cartilage is lengthening. A full-length bone results, complete with bone marrow.

Endochondral ossification can occur during embryonic development as increasingly firmer support is needed for the growing fetus, or can occur well into the teenage years, depending on the bone that is involved.

Ossification defects can occur during development of a fetus, with catastrophic effects. The use of angiotensin-converting enzyme (ACE) inhibitors as a means of controlling pregnancy-induced high blood pressure has been shown to decrease ossification, particularly of the skull. The resulting malformation can be lethal.

See also Bone histophysiology; Bone injury, breakage, repair, and healing; Osteology

OSTEOLOGY

Osteology is the study of bone and bone **tissue**. Osteology also often encompasses the study of the **connective tissues** associated with bone and of the physical properties of bone.

Bones are structurally different from most other forms of living tissue because they blend a capacity for great strength with a capacity for growth and repair. Although lightweight, bones have a great strength and capacity to bear weight many times greater than the bone itself. Bones resist compression, but do show some elasticity (ability to bend). Portions of bone are also dynamic living tissues that can adapt to stress. This dynamic flexibility allows bone to respond to the correcting applications of leg braces or the bony portions of teeth to respond to the application of orthodontic braces.

Because bone is a living tissue, osteologists study the role that **nutrition** and physiological processes (e.g. endocrine function) play in the development and maintenance of bone tissue. Another area of increasing interest involves study of the effects of aging on bone. The processes of aging often involve osteolysis, the destruction and loss of bone tissue through reabsorption, dissolution, and softening. The interconnectivity of physiological processes is made readily apparent by the relationship between proper nutrition, loss of **calcium**, and osteolysis. The disease **osteoporosis**, four times more common in women than men, results in a progressive depletion of bone that makes bones—especially those of the spine, wrist, and hips—fragile and at greater risk of breakage.

Some osteologists are concerned primarily with the mathematical measurement of bone and the determination of bone growth patterns in between different groups. Osteometry, the measurement of bone, allows anatomists and osteologists a vast data base upon which to make judgments regarding, age of bone and the conditions under which bone growth occurred.

The formal study of osteology is also important for archaeologists and anthropologists. Careful examination of bone can reveal a great deal about the anatomical and physiological differences in ancient peoples, and provide information about the cultures in which they lived. The condition of bone can even allow anthropologists to make scientific, data-based judgments concern ancient religious and cultural practices.

Based upon the examination of bone fragments, osteologists, archaeologists, anthropologists, and forensic pathologists are often able to reliably determine the age at **death**, sex, height (stature), and cause of death of the individual from whom the bone fragments remain. Wear patterns on bone can even provide a guide to the type of work the individual routinely performed.

Radiographs of bone and dental structures are still used as accepted means of post-mortem identification.

In addition to a reliance on traditional and scientifically accepted anatomical and physiological principles, osteopathic medical practitioners place additional emphasis on the alignment and mechanics of living bone structure. One of the essential tools of the osteopathic practitioner is a reliance on the manipulation of muscular skeletal structure and a means of diagnosis and treatment of disease.

See also Anatomical nomenclature; Anatomy; Arthrology (joints and movement); Bone histophysiology; Skeletal and muscular systems, embryonic development; Skeletal muscle; Skeletal system overview (morphology)

OSTEOPOROSIS

Osteoporosis is a condition in which bone mass, and therefore, bone strength, is decreased. This results in an increased risk of fracture. Primary osteoporosis occurs due to normal, predictable changes within the body during the **aging process**. Secondary osteoporosis occurs as a result of some other specific disease process that produces osteoporosis as one of its symptoms.

To understand osteoporosis, it is helpful to understand the basics of bone formation. Bone is formed on a protein base (collagen) by the deposition of minerals, particularly **calcium**. This laying down of bone is carried out by specialized cells called osteoblasts. The formation of new bone occurs most effectively along lines of stress/weight that are experienced by the bone. Other cells, osteoclasts, are responsible for reabsorbing (taking up) bone. These cells actually digest already-formed bone.

This active reabsorbtion-formation cycle within bone occurs throughout life, so that old bone is always being replaced by new bone. When the reabsorbtion phase is accelerated, or the formation phase is slowed, less calcified bone exists. This is the state which results in the weakened bone structure present in osteoporosis.

A decrease in the rate of bone mineralization is a predictable effect of aging. For example, in infancy, the turnover rate of calcium in bone is 100% by adulthood, this turnover rate falls to only 18% per year.

Women are particularly prone to osteoporosis because of several factors. Women have less bone mass than men to

begin with, so the threshold level at which osteoporosis may cause fractures is reached more quickly. It is believed that the bone formation phase is encouraged in some way by the presence of **estrogen**. In women, estrogen production drops off drastically following **menopause** (the cessation of the menstrual period). This change in the chemical environment within the bodies of older women apparently results in a decrease in the bone formation phase. With bone reabsorbtion continuing at its normal pace, but without the normal pace of bone replacement occurring, bone mass decreases.

Because the pattern of bone formation occurs in response to weight/stresses borne by the bone, disuse osteoporosis occurs in individuals who are on bed rest for prolonged periods of time, as well as in individuals experiencing the relative weightlessness of space flight.

Other causes of osteoporosis include many diseases which alter the hormonal/chemical environment of the body, including thyroid disease, disease of the parathyroid (a gland responsible for calcium levels within the body), gastrointestinal diseases (which can alter the ability of the body to absorb calcium in the diet), diseases which decrease the amount of estrogen produced, and certain **liver** diseases.

Alcohol and some drugs can also affect calcium levels in the body, thus producing osteoporosis. Some of these drugs include thyroid medications, steroid preparations, anti-seizure medications, and certain **chemotherapy** (anti-cancer) agents.

Congenital diseases (diseases present at birth) of connective **tissue** (a group of tissues of the body which includes bone) can cause abnormalities of bone structure, and therefore osteoporosis. Such diseases include osteogenesis imperfecta (brittle bone disease) and Marfan's syndrome.

The importance of osteoporosis in terms of the misery it causes and its economic impact is underscored by these statistics. About one-third of all women over the age of 70 experience hip fracture. Of those elderly people who fracture a hip, about 15% die of complications secondary to that hip fracture. A large percentage of those who survive are unable to return to their previous level of activity, and many times a hip fracture precipitates a move from self-care to a supervised living situation or nursing home. The yearly cost of osteoporotic injury in the United States is greater than $10 billion.

See also Bone injury, breakage, repair, and healing; Posture and locomotion; Skeletal system overview (morphology)

OTIC EMBRYOLOGICAL DEVELOPMENT •

see EAR, OTIC EMBRYOLOGICAL DEVELOPMENT

OVARIAN CYCLE AND HORMONAL REGULATION

During their reproductive years, nonpregnant females usually experience a cyclical sequence of changes in their ovaries and uterus. Each cycle takes about one month and involves both **oogenesis**, the process of formation and development of the egg or ovum, and preparation of the uterus to receive a fertilized ovum. **Hormones** secreted by the **hypothalamus**, anterior pituitary gland and ovaries control the principal events. The ovarian cycle is a series of events in the ovaries that occur during and after the maturation of the oocyte. The uterine (menstrual) cycle is a concurrent series of changes in the endometrium of the uterus to prepare it for the arrival of a fertilized ovum that will develop in the uterus until birth. If **fertilization** does not occur, the stratum functionalis of the endometrium is shed. The general term female reproductive cycle encompasses the ovarian and uterine cycles, the hormonal changes that regulate them and also the related cyclical changes in the **breasts** and cervix.

The ovarian and uterine cycles are controlled by chemical messengers or hormones. Gonadotropin releasing hormone (GnRH) is secreted by the hypothalamus and stimulates the release of **follicle-stimulating hormone (FSH) and luteinizing hormone (LH)** from the anterior pituitary gland. FSH, in turn, initiates follicular growth and the secretion of estrogens by the growth follicles. LH stimulates the further development of ovarian follicles and their full secretion of estrogens, brings about ovulation, promotes formation of the corpus luteum and stimulates the production of estrogens, **progesterone**, relaxin and inhibin by the corpus luteum.

Estrogens are hormones having several important functions. They promote the development and maintenance of female reproductive structures, secondary sex characteristics and the breasts. The secondary sex characteristics include the distribution of **adipose tissue** in the breasts, **abdomen**, and hips; also voice pitch, a broad pelvis and the pattern of **hair** growth on the head and body. Estrogens increase protein anabolism and lower **blood** cholesterol level. Moderate amount of estrogens in the body inhibit both the release of GnRH by the hypothalamus and secretion of LH and FSH by the anterior pituitary gland. At least six different estrogens are present in the **plasma** of human females but only three are present in significant quantities: B-estradiol, estrone and estriol. In nonpregnant females the principle **estrogen** is B-estradiol, which is synthesized from cholesterol in the ovaries.

Progesterone is secreted mainly by cells of the corpus luteum and acts synergistically with estrogens to prepare the endometrium for the implantation of a fertilized ovum and the **mammary glands** for milk secretion. High levels of progesterone also inhibit the secretion of GnRh and LH. A small quantity of the hormone relaxin produced by the corpus luteum during each monthly cycle, relaxes the uterus by inhibiting contractions. This is probably to facilitate the implantation of an ovum which is perhaps more likely to occur in a relaxed uterus. During **pregnancy**, the **placenta** produces much more relaxin and continues to relax the uterine **smooth muscle**. At the end of pregnancy, relaxin also increases the flexibility of the **pubic symphysis** and may help dilate the uterine cervix, both of which ease delivery of the baby. Inhibin is secreted by granulosa cells of growing follicles and by the corpus luteum of the ovary. It inhibits secretion of FSH and to a lesser extent, LH.

The duration of the reproductive cycle is divided into four phases: the menstrual phase, the preovulatory phase, ovu-

lation, and the postovulatory phase. The menstrual phase lasts for about five days and by convention the first day of **menstruation** marks the first day of a new cycle. The endometrium is shed and the discharge occurs because the declining levels of hormones, especially progesterone, stimulates the release of **prostaglandins** that cause the uterine spiral arterioles to constrict. As a result the cells they supply become oxygen deprived and die and the stratum functionalis sloughs off. During this phase, some 20 secondary follicles in each ovary begin to enlarge and continue to do so through the preovulatory phase, the time between menstruation and ovulation, under the influence of FSH. By about day six, one follicle has outgrown the others and becomes the dominant follicle. Estrogens and inhibin secreted by the follicle decrease the secretion of FSH and the other follicles stop growing. The mature dominant follicle, or Graafian follicle, continues to enlarge until it is ready f or ovulation. It continues to produce estrogen under the influence of LH. At day 14, the follicle ruptures and releases an oocyte into the pelvic cavity. This process is known as ovulation. After ovulation the mature follicle collapses.

The postovulatory phase of the female reproductive cycle is the most constant in duration, lasting approximately from day 15 to 28 and represents the time between ovulation and the onset of the next menses. In the ovary, after ovulation, the LH stimulates the remnants of the mature follicle to develop into the corpus luteum, which secretes increasing quantities of progesterone and some estrogens. This is called the luteal phase of the ovarian cycle. Subsequent events in the ovary that ovulated an oocyte depend on whether or not the oocyte becomes fertilized. If the oocyte is not fertilized, the corpus luteum has a lifespan of only two weeks, after which it degenerates into a corpus albicans. As the levels of progesterone, estrogens and inhibin decrease during this phase, GnRH, FSH and LH release increases because of the lack of feedback suppression by the ovarian hormones. Then follicular growth resumes and a new ovarian cycle begins.

If, however, the oocyte is fertilized and begins to divide, the corpus luteum persists past its normal two week lifespan. It is prevented from degenerating by the human chorionic gonadotropin (hCG), a hormone produced by the chorion of the embryo as early as 8–12 days after fertilization. HCG acts like LH in stimulating the secretory activity of the corpus luteum and the presence of hCG in maternal blood or urine is an indicator of pregnancy.

See also Gonads and gonadotropic hormone physiology; Hormones and hormone action; Reproductive system and hormones (female); Sex hormones

OVARY • *see* REPRODUCTIVE SYSTEM AND ORGANS (FEMALE)

OXYGEN TRANSPORT AND EXCHANGE

Oxygen transport and exchange provide necessary oxygen to the body's cells through the process of respiration. Oxygen enters the **respiratory system** through the mouth and **nose**, and travels through the **pharynx**, **larynx**, trachea, and **bronchi** during inspiration (inhalation) to reach the **lungs**. The bronchi enter into the lungs and branch to form bronchioles. The bronchioles further divide to form alveolar ducts. These ducts lead to tiny sacs called **alveoli** that are surrounded by pulmonary **capillaries** where oxygen exchange takes place. There are approximately 300 million alveoli that greatly increase the surface area of the lungs. Oxygen exchange occurs by the process of diffusion across the alveolar-capillary wall. During diffusion, oxygen travels down its concentration gradient into the **blood** and is exchanged for carbon dioxide. The partial pressure of oxygen within the body is denoted by pO_2 and is measured in millimeters of mercury (mm Hg). The pO_2 of oxygen in the alveoli is 105 mm Hg and the pO2 of deoxygenated blood entering the pulmonary capillaries at rest is 40 mm Hg. Because of this difference in pO_2, oxygen is exchanged for carbon dioxide until the pO2 of the capillaries equals the pO_2 of the alveoli. At **equilibrium** the pO_2 of oxygen equals 105 mm Hg in the now oxygenated blood of the capillaries. Likewise, carbon dioxide will diffuse down its concentration gradient from the deoxygenated blood of the capillaries, with a pCO_2 of 45 mm Hg, into the alveoli until it reaches equilibrium. Once equilibrium is reached, the pCO_2 in the capillaries will be 40 mm Hg, equal to the pCO_2 in the alveoli.

The rate of oxygen exchange is dependent upon the partial pressure difference, surface area of gas exchange, diffusion distance, and the rate and depth of respiration. In order for diffusion to take place, there must be a difference in the partial pressure of oxygen between the alveoli and the pulmonary capillaries. At higher altitudes, the partial pressure of atmospheric oxygen decreases, which decreases the pO_2 in the alveoli and thus impairs the diffusion process. Oxygen exchange is largely dependent on the available surface area of the alveoli. For example, patients who have pulmonary disease have their alveolar-capillary membrane destroyed, thereby decreasing the surface area available for oxygen exchange. The effectiveness of oxygen exchange is in part due to the short distance that oxygen has to diffuse across the alveolar-capillary membrane. Additionally, the diameter of the capillaries is only wide enough to allow one red blood cell to pass through at a time. If the capillaries swell due to a buildup of fluid, the diffusion distance is increased and oxygen exchange is hindered. Oxygen exchange is also dependent on the amount of oxygen that is available for diffusion. The respiratory rate and depth can affect the amount of oxygen available for exchange.

Once the red blood cells have received the oxygen, the oxygenated blood travels from the pulmonary capillaries to the pulmonary **veins**, and to the **heart** where it is pumped through the systemic **arteries** to the systemic capillaries and then to tissues. Oxygen is exchanged from the oxygenated systemic capillaries to the deoxygenated tissues. The pO_2 of the oxygenated systemic capillaries is 105 mm Hg and the pO_2 of the tissues is 40 mm Hg; therefore, oxygen diffuses down its concentration gradient until it reaches equilibrium. During oxygen diffusion, carbon dioxide is also being exchanged from the tissues to the systemic capillaries until equilibrium is

reached. Diffusion of carbon dioxide occurs because the pCO_2 in the tissues is 45 mm Hg while the pCO_2 in the systemic capillaries is 40 mm Hg. Once oxygen and carbon dioxide exchange is complete, the deoxygenated blood returns to the heart where it is pumped to the lungs as respiration continues.

About 98% of oxygen is transported by molecules of **hemoglobin** in red blood cells. One molecule of hemoglobin contains four heme groups and one globin protein. A heme group is an iron containing pigment that is able to bind one oxygen molecule. Only 2% of oxygen dissolves in the blood **plasma** because oxygen is relatively insoluble in water. The most important factor that determines the ability of oxygen to bind to hemoglobin is the partial pressure of oxygen in the blood. However, the oxygen bound to hemoglobin has no effect on the pO_2 of the blood, but rather is determined by the amount of dissolved oxygen. Hemoglobin is able to bind dissolved oxygen from the blood and maintain a low partial pressure of oxygen. However, the higher the pO2, the greater affinity hemoglobin has for oxygen. When the pO_2 is 105 mm Hg, hemoglobin becomes fully saturated. When the pO_2 is 40 mm Hg, hemoglobin will contain about 75% of its total oxygen capacity. The hemoglobin molecule is said to be 75% saturated, meaning that three of the four heme groups will have a bound oxygen molecule. Therefore, under resting conditions, the tissues only utilize 25% of the available oxygen from hemoglobin.

Other factors that influence oxygen's ability to bind to hemoglobin include acidity, partial pressure of carbon dioxide, temperature, and 2,3-biphosphoglycerate (BPG). In acid environments, oxygen has a decreased ability to bind to hemoglobin; this is called the Bohr effect. Carbon dioxide can displace oxygen from hemoglobin; thus as pCO_2 increases, hemoglobin has a decreased affinity for oxygen. Additionally, high pCO_2 can increase the blood acidity due to the conversion of carbon dioxide to carbonic acid in blood. An increase in temperature can also increase the release of oxygen from hemoglobin. BPG is a byproduct of glycolysis found in red blood cells that decreases hemoglobin's ability to bind oxygen.

See also Breathing; Respiration control mechanisms

OXYTOCIN AND OXYTOXIC HORMONES

Oxytocin is a small protein that is manufactured in the **brain** of humans and other mammals. It is secreted in the brain, ovary and testes, were it functions in a number of activities that are vital to the birth process and for milk production to nourish the newborn.

In females, as befits its origin from the Greek for "swift birth," oxytocin stimulates the smooth muscles of the uterus to contract. This occurs during labor as part of the birth process. Indeed, when uterine contractions are proving insufficient to propel the baby down the birth canal, physicians will sometimes administer oxytocin (also known as pitocin) to complete delivery. Oxytocin also causes contraction of muscle in the **mammary glands**, which ejects milk from the breast to the suckling infant. Studies in mammals have established that approximately 80% of breast milk is available only after ejection. Thus, proper **nutrition** is dependent on the presence of oxytocin.

Oxytocin has other functions as well. In mammals other than humans, oxytocin also plays an important role in maternal behavior, such as nest building and care for infants. In humans its release in the brain has a calming effect. More recent research has indicated that oxytocin may also be associated with the ability of humans to "bond" with others in psychologically healthy and meaningful relationships. So, in addition to stimulating birth and nutrition of the newborn, oxytocin may also facilitate the emotional bond between mother and child.

Oxytocin also has a role to play in male **physiology**. It is detected during ejaculation. Hence, it has been surmised that oxytocin in men facilitates **sperm** transport and perhaps aspects of sexual behavior that lead to mating.

Other **hormones** are oxytocin-like in their structure or action. Prolactin is an ancient hormone that originally functioned to maintain the salt and water balance in vertebrates. With evolutionary changes, prolactin assumes a variety of functions. In conjunction with **estrogen**, prolactin controls the production of milk (whose subsequent ejection from the breast is controlled by oxytocin). Vasopressin, which has an almost identical amino acid composition as oxytocin, is an antidiuretic hormone. Nonetheless, its function is quite different. Vasopressin controls the reabsorption of water by the **kidneys** and regulates the osmotic content of the **blood**.

See also Parturition

P

PAIN

Pain is one of the three existing physiologic types of somatic senses, the other two being the mechanoreceptive somatic sense and the thermo-receptive sense. The pain sense is activated by **tissue** damage, and therefore constitutes an important protective mechanism of tissue integrity. According to pain intensity, pain is classified as acute pain, such as that caused by cuts or **burns**, or chronic pain, usually caused by progressive tissue destruction, such as chronic inflammation or infections.

Skin and other tissues contain embedded free nerve ends, known as pain receptors, which are excited by three different types of stimuli: thermal, mechanical, and chemical. Acute pain is usually caused by thermal stimuli (i.e., burns) or mechanical stimuli (bruises, cuts, etc), whereas the chronic pain may be caused by any of the three types of stimuli. Pain receptors conduct pain signal through two different pathways into the **central nervous system**, one of them conducting acute pain signals and the other chronic pain signals. The **brain** is able to identify with more precision the area of origin of acute pain than the area affected by chronic pain.

Pain signals are transmitted to the brain by entering the **spinal cord** where one of two pathways is used to reach the brain: the neospinothalamic tract or the paleospinothalamic tract. The neospinothalamic tract transmits acute pain signals to the brain, mainly to the **thalamus** and to other basal areas, and from the thalamic areas the signals are driven to the somatosensory cortex as well. The paleospinothalamic tract transmits mainly chronic pain signals, but it can also transmit some acute pain signals, with its fibers ending mostly in laminae II and III of the dorsal horns, known as substantia gelatinosa. After passing this area, the last conducting neural cells join the fibers of the first pathways (for acute pain), carrying the signal into the brain stem and thalamus. Once the reticular formation and the thalamus receive the signal, conscious perception of pain occurs. The somatic sensory cortical areas in turn, interpret the quality of the perceived pain. Pain detection leads to an overall brain excitability that triggers the production of endogenous analgesic substances similar to opiates, known as enkephalins and **endorphins**, such as beta-endorphin, met-enkephalin, dynorphin, and leu-enkephalin. **Neurons** in multiple areas of the brain do present synaptic receptors for these naturally occurring painkillers. The pharmaceutical drugs used to treat pain are usually molecules similar to the endogenous endorphins and enkephalins, able to occupy the same synaptic receptors.

See also Anesthesia and anesthetic drug actions; Basal nuclei; Brain stem function and reflexes

PALADE, GEORGE EMIL (1912-)
Romanian-born American cell biologist

George Palade entered the science of cell biology at a time when techniques such as electron microscopy and sedimentation of discrete bits of **cell structure** were beginning to reveal the minute structure of the cell. He not only advanced these techniques, but also, by investigating the ultrastructure or fine structure of animal cells, identified and described the function of **mitochondria** as the powerhouse of the cell and of ribosomes as the site of protein manufacture. For his research in the function and structure of such cell components, he shared the 1974 Nobel Prize in physiology or medicine with two other cell researchers, **Albert Claude** and **Christian R. Duvé**.

George Emil Palade was born on November 19, 1912, in Jassy, in northeastern Romania. One of three children, Palade came from a professional family—his father, Emil, was a philosophy professor at the University of Jassy, while his mother, Constanta Cantemir, taught elementary school. Palade's two sisters, Adriana and Constanta, would grow up to be a professor of history and a pediatrician, respectively.

Attending school in Buzau, Palade entered the University of Bucharest in 1930 as a medical student. Ten years later he received his degree, having completed his internship as well as a thesis on the microanatomy of the porpoise kidney. Having earned his medical degree, Palade chose to focus on research instead of practicing medicine. His particular interest was **histology**, or the microscopic structure of plant and animal **tissue**. With the advent of the World War II, Palade was drafted into the army and stationed at the University of Bucharest Medical School as an assistant professor of **anatomy**. In 1941, he married and he and his wife eventually had two children.

In 1945, after being discharged from the army, Palade obtained a research position at New York University. While there, he met the eminent cell biologist Albert Claude, who had pioneered both the use of the electron microscope in cell study and techniques of cell fractionation (the separation of the constituent parts of cells by centrifugal action). The older scientist invited Palade to join the staff at the Rockefeller Institute (now Rockefeller University), and in 1946 Palade accepted a two-year fellowship as visiting investigator. In 1947, the communist-led government in Romania declared the country a people's republic and forced the abdication of King Michael. Palade, who had always planned to return home to work, now opted to remain in the United States. He became a United States citizen in 1952 and a full professor of cytology at Rockefeller in 1958.

At the Rockefeller Institute, Palade's first achievements came in the preparation of cell tissue for both the fractionation process and electron microscopy. In the former, collaborating with W. C. Schneider and George Hogeboom, Palade introduced as a fixative the use of gradient sucrose, and in the latter, buffered osmium tetroxide. However, his accomplishments soon went beyond improvements in methodology. Claude left the institute in 1949, and in the next decade Palade and his collaborators, building on Claude's work, reported groundbreaking descriptions of the fine appearance of the cell and of its biochemical function. Concentrating on the cytoplasm—the living material in the cell outside the nucleus—Palade was first attracted to larger organelles (bodies of definite **structure and function** in the cytoplasm) which Claude had earlier called "secretory granules." Palade showed that these tiny sausage-shaped structures, mitochondria, are the site where biochemical energy for the cell is generated. Animal cells typically contain a thousand such mitochondria, each creating adenosine triphosphate—ATP, a high-energy phosphate molecule—through enzymic (enzyme-catalyzed) oxidation or breakdown of **fat** and sugar. The **ATP** is then released into the cytoplasm where it powers energy-requiring mechanisms such as nerve impulse conduction, **muscle contraction**, or **protein synthesis**.

Using the high-power electron microscope (a device that utilizes electrons instead of light to form images of minute objects), Palade next revealed a delicate tracery, subsequently termed the endoplasmic reticulum by his collaborator, Keith R. Porter. The endoplasmic reticulum is a series of double-layered membranes present throughout all cells except mature **erythrocytes**, or red **blood** cells. Its function is the formation and transport of fats and proteins. By far

Palade's most significant work was with so-called microsomes, small bodies in the cytoplasm that Claude had earlier identified and shown to have a relatively high **ribonucleic acid** (**RNA**) content. RNA is the genetic messenger in protein synthesis. Palade observed these microsomes both as free bodies within the cytoplasm, and attached to the endoplasmic reticulum. In 1956, using a high-speed centrifuge, Palade and his colleague Philip Siekevitz were able to isolate microsomes and observe them under the electron microscope. They discovered that these microsomes were made of equal parts of RNA and protein.

Palade assumed that these RNA-rich microsomes were in fact the factories producing protein to sustain not only the cell but also the entire organism. The microsome was renamed the ribosome, and Palade and his team went to work to investigate the pathway of protein synthesis in the cell. Palade and Siekevitz began a series of experiments on ribosomes of the **liver** and **pancreas**, employing autoradiographic tracing, a sophisticated process similar to x-ray photography in which a picture is produced by radiation. Investigating in particular exocrine cells (those that secrete externally) of the guinea pig pancreas, the team was able, by 1960, to show that ribosomes do in fact synthesize proteins that are then transported through the endoplasmic reticulum. Further research with Lucien Caro, J. D. Jamieson, C. Redman, David Sabatini, and Y. Tashiro elucidated the function of the larger ribosomes attached to the endoplasmic reticulum, establishing them as the site where **amino acids** assemble into polypeptides (chains of amino acids). Palade's team also traced the transportation network as well as the function of the Golgi complex, tubelike structures where proteins are sorted before final transport to the cell surface for export.

Having completed his work on protein synthesis, Palade turned his attention to cellular transport—the means by which substances move through cell membranes. Working with Marilyn G. Farquhar, Palade demonstrated by electron micrography (images formed using an electron microscope) that molecules and ions were engorged by sacs or vesicles that move to the surface from within the cell. These vesicles actually merge with the outer membrane for a time, and then swallow up and bring the substances inside the cell. This vesicular model was in distinct contrast to the then current pore model whereby it was thought that molecules simply entered the cell through pores in the membrane.

Following the death of his wife in 1969, Palade remarried. In 1972, he left the Rockefeller Institute and became a full professor of cell biology at Yale University, continuing his research in cell morphology and function, but also turning to practical clinical uses of his discoveries. His later work is an attempt to establish links between defects in cellular protein production and various illnesses. In 1974, Palade shared the Nobel Prize in physiology or medicine with his former mentor, Albert Claude, and with Christian R. de Duvé, for their descriptions of the detailed microscopic structure and functions of the cell. He was also the recipient of the Passano Award in 1964, the Albert Lasker Basic Medical Research Award in 1966, the Gairdner Foundation Special Award in

1967, and the Horowitz Prize in 1970. In addition, Palade is the founding editor of the *Journal of Cell Biology*.

In 1990, Palade left Yale to become the dean for scientific affairs and to serve as a professor-in-residence in cellular and molecular medicine at the University of California, San Diego.

See also Biochemistry; Cell membrane transport; Mitochondria and cellular energy; Molecular biology; Protein metabolism

PALATE EMBRYONIC DEVELOPMENT · *see* FACE, NOSE, AND PALATE EMBRYONIC DEVELOPMENT

PALATE (HARD AND SOFT PALATE)

The palate is essentially the roof of the mouth. The front portion of the palate is constructed of bone (specifically two bones called the maxilla and the palatine) covered with a mucous membrane. Together these form the hard palate. The hard palate can be felt by running the tongue over the roof of the mouth. Further back in the mouth, behind the hard palate, lies the soft palate. The soft palate is made up of muscular **tissue** that is covered by **epithelial tissue**. A projection of tissue known as the uvula hangs down from the middle of the soft palate over the root of the tongue. It is thought that the uvula functions to keep food from straying down the **breathing** passage during **swallowing**. In singers, the uvula has been claimed to function in the generation of the vibrato, or wavy up-and-down sound.

The hard and soft palate separates the oral cavity from the nasal cavity. In other animals the roof of the mouth is actually the base of the **skull**.

The presence of the palate makes it possible to breathe and chew at the same time. When food is swallowed, the soft palate rises up and blocks off the entrance to the rear nasal passage. When food is not being swallowed, this passage is open, making it possible to breathe through the mouth and through the **nose**. As well, prior to swallowing food is pressed up against the palate and pushed to the back of the throat using the tongue.

The palate also functions in speaking and singing. When sound emerges from the chest, the sound waves that have been produced by the **vocal cords** bounce off the hard palate and out the mouth. The hard palate directs and resonates.

Formation of the palate occurs during development of the fetus. Improper formation of the hard palate occurs in one of every 500–1,000 babies. This condition, called cleft palate, is correctable by surgery. Its cause is still unresolved. A combination of inherited traits and some environmental factors in the mother's womb are suspected of causing the abnormality.

The uvula is implicated in snoring and **sleep** apnea (interrupted breathing during sleep). Treatment involves removal of excess flesh from the uvula.

See also Embryonic development, early development, formation, and differentiation; Face, nose, and palate embryonic development; Gustatory structures; Larynx and vocal cords; Swallowing and dysphagia

PANCREAS

The pancreas in humans is a solid, elongated, flattened gland about 10 in. (25 cm) long, lying behind the stomach and attached to the back of the abdominal cavity. Its "head" is just to the right of the mid-line and its "body" and "tail" point slightly upwards and lie just beneath the extreme edge of the left side of the ribs. The head is closely attached to the first part of the small intestine, into which the stomach empties partially digested solid and liquid food. It is to this that the pancreas adds its digestive juices, containing **enzymes**. The tube draining the **liver** of its **bile** (the bile duct) lies just behind the head of the pancreas and usually joins the bowel at the same place where the fluids from the pancreas enter the bowel.

The pancreas is composed of two major types of tissues: exocrine **tissue**, the acini, which secrete digestive enzymes into the duodenum through the pancreatic duct, and endocrine tissue, the islets of Langerhans, which produce and secrete the **hormones** insulin and glucagon directly into the **blood**. Endocrine tissue contains alpha, beta and delta cells, which are distinguished from one another by their morphology and staining characteristics. Beta cells produce insulin and alpha cells produce glucagon, two hormones regulating blood glucose levels. Delta cells secrete the hormone somatostatin, which inhibits insulin and glucagon secretion.

A deficiency of the hormone insulin in the body results in **diabetes mellitus**, which affects about 13 million individuals in the United States. Beta cells are usually present in the pancreas of a person who has severe diabetes, but they have a hyalinized appearance, contain no secretory granules and do not exhibit staining reactions for insulin. Diabetes is characterized by a high blood glucose (sugar) level and glucose spilling into the urine due to a deficiency of insulin. As more glucose concentrates in the urine, more water is excreted, resulting in extreme **thirst**, rapid weight loss, drowsiness, fatigue, and possibly dehydration. Because the cells of the diabetic cannot use glucose as fuel, the body uses stored protein and **fat** for energy, which leads to a build-up of acid (**acidosis**) in the blood. If this condition is prolonged, the person can fall into a diabetic **coma**. A diabetic coma can be simply diagnosed as it is characterized by deep, labored **breathing** and a sweet-smelling, "fruity" breath.

There are two types of diabetes. In Type I diabetes, formerly called juvenile-onset diabetes, the pancreas cannot produce insulin. People with Type I diabetes must have daily insulin injections. It is necessary to avoid taking too much insulin as it can lead to insulin **shock**. Insulin shock begins with a mild hunger, which is quickly followed by sweating, shallow breathing, dizziness, palpitations, trembling, and mental confusion. As the blood sugar falls, the body tries to compensate by breaking down fat and protein to make more sugar. Eventually, low blood sugar leads to a decrease in the sugar supply to the **brain**, resulting in a loss of consciousness. Eating a sugary food can prevent insulin shock until appropriate medical measures can be taken. Type II diabetes, formerly called adult-onset diabetes, can occur at any age. In this condition, the pancreas can produce insulin, but the cells do not respond to it.

See also Cell structure; Digestion; Endocrine system and glands

PARAGANGLIA

Paraganglia are groups of cells that originate from the neural crest tissues. They are described as two different types. One type is found inside the adrenal medulla and is responsible for the secretion of **epinephrine** and norepinephrine. The other type is located outside the adrenal glands and has an entirely different function. It is believed the second type is responsible for the detection of oxygen and carbon dioxide levels in the **blood.**

Other paraganglia are found in small pockets or groups mostly in the head and neck region. Several types have been identified. Type I are cells called "true paraganglia" and are the type located along nerve fibers or near ganglia of the **central nervous system**. Type II are "free paraganglia" and are found within **adipose tissue** of the **heart**. They do not show any definite connection to specific structures. Type III, the "interganglionic paraganglia" are located within the ganglia themselves. Finally, the type IV are the "Intramyocardic paraganglia" which are immunoreactive cells that lie near the myocardiocyte bundles (excitable myocardial cell fibers that transmit action potentials).

One of the most interesting functions of the paraganglia is their role in triggering signals to the **brain** via the vagus nerve. When a person is sick, macrophages (a type of **immune system** cell) releases cytokines that let the hypothalmus part of the brain stimulate and immune reaction. The cytokines are too big to pass the blood-brain barrier, but, instead, they attach to pockets of paraganglion cells located along the vagus nerve. The paraganglioic cells have receptors for interleukin-1, one of the types of cytokines carried by the blood. In turn these cells release **neurotransmitters** the vagus nerve and help send the signal to the brain to stimulate the body's immune response.

See also Neurotransmitters; Nerve impulses and conduction of impulses

PARASYMPATHETIC NERVOUS SYSTEM

The **autonomic nervous system** (ANS) is divided into two subsystems, the parasympathetic and sympathetic systems. Parasympathetic fibers innervate **smooth muscle**, cardiac muscle, and glandular **tissue**. In general, stimulation via

parasympathetic fibers slows activity and results in a lowering of metabolic rate and a concordant conservation of energy.

Most target **organs** and tissues are innervated by neural fibers from both the parasympathetic and sympathetic systems. The systems can act to stimulate organs and tissues in opposite ways (antagonistically). For example, parasympathetic stimulation acts to decrease **heart** rate. In contrast, sympathetic stimulation results in increased heart rate. The systems can also act in concert to stimulate activity (e.g., both increase the production of **saliva** by salivary glands, but parasympathetic stimulation results in watery as opposed to viscous or thick saliva). Although they share a number of common features, the classification of the parasympathetic and the sympathetic systems of the ANS is based both on anatomical and physiological differences between the two divisions.

Both the parasympathetic and sympathetic systems contain myelinated preganglionic nerve fibers that usually connect with (**synapse** to) unmyelinated postganglionic fibers, via a cluster of neural cells termed ganglia. The preganglionic fibers of the parasympathetic system derive from the neural cell bodies of the motor nuclei of the occulomotor (cranial nerve: III), facial (VII), glossopharyngeal (IX), and vagal (X) **cranial nerves**. There are also contributions from cells in the sacral segments of the **spinal cord**. These cranio-sacral fibers generally travel to a ganglion that is located near or within the target tissue. Because of the proximity of the ganglia to the target tissue or organ, the postganglionic fibers are much shorter.

Parasympathetic stimulation of the pupil from fibers derived from the occulomotor (cranial nerve: III), facial (VII), glossopharyngeal (IX) nerves constricts or narrows the pupil. This reflexive action is an important safeguard against bright light that could otherwise damage the **retina**. Parasympathetic stimulation also results in increased lacrimal gland secretions (tears) that protect, moisten, and clean the **eye**.

The vagus nerve (cranial nerve X) carries fibers to the heart, **lungs**, stomach, upper intestine, and ureter. Fibers derived from the sacrum innervate reproductive organs, portions of the colon, bladder, and rectum.

With regard to specific target organs and tissues, parasympathetic stimulation acts to decrease heart rate and decrease the force of contraction. Parasympathetic stimulation also reduces the conduction velocity of cardiac muscle fibers.

Parasympathetic stimulation of the lungs and smooth muscle surrounding the **bronchi** results in bronchial constriction or tightening. Parasympathetic stimulation can also result in increased activity by glands that control bronchial secretions.

Parasympathetic stimulation usually causes a dilation of arterial **blood** vessels, increased **glycogen** synthesis within the **liver**, a relaxation of gastro-intestinal sphincters (smooth muscle valves or constrictions), and a general increase in gastrointestinal motility (the contracts of the intestines that help food move through the system.

Parasympathetic stimulation results in a contracting spasm of the bladder. Accompanied by a relaxation of the sphincter, parasympathetic stimulation tends to promote urination.

The chemical most commonly found in both pre and postganglionic synapses in the parasympathetic system is the neurotransmitter acetylcholine.

See also Muscular innervation; Nerve impulses and conduction of impulses; Nervous system overview; Nervous system, embryological development; Neural plexi; Neurology; Neurons; Neurotransmitters

PARATHYROID GLANDS AND HORMONES

Parathyroid glands are found on the posterior surface of the thyroid gland. Normally, there are four parathyroids, averaging 120 mg in total mass but as many as 5% of normal individuals have more than four. The two superior parathyroids are found near the upper poles of the thyroid and the two inferior glands are near the lower poles. Parathyroid glands can be found in aberrant locations such as the tracheoesophageal groove, the retroesophageal space and the anterior **mediastinum**.

Parathyroid glands produce and secrete parathyroid hormone (PTH) that, together with vitamin D is the principle regulator of **calcium** ion (Ca^{+2}) in **extracellular fluid**. PTH is synthesized in the parathyroids as a "preprohormone" which is cleaved to give a prohormone. Further cleavage yields the 84 amino acid hormone, PTH.

Maintenance of extracellular Ca^{+2} within narrow limits is essential to the proper function of many tissues. It is responsible for cardiac and **skeletal muscle** contraction and relaxation. Excess calcium can lead to **heart** arrhythmias, muscle twitches and cramps. Calcium concentration is key to the secretory activity of practically all endocrine and exocrine glands. It is a primary factor in **blood** clotting and an important regulator of many cellular enzyme activities.

Disturbance in calcium homeostasis can lead to a wide variety of disease states including **hyperthyroidism**, hypothyroidism, **osteoporosis**, Paget's disease, renal disease, rickets and pancreatitis.

PTH regulates extracellular Ca^{+2} within very narrow limits. When circulating Ca^{+2} levels fall, the parathyroids are stimulated to secrete PTH. Conversely, when Ca^{+2} levels rise above **equilibrium**, there is evidence for intracellular degradation of synthesized PTH. Secretion of PTH can be inhibited by high extracellular magnesium ion concentrations. The presence of the active metabolite of vitamin D (dihydroxycholecalciferol) also suppresses PTH secretion and action.

PTH acts directly on kidney and bone. This action results from the binding of PTH to specific membrane-bound receptors on target cells of both **organs**. In the kidney, PTH increases the reabsorption of calcium (and magnesium) ions from the glomerular filtrate into the blood, increases excretion of phosphate and bicarbonate ions via urine and activates the enzyme that forms dihydroxycholecalciferol. In bone, PTH causes release of calcium and phosphate into the extracellular fluid by acting directly on the osteoblasts.

The hormone acts indirectly on the small intestine to maintain normal concentrations of serum Ca^{+2}. The stimulation of dihydroxycholecalciferol causes increased absorption of calcium by the epithelial cells of the intestinal lumen into the bloodstream.

See also Calcium and phosphate metabolism; Electrolytes and electrolyte balance; Hormones and hormone action; Thyroid histophysiology and hormones

PARTURITION

Labor (parturition) refers to the series of events that expel the fetus out of the mother's body by means of uterine contractions. The musculature of the pregnant uterus is arranged in three strata: an external hood-like layer, which arches over the fundus and extends into the various ligaments; an internal layer consisting of sphincter-like fibers around the orifices of the tubes and the internal Os; and the main portion of the uterine wall, formed by a middle layer, which consists of an interlacing network of muscle fibers. The uterus has pacemakers to produce the rhythmic coordinated contractions of labor.

Parturition results from a combination of several different factors that progressively increase the excitability of the uterine musculature. The **placenta** at term begins to secrete a higher ratio of **estrogen** (which excites uterine activity) to **progesterone** (which inhibits uterine activity). The fetal size produces a stretching of uterine musculature that also increases its excitability. The fetal presenting part (usually the head) mechanically presses downward on the maternal pelvis, stretching the cervix, which in turn, increases uterine activity. Relaxin, produced by the ovary, increases cervical dilation and relaxation of pelvic ligaments. Its levels increase just prior to parturition. The posterior pituitary gland increases secretion of the hormone **oxytocin**, while at the same time, uterine muscle cells increase the number of oxytocin receptors, increasing the uterus' sensitivity to oxytocin. Thus, the fetal presenting part stretches the cervix, which causes a greater increase in uterine activity, due to both mechanical pressure and biochemical action (signals from the cervix to the **hypothalamus**, causing an increased oxytocin production). As a result of these factors, the rhythmic contractions of the uterus become strong enough to begin pushing the fetus into the birth canal.

Normal labor consists of three stages. The first stage is the dilation stage, from the onset of labor to complete dilation of the cervix. The cervix dilates up to 10 cm in diameter. The first stage is divided in two substages, the latent phase that encompasses cervical effacement (thinning) and early dilatation. The second is the active phase, during which more rapid cervical dilatation occurs (starting from 3–4 cm). During the dilation stage, the amniotic fluid is expelled ("water breaks").

The second tage of labor is the expulsion stage, and encompasses complete cervical dilation throughout the delivery of the infant. The placental stage (stage three) begins after the delivery of the infant and ends with the expulsion of the placenta. Stage three occurs within 5–30 minutes after delivery of the infant.

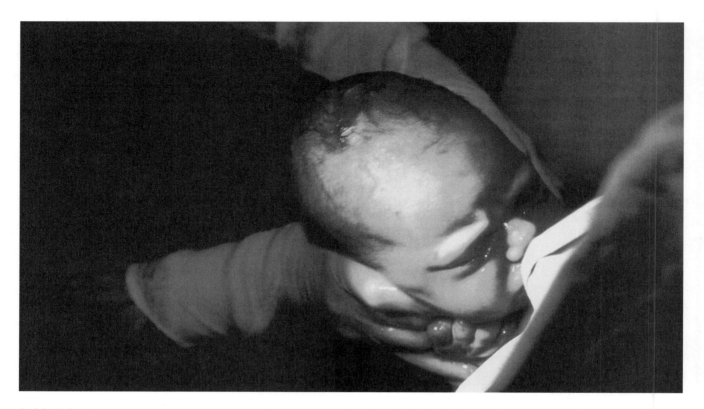

An infant's head emerges from the mother during birth. © SIU, National Audubon Society Collection/ Photo Researchers, Inc. Reproduced with permission.

The mechanism of labor consists in several changes of position (cardinal movements) of the fetus as it passes through the birth canal. This accommodation is accomplished by the forceful contractions of the uterus. In other words, the fetal head adapts to the maternal bony pelvis, and the lower most presenting part of fetus (usually the occipital portion of the head) rotates toward the largest pelvic segments. The head-first, or vertex, presentation of the infant occurs in 95% of term labors. The cardinal movements include engagement, the descent of biparietal diameter of fetal head below the pelvic inlet, descent of the presenting part through the birth canal (mostly occurring during the latter portion of the first stage and the second stage of labor), flexion of the fetal head to present the smallest diameter of the fetal head to the maternal pelvis, internal rotation that facilitates the presentation of the optimal diameters of the fetal head to the bony pelvis, extension of the fetal head as it reach the vaginal introitus, external rotation of the fetal head after delivery to "face forward" relative to his shoulders, and finally, expulsion, the final delivery of the fetus from the birth canal.

See also Pregnancy; Pregnancy, maternal physiological and anatomical changes; Prenatal growth and development

PASSIVE TRANSPORT · *see* CELL MEMBRANE

TRANSPORT

PATHOLOGY

Pathology is the scientific study of disease processes that affect normal **anatomy** and **physiology**. Anatomical and physiological changes are pathological changes when they result from an underlying disease process or abnormality.

Pathologists play an increasingly important role in diagnosis, research, and in the development of clinical treatments for disease. A specialized branch of pathology, **forensic pathology**, offers a vast array of molecular diagnostic techniques (including **DNA** fingerprint analysis) toward identification of remains, gathering of evidence, and identification of suspects.

Modern pathology labs rely heavily on **molecular biology** techniques and advances in biotechnology. During the last two decades, there have been tremendous advances in linking changes in cellular or **tissue** morphology (i.e., gross appearance) with genetic and/or intracellular changes. In many cases, specific molecular tests can definitively identify disease process, and of critical importance to the treatment of disease, and make a correct diagnosis at an earlier stage in the disease process.

Pathologists attempt to relate observable changes to disease process. Whether the changes are evident morphologically—or are distinguishable only via sophisticated molecular tests—the goal is to determine the existence and/or etiology of disease (the cause of disease). Once the etiologic agents are identified, the general goal of research is to document and gather evidence of the pathogenesis of disease (i.e., the mechanisms by which etiologic agents cause disease).

On a daily basis, pathologists perform a broad spectrum of tests on clinical samples to determine anatomical and physiological changes associated with a number of disease processes, including the detection of cancerous cells and **tumors**.

Major branches of pathology include the study of anatomic, cellular, and molecular pathology. Specific clinical studies often focus on transplantation pathology, neuropathology, immunopathology, virology, parasitology, and a number of clinical subspecialties (e.g. pediatric pathology).

See also Histology and microanatomy; History of anatomy and physiology: The Classical and Medieval periods; History of anatomy and physiology: The Renaissance and Age of Enlightenment; History of anatomy and physiology: The science of medicine; Human genetics; Immunology; Medical training in anatomy and physiology

PAVLOV, IVAN PETROVICH (1849-1936)
Russian physiologist

Ivan Petrovich Pavlov's research on mammalian **digestion** earned him the Nobel Prize and his research regarding conditioned **reflexes** brought him international recognition. The colloquial expression "Pavlov's dog" refers to Pavlov's famous experiments in which he taught a dog to salivate at the sound of a bell by associating the bell with feeding. This research helped spawn a physiologically oriented school of psychology that focused on the influence of conditioned **reflexes** on learning and behavior. Because of his contribution to the fields of psychology and physiology, Pavlov became one of Russia's most revered scientists in his day—even tolerated by the communist Soviet regime, of which he was openly critical.

Pavlov was born in Ryazan, Central Russia, on September 26, 1849. His father, Pyotr Dmitrievich Pavlov, was a priest who rose through the ranks and eventually headed one of the most influential parishes in the area. A devoted reader and scholar, Pyotr taught his son at an early age to read all worthwhile books at least twice so that he would understand them better—a bit of fatherly advice that helped shape Pavlov's intense dedication to his work. Pavlov's mother, Varvara Ivanova, also came from a family of clergy and had ten children after Pavlov, six of whom died in childhood.

The family expected the young Pavlov to follow the family tradition of entering the clergy. Accordingly, Pavlov attended Ryazan Ecclesiastical High School and the Ryazan Ecclesiastical Seminary. During his studies at the seminary, Pavlov became seriously interested in science, physiology in particular, and was greatly influenced by a radical philosopher named Dmitri Pisarev who espoused many of evolutionist Charles Darwin's theories.

In 1870, after the government decreed that divinity students could attend nonsectarian universities, Pavlov decided to leave the seminary and attend St. Petersburg University to study the natural sciences. At St. Petersburg, Élie de Zion, a professor of physiology, made a formidable impression on Pavlov. By all accounts, the two scientists had a mutual admiration for one another. According to Boris Babkin in his book *Pavlov: A Biography,* Pavlov said of his early mentor, "Never can such a teacher be forgotten," and in turn Zion called Pavlov a "skilled surgical operator."

Upon graduation from St. Petersburg University in 1875, Pavlov followed Zion to the Military Medical Academy in St. Petersburg, where Zion had been appointed chair of physiology. Pavlov became an assistant in Zion's laboratory and worked toward his medical degree. But Zion was soon dismissed because he was Jewish, and Pavlov, intolerant of his mentor's dismissal, left the Medical Academy in favor of the Veterinary Institute where he spent the next two years studying digestion and circulation. In 1877, Pavlov traveled to Breslau, Germany (later Poland), to study with Rudolf Heidenhain, a specialist in digestion. After receiving his medical degree from the Military Medical Academy in 1879, Pavlov went on to earn his postdoctoral degree and was honored with a Gold Medal for his doctoral dissertation in 1883.

Upon graduation, Pavlov was one of ten students awarded a government scholarship for postgraduate studies abroad. Pavlov returned to Germany to work with Carl Ludwig on cardiovascular physiology and **blood** circulation; he also collaborated with Heidenhain again on further digestion research. Another mentor, Sergei Botkin, eventually asked Pavlov to direct an experimental physiological laboratory. This lab, devoted to the school of "scientific medicine," focused on the physiological and pathological relations in an organism. Under Botkin's guidance, Pavlov first developed his interest in "nervism," the pathological influence of the **central nervous system** on reflexes.

Pavlov returned to St. Petersburg University and began his exhaustive research on digestion that eventually gained him worldwide recognition in scientific circles. Focusing on the physiology of digestion and gland secretions, Pavlov devised an ingenious experiment in which he severed a dog's gullet, forcing the food to drop out before it reached the animal's stomach. Through this sham feeding, he was able to show that the sight, **smell**, and **swallowing** of food was enough to cause secretion of the digestive acidic "juices." He demonstrated that stimulation of the vagus nerve (one of the major **cranial nerves** of the **brain**) influences secretions of the gastric glands. In 1904 Pavlov received the Nobel Prize in physiology or medicine for these pioneering studies on the physiology of the digestive system.

Pavlov's work on blood circulation earned him a professorship at the Military Medical Academy. In 1895 Pavlov became the chairman of the physiology department at the St. Petersburg Institute for Experimental Medicine, where he spent the greater part of his remaining career. Ironically, by the time Pavlov received the Nobel Prize, his work delineating the central nervous system's effects on digestive physiology was soon to be overshadowed by subsequent investigations by William Bayliss and others, who demonstrated that chemical (hormone) stimulation induces digestive secretions from the

pancreas. Ever curious, Pavlov himself conducted experiments that also confirmed this discovery.

Although Pavlov's "nervism" theory was relegated to secondary importance in the study of digestion, his experiments profoundly influenced biological research. Pavlov strongly believed that a healthy laboratory animal subject free from disease and the influence of pharmaceuticals was imperative to his work. "It has become abundantly clear that the usual simple cutting of an animal, the so-called acute test, is a source of many errors," Pavlov said, as noted by Alexander Vucinich in his book *Science in Russian Culture*. Among the laboratory techniques advanced by Pavlov were the use of aseptic surgical procedures on laboratory animals and the development of chronic, or long-lasting, experiments on the same animal. Pavlov believed in minimizing an animal's **pain** for both moral and scientific reasons and led the way for the humane treatment of laboratory animals.

Ironically, Pavlov's most famous studies were conducted after he received the Nobel Prize. Concentrating on the neural influences of digestion, Pavlov set out to determine whether he could turn normally "unconditioned" reflexes or responses of the central nervous system into conditioned reflexes. Pavlov noticed that the laboratory dogs would sometimes salivate merely at the approach of lab assistants who often fed them. Through careful repeated experiments, Pavlov demonstrated that if a bell is rung each time a dog is given food, the dog eventually develops a "conditioned" reflex to salivate at the sound of the bell, even when food is not present. Thus, Pavlov showed that the unconditioned reflexes—gastric activity and salivation—could become conditioned responses triggered by a stimulus (the bell) not previously associated with the physiological event (eating).

Pavlov traced this phenomenon to the **cerebral cortex** and continued to study the brain's role in conditioned reflexes for the remainder of his life. Although this research led to a proliferation of studies of conditioned reflexes in physiology, the conditioned reflex theory became a popular subject in the fields of psychiatry, psychology, and education.

Pavlovian psychology contends that a person's behavioral development and learning are profoundly affected by conditioned nervous responses to life events, similar to the dog's learned response to the bell. This behavioral theory created a schism in the field of psychology, with Pavlovian psychologists opposing the views of Sigmund Freud, who theorized that an individual's thought processes—especially the unconscious—were the driving forces of human behavior. Eventually, Freudian psychology usurped Pavlovian psychology in popularity to become the primary approach to mental health treatment outside of Russia. However, Pavlov maintained his devotion to the importance of conditioned reflexes in human behavior, believing that human language was probably the most intricate example of such conditioning. Pavlov also applied his theory to the treatment of psychiatric patients in which he placed patients in a quiet and isolated environment in order to remove any possible physiological or psychological stimuli that might negatively affect their mental health.

Pavlov's life spanned three distinct Russian political eras, which sometimes intruded upon his personal and professional life. He was born during the reign of Czar Nicholas I, an oppressive feudal monarch who sought to retain aristocratic rule at any price. Pavlov saw this oppressive regime give way to a new ideology of reform, known as post-Emancipation Russia, which heralded technological advancements, but remained mired in turmoil on both the social and political level. Shortly after the Bolshevik Revolution in 1917 that attempted to impose a socialist structure on society, Pavlov became a staunch and vocal opponent of the new and often hostile regime. Years earlier, Pavlov had shown a willingness to oppose authority when he resigned from the Medical Military Academy to protest the dismissal of his mentor, Zion, because he was Jewish.

By the time of the Bolshevik Revolution, Pavlov had achieved international recognition and was living a comfortable life. He had overcome the extreme economic hardships he faced early in his career when he had struggled to support his growing family on the meager salary of a lab assistant. In 1881, he married Seraphima Vasilievna Karchevskaya, a naval doctor's daughter and friend of Russian novelist Fyodor Dostoevsky. They eventually had four sons and a daughter.

Pavlov was willing to risk his hard-earned success by opposing the new communist regime. His religious background caused him to become enraged when all clergymen's sons were expelled from the Medical Academy, and in 1924, he resigned from the Medical Academy as the chair of physiology in protest. The new Soviet government, however, was intent on accommodating a person they considered to be a shining example of Russian science, who, in addition to the Nobel Prize, had been awarded the Order of the Legion of Honor of France and the Copley Medal of the Royal Society of London. Vladimir Lenin, who emerged as the most powerful leader of the revolution, signed a decree guaranteeing Pavlov's personal freedom and his right to continue his research and even to attend church. These special privileges were in stark contrast to countless other scientists whose work was suppressed by the government. Pavlov, however, continued to speak out, once refusing extra rations of food unless all his laboratory assistants received the same privileges.

The rulers in the new Soviet republic believed that Pavlov's work with conditioned reflexes could be adapted for political purposes. For example, they hypothesized that the masses could be conditioned just as Pavlov had conditioned the dog. In a sense, they saw the opportunity to develop a type of mass mind control in which the Soviet system would appear to have complete power, even over those who were originally reluctant to follow the communist way. To appease their favorite scientist, in 1935 the government built Pavlov a spacious laboratory equipped with the latest scientific technology which the scientist called "the capitol of conditioned reflexes."

Although Pavlov was known in the last five years of his life to publicly praise the government for their efforts to foster education and science, he repeatedly denounced the Soviet "social experiment." He died of pneumonia on February 27, 1936, in Leningrad.

See also Brain stem function and reflexes; Brain: Intellectual functions; Mastication; Psychopharmacology; Reflexes; Saliva

PENIS • *see* GENITALIA (MALE EXTERNAL)

PERICARDIUM

The pericardium refers to a thin membranous sac that surrounds the **heart** and the roots of the main **blood** vessels leading from the heart. Compositionally, it contains three layers. These layers are the fibrous pericardium, visceral pericardium (also known as the epicardium), and the parietal pericardium. Fluid is found in between the visceral pericardium and the parietal pericardium layers. This area is called the pericardial cavity. The fluid acts as a lubricant and as a **shock** absorber, reducing the friction between the membranes.

The fibrous pericardium is the outer fibrous sac that covers the heart. It is a protective layer. The next layer is the epicardium, the outer layer of the heart wall. It is composed of connective **tissue** that is covered by epithelium. The epicardium is also a protective layer. Between these two layers is the parietal pericardium. It is an insulating layer.

The pericardium has several functions. Its fibrous construction acts to hold the heart within the chest cavity. Also the membrane layers and the intervening fluid acts as a shock absorber to limit the internal motion of the heart generated by the pumping action of the organ. It also acts as a glove to prevent the heart from becoming too large when the volume of blood flowing through the heart increases.

The pericardium can become inflamed, condition termed pericarditis. Men between the ages of 20 and 50 are most at risk for pericarditis. The inflammation produces an increase in the amount of the fluid. This increased volume impinges on the heart, and the pressure can adversely affect the functioning of the heart. As a consequence of the increased fluid pressure, a sharp and piercing **pain** develops in the chest that becomes more severe when a breath is taken during inspiration.

Pericarditis can be caused by an **infection**, heart attack, **cancer**, radiation treatment, or as a result of injury or surgery. Sometimes pericarditis can be a secondary ailment accompanying rheumatoid arthritis, lupus, and failure of the kidney. Treatment for pericarditis can involve the administration of antibiotics, if the cause is an infection, and drugs to reduce inflammation and relieve pain. In more extreme cases, a needle may be inserted into the pericardium to draw of excess fluid (aspiration).

See also Cardiac cycle; Cardiac disease; Cardiac muscle; Connective tissues; Inflammation of tissues

PERIPHERAL NERVOUS SYSTEM

The peripheral nervous system (PNS) is one component of the overall nervous system network in the human body. The other component is the **central nervous system** (CNS). The two are interconnected, and function to both sense and react to environments inside and outside the body.

The PNS is made up of **neurons**; the cellular pathways for the various signals on which the proper operation of the nervous system relies. There are two types of neurons operating in the PNS. The first are the sensory neurons that run from the myriad of sensory receptors throughout the body. Sensory receptors provide the connection between the stimulus (such as heat, cold and **pain**) and the CNS. As well, the PNS also consists of motor neurons. These neurons connect the CNS to various muscles and glands throughout the body. These muscles and glands are also known as effectors. That is, they are the places where the responses to the stimuli are translated into action.

An example of the above is the sequence of events that occurs when a hot object, such as a frying pan, is touched. Sensory receptors detect the heat and associated pain. These signals are conveyed to the **brain** by the sensory neurons. Interpretation of the signals occurs in the brain. A response signal is ferried to the appropriate muscles by motor neurons. Finally the particular effectors are stimulated to perform the action, in this example, contraction of the muscles, so as to pull the hand away from the source of the heat and pain. This cycle is completed within a second.

The PNS is divided into two subsystems: the sensory-somatic nervous system and the **autonomic nervous system**. The sensory-somatic nervous system is the sensory gateway between the environment outside of the body and the central nervous system. Responses tend to be conscious. The autonomic nervous system is the gateway between the internal environment and the central nervous system. The control of activity, typically automatic, of many **organs** and muscles is the responsibility of the autonomic nervous system.

The sensory nervous system comprises twelve pairs of **cranial nerves** and thirty-one pairs of spinal nerves. Some pairs are exclusively sensory neurons, such as the pairs involved in **smell**, vision, hearing, and balance. Other pairs are strictly made up of motor neurons, such as those involved in the movement of the eyeballs, **swallowing**, and movement of the head and shoulders. Still other pairs consist of a sensory and a motor neuron working in tandem, such as those involved in **taste**, and other aspects of swallowing. All of the spinal neuron pairs are mixed; that is, they contain both sensory and motor neurons. This allows the spinal neurons to properly function as the conduit of transmission of the signals of the stimuli and the subsequent response.

The autonomic nervous system consists of both sensory and motor neurons. These connect the brain components of the central nervous system (in particular the **hypothalamus** and **medulla oblongata**) and various internal organs (such as the **heart, lungs,** and glands). In yet another subdivision, the autonomic nervous system consists of three subsystems; the **sympathetic nervous system**, the **parasympathetic nervous** system and the enteric nervous system. The sympathetic nervous system is involved with the response of the body to emergencies—a quite general reaction commonly known as the **"fight or flight" reaction**. Release of a chemical called noradrenaline (also called norepinephrine) alternately stimulates or inhibits the functioning of a myriad of glands and muscles. Examples include the acceleration of the heartbeat, raising of **blood** pressure, shrinkage of the pupils of the eyes, shrinkage of the air passages in the lungs, and the redirection of blood away from the skin to muscles, brain and the heart.

The parasympathetic nervous subsystem operates to return the body to it's normal levels of function following the sudden alteration by the sympathetic nervous subsystem; the so-called "rest and digest" state. Examples include the restoration of resting heartbeat, blood pressure, pupil diameter, and flow of blood to the skin.

The enteric nervous system is made up of nerve fibers that supply the "viscera" of the body; the **gastrointestinal tract, pancreas** and gall bladder.

See also Homeostatic mechanisms

PERISTALSIS • *see* INTESTINAL MOTILITY

PERITONEUM

The peritoneum is the largest serous membrane in the body. It completely surrounds the body cavity and contains the viscera and other body **organs**. In males, it is a completely closed sac, but in women, the peritoneum is perforated by the ovaries and fallopian tubes. The peritoneum is divided into two subdivisions, the parietal and visceral. The peritoneal cavity is the space that lies between these two portions of the membrane.

This important membrane has an interesting embryonic development. From early embryonic tissues a whole series of supportive tissues grow to contain and support the internal organs. The peritoneum is multi-layered. The outer layer of the visceral peritoneum forms the serosa of most of the intestinal tract. Other layers emerge into the abdominal cavity and attach to and support the organs including the intestinal tract. These connective and supportive layers are called the mesenteries. These specialized tissues keep the intestines in place and carry **blood** vessels and nerve tracts to the organs.

The mesenteries are free membranes connected to the entire digestive system. They are composed of flattened endothelium cells. A thin film of serous fluid lubricates the entire surface. This makes it possible for the organs to slide easily against one another within the body cavity as it moves. The outer surface of the peritoneum is attached to the body wall and is called the smooth surface. Organs slide easily along the cavity wall. The wall of the peritoneum reflects inwardly at various points along its wall. These reflections penetrate deeply into the body cavity and enclose the stomach, **liver**, spleen, and **heart**. The offshoots of this important membrane form the greater and small omentum that overlie the abdominal (stomach) wall under the linea alba.

See also Abdomen

PFLÜGER, EDUARD FRIEDRICH WILHELM (1829-1910)
German physician

E.F.W. Pflüger, the most prominent general physiologist of his time, founded *Archiv für die gesamte Physiologie* (Archives of All Physiology), a scholarly journal better known simply as *Pflügers Archiv*, in 1868. He contributed significantly to most subfields of physiology, especially physiological chemistry, embryological physiology, **metabolism**, and respiratory physiology.

Pflüger was born in Hanau am Main, Germany, the son of an activist leftist politician from whom he probably inherited his combative temperament. Both Pflügers got into trouble in 1849, as the 1848 revolutions failed, but while the father was unrepentant, the son, then a student at the University of Heidelberg, lost interest in politics, quit his pursuit of a law degree, and began instead to study medicine. He was a medical student at the University of Marburg from 1850 to 1851, earned his first M.D. at the University of Giessen in 1851, became the protégé of **Johannes Müller** and Emil du Bois-Reymond (1818–1896) at the University of Berlin also in 1851, and earned his second M.D. there in 1855.

Under du Bois-Reymond, Pflüger became an expert on the relatively new science of electrophysiology, which had arisen quite a while after the discovery of electric charges in nerve impulses by **Luigi Galvani**. Pflüger's 1859 book on this topic gained him the coveted professorship of physiology at the University of Bonn, where he succeeded Hermann von Helmholtz (1821–1894). He held that chair for the rest of his life and died in Bonn while lying in bed, editing page proofs for his *Archiv*.

Naturally belligerent, Pflüger mercilessly criticized other physiologists, who almost all feared, respected, and despised him. He was close to his wife and children but had hardly any friends outside his family. Even though he was absolutely devoted to science and believed that science progresses through controversy, he sometimes picked fights just for the sake of fighting. He struggled long and bitterly against Carl von Voit (1831–1908) about whether protein could produce **glycogen**, but finally had to admit that Voit was right. Hermann Senator (1834–1911), Hermann Munk (1839–1912), and Emil Fischer (1852–1919) are among the other prominent physiologists who felt Pflüger's wrath in public.

Pflüger was obsessed with work. He investigated the formation of eggs in the ovary; the exchange of oxygen and carbon dioxide in respiration; the polarity of nerves; the relative energy requirements of various types of cells; the **digestion** and absorption of proteins, fats, and **carbohydrates**; the metabolism of nitrogen compounds; the natural course of diabetes; the cardiological functions of the vagus nerve; the production of body heat; the comparative **embryology** of the female reproductive **organs**; the gas content of **blood**; the etiologies of dyspnea and apnea; and the neurological properties of the salivary glands. In one famous experiment, he replaced the blood of a frog with physiological saline solution and observed no significant difference in gas exchange, thus demonstrating that the seat of respiration is the bodily tissues, not the blood. He also invented several physiological instruments, including the lung catheter, the blood-gas evacuation pump, and the pneumonometer.

See also Biochemistry; Electrolytes and electrolyte balance; Embryology; Glycogen and glycolysis; Nerve impulses and conduction of impulses; Ovarian cycle and hormonal regulation; Protein metabolism; Respiratory system

PHAGOCYTOSIS · *see* CELL MEMBRANE TRANSPORT

PHALANGES

The phalanges are the bones that are found in the fingers and the toes. The phalanges of the hand are collectively known as the *phalanges digitorum manus*. The phalanges of the foot are collectively known as the *phalanges digitorum pedis*.

In the hand, the thumb has two phalanges: the distal phalange (farther from the wrist) and the proximal phalange (closer to the wrist or base of the finger). The remaining fingers add an intervening third phalange appropriately named the middle phalange. In total there are fourteen phalanges in the hand.

The phalanges of the hand are thickest at the end nearest the wrist, and narrow along their length. They are flat along their top and bottom surfaces, and their sides are rough in order to make the attachment of the flexor **tendons** easier. There is a great deal of articulation (movement of one bone relative to its neighbor) in the phalanges. This is one feature that allows the hand to be so dexterous. Often the phalanges are concave at the end that articulates with the next phalange. The convex portion of next phalange allows a smooth fit of one phalange with another.

Formation of the bones of the phalanges begins at about the eight week of fetal development for some of the phalanges. However, full development of the extreme phalanges is not complete until more than a year after birth, and development of all the phalanges is not complete until the end of the teenage years.

The number and arrangement of the phalanges of the foot is similar to those of the hand. The largest toe has two phalanges and the other toes have three phalanges. However, in the toe the phalanges are smaller, shorter and more compressed than those of the hand.

Other animals besides humans have phalanges. In the horse and the ox, as examples, the distal phalange is the bone of the hoof. In these animals the distal phalange is a single

unit. There are not multiple distal phalanges, because horse and oxen do not, of course, have fingers.

Like all bones, phalanges are subject to breakage. Healing of a broken hand or foot bone involves the immobilization of the bone. In the hand, this can often take the form of a splint on the particular digit that has been damaged. In the foot, however, the entire foot is usually immobilized in a cast.

Some forms of disease strike preferentially at phalanges. Rheumatoid arthritis strikes at proximal phalanges while osteoarthritis generally affects distal phalanges. In either case, arthritis can debilitate the fingers and, over a period of time greatly decrease dexterity and function.

See also Anatomical nomenclature; Bone injury, breakage, repair, and healing; Joints and synovial membranes

PHARMACOGENETICS

One of the newest subspecialties of biological science is pharmacogenetics, the science that deals with the relationship between inherited genes and the ability of the body to metabolize drugs. Medicine today relies on the use of therapeutic drugs to treat disease, but one of the longstanding problems has been the documented variation in patient response to drug therapy. The recommended dosage is usually established at a level shown to be effective in 50% of a test population, and based on the patient's initial response, the dosage may be increased, decreased, or discontinued. In rare situations, the patient may experience an adverse reaction to the drug and be shown to have a pharmacogenetic disorder. The unique feature of this group of diseases is that the problem does not occur until after the drug is given, so a person may have a pharmacogenetic defect and never know it if the specific drug required to trigger the reaction is never administered.

Consider the case of a 35-year-old male who is scheduled for surgical repair of a hernia. The patient is otherwise in excellent health and has no family history of any serious medical problems. After entering the operating theater, an inhalation anesthetic and/or muscle relaxant is administered to render the patient unconscious. Unexpectedly, there is a significant increase in body temperature, and the patient experiences sustained **muscle contraction**. If this condition is not reversed promptly, it can lead to **death**. Anesthesiologists are now familiar with this type of reaction. It occurs only rarely, but it uniquely identifies the patient as having malignant **hyperthermia**, a rare genetic disorder that affects the body's ability to respond normally to anesthetics. Once diagnosed with malignant hyperthermia, it is quite easy to avoid future episodes by simply using a different type of anesthetic when surgery is necessary, but it often takes one negative, and potentially life threatening, experience to know the condition exists.

An incident that occurred in the 1950s further shows the diversity of pharmacogenetic disorders. During the Korean War, service personnel were deployed in a region of the world where they were at increased risk for **malaria**. To reduce the likelihood of acquiring that disease, the antimalarial drug primaquine was administered. Shortly thereafter, approximately 10% of the African-American servicemen were diagnosed with acute anemia and a smaller percentage of soldiers of Mediterranean ancestry showed a more severe hemolytic anemia. Investigation revealed that the affected individuals had a mutation in the glucose 6-phosphate dehydrogenase (G6PD) gene. Functional G6PD is important in the maintenance of a balance between oxidized and reduced molecules in the cells, and, under normal circumstances, a mutation that eliminates the normal enzyme function can be compensated for by other cellular processes. However, mutation carriers are compromised when their cells are stressed, such as when the primaquine is administered. The system becomes overloaded, and the result is oxidative damage of the red **blood** cells and anemia. Clearly, both the medics who administered the primaquine and the men who took the drug were unaware of the potential consequences. Fortunately, once the drug treatment was discontinued, the individuals recovered.

Drugs are essential to modern medical practice, but, as in the cases of malignant hyperthermia and G6PD deficiency, it has become clear that not all individuals respond equally to each drug. Reactions can vary from positive improvement in the quality of life to life threatening episodes. Annually, in the United States, there are over two million reported cases of adverse drug reactions and a further 100,000 deaths per year because of drug treatments. The Human Genome Project and other research endeavors are now providing information that is allowing a better understanding of the underlying causes of pharmacogenetic anomalies with the hope that eventually the number of negative episodes can be reduced.

In particular, research on one enzyme family is beginning to revolutionize the concepts of drug therapy. The cytochrome P450 system is a group of related **enzymes** that are key components in the metabolic conversion of over 50% of all currently used drugs. Studies involving one member of this family, CYP2D6, have revealed the presence of several polymorphic genetic variations (poor, intermediate, extensive, and ultra) that result in different clinical phenotypes with respect to drug **metabolism**. For example, a poor metabolizer has difficulty in converting the therapeutic drug into a useable form, so the unmodified chemical will accumulate in the body and may cause a toxic overdose. To prevent this from happening, the prescribed dosage of the drug must be reduced. An ultra metabolizer, on the other hand, shows exceedingly rapid breakdown of the drug to the point that the substance may be destroyed so quickly that therapeutic levels may not be reached, and the patient may therefore never show any benefit from treatment. In these cases, switching to another type of drug that is not associated with CYP2D6 metabolism may prove more beneficial. The third phenotypic class, the extensive metabolizers, is less extreme than the ultra metabolism category, but nevertheless presents a relatively rapid turnover of drug that may require a higher than normal dosage to maintain a proper level within the cells. And, finally, the intermediate phenotype falls between the poor and extensive categories and gives reasonable metabolism and turnover of the drug. This is the group for whom most recommended drug dosages appear to be appropriate. However, the elucidation of

the four different metabolic classes has clearly shown that the usual "one size fits all" recommended drug dose is not appropriate for all individuals. In the future, it will become increasingly necessary to know the patient's metabolic phenotype with respect to the drug being given to determine the most appropriate regimen of therapy for that individual.

At the present time, pharmacogenetics is still in its infancy with its full potential yet to be realized. Based on current studies, it is possible to envision many different applications in the future. In addition to providing patient specific drug therapies, pharmacogenetics will aid in the clinician's ability to predict adverse reactions before they occur and identify the potential for **drug addiction** or overdose. New tests will be developed to monitor the effects of drugs, and new medications will be found that will specifically target a particular genetic defect. Increased knowledge in this field should provide a better understanding of the metabolic effects of food additives, work related chemicals, and industrial byproducts. In time, these advances will improve the practice of medicine and become the standard of care.

See also Pharmacology; Human genetics

PHARMACOLOGY

Pharmacology is the study of the changes produced in living animals by chemical substances. There is a great emphasis on the mechanisms underlying the actions of drugs, and both physicians and patients acknowledge the role of drugs in modern therapy. Drug treatment is recognized as the primary modality in the prevention and alleviation of diseases. For hundreds of years most drugs were impure mixtures of only vaguely known composition, and primarily of plant and animal origin. A physician was required to know only what effects might be anticipated from a particular preparation. How the mixture produced the effects was beyond the scope of knowledge of that period. Over the past 60 years the situation has changed and today the physician should know the overt expected effects of a drug, its precise mechanism of action, its elimination routes, its side effects and its potential toxicity. Systematic investigations contributing to such knowledge of the effects of drugs are most often based on results from animal experimentation and the use of isolated and purified active substances and these developed in the mid-nineteenth century. Pharmacologists today draw on a number of biomedical disciplines, including **physiology, pathology, biochemistry**, and bacteriology, to understand the interaction of drugs with body constituents as completely as possible.

There are several reasons for considering pharmacology as one of the increasingly important basic sciences of medicine. Large numbers of drugs are used in the practice of medicine. They cannot be applied intelligently or even safely without some insight into their mode of action, side effects, toxicity, and **metabolism**. As powerful new drugs are introduced, the necessity for adequate pharmacological knowledge on the part of the podiatrist becomes increasingly mandatory. Pharmacology embraces a number of sciences, including pharmacodynamics (the study of the action of drugs on a living body), therapeutics (use of drugs and the method of administration in the treatment of diseases), *materia medica* (the study of the source, composition, characteristics, and preparation of drugs), toxicology (the study of poisons and their action and also the ways of treating poisoning), pharmaceutical chemistry (chemistry in relation to drugs), and pharmacy (the preparation and dispensing of drugs for medical use).

As recently as the 1920s, relatively few medically beneficial active ingredients existed. Most active ingredients were used in only partially purified form, generally from plant or animal sources. Since then, vastly improved tools and methods have developed and it is now possible to identify which compounds in the earlier crude mixtures produce beneficial or undesirable effects. Techniques of synthetic organic chemistry have been used to define the chemical structures of drugs, to synthesize analogs and to test them for pharmacological activity. A significant breakthrough occurred in the 1940s when it was discovered that microorganisms could produce compounds called antibiotics, which were capable of killing other microorganisms. During the 1940s and 1950s, pharmacology depended greatly on organic chemistry to provide the background necessary to synthesize new drugs. Developments in biochemistry during the 1960s through to the present have changed the emphasis to molecular pharmacology and qualitative and quantitative molecular mechanisms of cellular processes. It is now possible to identify, isolate and characterize in the body fluid or **tissue** certain constituent changes responsible for the beneficial effects of drugs. Most recently, advances in genetic engineering and **molecular biology** have accelerated research on the nature of drug sites of action. Such knowledge may provide a sound basis for rational drug use in medial therapy as well as provide a foundation for the design of improved drugs with minimal side effects.

In the field of rational drug design, it is important to realize that each patient is unique. Some respond adequately to a selected drug while others may experience strong side effects. **Pharmacogenetics** is that area of pharmacology concerned with the unanticipated or unusual responses to drugs that may have a hereditary basis for their action. These effects must be distinguished from toxic side effects that can generally be anticipated or result from allergic manifestations. Today, pharmacogenetics is a vastly developing field in the wake of the human genome project. It is now being anticipated that in the future it will be possible to design drugs specifically for individual patients on the basis of an understanding of their **genetic code**.

See also Drug effects on the nervous system; Drug interactions

PHARYNX AND PHARYGEAL STRUCTURES

The pharynx is a tube made of fibers and muscle enclosing a musculomembranous lined passage continuous with the nasal cavity (nasal pharynx), oral cavity (oral pharynx), and the **larynx** (laryngeal pharynx). Accordingly, The pharynx is multifunctional. As a component of the digestive system, it aids

food in beginning its passage to the stomach. It also helps route air to the larynx, trachea, and the **lungs**. Finally, the pharynx acts as a resonating chamber for the sounds produced in the larynx.

The outer walls of the pharynx are made of three circularly arranged muscles, which are called constrictors. The superior pharyngeal constrictor overlaps with the middle pharyngeal constrictor, which itself overlaps with the inferior pharyngeal constrictor. The arrangement has been described as three rings stacked one inside the other.

The inner walls of the pharynx are made of three muscles that are oriented in an up and down direction. These muscles are the stylopharyngeus, palatopharyngeus, and salptngopharyngeus. The first of this trio of muscles arises from a region outside the pharynx called the styloid process. It passes through the superior and middle constrictors into the pharynx. The second muscle originates in the soft region underneath the mouth and from the base of the tongue, and runs down the length of the pharynx. The last muscle originates at **cartilage** locate at the opening of the vocal tube.

The muscles of the pharynx function in eating. Successive contraction of the constrictors forces the ball of food down into the esophagus. The contraction of the longitudinal muscles causes the pharynx to rise. This acts to raise the pharynx up and around the chewed bolus (ball) of food. Between swallows the inferior constrictor may close to guard the entry to the esophagus and prevent unwanted entry of air. This muscle likely functions in a similar manner as the sphincter of the intestinal tract.

The pharynx is well supplied with **blood**, via the ascending pharyngeal **arteries** and branches of the superior and inferior thyroid arteries. Also, the area is well infused with nerves. This is to be expected, as the operation of the pharynx must be highly coordinated with eating, **swallowing**, and **breathing**.

The pharynx also contains specialized lymphatic **tissue** that acts as a barrier to microbial pathogens (disease-causing organisms). There are three types of this tissue. The nasopharyngeal **tonsils**, also known as the adenoids, are located in nasal pharynx. The palatine tonsils, which are familiar to many people simply as tonsils, are located in the oral pharynx. Finally, the lingual tonsils are located on the underside of the tongue.

See also Cranial nerves; Mastication; Palate (hard and soft palate)

PHOSPHATES AND PHOSPHATE METABOLISM • *see* CALCIUM AND PHOSPHATE METABOLISM

PHYSIOLOGY

Physiology is the study of how various biological components work independently and together to enable organisms, from animals to microbes, to function. This scientific discipline covers a wide variety of functions from the cellular and sub-

Just as a pole vault may be depicted as the coordinated culmination of a series of smaller steps and movements, so too can many physiological processes be depicted as a series of intricate and coordinated steps. © Gilbert lundt/CORBIS. Reproduced by permission.

cellular level to the interaction of organ systems that keep more the complex biological machines of humans running.

Physiological studies are aimed at answering many questions. Physiologists investigate topics ranging from precise mloecular studies of how food is digested to more general studies of how thought processes relate to electrical and biochemical patterns found in the **brain** (a branch of this discipline known as neurophysiology). It is often physiology-related investigations that uncover the origins of diseases.

Human (or mammalian) physiology is the oldest branch of this science dating back to at least 420 B.C. and the time of **Hippocrates**, the father of medicine. Modern physiology first appeared in the seventeenth century when scientific methods of observation and experimentation were used to study **blood** movement, or circulation, in the body. In 1929, American physiologist W. B. Cannon coined the term homeostasis to describe one of the most basic concerns of physiology: how the varied components of living things adjust to maintain a constant internal environment conducive to optimal functioning.

With the steady advance of scientific technology—from the simple microscope to ultra high-tech computerized scanning devices—the field of physiology grew in scope. No longer confined to investigating the functioning components of life that could be observed with the naked eye, physiologists began to delve into the most basic life forms, like **bacteria**. They could

also study organisms' basic molecular functions, including electrical potentials in cells that help control **heart** rhythums.

The branches of physiology are almost as varied as the countless life forms that inhabit the earth. Viral physiology, for example, focuses on how viral particles grow, reproduce, and excrete by-products. However, the more complex an organism, the more avenues of research open to the physiologist. Human physiology, for instance, is concerned with the function of large scale integration founrd in **organs**, organ systems, and sensory pathways (e.g., sight and **smell**).

Physiologists also observe and analyze how certain body systems, like the circulatory, respiratory, and nervous systems, work independently and in concert to maintain life. Cellular physiology, an important branch of cell biology, focuses on the functions of the cell, especially the relationship between **structure and function**. Many branches of physiology are often important components included in the studies of other disciplines including **biochemistry**, biophysics, **immunology**, and endocrinology.

See also History of anatomy and physiology: The Classical and Medieval periods; History of anatomy and physiology: The Renaissance and Age of Enlightenment; History of anatomy and physiology: The science of medicine; Homeostatic mechanisms

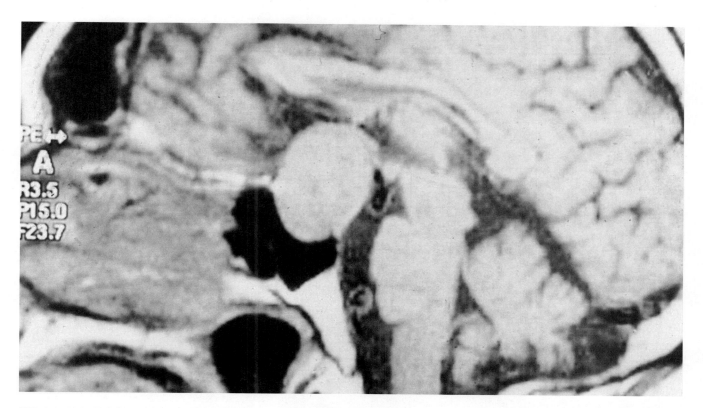

MRI pictograph depicting the relationship of the pituitary gland (center) to other cerebral structures. *Photograph by Bates M.D. Custom Medical Stock Photo. Reproduced by permission.*

PINEAL GLAND • *see* EPITHALAMUS

PITUITARY GLAND AND HORMONES

The pituitary gland is sometimes called the "master" gland of the **endocrine system**, because it controls the functions of many other **endocrine glands**. It is a small, oval gland, no larger than a pea, and is located at the base of the **brain**. It is attached to the hypothalumus (a part of the brain that affects the pituitary gland) by nerve fibers.

Anatomically, the pituitary gland consists of three sections: the anterior lobe, the intermediate lobe and the posterior lobe. The anterior and posterior lobes are derived from different embryological sources. The anterior lobe, or adenohypophysis, grows upward from the pharyngeal **tissue** at the roof of the mouth. The posterior lobe, or neurohypophysis, grows downward from neural tissue. It is structurally continuous with the **hypothalamus**, to which it remains attached by the hypophyseal, or pituitary, stalk. The intermediate lobe also originates in the **pharynx**, like the anterior lobe, but in humans it is greatly reduced in **structure and function**. The hypothalamus controls almost all secretions of the pituitary gland. The posterior lobe is controlled by nerve fibers that originate in hypothalamic **neurons** and the anterior lobe by substances that are transported from the hypothalamus by tiny **blood** vessels.

The tissues in the anterior lobe consist of extensive vascular areas interspersed among glandular cells that secrete a number of different **hormones**. It was formerly believed that various **enzymes** stimulated a master molecule, which then produced these hormones. Present evidence indicates that each hormone is individually synthesized by a specific type of glandular cell. Three types of such cells exist in the anterior pituitary gland: acidophils, basophils, and chromophobes.

The anterior lobe secretes important hormones that influence bodily functions by stimulating target **organs**. Adrenocorticotropic hormone (ACTH) is a peptide of 39 **amino acids** and controls the secretion of steroid hormones by the adrenal cortex. These affects glucose, protein, and **fat metabolism**. ACTH also stimulates the production of androgens, like **testosterone**, and in the foetus it stimulates the adrenal cortex to synthesize a precurser to **estrogen** called dehydroepiandrosterone sulfate (DHEA-S), which helps prepare the mother for giving birth. Another important hormone produced by the anterior pituitary lobe is thyroid stimulating hormone (TSH), also known as thyrotropin. This controls the rate of thyroxine synthesis by the thyroid, which is the principal regulator of body metabolic rate.

Human growth hormone (GH), or somatotropin, is a protein of 191 amino acids and GH-secreting cells in the anterior lobe are stimulated to synthesize and release GH by the intermittent arrival of growth hormone releasing hormone (GHRH) from the hypothalamus. GH promotes bodily growth by binding to receptors on the surface of **liver** cells, stimulating them to release growth factors promoting bone elongation. Oversecretion of GH can cause gigantism if it occurs before growth of the long bones is complete, or acromegaly if it begins during adulthood. Undersecretion can lead to dwarfism if experience during childhood, and decreased endocrine function accompanied by lethargy and loss of sexual capacity in the adult.

Other hormones secreted by the anterior lobe include prolactin, which regulates the formation of milk after the birth of an infant and also three separate gonadotropic hormones, follicle-stimulating hormone (**FSH**), **luteinizing hormone** (LH), and luteotropic hormone which control the growth and reproductive activity of the **gonads**. The release of each of the hormones from the anterior lobe is controlled by a specific substance secreted by nerve cells in the hypothalamus. These substances, called releasing factors, are transmitted by nerve fibers to tiny **capillaries** in the hypophyseal stalk. They move through blood vessels to the anterior lobe, where each one is responsible for the release of a specific pituitary hormone.

The posterior lobe releases two hormones into the body's circulation, both of which are synthesized by nerve cells in the hypothalamus. They are transported by nerve fibers to nerve endings in the posterior lobe, where they are released. The hormones are antidiuretic hormone (ADH or vasopressin), a very small peptide of nine amino acids, which acts on the collecting ducts of the kidney to facilitate the reabsorption of water into the blood. This acts to reduce the volume of urine formed, hence it is named antidiuretic hormone. The other hormone secreted by the posterior lobe is **oxytocin**, also a peptide of none amino acids, which aids in the release of milk from **mammary glands** and causes uterine contractions. Oxytocin is often given to mothers to hasten birth. The only hormone that is synthesized by the intermediate lobe is the melanocyte-stimulating hormone, which appears to control skin pigmentation.

See also Endocrine system and glands

PLACENTA AND PLACENTAL NUTRITION OF THE EMBRYO

The placenta (from Greek *plakuos,* flat cake) is an organ created from the zygote that links two individuals, the mother and the fetus. Throughout the **pregnancy**, placental mass maintains a dynamic relationship with the weight of developing fetus. The placenta serves to attach the embryo-fetus to the uterine wall and to exchange nutrients, wastes, and gases between the maternal **blood** and the embryo-fetal blood. It also provides endocrine and immune support for the developing fetus.

The placenta forms by the differentiation of the trophoblast (the outer cell mass of the early embryo that gives rise to the placenta) in two different layers, the cytotrophoblast (mitotically active) and the syncytiotrophoblast (lacunas forming **tissue** able to fuse the placenta to the wall of the uterus).

The placental **anatomy** is comprised of the placental parenchyma, umbilical cord, and amnion, all of which have specific macroscopic and microscopic features.

The full-term human placenta is a disk-like round organ weighing approximately 1 lb. (approximately 450 g) at delivery that is formed by a chorionic plate and a basal plate. The chorionic plate is a fibrous disc into which two umbilical **arteries** (sometimes one) coming from the embryo-fetal district ramify as chorionic plate vessels, each penetrating the plate to enter the stem of a villous tree. A vein draining oxygenated blood accompanies each arterial vessel back toward a single umbilical vein. The chorionic plate is covered by the amniotic membrane. The basal plate (maternal surface) is composed of 10–40 islands of parenchyma called cotyledons, subdivided by an incomplete system of septa. Cotyledons and septa are generated during the placental growth as the gestation progresses due to the relative lack of space in the uterine cavity. The basal plate is composed of a mixture of trophoblastic and decidual cells embedded in extracellular debris as well as blood clot. The placental parenchyma is composed of a highly vascularized stromal compartment lined or covered by trophoblast. The stroma is **mesoderm** filled with vascular and lymphatic vessels. The trophoblast expresses unique **antigens** (HLA-G and Fas ligand) that promote its graft-like immunological acceptance (tolerance) by the antigenically dissimilar maternal host.

From the microscopic point of view, it is essential to refer to the placental villous system development. At two weeks gestation, in the lacunar spaces, formed by the rapid expansion of the syncyziotrophoblast, finger-like columns of the proliferating cytotrophoblasts form the primary villi. At 3 weeks gestation, extra-embryonic mesenchyme, the future connective tissue of the placenta, grows into the center of the villi that are now called secondary villi. Successively, the mesenchyme differentiates into blood vessels and cells, forms an arteriocapillary network fused with placental vessels, and yields, at 4 weeks, the formation of the tertiary villi. Five types of villi have been detected on the basis of their size, stromal characteristics and vessel structure: (1) Stem villi give mechanical support to the villous trees. They have a dense fibrous stroma with relatively large vessels that are centrally located; (2) long slender mature intermediate villi arise from the ends of stem villi. They have a loose stromal core, with small arterioles and **capillaries**; (3) terminal villi are the final branches of the villous tree. They have an extensive system of capillaries and an inhomogeneous thickness of the trophoblast. In some area, trophoblast is reduced just to a thin layer known as vasculosyncytial membrane. The proportion of villus surface occupied by the vasculosyncytial membrane increases with the gestational age; (4) immature intermediate villi, usually present at the central area of the lobule, are recognized because of the presence of several macrophages (Hofbauer cells); (5) mesenchymal villi are immature villi present at the beginning of the gestation usually confined on the surface of immature intermediate villi as the pregnancy progresses.

In all mammalian species, **nutrition** of the conceptus is initially histiotrophic with the growing and rapidly dividing trophoblast deriving nutrients from simple diffusion and from phagocytosing maternal tubal and uterine secretions; following implantation and establishment of the placenta there is a transition to hemotrophic nutrition, with exchange between maternal and embryofetal blood circulations.

A possible histiotrophic pathway during early pregnancy has been extensively described. It involves the secretions from the uterine glands that are known to enter the intervillous space. The maternal secretions are taken up by the trophoblast and locally digested or diffuse throughout the villus mesenchyma along the stromal channels into the coelomic fluid. Finally, they are absorbed by the epithelia of the yolk sac and pass to the embryo though the vitelline circulation (the earliest embryonal circulation). At early stages of pregnancy, the yolk sac has a determinant role in delivering the nutrients because its vascularization is much more developed than that of the villous system. Histiotrophic nutrition benefits the embryo especially during embryogenesis because it reduces the risk to high oxygen exposure. Support for this theory comes from the observations that known teratogens as well as thalidomide and ethanol disrupt normal developmental patterns through the formation of elevated levels of free radicals (reactive oxygen species).

Hemotrophic nutrition takes place when the two different sides, both maternal and fetal, create a complex circulation. The maternal or intervillous placental circulation proceeds as the trophoblastic shell invades the uterine stroma to reach the maternal capillaries. **Fetal circulation** arises from the stroma of the villi when they begin to organize a primitive capillary system that connects to the intrafetal circulation. Toward the end of the first trimester both fetal and maternal circulation results in a hemochorial arrangement where maternal and fetal blood ideally are separated by a barrier. Maternal nutrients cross the intervillous space, trophoblast cells, the fibrous core of the villus, and the endothelial cells of the fetal capillaries to enter the fetal blood. Wastes from the fetus move in the opposite direction.

Nutrition of the embryo and fetus is mediated by very complex systems of transport, the maternal-fetal exchange. Placenta has several mechanisms of transport including simple diffusion (small molecules like oxygen, water, CO_2 move along concentration gradients from higher to lower concentrations), facilitated diffusion where carrier molecule dependent diffusion also moves molecules such as glucose along concentration gradients from higher to lower concentrations) dependent on local concentration gradients, needs a carrier molecule), and active transport against a concentration gradient. Active transport utilizes **ATP** and is therefore ATP dependent. Active transport is required to move **amino acids, calcium,** iron, and some water-soluble **vitamins.**

Glucose is the major metabolic "fuel" for the embryo and fetus (and placenta). Specific glucose transporters named GLUT 1 (much more widespread) and GLUT 3 (more vessels specific) are present on **plasma** membranes of placental barriers facing the maternal and fetal sides. Proper glucose uptake is achieved through a lower fetal glucose concentration, which increases the transplacental glucose gradients. With the advance of gestational age, in normal pregnancy, the glucose concentration decreases in the fetal blood despite the fact that in the maternal blood it remains constant.

The transport of aminoacids across the placenta is active and involves three fundamental steps. First, the uptake from maternal circulation across the microvillus membrane. Second, the transport throughout the trophoblast cytoplasm. Third, the transport across the basal membrane and into the umbilical circulation. A number of different variations of amino acid transport systems exist. For example, neutral aminoacids as well as glycine and alanine are transported by means of at least six different systems. It is worth noting how the serine is not transported from the maternal to the fetal circulation and some of its production occurs in the fetal **liver** derived from the glycine. Also is important the exchange of glutamine and glutamate performed by the placenta-fetal liver interaction. In particular, the fetal liver takes up glutamine and releases glutamate.

The placenta has a considerable capacity for lipid uptake and transport. Several lipase activities are involved in such a mechanism. Most of the transport involves the breakdown of **triglycerides** to free fatty acids and glycerol and re-esterification with intracellularly generated glycerol phosphate on the fetal side. Again, a direct passage of free fatty acids is also possible, and, although the fetus has an internal (endogenous) source of cholesterol, there is evidence that yolk sac and placenta take up maternal cholesterol in the form of LDL and HDL Fetal lipid metabolism is essential for the rapid fetal growth and cholesterol, in particular, is needed for a normal development.

See also Embryonic development: early development, formation, and differentiation; Glucose utilization, transport and insulin; Metabolic waste removal; Parturition; Prenatal growth and development; Vascular exchange

PLASMA AND PLASMA CLEARANCE

Blood consists of blood cells, **platelets,** and other particles suspended in a clear, pale-red liquid called plasma. Plasma, which makes up 50%–60% of blood by volume, is a complex solution of water, proteins, inorganic salts, urea, **uric acid,** creatinine, ammonia salts, **amino acids,** sugar, **fat, hormones,** and many other substances.

As blood passes through the **capillaries,** some plasma leaks out through the capillary walls. This fluid bathes the cells of the surrounding tissues and is called **tissue** fluid. If tissue fluid accumulates, the tissues swell, causing **edema;** normally, however, it is drawn off by the venules (small **veins** exiting the capillary bed) and by the vessels of the lymphatic system. Once in the lymphatic system, the fluid is called lymph. Lymph is eventually returned to the bloodstream, helping balance plasma loss from the capillaries.

Many proteins are dissolved in plasma, including fibrinogen, albumin, and the globulins. These proteins are formed mostly in the **liver** and perform a wide variety of functions. They aid in the transport of iron, copper, **lipids,** hormones, **bilirubin,** fatty acids, drugs, and other substances;

function as antibodies, clotting factors, and **enzymes**; correct the **acid-base balance** of the plasma; form a protein reserve that the body can access during starvation; and by their mere presence create inward-acting osmotic pressure that restrains plasma from leaking out of the capillaries too quickly.

The rate at which drugs are removed or cleared from plasma is important in medicine because it affects how long drugs remain active in the body. The volume of plasma that is completely cleared of a substance in a given time is called the plasma clearance of that substance. For example, if 130 milliliters (ml) per minute of plasma are being completely cleared of a given drug by an individual's **kidneys** every minute, then the plasma clearance of that drug, for that individual at that time, is 130 ml/min. When a substance is being cleared from the plasma by more than one process—metabolism and urinary excretion, for example—the calculation of plasma clearance is more complex.

See also Blood coagulation and blood coagulation tests; Lymphatic system; Osmotic equilibria between intercellular and extracellular fluids

PLATELETS AND PLATELET COUNT

Platelets, also called thrombocytes, are small disk-shaped cytoplasmic fragments produced in the bone marrow and involved in the process of **blood** clotting. Platelets are not true cells because they do not contain nuclei. When a blood vessel is injured, platelets first adhere to the site of the injury. Then, during the activation phase of clotting, the platelets bind to each other. Platelets react with a protein in the blood called fibrinogen to form fibrin, a thread-like substance that helps create a web-like structure of blood cells and platelets at the injury site, ultimately forming a clot to stop the bleeding. There are normally between 150,000–450,000 platelets in each microliter of blood. Low platelet counts or abnormally shaped platelets are associated with bleeding disorders. High platelet counts sometimes indicate disorders of the bone marrow.

A platelet count is a diagnostic test that determines the number of platelets in the patient's blood. The primary functions of a platelet count are to assist in the diagnosis of bleeding disorders and to monitor patients who are being treated for any disease involving bone marrow failure. Patients who have **leukemia**, polycythemia vera, or aplastic anemia are given periodic platelet count tests to monitor their health.

An abnormally low platelet level (thrombocytopenia) is a condition that may result from increased destruction of platelets, decreased production, or increased usage of platelets. In idiopathic thrombocytopenic purpura (ITP), platelets are destroyed at abnormally high rates. Hypersplenism is characterized by the collection (sequestration) of platelets in the spleen. Disseminated intravascular coagulation (DIC) is a condition in which blood clots occur within blood vessels in a number of tissues. All of these diseases produce reduced platelet counts.

Abnormally high platelet levels (thrombocytosis) may indicate either a benign reaction to an **infection**, surgery, or

certain medications; or a disease like polycythemia vera, in which the bone marrow produces too many platelets too quickly.

See also Blood coagulation and blood coagulation tests

PLEURA

The pleura are the thin, double-layered, moist membranes that line the chest wall and cover the **lungs**.

Parietal pleura, the pleurae lining the chest, doubles back over at the root of the lung where it becomes known as the visceral pleura. The two membranes are really one continuous sheet of **tissue**. They enclose the narrow pleural cavity, a thin, slit-like space containing a friction-reducing fluid that lubricates the membranes and aides **breathing**. In normal health, this cavity or "pleural space" goes unnoticed. The cavity becomes apparent, however, when a lung collapses or air or liquid collects between membranes.

The parietal and visceral pleurae enclose each lung in a different sac. The parietal pleurae consist of one elastic layer and rest on the fibrocollagenous tissue attached to the ribs. Parietal pleurae are divided into four categories: costal, diaphragmatic, mediastinal, and cervical. They are all supplied by intercostal and phrenic nerves that allow this type of pleura to sense **pain**. The visceral pleura are made up of five elastic layers, one of which is irregular. Visceral pleura help form pulmonary ligaments and connect to the autonomic nerves.

Diseases of the pleura include pleurisy and pleural effusion. Pleurisy is an inflammation of the pleura characterized by a sharp, stabbing chest pain that becomes worse with deep breathing or coughing. Excess liquid in the pleural space results in pleural effusion, and can be the result of disease or trauma. **Tumors** of the pleura (mesothelioma) are seen most frequently in asbestos workers.

See also Respiratory system

POISON AND ANTIDOTE ACTIONS

A poison is a compound that produces a deleterious change on or in the body. There are a large number of known poisons, with a myriad of effects. Toxicity is a general term used to indicate adverse effects produced by poisons. These adverse effects can range from slight symptoms such as headaches or **nausea**, to severe symptoms such as **coma**, convulsions, and **death**.

The hallmark of a poison is that it changes some aspect of a body function, often the speed of a function. Examples of this include increased **heart** rate or sweating, or decreasing breath (sometimes to the point of death). A poison may have wide-ranging effects in the body, may damage only a particular region or organ, or may do both. An example of the latter is an insecticide called Parathion. It inactivates a particular enzyme that functions in communication between nerves. The enzyme is very widespread in the body, and thus many varied effects are seen.

Toxicity is based on the number of exposures to a poison and the time it takes for toxic symptoms to develop. Two common types are acute and chronic toxicity. Acute toxicity is due to short-term exposure and happens within a relatively short period of time, whereas chronic toxicity is due to long-term exposure and happens over a longer period.

Some poisons produce a mild reaction. Poison ivy, poison oak and poison sumac all contain a sticky sap comprised of a compound called toxicodendrol. For individuals who are allergic to the compound—more than half the population—a red, blistering rash called rhus dermatitis results upon contact with the plant. There are no antidotes per se, as the rash cannot be reversed. Antihistamines or drying agents such as calamine provide comfort and lessen the rash.

The toxins produced by **bacteria** are far more potent poisons than toxicodendrol. The effects of bacterial toxins are varied, ranging from the vomiting and **diarrhea** associated with toxins of *Escherichia coli* and Shigella, to the paralysis and death caused by the toxin produced by *Clostidium botulinum*. If detected early enough, relief is brought by the injection of an antitoxin, which neutralizes the toxin that has not yet bound to its target. This antidote is ineffective on toxin that has already bound to host **tissue**.

Plants are another source of poisons. Very many plants, if ingested, can cause vomiting, depression, tremors or convulsions, stomach **pain**, kidney or **liver** failure, coma, or death, to name just a few symptoms. The antidote depends on the type of plant. Treatment with ipecac to induce vomiting is a common antidote. But in some cases, an antidote does not exist. Then, stabilization of the patient and medical monitoring to prevent further damage is the course of action.

Compounds that are effective in one setting, or drugs that are therapeutic at certain concentrations, can be poisonous if used in an inappropriate way or at too high a concentration. As examples, bleach and other household detergents and cleaning agents are poisonous if ingested. And, while two aspirin are effective for treatment of a **headache**, 30 aspirin at one time are poisonous. The list of potential poisons is thus, numerous.

See also Drug effects on the nervous system; Homeostatic mechanisms

PONS

The pons is a small primitive structure in the mammalian **brain**. By primitive, researchers mean that it is found in mammals considered to have evolved before *Homo sapiens*. The structure itself is small and appears insignificant compared to other more complex structures of the human brain. However, it is a very important structure for motor functions (movement) of many regions of the body. Its primary function seems to be one of integrating signals from the cerebrum to the **cerebellum** and sending them to the proper parts of the body.

Another function involves the formation of several of the **cranial nerves** for the head.

When looking at fetal development of the brain, the developing pons is found at the point where the brainstem flexes or **bends** back on itself and eventually comes in contact with the growing cerebellum. As a result of the bending, the fourth ventricle lies just next to the adult pons. The ventricle is the remnant of the initial space between the cerebellum and the pons. This spot is simply called the pons curvature.

At one time the pons was termed the pons Varolii, but as the trend to drop personal names from anatomical structures continues, this name also name has been dropped.

The pons is anterior (front of body) swelling of **tissue** on an area called the hindbrain or metencephalon. It is located above (superior) to the medulla oblangata, below (inferior) to the crura cerebri, and in the middle or between the two hemispheres (sides) of the cerebellum. It is a tiny structure being only about an inch long and thick and about and an inch and one half wide. The pons is directly connected to the medulla and the midbrain. Its sides are free and it is separated from the cerebellum by the cerebro-spinal fluid in the fourth ventricle.

As mentioned, the pons receives motor signals from the cerebellum and uses the Purkinje motor **neurons** to send the impulses to the proper regions of the body. These nerve tracts are seen as white fibers from side to side throughout the entire structure. The reticular (a latticework type pattern) formation of the medulla extends into the top portion of the pons. The reticular formation aids in the regulation of **sleep** and other daily cycles of the brain.

Several of the cranial nerves have their nuclei located in the pons. The nuclei lead to fibers that emerge from the pons and out to regions of the face. Cranial nerves V (trigeminal), VI (abducens), VII (facial), VIII (vestibulucochlear) come from the junction of the pons with the medulla.

The **gray matter** of the pons is the unmyelinated portion of the structure. Its function is to relay signals from the cerebrum to the cerebellum. Without this relay, motor functions necessary for **breathing**, **posture**, **swallowing**, and bladder control would be difficult if not impossible.

See also Cerebral morphology; Motor functions and controls

PORTER, GEORGE (1920-   )
English chemist

Sir George Porter shared the Nobel Prize in chemistry in 1967 with his former teacher, Ronald G. W. Norrish, and Manfred Eigen for their contributions to the study of rapid chemical reactions. Porter's efforts included research on flash photolysis, which has been used widely in the fields of organic chemistry, **biochemistry**, and photobiology. Flash photolysis is a technique used in studies of **ATP** and muscle cell **physiology** that enables investigators to exam the rapid (millisecond) actions of cross-bridging, and the role of ATP (and ATPases) during **muscle contraction**.

Porter was born on December 6, 1920, to John Smith Porter and Alice Ann (Roebuck) Porter in Stainforth, West Yorkshire, where he received his early education at Thorne Grammar School. With the award of an Ackroyd Scholarship, he entered Leeds University in 1938 to study chemistry and received his Bachelor of Science degree in 1941. While at Leeds he also studied radio physics and electronics, and he drew upon this background while serving in the Royal Navy Volunteer Reserve as a radar specialist during World War II. At the end of the war, Porter entered Emmanuel College at Cambridge University to do graduate work. There he met and studied under Norrish, who had pioneered research in the area of photochemical reactions in molecules. Porter received his doctorate degree from Cambridge in 1949.

Using very short pulses of energy that disturbed the **equilibrium** of molecules, Porter and Norrish developed a method to study extremely fast chemical reactions lasting for only one-billionth of a second. The technique is known as flash photolysis. First, a flash of short-wavelength light breaks a chemical that is photosensitive into reactive parts. Next, a weaker light flash illuminates the reaction zone, making it possible to measure short-lived free radicals, which are especially reactive atoms that have at least one unpaired electron. Flash photolysis made it possible to observe and measure free radicals for the first time and to study the sequence of the processes of reactants as they are converted into products. When Porter won the Nobel Prize in 1967, he was praised, along with Norrish and Eigen, for making it possible for scientists around the world to use their techniques in a wide range of applications, opening many passageways to scientific investigation in physical chemistry. In his own work, Porter was able to apply his methods from his early work with gases to later work with solutions. He also developed a method to stabilize free radicals, which is called matrix isolation. It can trap free radicals in a structure of a supercooled liquid (a glass). Porter also made important contributions in the application of laser beams to photochemical studies for the purpose of investigating biochemical problems. Some practical applications of photochemical techniques include the production of fuel and chemical feedstocks.

In 1949, Porter became a demonstrator in chemistry at Cambridge University and an assistant director of research in the Department of Physical Chemistry in 1952. While he was at the British Rayon Research Association as assistant director of research in 1954, Porter used his method of flash photolysis to record organic free radicals with a lifetime as short as one millisecond. Also at the Rayon Association, he worked on problems of light and the fading of dye on fabric.

Porter was appointed professor of physical chemistry at Sheffield University in 1955, and in 1963, he became the head of the chemistry department and was honored as Firth Professor. During his years at Sheffield, Porter used his flash photolysis techniques to study the complex chemical interactions of oxygen with **hemoglobin** in animals. He also investigated the properties of chlorophyll in plants with the use of his high-speed flash techniques. He was able to improve his techniques to the degree that he could examine chemical reactions that were more than a thousand times faster than with the use

of flash tubes. Porter also studied chloroplasts and the primary processes of photosynthesis.

In 1966, Porter also became Fullerian Professor of Chemistry at the Royal Institution in London and the Director of the Davy Faraday Research Laboratory. He left there to take the position of chair for the Center for Photomolecular Sciences at Imperial College in London in 1990. During his career, Porter received many other honors and awards in addition to the Nobel Prize. He was knighted in 1972, and he has been granted numerous honorary doctorate degrees and awarded prizes from British and American scientific societies, including the Robertson Prize of the American National Academy of Sciences and the Rumford Medal of the Royal Society, both in 1978.

Porter has been active outside scientific circles in the promotion of science to the general public. His concern about communication between scientists and the rest of society induced him to participate as an adviser on film and television productions. He has been praised for his activities in educating young people and people in non-scientific fields about the value of science. He was an active participant during his service with the Royal Institution in a science program series for British Broadcasting Company television (BBC-TV) called *Young Scientist of the Year*. Another BBC-TV program in which he participated was called *The Laws of Disorder* and *Time Machines*. Porter has also served on many policy and institutional committees that are involved in promoting science and education in Europe, England, and America.

Porter, an active contributor to scientific journals and an important scientific advisor to industry, is also praised for being a tireless promotor of science education. Porter married in 1949. He and his wife have two sons. In addition to sailing, Porter spends some leisure time vacationing on the coast of Kent with his family.

See also Enzymes and coenzymes; Muscular system overview; Myology

PORTER, RODNEY (1917-1985)
English biochemist

Rodney Porter was a biochemist who spent most of his professional life investigating the chemical structure and functioning of antibodies, a class of proteins which are also called immunoglobulins. Since 1890, scientists had known that antibodies are found in the **blood** serum and provide immunity to certain illnesses. However, when Porter began his research in the 1940s, little was known about their chemical structure, or how **antigens** (substances that cause the body to produce antibodies) interacted with them. Using the results of his own research as well as that of Gerald M. Edelman, Porter proposed the first satisfactory model of the immunoglobulin molecule in 1962. The model allowed the development of more detailed biochemical studies by Porter and others that led to a better understanding of the way in which antibodies worked chemically. Such understanding was key to research on the prevention and cure of a number of diseases and the solution to problems related to organ transplant rejection. For his work, Porter shared the 1972 Nobel Prize in physiology or medicine with Edelman.

Rodney Robert Porter was born in Newton-le-Willows, near Liverpool in Lancashire, England. His mother was Isobel Reese Porter and his father, Joseph L. Porter, was a railroad clerk. "I don't know why I became interested in [science]," Porter once told the *New York Times*. "It didn't run in my family." He attended Liverpool University, where he earned a B.S. in **biochemistry** in 1939. During World War II, he served in the Royal Artillery, the Royal Engineers, and the Royal Army Service Corps, and participated in the invasions of Algeria, Sicily, and Italy. After his discharge in 1946, he resumed his biochemistry studies at Cambridge University under the direction of Frederick Sanger.

Porter's doctoral research at Cambridge was influenced by Nobel laureate Karl Landsteiner's book, *The Specificity of Serological Reactions,* which described the nature of antibodies and techniques for preparing some of them. Antibodies, at the time, were thought to be proteins that belonged to a class of blood-serum proteins called gamma globulins. From Sanger, who had succeeded in determining the chemical structure of insulin (a protein that metabolizes **carbohydrates**), Porter learned the techniques of protein chemistry. Sanger had also demonstrated tenacity in studying problems in protein chemistry involving amino acid sequencing that most believed impossible to solve, and he was a model for the persistence Porter would show in his later work on antibodies.

Fortunately, Porter chose rabbits to experiment on for his research. Although this was not known at the time, the antibody system is not as complex in this animal as it is in some. The most important antibody, or immunoglobulin, in the blood is called IgG, and contains more than 1,300 **amino acids**. The problem of discovering the active site of the antibody—the part that combines with the antigen—could be solved only by working with smaller pieces of the molecule. Porter discovered that an enzyme from papaya juice, called papain, could break up IgG into fragments that still contained the active sites, but were small enough to work with. He received his Ph.D. for this work in 1948.

Porter remained at Cambridge for another year, then in 1949, he moved to the National Institute for Medical Research at Mill Hill, London. There, he improved methods for purifying protein mixtures and used some of these methods to show that there are variations in IgG molecules. He obtained a purer form of papaya enzyme than had been available at Cambridge and repeated his earlier experiments. This time the IgG molecules broke into thirds, and one of these thirds was obtained in a crystalline form which Porter called fragment crystallizable (Fc).

Obtaining the Fc crystal was a breakthrough; Porter now was able to show that this part of the antibody was the same in all IgG molecules, because a mixture of the different molecules would not have formed a crystal. He also discovered that the active site of the molecule (the part that binds the antigen) was in the other two-thirds of the antibody. These he called fragment antigen-binding (or FAB) pieces. After Porter's research was published in 1959, another research

group, led by Gerald M. Edelman at Rockefeller University in New York, split the IgG in another way—by separating amino acid chains rather than breaking the proteins at right angles between the amino acids as Porter's papain had done.

In 1960, Porter was appointed professor of **immunology** at St. Mary's Hospital Medical School in London. There he repeated Edelman's experiments under different conditions. After two years, having combined his own results with those of Edelman, he proposed the first satisfactory structure of the IgG molecule. The model, which predicted that the FAB fragment consisted of two different amino acid chains, provided the basis for far-ranging biochemical research. Porter's continuing work contributed numerous studies of the structures of individual IgG molecules. In 1967, Porter was appointed Whitley Professor of Biochemistry and chairman of the biochemistry department at Oxford University. In his new position, Porter continued his work on the immune response, but his interest shifted from the structure of antibodies to their role as receptors on the surface of cells. To further this research, he developed ways of tagging and tracing receptors. He also became an authority on the structure and **genetics** of a group of blood proteins called the complement system, which binds the Fc region of the immunoglobulin and is involved in many important immunological reactions.

Porter married Julia Frances New in 1948. The couple had five children and lived in a farmhouse in a small town just outside of Oxford. Porter was killed in an automobile accident a few weeks before he was to retire from the Whitley Chair of Biochemistry. He had been planning to continue as director of the Medical Research Council's Immunochemistry Unit for another four years; he had also intended to continue his laboratory work, attempting to crystallize one of the proteins of the complement system. Porter's awards in addition to the Nobel Prize include the Gairdner Foundation Award of Merit in 1966 and the Ciba Medal of the Biochemical Society in 1967.

See also Immune system

POSTERIOR • *see* ANATOMICAL NOMENCLATURE

POSTURE AND LOCOMOTION

Walking on two legs, or bipedal locomotion, was adopted by man's ancestors millions of years ago, and was among the earliest features distinguishing the hominid line from the apes. Bipedal locomotion was an adaptation to life on the African plains, where food sources were widely scattered. It allowed the ability to move from one place to another while holding things at the same time. It also provided a higher vantage point from which to scan the area for predators.

While advantageous in many ways, the upright posture is inherently unstable, like a two-legged stool. Over time, **evolution** has brought changes to the human musculoskeletal system, especially the pelvis and lower limbs, to help balance and support the mass of the body with a minimum expenditure of energy. These include powerful and bulky leg muscles, and a wide pelvic girdle to support the body's weight and transmit it to the lower limbs.

Humans employ a plantigrade gait, in which the entire sole of the foot is placed on the ground. The heel strikes the ground first, followed by the rest of the sole. The ankle and toes are then flexed, pushing off to provide forward motion. Meanwhile, the other leg is off the ground. It raises, swings forward, and is planted on the ground to continue the cycle.

Because the body's center of gravity generally falls between the two feet, every time we lift a foot to take a step, we create even more instability for our musculoskeletal system to handle. It does this by shifting the body's weight over the supporting leg using the hip abductor muscles.

See also Lower limb structure; Motor functions and controls; Muscular system overview; Skeletal system overview; Sports physiology; Vertebral column

PREGNANCY

Pregnancy is a state in which a woman carries a fertilized egg inside her body. After the egg is fertilized by a **sperm** and then implanted in the lining of the uterus, it develops into the **placenta** and embryo, and later into a fetus. Pregnancy usually lasts 40 weeks, beginning from the first day of the woman's last menstrual period, and is divided into three trimesters, each lasting three months.

At the end of the first month, the embryo is about a third of an inch long, and its head and trunk—plus the beginnings of arms and legs—have started to develop. The embryo receives nutrients and eliminates waste through the umbilical cord and placenta. By the end of the first month, the **liver** and digestive system begin to develop, and the **heart** starts to beat.

In the second month, the heart starts to pump and the nervous system (including the **brain** and **spinal cord**) begins to develop. The 1 in. (2.5 cm) long fetus has a complete **cartilage** skeleton, which is replaced by bone cells by month's end. Arms, legs and all of the major **organs** begin to appear. Facial features begin to form.

By the end of the third month, the fetus has grown to 4 in. (10 cm) and weighs a little more than an ounce (28 g). Now the major **blood** vessels and the roof of the mouth are almost completed, as the face starts to take on a more recognizably human appearance. Fingers and toes appear. All the major organs are now beginning to form; the **kidneys** are now functional and the four chambers of the heart are complete.

During the fourth month, the fetus begins to kick and swallow, although most women still can't feel the baby move at this point. Now 4 oz. (112 g), the fetus can hear and urinate, and has established sleep-wake cycles. All organs are now fully formed, although they will continue to grow for the next five months. The fetus has skin, eyebrows, and **hair**.

At five months, now weighing up to a 1 lb. (454 g) and measuring 8–12 in. (20–30 cm), the fetus experiences rapid growth as its internal organs continue to grow. At this point,

the mother may feel her baby move, and she can hear the heartbeat with a stethoscope.

After six months, even though its **lungs** are not fully developed, a fetus born during this month can usually survive with intensive care. Weighing 1–1.5 lb. (454–681 g), the fetus is red, wrinkly, and covered with fine hair all over its body. The fetus will grow very fast during this month as its organs continue to develop.

After seven months, there is a better chance that a fetus born during this month will survive. The fetus continues to grow rapidly, and may weigh as much as 3 lb. (1.3 kg) by now. Now the fetus can suck its thumb and look around its watery womb with open eyes.

After eight months, growth continues but slows down as the baby begins to take up most of the room inside the uterus. Now weighing 4–5 lb. (1.8–2.3 kg) and measuring 16–18 in. (40–45 cm) long, the fetus may at this time prepare for delivery next month by moving into the head-down position.

At nine months, adding 0.5 lb. (227 g) a week as the due date approaches, the fetus drops lower into the mother's **abdomen** and prepares for the onset of labor, which may begin any time between the 37th and 42nd week of gestation. Most healthy babies will weigh 6–9 lb. (2.7–4 kg) at birth, and will be about 20 in. (51 cm) long.

The first sign of pregnancy is usually a missed menstrual period, although some women may experience some bleeding in the beginning. A woman's **breasts** swell and may become tender as the **mammary glands** prepare for eventual breastfeeding. Nipples begin to enlarge and the **veins** over the surface of the breasts become more noticeable.

Nausea and vomiting are very common symptoms and are usually worse in the morning and during the first trimester of pregnancy. They are usually caused by hormonal changes, in particular, increased levels of **progesterone**. Women may feel worse when their stomach is empty, so many women eat several small meals throughout the day, and keep things like crackers on hand to eat even before getting out of bed in the morning.

Many women also feel extremely tired during the early weeks. Frequent urination is common, and there may be a creamy white discharge from the vagina. Some women crave certain foods, and an extreme sensitivity to **smell** may worsen the nausea. Weight begins to increase.

In the second trimester (13–28 weeks) a woman begins to look noticeably pregnant and the enlarged uterus is easy to feel. The nipples get bigger and darker, skin may darken, and some women may feel flushed and warm. Appetite may increase. By the 22nd week, most women have felt the baby move. During the second trimester, nausea and vomiting often fade away, and the pregnant woman often feels much better and more energetic. Heart rate increases as does the volume of blood in the body.

By the third trimester (29–40 weeks), many women begin to experience a range of common symptoms. Stretch marks may develop on abdomen, breasts, and thighs, and a dark line may appear from the navel to pubic hair. A thin fluid may be expressed from the nipples. Many women feel hot, sweat easily and often find it hard to get comfortable. Kicks

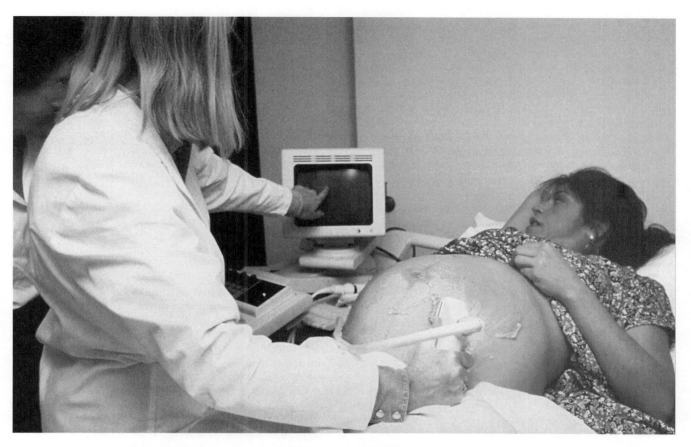

Ultrasound tests establish the position of the fetus *in utero* and allow non-invasive examination of the fetus' general morphological features. *Amy Etra/PhotoEdit. Reproduced by permission.*

from an active baby may cause sharp pains, and lower back-aches are common. More rest is needed as the woman copes with the added stress of extra weight. Braxton Hicks contractions may get stronger.

At about the 36th week in a first pregnancy (later in repeat pregnancies), the baby's head drops down low into the pelvis. This may relieve pressure on the upper abdomen and the lungs, allowing a woman to breathe more easily. However, the new position places more pressure on the bladder.

A healthy gain for most women is between 25–35 lb. (11–16 kg). Women who are overweight should gain less; and women who are underweight should gain more. On average, pregnant women need an additional 300–500 calories a day. Generally, women will gain 3–5 lb. (1.3–2.3 kg) in the first three months, adding one to two pounds a week until the baby is born. An average, healthy full-term baby at birth weighs 7.5 lb. (3.4 kg), and the placenta and fluid together weigh another 3.5 lb. (1.6 kg) The remaining weight that a woman gains during pregnancy is mostly due to water retention and **fat** stores. Her breasts, for instance, gain about 2 lb. (0.9 kg) in weight, and she gains another 4 lb. (1.8 kg) due to the increased blood volume of pregnancy.

Pregnancy is a natural condition that usually causes little discomfort provided the woman takes care of herself and gets adequate prenatal care. Due to technological advances, pregnancy is increasingly occurring among older women in the United States.

See also In vitro and in vivo fertilization; Parturition; Pregnancy, maternal physiological and anatomical changes; Prenatal growth and development

PREGNANCY, MATERNAL PHYSIOLOGICAL AND ANATOMICAL CHANGES

During **pregnancy**, the metabolic demands brought on by the fetus, **placenta**, and uterus, along with the increase of pregnancy **hormones** and, latter, the mechanical pressure of an expanding uterus, create alterations in a woman's body from both a physiological and anatomical point of view. These changes are extensive and may be systemic or localized. They are necessary to maintain a healthy fetus without compromising the mother's health.

After intercourse (coitus), millions of motile **sperm** make their way upward through the uterus and fallopian tubes.They are able to live in female genital tract for 72 hours but are usually highly fertile only for 24 hours. Accordingly, for **fertilization** (the fusion of the nuclear components in the head of a sperm with the nucleus of an ovum) to occur, it is usually necessary that an ovum be expelled from the ovary 24 hours prior to coitus. This allows the ovum to begin a process of division while traveling downwards along the fallopian tubes to reach the uterus. The process takes approximately three days. The dividing ovum develops an outer layer of trophoblast cells capable of phagocytizing nutrient material. Secreting proteolytic **enzymes** eat their way into endometrium of the uterus and allow implantation with the formation of a scar-like corpus luteum. The trophoblastic cells also secrete the human chorionic gonadotropin (HCG or hCG), a hormone that has almost the same effect as does Lutenizing hormone (LH). HCG keeps corpus luteum from degenerating and keeps it secreting large quantities of **progesterone**. The effects of these hormones result in a thickening of the endometrium and prevention of **menstruation**.

In addiction to hCG, the placenta secretes other important hormones. After approximately three months of pregnancy, the secretion of HCG gradually reduces, and from that time onward, progesterone and estrogens from the placenta are essential for the maintenance of pregnancy, and **growth and development** of fetus. The placenta also produces also a large quantity of human chorionic somatomammotropin. This hormone promotes growth of the fetus, increases use of fatty acid by the mother for energy and reduces use of glucose (i.e., an excess of glucose in maternal **blood** able to be used by the fetus as a source of energy), and promotes growth and development of breast **tissue** for **lactation**.

The most obvious physical changes in the mother are weight gain and altered body shape. Weight gain is due not only to uterus and its contents, but also to increased breast tissue, blood, and water volume. Deposition of **fat** and protein, and increased cellular water are added to the maternal stores. The average weight gain in pregnancy is about 27.5 lb. (12.5 kg). During normal pregnancy, approximately 2.2 lb. (1,000 g) of weight gain is attributable to protein. Half of this is found in the fetus and placenta, with the rest distributed as uterine contracting proteins, **plasma** protein, **hemoglobin**, and breast glandular tissue.

The expansion of total body fluid volume in pregnancy is accompanied by retention of 900–1000 meq of sodium and six to eight liters of water, distributed among the fetus, amniotic fluid, and extracellular and intracellular spaces. Plasma volume increases by 10–15% at six to twelve weeks of gestation, expands rapidly until 30–34 weeks, after which there is only a modest rise. At term, the total gain is about 1,100–1,600 mL and results in a plasma volume of 4,700–5,200 mL (30–50%) above that found in non-pregnant women). The increase is needed for the extra blood flow to the uterus, to facilitate maternal and fetal exchanges of respiratory gases, nutrients, and metabolites, and to compensate for maternal blood loss at delivery.

A greater increase in intravascular plasma volume compared to red cell mass results in the dilutional or physiologic anemia of pregnancy. This becomes most apparent at 30–34 weeks of gestation, when plasma volume peaks in relation to

red cell volume. Iron used in pregnancy prevents only maternal anemia, because iron is actively transported to the fetus from maternal compartment and fetal hemoglobin levels are maintained despite maternal anemia. Red blood cell (RBC) mass begins to increase at eight to ten weeks of gestation and steadily rises, in women taking iron supplements, by 20–30% (250–450 mL) above nonpregnant levels by the end of pregnancy. Among women not taking iron supplements, the red cell mass may only increase by 15–20%. Increased plasma erythropoietin induces the rise in red cell mass, which partially supports the higher metabolic requirement for oxygen during pregnancy.

During normal pregnancy, the cardiac output rises 30–50% (1.8 L/min) above baseline; this increase is evident by eight weeks of gestation. The elevation in cardiac performance results in part from changes in three important factors that determine cardiac output, the associated rise in blood volume, a decline in systemic vascular resistance, and a maternal **heart** rate that rises by 15–20 beats per minute. Physiologic changes of the heart and chest wall during pregnancy cause changes in the electrocardiogram that are unrelated to disease. The heart is rotated toward the left, resulting in a 15–20° left axis deviation. Marked variation in chamber volumes, especially left atrial enlargement, leads to stretching of the cardiac conduction pathways and predisposes the pregnant woman to common alterations in cardiac rhythm. In early pregnancy, increased cardiac output is primarily related to the rise in **stroke** volume (35%). In late pregnancy, heart rate is the major factor. The blood pressure (BP) typically falls early in gestation and is usually 10 mmHg below baseline in the second trimester, declining to a mean of 105/60 mmHg. The fall in BP is induced by vasodilation accompanying a reduction in systemic vascular resistance. In the third trimester, the diastolic blood pressure gradually increases and may normalize to nonpregnant values by term.

One of the earliest set of symptoms of pregnancy are **nausea** and vomiting, a syndrome known as "morning sickness." Morning sickness typically begins between four to eight weeks of gestation and disappears between 14–16 weeks. The exact etiology is not known, but appears correlated to the high levels of progesterone, hCG, and their effect on the relaxation of **smooth muscle** cells of stomach. The enlarging uterus causes a gradual cephalic displacement of stomach and intestines. The stomach at full term has attained vertical position rather than its normal horizontal one. These mechanical forces lead to increased intragastic pressure, as well as change in the angle of gastroesophageal junction, which in turn tends toward greater esophageal reflux. The large and small bowel move upward and laterally, and the **appendix** is displaced superiorly in the right flank area. These **organs** return to normal position in the early puerperium. In general, there is decreased gastrointestinal motility during pregnancy because of increasing levels of progesterone. As a result, gastric emptying time is prolonged and there is decreased esophageal tone and incompetence of the esophageal-stomach sphincter, leading to gastric reflux and heartburn, common complaints in pregnancy.

Transit time of food throughout intestinal tract may be so much slower that more water than normal is reabsorbed, leading to constipation. Gallbladder function is also delayed in pregnancy because of hypotonia of the smooth muscle wall due to progesterone activity. Emptying time is slowed and often incomplete. **Bile** can become thick, and bile stasis may lead to gallstone formation.

Early in pregnancy, in the mucosal vasculature throughout the respiratory tract occurs a capillary engorgement and increased amounts of secretions in the **nose**, oropharynx,larynx and trachea.This causes allergy-like symptoms, chronic colds, nasal congestion, voice changes, and can make **breathing** difficult. As the uterus enlarges, the subcostal angle increases chest circumference (by up to six cm) and the diameter slightly increases (by two cm) due to the elevation of the **diaphragm**. In pregnancy, respiration is more diaphragmatic than thoracic. There is a 30–40% of increase in tidal volume as pregnancy progresses. Inspiratory capacity increases five to ten percent, as does the respiratory rate (two to three breaths per minute increase). The functional residual capacity, expiratory volume, and residual volume are all decreased by 20%. The total lung capacity is also decreased by approximately 5%, with a resulting increase of about two to three respirations per minute.

During pregnancy, the **kidneys** enlarge approximately 0.4–0.6 in. (1–1.5 cm), with a concomitant increase in weight. At term, both kidneys are 50–60% higher than in the nonpregnant status. This is the result of an increase in interstitial volume, as well as distended renal vasculature. Renal plasma flow begins to increase early in the first trimester and increases by as much as 25–75% over non-pregnant levels at term. Similarly the **glomerular filtration** rate increases to 50% over the non-pregnant state. Both renal pelvis and ureters are dilated during pregnancy because of the relaxing effect of progesterone and the enlarging and dextorotation of the uterus, among other factors. Because progesterone also decreases bladder tone, there is an increased residual volume and with a dilated collecting system, urinary stasis can result. There is also loss of urinary control as pregnancy advances because the bladder is displaced upward and flattened in anterior posterior diameter by the enlarging uterus. Pressure from the uterus leads to a reduction in bladder capacity resulting in increased urinary frequency.

The effects of pregnancy on the female reproductive **anatomy** include an increase in vascularity in the vulva. An increase in vaginal transudation and the stimulation of vaginal mucosa produce an important vaginal discharge. The epithelium of the cervical canal everts onto the ectocervix. The uterus undergoes an enormous increase due to hypertrophy and hyperplasia of the myometrium. Its weight passes from the approximately 0.16 oz. (70 g) of the non-pregnant status to 2.5 oz. (1,100 g) at term. Similarly, the uterine cavity grows from a volume of about 0.34 fl. oz. (10 ml) to approximately 5.3 qt. (5 L) at term. Cardiac output to the uterus increases from 2% in the non-pregnant state to 15–20% at term. The enlarged uterus causes an increase in intra-abdominal pressure. As pregnancy progresses, the spine is positioned in a compensatory lumbar lordosis (abnormal forward curvature). The effects of progesterone and relaxin cause an increased laxity of the ligaments. The **pubic symphysis** separates at about

28–30 weeks. There is also an increased rate of PTH (parathyroid hormone), which leads to increased absorption of **calcium** from the intestine and a decreased loss of calcium by the kidneys, due to fetal needs. Despite the increase of PTH, the skeleton is well maintained by the calcitonin effect.

See also Anatomical nomenclature; Birth defects and abnormal development; Blood pressure and hormonal control mechanisms; Calcium and phosphate metabolism; Diuretics and antidiuretic hormones; Embryonic development, early development, formation, and differentiation; Ethical issues in embryological research; Ethical issues in genetics research; Fetal circulation; Fetal membranes and fluids; Follicle stimulating hormone (FSH) and luteinizing hormone (LH); Heart, embryonic development and changes at birth; Heart, rhythm control and impulse conduction; Homeostatic mechanisms; Hormones and hormone action; Infertility (female); Intestinal motility; Mammary glands and lactation; Parturition; Placenta and placental nutrition of the embryo; Plasma and plasma clearance; Prenatal growth and development; Pubic symphysis (gender differences); Reproductive system and organs (female)

PRENATAL GROWTH AND DEVELOPMENT

The age of an embryo or fetus can be measured from both the **fertilization** date, and the date of the last menstrual period. Usually in the embryonic period, the fertilization date is used, while, in the fetal period and in obstetrical practice, the date of the last **menstruation** is used.

The fusion of the male **sperm** and female oocyte leads to fertilization and formation of the zygote (or fertilized embryo). As the zygote traverses the maternal tubal and uterine cavity, the initial stages of embryonic development have been completed. During the first week, the first zygote's (mitotic) divisions produces a series of blastomeres and, because of further cellular divisions (cleavage), the so called morula. A morula contains about 10–30 cells. The morula stage is the final stage prior to formation of a fluid-filled cavity called the blastocele cavity. Once the cavitation is recognizable, the conceptus is called an early blastocyst. Inside the blastocyst is an inner cell mass or embryoblast (future embryo), and the outer cell mass or trophoectoderm (future **placenta**). The zygote inherits protein information from the oocyte cytoplasm and mitochondrial **DNA** (of maternal origin), in order to mantain the continuity of **metabolism** and cell divisions. As development advances, there is a switch to the zygotic genome control. A coordinate expression of several genes including cytoskeletal proteins, growth factors, and intercellular recognition molecules is essential for the correct implatation in the endometrium. Blastocyst implantation occurs at the end of the first week. At two weeks gestation, the development of the so called bilaminar embryonic disk takes place.

The embryoblast then splits into the epiblast and hypoblast. The epiblast is an external layer of primary ectoderma; the hypoblast is a inner layer of primary **endoderm**. A small space forms from successive coalescences of the epiblast, yielding the amnion and the amniotic cavity. Hypoblastic cells migrate along the inner surface of the cytotrophoblast to form the primary yolk sac. Such a cavity will be successively reduced and will be named secondary, or definitive, yolk sac. The extraembryonic **mesoderm** arises from a cell proliferation of the epiblast. It initially consists of a layer of cells between the cytotrophoblast and primitive yolk sac. As the trophoblast grows outward rapidly, cavitations appear within the extraembryonic mesoderm. The cavitations grow larger and eventually join to form a large cavity that completely surrounds the embryo, except at the stalk where it remains attached to the cytotrophoblast. This new cavity is called the extraembryonic coelom and will become the future chorionic cavity. The extraembryonic coelom divides the extraembryonic mesoderm into two layers named extraembryonic splanchnic mesoderm, which surrounds the yolk sac and extraembryonic somatic mesoderm, which lines the trophoblast and covers the amnion. The trophoblast develops into cytotrophoblasts. The outer layer of cytotrophoblasts evolves into syncytiotrophoblasts. Syncytiotrophoblasts are responsible for implantation. Cytotrophoblast evolves into chorionic villi.

Embryogenesis (3–8 weeks) involves three major processes: morphogenesis (generation of shape), pattern formation (biologic-spatial cell organization), and differentiation (specialization in specific phenotypes). Morphogenesis and pattern formation are regulated by bone morphogenetic proteins and homeobox genes. Embryonic **tissue** and organ development (cellular differentiation) is regulated by local interactions between cells and extracellular matrix. The interaction can occur by means of cell-cell contact, or via specialized molecules such as adhesion molecules, intergins, or growth factors. During embryogenesis tissues and **organs** develop.

An important event in the embryonic period is gastrulation. Gastrulation begins with epiblastic proliferations, and is a process able to convert the bilaminar embryonic disk into the three primary embryonic germ layers, **ectoderm**, endoderm, and mesoderm. Ectoderm, meaning outside, ia an embryonic layer that more or less surrounds the other germlayers. The mesoderm, or middle, is a germ layer that lies between the ectoderm and the mesoderm. The endoderm, meaning inside, is a germ layer that lies at the most interior of the embryo. During this period, the human embryo converts from a flat embryonic disc to a three dimensional embryo. This is accomplished by folds, the head fold of the body, the tail fold of the body (both seen in sagittal section), and the lateral folds of the body.

Next, the formation of the dorsal mesoderm, (**notochord** and paraxial mesodermal cords) occurs. **Somites** arise from the coalescence of mesodermal cords. The mesoderm is further organized into intermediate and lateral mesoderm. Essentially, the vast majority of the urogential apparatus, cardiovascular apparatus, branchial apparatus, and muscles and bones derive from mesoderm. The evolution of the ectoderm then occurs, into the neural tube that will produce the nervous system and the skin. Also, the evolution of the endoderm occurs, that essentially will develop into the digestive apparatus, respiratory apparatus, some parts of the **urogenital system**,

and branchial pouches (part of the branchial apparatus). By the end of eight weeks, the embryo is about one inch long. The arms and legs are starting to form, and the embryo is beginning to look more human. Fingers and toes are growing and facial features are becoming more prominent. The head seems big compared to the body because the **brain** is growing at a very rapid rate. At this point, the embryonic period ends and all essential structures are present.

The fetal period ranges from nine to 38 weeks. It is characterized by rapid body growth (from 0.14 to 5.7–9.1 oz [60 to 2,500–4,000 gr]). During this period, the embryo, now named the fetus, is starting to grow. Length velocity has a peak at 20 weeks about in. (10 cm) in four weeks. Weight velocity has a peak at 30–34 weeks. At the beginning of the fetal period, the fetus heartbeat may be heard during a Doppler exam (performed with an ultrasound machine). When the gross structure of the nervous system has been established, the fetus uses its nervous senses to react to its environment. The fetus grows rapidly and tissues and organs within the fetus continue to develop and look more like the adult structures. The fetal size is measured during this period by the distance beween "crown to rump;" also long bones and head diameter can be used to estimate fetal age.

At nine weeks, half the fetus' overall size is its head, but by twelve weeks, the body size has doubled. During this period the fetus begins making urine. Male and female fetuses now begin to look different. The primary **ossification** centers appears. The **liver** becomes the site of **hemopoiesis**. From 13 to 16 weeks, the head is relatively smaller and the limbs (legs) are longer. The face of the fetus grows, and features on the face are begin to be placed in their correct positions assuming a human shape. The fetus performs **swallowing** and **breathing** motions.

During the fifth month (17–20 weeks), the growth slows down but the fetus is taller and the limbs reach their final relative proportions. Fetal movements become coordinated and can be felt by the mother (quickening). Vernix caseosa, a white substance that covers the skin of the fetus, appears. It helps to protect the fetus from macerazion due to the amniotic fluid. Brown **fat** is produced and accumulated on the neck and in the perirenal spaces for heat production.

During the sixth month (21–24 weeks), the most consistent event is the secretion of surfactant by the **lungs**. At about seven months (25–28 weeks) the lungs can work and breathe air. At eight months (29–32 weeks) the papillary reflex is present. The skin is smooth. By nine months, the (33–38 weeks), the head is about one quarter of the total body size. The bone marrow becomes the definitive site of hemopoiesis.

The development of ultrasound scanning has allowed accurate growth measurements of various fetal structures to be made. By evaluating large groups of fetuses, standard tables, percentiles, and curves of fetal growth have been created. Assuming a normal fetal growth, biometric measurements are determined by gestational age, and the expected date of delivery can be predicted. The most common and reliable fetal measurements include gestational sac diameters, the embryonic crown-rump length, the biparietal diameter (measures the diameter between the two sides of the head), head circumference, femur length, and the abdominal circumference.

See also Amniocentesis; Anatomical nomenclature; Birth defects and abnormal development; Cell differentiation; Ear, otic embryological development; Embryology; Embryonic development, early development, formation, and differentiation; Ethical issues in embryological research; Ethical issues in genetics research; Eye: Ocular embryological development; Face, nose, and palate embryonic development; Fetal circulation; Fetal membranes and fluids; Gastrointestinal embryological development; Genetic code; Genetic regulation of eukaryotic cells; Genetics and developmental genetics; Heart, embryonic development and changes at birth; Human Genetics; Implantation of the embryo; *In vitro* and *in vivo* fertilization; Integumentary system, embryonic development; Limb buds and limb embryological development; Neonatal growth and development; Nervous system, embryological development; Placenta and placental nutrition of the embryo; Pubic symphysis (gender differences); Renal system, embryological development; Reproductive system and organs (female); Reproductive system and organs (male); Respiratory system embryological development; Skeletal and muscular systems, embryonic development; Urogenital system, embryonic development; Vascular system, embryonic development

PRIESTLEY, JOSEPH (1733-1804)

English chemist

Joseph Priestley pioneered research into the nature and composition of gases involved in physiological processes, including air inspired and expired during the respiratory cycle, and was the first scientist to report the discovery of oxygen.

Born into a poor family in a village near Yorkshire, England, Priestley lost his mother at an early age and was sent to live with his aunt, a devout Protestant. He was educated at religious schools that endorsed nonconformist beliefs; he never formally studied science, but he did excel as a scholar of languages, logic, and philosophy. Priestley became a country preacher, but eventually turned to teaching. While employed at the Warrington Academy in the 1760s, he argued that school curriculums should reflect contemporary discoveries, rather than following outdated classical models. In 1762, Priestley married Mary Wilkinson, the sister of one of his schoolmates.

In 1766, Priestley visited London, England, and met Benjamin Franklin, who, prior to the American Revolution, was trying to settle a dispute between the American colonies and the British government. Soon after meeting Franklin, Priestley took over a pastorate in Leeds, England. Next door to his home was a brewery, and Priestley's scientific curiosity was aroused by the layer of heavy gas hovering over the huge fermentation vats. Priestley began experimenting with this gas, which we know today as carbon dioxide. Finding that the gas was heavier than air and that it could extinguish flames, Priestley realized he had isolated the same gas Joseph Black had designated as fixed air. Conducting various experiments

with this gas, he found that when dissolved in water, a bubbly drink was produced. Priestley had invented soda water, or seltzer.

During this period, Priestley also wrote a history of optics and an immensely successful history of electrical research. He discovered that carbon conducts electricity, for example, and he learned that an electrostatic charge collects on the outer surface of a charged object. As the first scientist to predict a relationship between electricity and chemistry, Priestley anticipated the new field of electrochemistry. At this time, Priestley also gave the modern name rubber to a Brazilian tree-sap product that had just been introduced to Europe. It was Priestley who told draftsmen that the material could be used to erase, or "rub out," pencil marks on their drawings.

Soon Priestley turned to chemistry, particularly the study of gases. During the early 1770s, he developed new methods of collecting gases in the laboratory and prepared several gases unknown to chemists at the time. Priestley adapted a device called the pneumatic trough, filling it with liquid mercury instead of water to obtain samples of gases. In this way, he was able to isolate gases such as sulfur, dioxide ammonia, and hydrogen chloride. (Ammonia and hydrochloric acid were known earlier, but only as liquids.) Priestley also discovered nitrous oxide (N_2O) years before Sir Humphry Davy popularized the gas's properties. Other gases isolated and identified by Priestley include nitrogen dioxide and silicon fluoride. As a result of these accomplishments, Priestley was elected to the French Academy of Sciences.

In 1773, Priestley won a lucrative post as librarian and companion to Lord Shelburne (1737-1805), a liberal politician. Priestley and his employer both sympathized with the colonial American rebels, who were then ready to begin the American Revolution, and an essay on government published by Priestley in 1768 provided Thomas Jefferson (1743-1826) with ideas for writing the Declaration of Independence.

Priestley's most famous scientific research was done during his eight years with Lord Shelburne. Because most of the gases he had studied were created by heating various substances, Priestley obtained a large magnifying lens. In 1774, Priestley used this lens to discover oxygen. Although Swedish chemist Carl Wilhelm Scheele had discovered the gas just a few years earlier, Priestley's results were reported first, and he usually gets the credit for the discovery. Priestley found that mercuric oxide, when heated, breaks down to form shiny globules of elemental mercury, while giving off a gas with unusual properties. A smoldering ember of wood, for example, burst into flames when exposed to the gas. In addition, a mouse trapped in a container of the gas became frisky and survived for a longer time than it would when trapped in ordinary air. Moreover, when Priestley inhaled the gas, he reported feeling "light and easy Priestley realized that this same gas was produced by plants, enabling them to restore "used-up" air to its original freshness.

In keeping with the scientific theory accepted at that time, Priestley named the gas dephlogisticated air because it absorbed phlogiston so readily. Phlogiston was thought to be the substance that gives materials their ability to burn.

Supposedly, during combustion, phlogiston is released from burning material and absorbed by the surrounding air or gas. When Priestley reported his findings, he unknowingly gave Antoine-Laurent Lavoisier the key to a new theory of combustion that contradicted the phlogiston theory. It was Lavoisier who later expanded on Priestley's work, re-named the gas oxygen, and explained how substances burn by combining chemically with oxygen.

While conducting his experiments, Priestley continued to speak out aggressively on political and religious issues. He not only supported the American colonists' war with England, but also sympathized with supporters of the French Revolution. Priestley's religious allegiance had shifted toward the Unitarian Church, which was also unpopular in England at the time. Priestley eventually settled in Birmingham, England, where he served as a chapel minister and joined the Lunar Society, a club of respected scientists and inventors, but in 1791 an angry "Church and King" mob retaliated against France's supporters and burned down Priestley's home and laboratory, destroying much of his research. Priestley escaped to London, but even there, his beliefs were barely tolerated. After the French people beheaded their king, declared war on England, and offered to make Priestley a citizen, Priestley gave in to public outrage and left for America. There he became a personal friend of Jefferson and of other politicians.

See also Anesthesia and anesthetic drug actions; Breathing; Gaseous exchange; Hemoglobin; Hypoxia; Oxygen transport and exchange; Respiration control mechanisms; Respiratory system

PROGESTERONE

Progesterone is a hormone produced primarily by females. **Hormones** are a group of chemicals secreted by special glands throughout the body. Hormones are carried by the bloodstream to target tissues, where they stimulate the tissues to carry out specific functions. Progesterone is essential for the proper functioning of the female menstrual cycle and formation of the **placenta** during **pregnancy**.

The chemical name for progesterone is pregn-4-ene-3,20-dioneand its symbol is $C_{21}H_{30}O_2$. The main site of production is in the cells of the corpus luteum of the ovaries. There are two types of cells in the ovaries that produce progesterone. In one type of cell, the small luteal cells, luteinizing hormone (LH) stimulates the secretion of progesterone. Large luteal cells also produce progesterone, but their production is controlled by a chemical called prostaglandin (PG)F2alpha which causes the **death** of these particular cells.

The production of progesterone is limited by the synthesis of cholesterol. The **mitochondria** of cells transport cholesterol across their membranes (steroidogenesis) and the rate at which this happens is regulated by the speed at which cholesterol is transported across the inner and outer mitochondr-

ial membrane. The initiation of progesterone production begins in this area of the mitochondrion.

The effect of progesterone is most keenly observed in the uterus where the hormone aids in the successful implantation of a fertilized egg. It stimulates the growth of the placenta where additional progesterone is produced. In this case progesterone prevents spontaneous abortion. The hormone also targets tissues of the **mammary glands** and prepares them to secrete milk. If no **fertilization** occurs, the synthesis of progesterone is reduced and the onset of **menstruation** occurs. Progesterone is also produced in the adrenal cortex where it is used for the synthesis of **testosterone** and other steroids.

Today progesterone is synthesized in laboratories and is a useful medication. Some ovarian disorders are regulated by the consumption of synthetic progesterone. It has been successfully used in combination with other hormones for birth control.

See also Ovarian cycle and hormonal regulation

PRONATION • *see* ANATOMICAL NOMENCLATURE

PROSTAGLANDINS

Prostaglandins are a type of chemical that can be found in almost all tissues of mammals. Prostaglandins exist in cyclic form, and are composed of derivatives of fatty acids. They exert their functions both locally (in the basic vicinity of their production), as well as traveling through the bloodstream to act on distant target **organs** (in an endocrine manner). In fact, some prostaglandins are the "middle man" messengers between **hormones** and other components in cells. There are many types of prostaglandins, including those labeled PGA1, PGE1, PGE2, and PGI1.

Prostaglandins work on a number of different target tissues, including the uterus, ovaries, and fallopian tubes. Although the actions of prostaglandins are diverse depending upon the cell type in which they interact, they are best known for producing **smooth muscle** contraction. Prostaglandins are known to be involved in the initiation and maintenance of uterine contractions during labor, as well as in the production of the uncomfortable cramps occurring during menstrual periods. Prostaglandins are also involved in regulating the size of the **blood** vessels of the kidney. Prostaglandins affect the actions of the blood cells responsible for clotting (**platelets**). Prostaglandins are involved in the production of fever. Prostaglandins are also involved in the process of inflammation, and anti-inflammatory medications such as aspirin and ibuprofen work by interfering with the production of prostaglandins.

See also Endocrine system and glands

PROSTATE GLAND

The prostate gland is part of the male reproductive system. It is a small, muscular organ about the size of a walnut, located at the base of the bladder and in front of the rectum.

The prostate, together with the seminal vesicles and the Cowper's glands, comprise the male sexual accessory glands. They produce over 95% of the ejaculate by means of which the **sperm** is transported into the female reproductive tract.

Specific compounds in the secretions of each gland contribute to **fertilization**, for example by maintaining the optimal fluid viscosity, encouraging the viability of the sperm cells, or insulating them from the female **immune system**. The prostate itself produces between about 15 and 30% of the ejaculate. Its secretions include citric acid, which maintains the osmotic **equilibrium** between the fluid and the sperm cells so that they do not become either engorged or dehydrated. It also contains zinc, a bactericide that may help prevent urinary tract infections.

Sperm cells from the testes plus the fluid from the seminal vesicles enter the prostate via a tube called the vas deferens. In the prostate, the additional secretions are added, and the muscular gland pumps the fluid into the urethra during ejaculation.

The prostate surrounds the urethra like a collar, so a swollen prostate is a common cause of urinary problems in men, especially as they age. About one in four will eventually require surgery for benign prostate hyperplasia, and prostate **cancer** is now the most common malignancy in men. Early diagnosis of prostate cancer is now possible by means of a **blood** test for prostate-specific antigen (PSA). PSA is a protein that is normally found in the ejaculate, and enters the blood when the prostate is damaged.

See also Osmotic equilibria between intercellular and extracellular fluids; Reproductive system and organs (male); Semen and sperm; Sexual reproduction; Urogenital system

PROTEIN METABOLISM

In humans, there is a large amount of protein turnover. Approximately one pound of protein is degraded and re-synthesized each day. Most of the **amino acids** liberated by protein degradation are used again in the re-synthesis of new protein but about ten percent of the total amino acid content is lost as it is converted to other important molecules involved in nervous system function, pigments, various **hormones**, and a variety of other essential activities. Amino acids are also used as fuel and, when present in excess in the diet, can be converted to **fat** for storage of excess calories.

Because of the universal importance of protein molecules to living cells, both plant and animal tissues can provide dietary protein. During **digestion**, the long chains of amino acids that make up complex protein molecules are disassembled to produce the twenty different single amino acids. These are taken up by cells in the digestive system, mostly in the

small intestine, and released to the **blood** where they are transported around the body. Amino acids from the diet are used in three ways. They are uniquely used in the synthesis of new protein and, in a well-fed body, cells are actively synthesizing the structural and enzymatic proteins required for healthy functioning. The synthesis of these proteins is closely regulated by the expression of particular genes. It is this selective regulation that determines which proteins are to be synthesized, and in a more global sense, the characteristics, abilities, and activities of each individual cell.

When present in excess, amino acids are used as fuel. The different carbon skeletons of the twenty different amino acids are each metabolised through a more or less unique series of reactions. Thus, the degradation of each amino acid occurs by means of a specific pathway. However, the end products of these pathways are the same as various intermediates in the breakdown of glucose. Thus, overall, amino acid degradation results in the production of acetyl-CoA or its precursors and several of the organic acids involved the tricarboxylic acid (TCA) cycle. This means that, like carbohydrate, the carbon atoms that make up the amino acids can be converted to CO_2 with the production of energy needed to support the life of the cell and the organism.

Excess amino acids can also be converted to fat. Again the picture is similar to that for carbohydrate in that carbon structures derived from the amino acids can be converted to citrate, a TCA cycle compound, and the first intermediate in the pathway of fat synthesis. Because the **liver** is the major site of fat synthesis, excess amino acids are taken up by the liver, converted to fat, packaged, transported, and stored as fat in **adipose tissue**.

Both carbohydrate and fat can be stored by cells, and by the organism, for use at a later time. **Glycogen** represents the storage form for carbohydrate and is present in many types of cells, particularly in the liver. **Triglycerides** represent the storage form for fatty acids synthesized in the liver and stored in adipose **tissue**. There is, however, no storage form for amino acids. They are either converted into protein or they are converted into other compounds. As a consequence, during the fasting state the body begins to break down protein to obtain the amino acids for the synthesis of new protein molecules needed to maintain or change metabolic activities. Each individual kind of protein molecule has a particular rate of turnover. Some proteins are degraded rapidly, such that half of the total amount of the enzyme in a single cell is broken down every 15 minutes or so. Others are degraded more slowly, where the time it takes to degrade half is perhaps an hour, a few hours, or in some instances, several days or weeks. Also during fasting, the amino acids liberated by protein breakdown also assist in energy production. This occurs both at the level of the individual cell in which protein degradation occurs and in whole body **metabolism**. It occurs in the following way: sugars, particularly glucose, are sources of TCA cycle intermediates and essential for the production of energy as **ATP**. Oxaloacetate (OAA) is a critical intermediate in the TCA cycle and the first step in the cycle involves the combination of OAA and acetyl-CoA to form citrate. During the breakdown of amino acids, the carbon skeletons of many of the amino acids are converted to one of the intermediates of the TCA cycle. Because of its cyclic character, once these intermediates enter the cycle they are easily converted to OAA. The production of OAA from amino acids means that the cell no longer needs to use as much glucose to maintain adequate levels of OAA in the TCA cycle. This, in turn, means that blood glucose is used more sparingly.

In the fasting state, a significant portion of the amino acids produced by the breakdown of protein in peripheral tissues, such as muscle, is released to the blood. Because of its very rich blood supply, the liver has excellent access to these circulating amino acids. These free amino acids are used for two major purposes. The first purpose is, just as in peripheral tissue, for the support of the synthesis of proteins needed by the liver to maintain its own structures and processes. The second purpose is for the synthesis of additional glucose for use by other tissues by a process called gluconeogenesis that takes place in the liver. Glucose can be synthesized from several key intermediates in metabolism. One of these is malate, one of the components of the TCA cycle. Just as for OAA, all of the TCA cycle intermediates can be converted to malate. Since the carbon skeletons of many of the amino acids are converted into TCA cycle intermediates, they also serve as starting material for the synthesis of glucose. This newly synthesized glucose can be released to the blood for use by the **central nervous system** and by other tissues.

See also Biochemistry

PROTEIN SYNTHESIS

Protein synthesis represents the final stage in the translation of genetic information from **DNA**, via messenger **RNA** (mRNA), to protein. It can be viewed as a four-stage process, consisting of amino acid activation, translation initiation, chain elongation and termination. The events are similar in both prokaryotes, such as **bacteria**, and higher eukaryotic organisms, although in the latter there are more factors involved in the process.

To begin with, each of the 20 cellular **amino acids** are combined chemically with a transfer RNA (tRNA) molecule to create a specific aminoacyl-tRNA for each amino acid. The process is catalyzed by a group of **enzymes** called aminoacyl-tRNA synthetases, which are highly specific with respect to the amino acid that they activate. The initiation of translation starts with the binding of the small subunit of a ribosome, (30S in prokaryotes, 40S in eukaryotes) to the initiation codon with the nucleotide sequence AUG, on the mRNA transcript. In prokaryotes, a sequence to the left of the AUG codon is recognized. This is the Shine-Delgrano sequence and is complementary to part of the small ribosome subunit. Eukaryotic ribosomes start with the AUG nearest the 5'–end of the mRNA, and recognize it by means of a "cap" of 7–methylguanosine triphosphate. After locating the cap, the small ribosome subunit moves along the mRNA until it meets the first AUG codon, where it combines with the large ribosomal subunit.

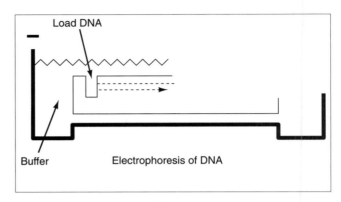

The method of submarine DNA electrophoresis. *Illustration by Hans & Cassidy.*

In both prokaryotes and eukaryotes, the initiation complex is prepared for the addition of the large ribosomal subunit at the AUG site, by the release of initiation factor (IF) 3. In bacteria, the large 50S ribosomal subunit appears simply to replace IF–3, with IF–1 and IF–2. In eukaryotes, another factor eIF–5 (eukaryotic initiation factor 5), catalyses the departure of the previous initiation factors and the joining of the large 60S ribosomal subunit. In both cases the release of initiation factor 2 involves the hydrolysis of the GTP bound to it. At this stage, the first aminoacyl-tRNA, Met-tRNA, is bound to the ribosome. The ribosome can accommodate two tRNA molecules at once. One of these carries the Met-tRNA at initiation, or the peptide-tRNA complex during elongation and is thus called the P (peptide) site, while the other accepts incoming aminoacyl-tRNA and is therefore called the A (acceptor) site. What binds to the A site is usually a complex of GTP, elongation factor EF-TU and aminoacyl-tRNA. The tRNA is aligned with the next codon on the mRNA, which is to be read and the elongation factor guides it to the correct nucleotide triplet. The energy providing GTP is then hydrolysed to GDP and the complex of EF-TU:GDP leaves the ribosome. The GDP is released from the complex when the EF-TU complexes with EF-TS, which is then replaced by GTP. The recycled EF-TU: GTP is then ready to pick up another aminoacyl-tRNA for addition to the growing polypeptide chain. On the ribosome, a reaction is catalysed between the carboxyl of the P site occupant and the free amino group of the A site occupant, linking the two together and promoting the growth of the polypeptide chain. The peptidyl transferase activity which catalyses this transfer is intrinsic to the ribosome. The final step of elongation is the movement of the ribosome relative to the mRNA accompanied by the translocation of the peptidyl-tRNA from the A to the P. Elongation factor EF-G is involved in this step and a complex of EF-G and GTP binds to the ribosome, GTP being hydrolysed in the course of the reaction. The de-acylated tRNA is also released at this time.

The end of polypeptide synthesis is signalled by a termination codon contacting the A site. Three prokaryotic release factors (RF) are known: RF–1 is specific for termination codons UAA and UAG, while RF–2 is specific for UAA

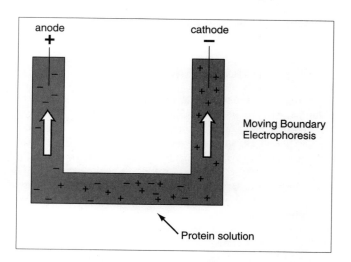

The simplest electrophoresis approach is the moving boundary technique. As shown, the charged molecules (proteins) to be separated are electrophoresed upward through a buffer solution toward electrodes immersed on either side of a U-shaped tube. This technique separates the biomolecules on the basis of their charges—positively charged molecules migrate toward the negative electrode (cathode) and negatively charged particles move to the positive electrode (anode). *Illustration by Hans & Cassidy.*

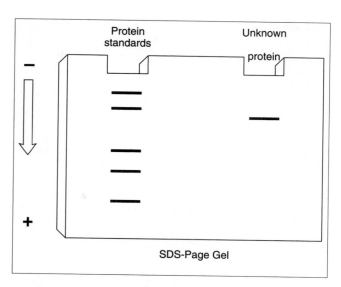

The SDS-PAGE technique of electrophoresis allows researchers to study parts of proteins and protein-protein interactions. If a protein has different subunits, they will be separated by SDS treatment and will form separate bands. *Illustration by Hans & Cassidy.*

and UGA. RF–3 stimulates RF–1 and RF–2, but does not in itself recognize the termination codons. RF–3 also has GTPase activity and appears to accelerate the termination at the expense of GTP. Only one eukaryotic release factor is known and it has GTPase activity.

At any one time, there can be several ribosomes positioned along the mRNA and thus initiation, elongation and termination proceed simultaneously on the same length of mRNA. The three dimensional structure of the final protein begins to appear during protein synthesis before translation is completed. In many cases, after the synthesis of the amino acid chain, proteins are subjected to further reactions which convert them to their biologically active forms, e.g., by the attachment of chemical groups or by removal of certain amino acids—a processes known as post-translational modification.

See also Genetic code

PRUSINER, STANLEY (1942-   )

American physician

Stanley Prusiner performed seminal research in the field of neurogenetics, identifying the prion, a unique infectious protein agent containing no **DNA** or **RNA**.

Prusiner was born on in Des Moines, Iowa. His father, Lawrence, served in the United States Navy, moving the family briefly to Boston where Lawrence Prusiner enrolled in Naval officer training school before being sent to the South Pacific. During his father's absence, the young Stanley lived with his mother in Cincinnati, Ohio. Shortly after the end of

World War II, the family returned to Des Moines where Stanley attended primary school and where his brother, Paul, was born. In 1952, the family returned to Ohio where Lawrence Prusiner worked as a successful architect.

In Ohio, Prusiner attended the Walnut Hills High School, before being accepted by the University of Pennsylvania where he majored in chemistry. At the University, besides numerous science courses, he also had the opportunity to broaden his studies in subjects such as philosophy, the history of architecture, economics, and Russian history. During the summer of 1963, between his junior and senior years, he began a research project on **hypothermia** with Sidney Wolfson in the department of surgery. He worked on the project throughout his senior year and then decided to stay on at the University to train for medical school. During his second year of medicine, Prusiner decided to study the surface fluorescence of brown **adipose tissue** (fatty **tissue**) in Syrian golden hamsters as they arose from hibernation. This research allowed him to spend much of his fourth study year at the Wenner-Gren Institute in Stockholm working on the **metabolism** of isolated brown adipocytes. At this he began to seriously consider pursuing a career in biomedical research.

Early in 1968, Prusiner returned to the United States to complete his medical studies. The previous spring, he had been given a position at the National Institutes of Health (NIH) on completing an internship in medicine at the University of California San Francisco (UCSF). At the NIH, he worked on the glutaminase family of **enzymes** in *Escherichia coli* and as the end of his time at the NIH began to near, he examined the possibility of taking up a postdoctoral fellowships in neurobiology. Eventually, however, he decided that a residency in **neurology** was a better route to developing a rewarding career in research as it offered him direct contact with patients and therefore an opportunity to learn about both the normal and abnormal nervous system. In July 1972, Prusiner began a residency at UCSF in the department of neurology. Two months later, he admitted a female patient who was exhibiting progressive loss of memory and difficulty performing some routine tasks. This was his first encounter with a Creutzfeldt-Jakob disease (CJD) patient and was the beginning of the work to which he has dedicated most of his life.

In 1974, Prusiner accepted the offer of an assistant professor position from Robert Fishman, the Chair of Neurology at UCSF, and began to set up a laboratory to study scrapie, a parallel disease of human CJD found in sheep. Early on in this endeavor, he collaborated with William Hadlow and Carl Eklund at the Rocky Mountain Laboratory in Hamilton, Montana, from whom he learned much about the techniques of handling the scrapie agent. Although the agent was first believed to be a virus, data from the very beginning suggested that this was a novel infectious agent, which contained no nucleic acid. It confirmed the conclusions of Tikvah Alper and J.S. Griffith who had originally proposed the idea of an infectious protein in the 1960s. The idea had been given little credence at that time. At the beginning of his research into prion diseases, Prusiner's work was fraught with technical difficulties and he had to stand up to the skepticism of his colleagues. Eventually he was informed by the Howard Hughes Medical Institute (HHMI) that they would not renew their financial support, and by UCSF that he would not be promoted to tenure. The tenure decision was eventually reversed, however, enabling Prusiner to continue his work with financial support from other sources. As the data for the protein nature of the scrapie agent accumulated, Prusiner grew more confident that his findings were not artifacts and decided to summarize his work in a paper, published in 1982. There he introduced the term "prion," derived from "proteinaceous" and "infectious" particle and challenged the scientific community to attempt to find an associated nucleic acid. Despite the strong convictions of many, none was ever found.

In 1983, the protein of the prion was found in Prusiner's laboratory and the following year, a portion of the amino acid sequence was determined by Leroy Hood. With that knowledge, molecular biological studies of prions ensued and an explosion of new information followed. Prusiner collaborated with Charles Weissmann on the molecular cloning of the gene encoding the prion protein (PrP). Work was also done on linking the PrP gene to the control of scrapie incubation times in mice and on the discovery that mutations within the protein itself caused different incubation times. Antibodies that provided an extremely valuable tool for prion research were first raised in Prusiner's lab and used in the discovery of the normal form of PrP protein. By the early 1990s, the existence of prions as causative agents of diseases like CJD in humans and bovine spongiform encephalopathy (BSE) in cows, came to be accepted in many quarters of the scientific community. As prions gained wider acceptance among scientists, Prusiner received many scientific prizes. In 1997, Prusiner was awarded the Nobel Prize in physiology or medicine.

See also Human genetics; Infection and resistance

PSYCHOPHARMACOLOGY

Basic **neurotransmitters** influence whether or not an **action potential** will be produced in the post-synaptic cells they affect. Glutamine, an excitatory neurotransmitter, increases the likelihood of an action potential, whereas GABA, an inhibitory neurotransmitter, decreases the likelihood. Beyond these basic neuotransmitters, however, there are several neurotransmitters that have modulatory effects. By acting on post-synaptic membranes, they can dampen or intensify the effects of the basic neurotransmitters. Because their effects are diffuse rather than concentrated in a small part of the **brain**, they have widespread effects on mood, motivation, memory, and other aspects of behavior. Therefore, many drugs that are used to treat disturbances in behavior work by interacting with these modulatory neurotransmitters.

Depression, which affects many people at some time in their lives, is treated with three different classes of drugs, which are classified according to their molecular **structure and function**. One group is the MAO inhibitors. Monoamine oxidase is an enzyme that breaks down norepinephrine and serotonin, two of the modulatory neurotransmitters. MAO inhibitors prevent this destruction. The tricyclic antidepressants block reupake of norepinephrine and serotonin by the pre-synaptic neuron. Selective serotonin reuptake inhibitors such as Prozac specifically prevent the reuptake of serotonin. All three classes of drugs increase the potential for modulatory neurotransmitters to act on the post-synaptic **neurons**.

Schizophrenia, a devastating mental illness, involves a break with reality. The boundary between what is real and what is not becomes blurred, affecting thought, perception, and mood. Neuroleptics, or antipsychotic drugs, are used to treat schizophrenia. Their mode of action is to block dopamine receptors on the post-synaptic neuron so that dopamine, which is believed to be excessive in some parts of a schizophrenic's brain, cannot exert its full effect. The more effectively a neuroleptic binds to the dopamine receptors, preventing dopamine from doing so, the less of it is needed to decrease the symptoms of schizophrenia. By contrast, drugs that increase dopamine activity in the brain cause the symptoms to become worse. Parkinson's disease is caused by a shortage of dopamine. When they are treated with its precursor, L-dopa, some Parkinson's patients experience psychosis.

Anxiety disorders are treated by benzodiazepines, the best known of which is Valium. These drugs bind to benzodiazepine receptors in the brain, particularly in the **cerebral cortex**, **hippocampus** and amygdala. These receptors then increase the action of GABA receptors. Since GABA is an inhibitory neurotransmitter, the effect is to decrease the rate of action potentials in the affected areas. This occurs because GABA-activated receptors cause chloride ions to enter the cell, hyperpolarizing it.

In obsessive compulsive disorder (OCD) people involuntarily engage in repetitive acts without an objective reason for doing so. Clomipramine, used to treat OCD, works by blocking serotonin uptake, causing less response to serotonin.

The psychotropic medications of today are effective in many cases, but they resemble blunt instruments. Many have undesirable side effects, and finding which ones will be effective for an individual is often a matter of trial and error. As knowledge of the genetic and biochemical bases of psychiatric disorders increases, it will be possible to design drugs that will exert their effects more specifically and can be chosen to suit an individual patient, increasing the precision of treatment.

See also Nervous system overview

PTH · *see* PARATHYROID GLANDS AND HORMONES

PUBERTY

Puberty is the period of sexual maturity when sexual **organs** mature and **secondary sexual characteristics** develop. Puberty is also the second major growth period of life—the first being infancy. A number of **hormones** under the control of the **hypothalamus**, pituitary, ovaries, and testes regulate this period of sexual growth, which begins for most boys and girls between the ages of nine and fifteen. The initial obvious sign of female puberty is the beginning of breast development, whereas the initial obvious sign in males is testicular enlargement. Since early signs of female puberty are more noticeable, it is sometimes assumed that female puberty precedes male puberty by quite a bit. However, males usually start puberty just a few months after females, on average. In males, puberty is marked by testicle and penile enlargement, **larynx** enlargement, pubic **hair** growth, and considerable growth in body height and weight. In females, puberty is marked by hip and breast development, uterine development, pubic hair growth, **menstruation**, and increases in body height and weight. Because of the extensive growth that occurs at this time, a balanced, nutritious diet with sufficient calories is important for optimal growth. Although puberty was originally used to classify the initial phase of early fertility, the term is also used to include the development and growth, which culminates in fertility. In this sense, puberty usually lasts two to five years and is accompanied by the psychological and emotional characteristics called adolescence.

Puberty marks the physical transition from childhood to adulthood. While the changes that accompany this time are significant, their onset, rate, and duration vary from person to person. In general, these changes are either sexual or growth related. The major control center for human pubertal development is the hypothalamus for both sexes, but puberty is accompanied by additional growth of the adrenal glands, as well. The added adrenal **tissue** secretes the **sex hormones**, androgens or estrogens, at low levels. The adrenal sex hormones are thought to initiate the growth of pubic and axillary (under-arm) hair. This adrenal maturation is called adrenarche.

It is not known exactly what triggers puberty to begin. However, the hypothalamus sends out gonadotropin hormones responsible for **sperm** and egg maturation. One theory holds that normal **brain** growth towards the end of childhood includes significant hypothalamic changes. Hypothalamic receptors are thought to become more sensitive to low levels of circulating sex steroids. These changes enable the neuroendocrine system to initiate spermarche (sperm maturation) and menstruation in puberty. However, these early hormonal fluctuations begin at night and remain a nocturnal **pulse** for some time before they are detectable while awake. Some behavioral changes are related to pubertal hormonal changes, as well. The increase in **testosterone** is associated with more aggressive behavior in males. And libido (sex drive) increases occur for some teenagers in association with **estrogen** and testosterone increases. These effects are also carried out through sex hormone receptors on the hypothalamus.

Major pubertal hormones secreted by the hypothalamus include gonadotropin releasing hormone (GRH) and growth hormone releasing hormone (GHRH). Both target the anterior pituitary gland, which in turn releases gonadotropins and growth hormone (also known as somatotropin). GRH is released in a pulsative fashion. This pulsation triggers release of the gonadotropins, **luteinizing hormone (LH) and follicle stimulating hormone** (FSH). LH stimulates testosterone release by the testes, and FSH is required for early stages of sperm maturation. GHRH is released on a daily basis throughout life, but growth hormones have an enhanced effect during puberty when they are combined with sex hormones.

The age of onset of puberty varies but can be between the ages of 9 and 14 in boys. However, individuals can mature as late as 20. When all of a male's organs and endocrine functions are normal but testicular development never occurs, he is said to display eunuchoidism. This name originates from China where servile classes of eunuchs were created by removing their testicles. Because of their lack of testosterone, they were less aggressive. Puberty that begins before the age of eight is called precocious. Precocious puberty can result from neurological disorders of the posterior hypothalamus or pituitary disorders such as **tumors** or infections.

The initial sign of male puberty is testicular enlargement. The testes secrete testosterone, which stimulates many primary and secondary sexual characteristics. Testosterone causes the **prostate gland** and seminal vesicles to mature. The seminal vesicles begin to secrete fructose, which is the primary nutrient sperm require. During puberty, primitive male **germ cells** begin to mature into primary spermatocytes. This early step in sperm maturation is testosterone-independent. However, the final stage of sperm maturation into spermatozoa is testosterone-dependent. Testicular size may double or quadruple at the start of puberty, but the rate of testicular growth is greatest in the middle of puberty. By the end, they will have doubled in size again. There is great variability in the final testicular size from man to man, but this difference has no affect on sexual ability.

The general progression of male genital area development is the onset of testicular enlargement, onset of penile enlargement, and the appearance of pubic hair (pubarche). The scrotal skin also becomes darker and more wrinkled. Penile enlargement usually begins about a year after testicular growth begins. The penis first becomes longer, and then becomes broader. Initial ejaculations usually occur later during **sleep**. Sperm count is low, at first.

Facial hair growth and a deepening voice are two secondary sexual characteristics, which develop about two years after pubic hair appears in males. Facial hair begins on the upper lip, becomes more confluent, extends to side-burns, and then grows on the chin. Hair also begins to appear on a pubertal boy's chest and **abdomen**. The voice deepens by dropping in pitch due to enlargement of the **vocal cords** in the larynx, voice box. In addition, other body hair grows, and the areola (pigmented ring around the nipple) enlarges.

Boys grow considerably in both height and mass during puberty. On average, boys will grow about 3.7 in/year (9.5 cm/year) at the peak year of their growth spurt. Boys average 4 ft 7 in (1.4 m) in height prior to the onset of puberty and grow an additional 15 in (38 cm) taller during their pubertal growth spurt. At the end of puberty, the average male height is 5 ft 10 in (1.8 m). The initial growth occurs in the leg bones increasing leg length. Then the torso lengthens causing an increase in sitting height. Between leg growth and torso growth, the arms, shoulders, and hips of boys grow considerably, as well. Muscle mass also increases-particularly in the shoulders. A temporary drop in subcutaneous **fat** occurs in the arms during this time with fat levels returning to normal at the end of puberty.

At the beginning of puberty, a girl's face rounds out, her hips widen, and her **breasts** begin to develop. Breast development can occur as early as age eight, but starts between 10 and 14 for most girls. Full breast development may take two to five years. Pubic hair begins to grow shortly afterwards, followed by the first menstrual period, or menarche. Like male puberty, female puberty is initiated by hypothalamic hormones. GRH secreted from the hypothalamus triggers LH and FSH release from the anterior pituitary. The LH and FSH, in turn, stimulate ova maturation. GHRH is also released from the hypothalamus and stimulates growth hormone secretion from the pituitary.

Breast development is called thelarche and can be measured in stages. The initial accumulation of tissue pads the underside of the areola around the nipple. Before puberty, the areola is usually about 0.5 in (1.2 cm) in diameter. By the end of puberty, it can be about 1.5 in (3.8 cm) in diameter. The breast enlarges developing a smooth curve. Then a secondary mound of tissue grows under the areola. Usually by age 18, a girl's breasts have reabsorbed the secondary mound giving a rounded contour to the now adult shape.

Breast budding is followed by menarche between 12 and 14 for most girls. However, normal menarche may occur between 10 and 16. Menstruation occurs as part of the menstrual cycle, which lasts about 28 days. The initial hormonal cycles associated with the menstrual period usually begin months before menarche, so for a while a girl usually has hormonal cycles without menstruation. The menstrual cycle is divided into two halves, the follicular and the luteal phases. During the follicular phase, an immature egg follicle ripens

and estrogen levels rise. On around day 14, LH and FSH trigger the egg to travel into the adjacent fallopian tube. During the luteal phase, high **progesterone** and estrogen levels prevent another egg from beginning another cycle. After about eight days, if the egg is not fertilized, then the uterine lining is shed as menstrual **blood**. Menstruation can last one to eight days, but usually lasts three to five days. The amount of blood lost varies from slight to 2.7 oz (80 ml) with the average being 1 oz (30 ml) lost for the whole period.

A number of factors affect when menstruation begins. Normal menarche is associated with good **nutrition** and health. Girls who are malnourished or ill may have later menarche. In addition, girls who are particularly athletic or involved in strenuous physical activities such as ballet often start menstruating later. Once menarche occurs, cycles are usually irregular for up to two years.

The pubertal growth spurt, of height and weight, in girls usually occurs a year or two before boys, on average. Increases in height and weight are followed by the increases in hip size, breast size, and body fat percentage. The peak growth rate during this time is 3.2 in (8 cm) per year, on average. The average female is 4 ft 3 in (1.3 m) tall at the beginning of puberty and gains 13.5 in (34 cm) total during her pubertal growth spurt. At the end of puberty, the average female height is 5 ft 4.5 in (1.6 m) tall. Girls also increase body fat at the hips, stomach, and thighs.

Around the world, entry into adulthood is often marked ceremoniously in males and females. A rite of passage ceremony is held to honor this transition. This type of ceremony is usually held in less-industrialized countries where boys and girls are expected to assume adult roles at the end of puberty. The Arapesh of New Guinea build the young woman a menstrual hut at the home of her husband-to-be. Her girlish ornaments are removed, and the girl acquires "womanly" markings and jewelry. The ceremony marks the beginning of her fertility. Young Mano men of Liberia go through a ceremonial "death" at puberty. These young men used to be stabbed with a spear and thrown over a cliff to symbolize death and rebirth into adulthood. Actually, a protective padding kept the spear from penetrating them, and a sack of chicken blood was tied over the spot to appear as though the boy had been stuck. He was not tossed over the cliff, but a heavy object was thrown over instead to sound like he had been thrown. Pubertal Apache girls are sometimes showered with golden cattail pollen (considered holy) as part of a four-day ritual. And boys and girls in Bali, Indonesia, formally come of age when a priest files their six top teeth even so they will not appear fanged.

By comparison, industrialized countries seldom have pubertal rites of passage. In fact, puberty may not be discussed often. Instead, these teenagers are usually expected to continue their education for some time before they can settle down and have a family. The changes that accompany puberty often bring on new feelings, however. Adolescents begin to contemplate independence from their parents and assume more adult roles in their family. In addition, puberty is a time when some boys and girls begin to think about their sexuality and sexual activity. Because the human body undergoes such significant and seemingly rapid changes in puberty, it can be a frightening time if a boy or girl does not understand what they are experiencing. Studies have shown that boys and girls who have been told about pubertal changes are less frightened and have fewer emotional problems related to puberty than children who have not been informed about what to expect.

With sexual maturation comes fertility. Most people do not become sexually active during puberty, but those who do have the additional adult responsibility to respect the possibility of **pregnancy**. For teenagers who begin having intercourse, contraceptive options exist to prevent pregnancy. Another serious consideration, however, is the possibility of contracting a sexually transmittable disease (STD). Not all STDs are curable. Some are debilitating, and others are fatal. The key is protection. Most contraceptives do not protect against both pregnancy and STD's. However, condoms (used correctly) will protect against both.

Puberty is not a good time to have a poor diet. A diet of potato chips and ice cream or celery and water will not optimize healthy growth. They will both hinder it. Loading up on junk food or slimming down by fasting are both dangerous. During puberty, a lot of body mass is constructed, and the right nutritional building blocks are essential. **Calcium**, protein, **carbohydrates**, minerals, and **vitamins** are all important. Enough calories to fuel development are also needed. During puberty, adolescents need about 2,000–2,500 calories a day. Some girls become self-conscious of their developing bodies and try to minimize fatty tissue growth by fasting or making themselves throw up food they have eaten. Both of these mechanisms to stay thin are extremely dangerous, can have long-term detrimental effects on health, and should be avoided. Adolescents who can turn to a trustworthy adult with their questions or concerns about puberty may find this transition easier.

See also Adolescent growth and development

PUBIC SYMPHYSIS (GENDER DIFFERENCES)

The pubic symphysis is the **fibrocartilage** that connects the pubic bones of the pelvis. The structure and relative elasticity of the pubic symphysis is important to normal childbirth because the broader **cartilage** that exists in females allows a greater spreading of the pelvis and wider birth canal for the passage of the fetus during childbirth (**parturition**).

The pubic symphysis is a cartilaginous articulation of the pelvic innominate bones and it is the anterior landmark of the pelvis. Along with the sacroiliac **joints**, the posterior **synovial joints** that articulate the sacrum and innominate bones.

Along with the pubic bones, the public symphysis defines the subpubic arch. In females, the pubic symphysis is broader than it is in males and thus the angle of the subpubic arch is less acute (greater angle). In females the public angle usually measures about 80° about twenty degrees greater than the average male public angle that measure about 60°.

Although in absolute terms the male pelvis is larger, the relative broadness of the pubic symphysis and the wider pubic angle allows the female pubic cavity to exceed the volume of the male pelvis. This increased capacity has obvious reproductive and evolutionary advantages.

Radiological examination (examination via x rays and other **imaging** techniques) establish that in both females and males there is a slight narrowing and hardening of the pubic symphysis that is identifiable between 25 and 35 years of age. Differentiation of sex, based upon pelvic angle, can also be made in imaging examinations of the developing fetus.

The determination of the size and degree of calcification of the public symphysis is important in the forensic or archaeological examination of skeletal remains. Measurements of the pubic symphysis (along with other features such as the curvature of the sacrum) allow examiner to definitively determine whether a skeleton is female or male. Examination of the public symphysis also allows a less reliable estimate of the age at **death**.

See also Anatomical nomenclature; Cartilaginous joints; Joints and synovial membranes; Ossification; Reproductive system and organs (female)

PULMONARY CIRCULATION

The pulmonary **circulatory system** delivers deoxygenated **blood** from the right ventricle of the **heart** to the **lungs**, and returns oxygenated blood from the lungs to the left atrium of the heart. At its most minute level, the alveolar capillary bed, the pulmonary circulatory system is the principle point of gas exchange between blood and air that moves in and out of the lungs during respiration.

The pulmonary vascular circuit begins with pulmonary **arteries** that branch from the pulmonary trunk leaving the right ventricle of the heart. Venous blood collected from the systemic and **coronary circulation** collects in the left atrium and during the diastolic portion of the **cardiac cycle**, flows into the right ventricle. As the heart contracts during systole, the semilunar valves that comprise the pulmonary valve separating the pulmonary trunk from the right ventricle open to allow blood to be rapidly pumped into the pulmonary system. As the pressure drops, the pulmonary valve closes to prevent backflow into the heart.

At about the level of the fifth or sixth thoracic vertebrae, the main pulmonary arterial trunk divides (bifurcates) into the left and right pulmonary arteries that travel to the corresponding lung. Upon entering the lungs, the pulmonary arteries rapidly divide to form a complex branch of pulmonary arterioles and ultimately a fine capillary bed that surrounds and supplies the **alveoli**.

Within the alveoli, gas exchange takes place. The principle exchanges involve allow the uptake of oxygen by hemoglobin-carrying red blood cells and the discharge of carbon dioxide—a metabolic waste product—into the respiratory air. During respiration, there is a subsequent **gaseous exchange** between the air within the **respiratory system** and air in the environment that allows a continual supply of oxygen to cells, while at the same time venting toxic carbon dioxide.

Lung **capillaries** ultimately fuse into venules and pulmonary **veins**. Pulmonary veins located in the portioned lobes of the lung usually unite to form a single efferent outgoing (efferent) vein from each lobe of the lung. Eventually, the pulmonary veins from the middle and upper lobes of the right lung (along with three lobes) fuse (anastomose) to create a pair of right pulmonary veins—an inferior pulmonary vein and a superior pulmonary vein. Matched by paired superior and inferior left pulmonary veins, the fur pulmonary veins travel separately to individually enter the left atrium of the heart.

An imbalance of fluid within the pulmonary system can lead to excessive fluid levels within pulmonary interstitial spaces (the spaces between cells) that results in pulmonary **edema**.

Pulmonary embolisms occur when a clot or air blocks the flow of blood through the pulmonary arterial system. The larger the clot and more primary the pulmonary artery (i.e., the closer it is to the primary pulmonary artery supplying each lung), the more serious the potential is damage to lung **tissue**.

See also Fetal circulation; Systemic circulation; Vascular exchange; Vascular system (embryonic development); Vascular system overview

PULMONARY EMBOLISM · *see* EMBOLISM

PULSE AND PULSE PRESSURE

The pulse is tactile sensation of **blood** coursing through the arterial system that corresponds to the rhythmic beating of the **heart**. The pulse reflects changing pressures within the arterial system due to changes caused by the cyclic pumping of blood by the heart. Pulse pressure is the calculated difference between the highest and lowest pressure arterial pressures.

The actual pressures in the arterial system are too small to be discerned by touch. The throbbing pulse sensation is actually caused by a shock wave emanating from the heart as it contracts. The shock wave travels trough the arterial system and is recorded as the pulse. The number of pulse sensations recorded in minute is termed the pulse rate. In many cases, pulse rate is determined by recording or counting pulsations for ten seconds and then multiplying the observed pulsations by six to determine the pulse rate. Longer recording periods allow more accurate determinations.

Various pulse reading may be recorded. The radial pulse is usually recorded by placing the fingers against the radial artery near the wrist. The carotid pulse is recorded by pressing against the carotid artery at the side of the neck. The temporal pulse is recorded by light pressure against the temporal **arteries** near at the corners of the forehead. Other important pulse sites include a femoral pulse that can be recorded on the front (anterior) side of the hip and the pedal pulse that can be taken

by pressure against the dorsalis pedis artery on the dorsum of the foot near the extensor tendon of the great toe.

The pulse shock wave correlates with changing pressures in the arterial system. The systolic pressure reflects the highest arterial pressure; the diastolic pressure reflects the lowest arterial pressure. A blood pressure is usually recorded as a ratio of systolic to diastolic pressures, usually measured in pressure units of millimeters of Mercury (i.e., 120 mm Hg over 80 mm Hg). The pressure units reflect the pressure (force multiplied by area) needed to causes changes in the height of a column of mercury in an evacuated column. The pulse pressure is the mathematical difference between the systolic and diastolic pressures (e.g. with a blood pressure expressed as 120/80 the pulse pressure is 40 mm Hg. 120 mm Hg – 80 mm Hg = 40 mm Hg. Because pulse pressure represents the stress on the arterial system, and because "normal" blood pressure ratios can exists over a wide, the pulse pressure—especially an elevated (broad) or depressed (narrow) pulse pressure—may often be an earlier indicator of underling **cardiac disease** or abnormality than the blood pressure ratio itself.

See also Blood pressure and hormonal control mechanisms; Cardiac cycle; Heart defects; Heart, embryonic development and changes at birth; Heart, rhythm control and impulse conduction

PURKINJE (PURKYNĔ), JAN EVANGELISTA
Czech histologist and physiologist

Jan Evangelista Purkinje is credited with articulating a physiological phenomenon (now known as the Purkinje effect) that involves changes in visual perception. Specifically as the intensity of light decreases individuals perceive that red colored objects red objects fade faster than blue objects of equal brightness.

As a boy growing up in Bohemia (now part of the Czech Republic), Jan Purkinje showed great promise. His father, an estate manager who encouraged his son's interests, died when Purkinje was six years old. At the age of ten, Purkinje, an only child, was admitted to a Piarist monastery (established in 1597 to educate the poor) near the Austrian border. Purkinje became a choirboy and outstanding student at the monastery, quietly studying for the priesthood. Just before he was to be ordained a priest, Purkinje decided to take up the study of philosophy at Prague University. While there, he became interested in medicine. His research and tutoring during this time strengthened his physics and optics background.

In 1818, Purkinje presented his graduate thesis which described a visual phenomena now known as the Purkinje effect. He stated that as the intensity of light decreases, different colored objects that appeared to be the same brightness in highly intense light, appear to be unequally bright. In other words, as light intensity decreases, blue objects might appear to be brighter than red objects—even though they appeared to be the same brightness in the more intense light.

After graduating in 1819, Purkinje developed wide-ranging interests in the areas of experimental **pharmacology** and psychology, phonetics, **histology, embryology**, and physical anthropology. In 1823, the same year he took the position of Professor of Physiology and Pathology at the University of Breslau (now the Wrocaw province in Poland), Purkinje published another paper that recognized fingerprints as a way to identify individuals. He noted that fingerprints seemed to follow nine general patterns, he mentioned the ridge formations of the human palm.

In 1832, after obtaining a modern microscope, Purkinje began a new period of research. He found different ways to examine tissues under the microscope—fixing, sectioning, and staining. With his techniques, he saw structures that other observers hadn't noticed. For example, in 1837, Purkinje discovered large pear-shaped nerve cells in the outer layer of the **brain** that had several branches. These are now called Purkinje cells. He was the first to describe these cells as formations in the **central nervous system** of vertebrates and pointed out that they play an important role in nervous activity.

Two years later, as Purkinje was investigating the function of muscular **organs**, he discovered Purkinje fibers—special muscle fibers in the ventricles of the **heart**. Later it would be shown that Purkinje fibers have an important function; they conduct contraction signals to all parts of the heart.

That same year, Purkinje (no doubt influenced by his theological background) described the contents of animal embryos using the term protoplasm in its scientific sense. To him, the term meant "first formed," but eventually it took on a more general meaning: the living material inside a cell. Purkinje also conducted comparative studies of animal and plant **tissue**, observing that "granules"—now termed cells—were present in both. These observations laid the groundwork for **Matthias Schleiden** and **Theodor Schwann** who formulated their **cell theory** in 1839. Purkinje went on to open an independent physiological institute in Breslau—the first of its kind. Although such an institute was very rare until the mid-nineteenth century, it eventually become a regular department of medical schools.

After 1850, Purkinje returned to Prague. He devoted the remainder of his life to the cause of Czech nationalism and making science more accessible to his countrymen.

See also Eye and ocular fluids; Heart, rhythm control and impulse conduction; Muscle contraction; Muscular innervation; Nerve impulses and conduction of impulses; Nervous system overview; Sense organs: ocular (visual) structures

PURKINJE SYSTEM

The Purkinje system, or subendocardial plexus of the **heart** has two functions. The primary role of the fiber tracts is to carry electrical impulses to the ventricles for contraction. The sec-ond, and less important role, is to provide extra support for the myocytes (heart cells) of the ventricles. The latter function enables the ventricles or myocardial **tissue** to expand and contract while returning to their original shape during resting parts of the **cardiac cycle**.

The Purkinje system is one of the last anatomical systems to retain the name of its first describer. Purkinje gained most of his reputation for the identification of the large integrator **neurons** of the **brain**. Because the use of the term Purkinje cells can refer to either location in the brain or heart, it is best to use the phrase Purkinje fibers when discussing the conduction system of the heart.

The primary function of the Purkinje fibers is to carry action potentials (electrical impulses) throughout both ventricles of the heart. Usually this type of function is performed by neurons. However, the Purkinje fibers are not neurons or nervous tissue of any type. The fibers are actually made from specialized myocytes that are collected in small bundles or fibers. All muscle cells are excitable in that they are all capable of carrying and transmitting action potentials. The Purkinje fibers are made of specialized muscle cells that are not contractile in that they do not have well-defined sarcomeres (the contracting segment of **skeletal muscle** fibers). Under microscopic view, some small and sparse sarcomeres can be identified, but for the most part, this just indicates their origin and relationship to muscle tissue. However, the fibers carry action potentials from the atrioventricular bundle branches throughout the myocardium of both ventricles. The route of an **action potential** throughout the entire heart begins at the sinoatrial node to the atrioventricular node to the atrioventricular bundle branches. The branches divide at the septum and lead to the Purkinje fibers. The fibers relay the signal throughout the entire myocardial tissue of both ventricles.

One of the trickier aspects of heart **muscle contraction** is timing. Both the atria and ventricles cannot contract at the same time, or **blood** would not move successfully through the several paths into and out of the heart. The Purkinje system and the cells that are part of these fibers are slower in passing action potentials. For example, the sinoatrial node can conduct between 70–80 action potential per minute at a resting heart state. The Purkinje system conducts about 20–40 action potentials per minute. What this accomplishes is a delayed contraction of the ventricles so that filling from the atria is complete. Then the ventricles can contract with the greatest force possible. The delay in the Purkinje cells that are spread throughout both ventricles and up around the papillary muscles permits a sort of "milking" movement in the ventricles to coordinate effective emptying of blood to peripheral systems. This system provides a critical role in the coordination of contraction and timing of the pumping action of the ventricular portion of the heart.

See also Cardiac muscle; Heart, rhythm control and impulse conduction

R

RADIATION DAMAGE TO TISSUES

In addition to **burns** to integumentary (skin) and organ systems, certain types of radiation exposure may cause mutations (**DNA** damage and genetic alterations) or accelerate the types of mutations that occur spontaneously at a very low rate. Ionizing radiation was the first mutagen that efficiently and reproducibly induced mutations in a multicellular organism. Direct damage to the cell nucleus is believed to be responsible for both mutations and other radiation mediated genotoxic effects like chromosomal aberrations and lethality. Free radicals generated by irradiation of the cytoplasm are also believed to induce gene mutations even in the non-irradiated nucleus.

There are many kinds of radiations that can increase mutations. Radiation is often classified as ionizing or non-ionizing depending on whether ions are emitted in the penetrated tissues or not. X rays, gamma rays (γ), beta particle radiation (β), and alpha particle (α) radiation (also known as alpha rays) are ionizing form of radiation. On the other hand, UV radiation, such as that experience by exposure to Sunlight is non-ionizing. Biologically, the differences between types of radiation effects fundamentally involve the way energy is distributed in irradiated cell populations and tissues. With alpha radiation, ionizations occur every 0.2-0.5 nanometers (nm) leading to an intense localized deposition of energy. Accordingly, alpha radiation particles will travel only about 50 nm before expending of their energy. Primary ionization in x rays or gamma radiation occurs at intervals of 100 nm or more and traverses some centimeters deeper into tissues. This penetration leads to a more even distribution of energy as opposed to the more concentrated or localized alpha rays.

This principle has been used experimentally to deliver radiation to specific cellular components. A cumulative effect of radiation has been observed in animal models. This means that if a population is repeatedly exposed to radiation, a higher frequency of mutations is observed that is due to additive effect. Intensive efforts to determine the mutagenic risk of low

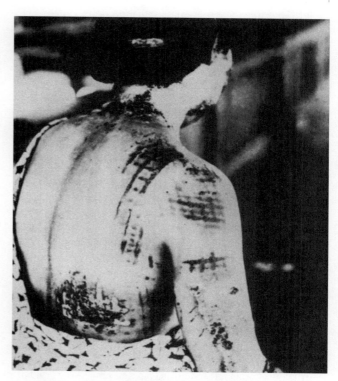

A woman recovering from radiation burns suffered after the atomic bomb blasts in Japan, 1945. *National Archives and Records Administration.*

dose exposure to ionizing radiation have been an ongoing concern because of the use of nuclear energy and especially because of the exposure to radon gas in some indoor environments. Radon is estimated by the United States Environmental Protection Agency to be the cause of more than 20,000 cases of lung **cancer** annually.

Investigation of radiation's mutagenic effects on different tissues, cells, and subcellular compartments is becoming possible by the availability of techniques and tools that allow the precise delivery of small doses of radiation and that provide better monitoring of effects. Reactive oxygen species released in irradiated cells are believed to act directly on nuclear DNA and indirectly by modifying bases that will be incorporated in DNA, or inactivating DNA repair **enzymes**. Novel microbeam alpha irradiation techniques have allowed researchers to investigate radiation-induced mutations in non-irradiated DNA. There is evidence that radiation induces changes in the cytosol that are transmitted to the nucleus and even to neighboring cells.

Critical lesions leading to mutations or killing of a cell include induction of DNA strand breaks, damaged bases, and production of abasic sites (where a single base is deleted), as well as large chromosomal deletions. Except for large deletions, most of these lesions can be repaired to a certain extent, and the lethal and mutagenic effect of radiation is assumed to result principally from incompletely or incorrectly repaired DNA. This view is supported by experimental studies which showed that mice given a single radiation dose, called acute dose, develop significantly higher level of mutations than mice given the same dose of radiation over a period of weeks or months.

See also Aviation physiology; Birth defects and abnormal development; Cell structure; Chemotherapy; Chromosomes; Human genetics; Inflammation of tissues; Molecular biology

RAMÓN Y CAJAL, SANTIAGO (1852-1934)
Spanish histologist

Santiago Ramón y Cajal is often regarded as the father of modern neuroanatomy. His studies of the fine structures of the cortex of the **brain**, the **spinal cord**, and nerve **tissue** led to a Nobel Prize in physiology or medicine n 1906.

Ramón y Cajal was born in Petilla de Aragón, Spain. His father, a country doctor, wanted young Santiago to pursue a medical career also and enrolled his son first, at the College of the Aesculpian Fathers and later, at the Institute at Huesca. As a youth, Ramón y Cajal was not really interested in medicine, however, and he left school to become apprenticed to a butcher, then to a shoemaker. At the age of 16, he returned to his formal education, entering the University of Zaragoza. There he was especially interested in **anatomy**. After graduation, Ramón y Cajal entered military service as a doctor in the Spanish army. After contracting **malaria** in Cuba, Cajal was discharged. Cajal then returned to Zaragoza, where he completed his medical degree in 1879.

Ramón y Cajal soon became involved in a controversy raging among anatomists at the time. The question was how nerve messages are transmitted through the body. One theory—the reticular theory—held that nerve messages travel through a complex network of nerve fibers in physical contact with each other. Cell bodies observed within this network were thought to play primarily a structural and supportive role. Ramón y Cajal was able to provide new evidence about this issue by developing new cell staining techniques. These stains showed more clearly than had ever been possible the detailed structure of nerve tissue. With this technique, he was able to see that nerve cells are distinct units whose extensions, axons and dendrites, are not in contact with each other, but are separated by narrow gaps (synapses). For this discovery, Ramón y Cajal shared the 1906 Nobel Prize in physiology or medicine with his contemporary **Camillo Golgi**. Ramón y Cajal continued working on nerve **structure and function** for another three decades. In 1891, he found that nerve messages enter a neuron by way of the dendrites and leave by way of the axon. Later studies dealt with the growth and degeneration of **neurons**. He also developed new stains that made possible even more detailed studies of nerve tissue.

Ramón y Cajal's first academic appointment was as Professor of Descriptive Anatomy at the University of Valencia in 1883. He then went on to become Professor of **Histology** at the University of Barcelona in 1887, and Professor of Histology and Pathological Anatomy at the University of Madrid in 1892. He served at Madrid until 1921, when he became director of the Cajal Institute, founded in his honor by King Alfonso XIII. Ramón y Cajal died in Madrid at the age of 82.

See also Nerve impulses and conduction of impulses; Nervous system overview; Neurotransmitters

RECIPROCAL INNERVATION

Reflexes that contract a muscle while relaxing the muscles that oppose it are said to feature reciprocal innervation (or reciprocal inhibition). One such reflex is the familiar knee-jerk reflex, in which a hammer-tap just below the kneecap briefly stretches the quadriceps muscle (thigh muscle) above the kneecap. This stimulates several hundred stretch-detecting **neurons** attached to the quadriceps muscle. Impulses travel up the fibers of these sensory neurons to the **spinal cord**, where each fiber forks into two branches. One branch connects directly to 100–150 motor neurons that return directly to the stretched quadriceps muscle. These motor neurons are stimulated to fire, causing the quadriceps muscle to contract and the tapped leg to extend in an involuntary kick. The other branch stimulates local-circuit neurons that travel a short distance inside the spinal cord and form inhibitory synapses on motor neurons connected to the hamstring muscles at the back of the upper leg. Inhibitory synapses are connections that discourage the postsynaptic (receiving) neuron from firing; the net effect of this arrangement is that the hammer-tap discourages the motor neurons of the hamstring muscles from firing. The back of the leg relaxes, allowing the thigh muscle to extend the leg.

The flexor reflex and crossed extension reflex also involve reciprocal innervation. In the flexor reflex, an injury-sensing neuron activated by an injury to the foot stimulates certain motor neurons in the spinal cord while inhibiting others so that the thigh is relaxed, the back of the leg contracted, and the foot bent up. The net effect is to raise the injured foot and leg away from the source of **pain**.

The crossed extension reflex builds on the flexor reflex by adding neurons that cross the spinal cord and reciprocally innervate the muscles of the uninjured leg. While the injured leg is reflexively raised, as described above, the uninjured leg is reflexively straightened, tending to keep the person upright.

See also Nerve impulses and conduction of impulses

REFERRED PAIN

Referred **pain** is the perception of pain in a region of the body that is not the actual source of that pain. One of the most common examples of this would be the radiating pain down the left arm experienced during a **heart** attack. What is the anatomical basis for this phenomenon? One explanation may lie in the number of pain receptors found on the skins surface and those of the internal **organs**, and the mechanism and hierarchy for how the **brain** is able to process the signals from these receptors. The number of cutaneous receptors is much greater then those of internal structures.

Pain sensation is caused by rapidly conducted action potentials (the electrochemical propagation of a nerve signal) from large-diameter myelinated axons, and more slowly propagated action potentials carried on smaller less myelinated axons. The former results in a sharper, more localized pain where as the latter causes a diffuse, or burning pain. Pain on the skin surface is highly localized due to the number of pain receptors and the presence of **mechanoreceptors**. Deep or visceral pain is much more diffuse. Referred pain may simply result from convergence of the two types of pain action potentials on the same afferent **neurons**. It is likely that the brain cannot distinguish the source of these two stimuli so the more rapidly propagated signal is given a preferential priority.

Other examples of referred pain often include the gall bladder referring pain to the top of the right shoulder, a **diaphragm** problem felt in the neck, and intestinal dysfunction felt in the middle or the low back. Referred pain can be a valuable diagnostic tool. Often the clinician is able to diagnose a condition or disease based on the typical location where the pain is referred.

See also Nerve impulses and conduction of impulses; Nervous system overview; Nervous system, embryological development; Neural damage and repair

REFLEXES

Reflexes are actions directed by the nervous system that do not require conscious direction or control. Reflexes are characterized as involuntary or voluntary. Reflexes designed to protect the body from injury are termed nociceptive reflexes. Such reflexes include the twitching or winking of the **eye**, coughing and sneezing to dislodge foreign bodies from the respiratory tract, and gagging (pharyngeal reflex) to as a mechanism to protect the respiratory and digestive tracts.

Under certain conditions it is possible for the body, as a result of experience, to develop rapid responses to certain stimuli so that rapid coordinate movements, such as those associated with running, are possible. These reflexes, however, are not the result of simplified neural pathways, but are the result of conditioning that allows the nearly automatic coordinated movements of many muscles.

Most reflexes require three different neural components within including a sensory neuron, a transmitting neuron, and a motor or effector neuron. The combined function of these **neurons** creates a reflex arc. Reflex arcs are most commonly associated with motor actions that allow a rapid avoidance of **pain**. For example, a quick jerk of an arm away from a hot object that avoids a burn is a classic example of a withdrawal reflex.

Although such actions do not require conscious thought, the transmitting neurons involved send messages to the brains pain perception neurons via the spinothalamic tract. In the case of a hot object, the voluntary muscle reflex arc begins with the mechanical receptors (thermal receptors) that provide stimulus to the transmitting neuron or neurons that, in turn, pass the neural impulse on to the motor neuron that causes the appropriate muscle movements.

The transmission of neural reflexes in reflex arcs uses the same electrical and chemical mechanisms of transmission as all other impulses. Within the neural cell body (axon), the impulse travels electrically as an **action potential**. At the **synapse**, the intercommunicating gap or space between neurons, the neural impulse is conducted through the release, diffusion, and binding of **neurotransmitters**. Accordingly, reflex impulses are limited to the same conduction speeds as any other form of neural impulse. Because reflexive actions are the result of simplified neural pathways as opposed to a special form of neural transmission, reflexes are also subject to rebound and fatigue.

One of the simplest forms of reflex involves a stretching response that involves only a sensory and a motor neuron. When stimulated by the extension of a particular muscle or the contraction of the muscle's antagonist muscle, the sensory neurons located in muscle spindles transmit a nerve impulse that, at the level of the **spinal cord**, directly stimulates the motor neurons that cause the muscle to contract. These contraction counter effects are vital to maintaining normal balance, and to prevent muscular damage from hyperextension.

Although simple in design, reflex pathways can achieve sophisticated results. In the case of the crossed extensor reflex, rapid contractions (e.g., jerking withdrawal movements) in one muscle or group of muscles also involve specific inhibition of other muscles or muscles groups required to maintain proper balance. If, for example, one places a foot on hot asphalt, jerking reflexive withdrawal of the left foot occurs via the inhibition of the quadriceps and extensors involved in advancing the foot and the rapid contraction of the hamstring muscles to withdraw the foot. This rapid shift from extension to flexion in one leg would then be uncoordinated with the already flexing right leg. Crossed extensor reflexes allow the rapid change from flexion to extension in the right leg, an reflex needed to maintain proper balance via inhibition of the right leg flexion muscles and the excitation of the extensor muscles.

Such stretching reflexes are also the involved in the automatic response to the contusion or stretching of knee (patellar) **tendons**. Because a muscle's agonist must contract as a muscle stretches, separate neural signals branch from the sensory neuron to via inhibitory neural pathways to inhibit the agonist in a process termed reciprocal inhibition. Tendons attached to muscles also respond to muscle stretching and contracting. In the deep tendon reflex, there is a process of reciprocal activation wherein receptors within the tendons send signals via transmitting neurons to agonist motor neurons that cause the agonist muscle or muscles to contract. This balancing of or complementation of **muscle contraction** and relaxation is a critical feature of normal body movement and coordination.

There are hundreds of defined reflexive actions, some normal and well known, other occurring only in conjunction with injury or diseases.

A blow to the Achilles tendon results in the plantar extension of the foot characterized by the familiar jerk of the ankle. The Babinski reflex involving the dorsiflexion (arching) of the big toe results from a scraping pressure drawn across the bottom of the foot.

The auditory reflex shuts both eyes as a familiar protective response to loud noise. Ciliary reflexes allow the pupil to rapidly accommodate to changing light conditions that could potentially damage the **retina** or optic nerve. Conjunctival reflexes cause the reflexive closure of the eyelids when the eye is touched.

Reflexes are often tested to assure normal neural functions and intact neural pathways. A common test involves the pupillary reflex that results in the constriction of the pupil when light is shown in the opposite eye.

See also Autonomic nervous system; Brain stem function and reflexes; Cough reflex; Nerve impulses and conduction of impulses; Nervous system overview; Neurology; Parasympathetic nervous system; Sympathetic nervous system; Yawn reflex

REICHSTEIN, TADEUS (1897-1996)

Polish organic chemist

It is now known that the **hormones** of the adrenal gland are essential to controlling many challenges to the human body, from maintaining a proper balance between water and salt to responding to stress. Tadeus Reichstein is one of those responsible for this knowledge; **Edward Kendall** and **Philip Hench** also played an important role in these efforts, and the three men shared the 1950 Nobel Prize in physiology or medicine. Reichstein's work has had effects throughout medicine—in the treatments of Addison's disease and rheumatoid arthritis, for example, and in the understanding of the fundamental biochemical processes of steroid hormone **metabolism**.

The eldest son of engineer Gustava Reichstein and his wife, Isidor, Reichstein was born near Warsaw in Poland. After moving first to Kiev in the Ukraine and then to Berlin, the family settled in Zürich and became Swiss citizens. Tadeus

attended the Eidgenössiche Technische Hochshule and graduated in 1920 with a chemical engineering degree. He worked briefly in a factory, then returned to the Eidgenössiche Technische Hochshule where he earned his doctorate in organic chemistry in 1922.

For several years thereafter Reichstein continued to work with his doctoral advisor, Hermann Staudinger, who would later win the 1953 Nobel Prize in chemistry. Reichstein's early work focused on identifying and isolating the chemical species in coffee that give it its flavor and aroma. This interest in plant products was to remain with Reichstein throughout his career. He had an early success when he discovered how to synthesize the newly discovered compound ascorbic acid (vitamin C). He published this method in 1933, and later that year Reichstein developed a second method of synthesis which is still widely used in the commercial production of this dietary supplement.

In 1934, Reichstein began work on what he originally believed to be a single hormone produced by the cortex or outer layers of the adrenal glands. He soon realized, however, that the adrenals were producing a milieu of active substances. His work began with more than a ton (about 1,000 kg) of adrenal glands that had been surgically removed from cattle. His first stage of purification resulted in about 2.2 lb. (1 kg) of biologically active extract. He established that the extract was biologically active by injecting it into animals whose adrenal cortices had been removed; if the compound was active it replaced what was missing as a result of the operation and allowed the animal to survive. The next stage of purification reduced the kilogram of extract to less than 1 oz. (25 g), only about one-third of which proved to be the critical hormone mixture. Instead of one hormone, this sample contained no fewer than twenty-nine distinct chemical species.

Reichstein isolated the twenty-nine species and then individually examined them. He identified the first four that were found to be biologically active, and later synthesized one of them. It was also Reichstein who demonstrated that these compounds were all steroids. Steroids are a group of chemicals which share a particular structure of four linked carbon-based rings; other important compounds having steroid structure include the **sex hormones**, cholesterol, and vitamin D.

Reichstein built on his earlier work with plant extracts to synthesize the steroid hormones. He and his colleagues developed several different methods to this end, though a process that used an animal waste product (ox **bile**) proved to be the most economical. One of the most important syntheses that Reichstein accomplished was that of aldosterone, which controls both water balance and sodium-potassium balance in the body. Reichstein's work was also critical to the eventual syntheses of desoxycorticosterone, which for many years was the preferred treatment for Addison's disease, and cortisone, which is used for treating rheumatoid arthritis. It was principally for this latter accomplishment that Reichstein shared the 1950 Nobel Prize in chemistry.

Reichstein moved to the University of Basel in 1938, where he was appointed director of the Pharmaceutical Institute; in 1946 he became head of the organic chemistry division. Here he turned his attention to plant glycosides, a group of compounds with wide-ranging biological effects. They are the basis for a number of widely used drugs, and one of these, digitalis, has proven useful in controlling the **heart** rate. Reichstein was able to identify both the plants and the parts of the plants that contained glycosides, and his contributions were critical for initiating many botanical studies. He was one of the first researchers to realize the value of the tropical rain forests to the pharmaceutical industry. His work has also been pivotal in the field of chemical taxonomy, where the identities of plants are determined through their chemical composition—a method which has a higher degree of certainty than identification through visible characteristics. This technique has had broad applications in the development of both natural insecticides and drugs.

Reichstein was presented with an honorary doctorate from the Sorbonne in 1947. He received the Marcel Benoist Award in 1947, the Cameron Award in 1951, and a medal from the Royal Society of London in 1968. He was a foreign member of both the Royal Society and the National Academy of Sciences.

Reichstein married Henriette Louise Quarles van Ufford in 1927, while still at the Eidgenössiche Technische Hochshule. They had one daughter. He retired from his academic posts in 1967, but continued to work in the laboratory until 1987. He died on in Basel, Switzerland, at the age of 99.

See also Adrenal glands and hormones; Hormones and hormone action

RENAL SYSTEM

The renal system is also known as the urinary system. Located in the retroperitoneal upper area of the abdominal cavity, the renal system is a collection of the **organs** and the structures that function in the removal of waste material from the body. In humans the renal system consists of the pair of **kidneys**, ureters, bladder, and the urethra.

The conversion of incoming nutrients to energy for the body's various processes generates waste. Much of this waste is soluble, that is, it can dissolve in water (and so in **blood**, which is largely comprised of water). This waste must be removed from the water and expelled from the body.

Waste removal takes place in the kidneys. These are the principle filtration units of the renal system. There are normally two kidneys in the human body. They are located on each side of the **abdomen** in the lower region of the back. Within the kidney, wastes are removed from the fluid. The wastes are concentrated into a waste product called urine.

Once produced, the urine passes from the body via the other components of the renal system. Tubes called ureters connect the kidneys to the urinary bladder. Each ureter is made of muscle and is between 16 and 18 in. (41–46 cm) long. Rhythmic contractions of the ureter force the urine downward into the bladder. The bladder is essentially a bag that holds the liquid until enough has accumulated that the need to urinate becomes evident. The bladder is also composed of muscle, which is capable of contraction. This forces the urine out the

bottom of the bladder and into a narrow tube called the urethra. The urethra leads to the outside, via the penis in the male or via the urethral opening superior to vaginal opening in the female.

The urethra is susceptible to **infection**, because of growth of microorganisms from the outside that infiltrate the urethral tubes. While usually treatable with antibiotics, chronic infections can develop that are quite resistant to treatment. These infections can be uncomfortable and can restrict urination, which may in turn influence the chemistry of the renal system.

Two other important functions of the renal system, which are also performed by the kidneys, is regulating and maintaining the balance of **electrolytes** and the pH of the fluid that has been processed. The electrolyte balance is maintained by regulating the amount and the composition of the fluid. Fluid passing through the kidneys is monitored and the proper concentrations of ions such as hydrogen, sodium, potassium, chloride, bicarbonate, sulfate, and phosphate are ensured. The production of ammonia also helps to keep the processed fluid at a pH of between 7.37 and 7.43. This range is critical for the proper function of the body.

The ability of the renal system to function depends upon the presence of two **hormones**. The first of these is called the antidiuretic hormone, or ADH. This hormone is produced by the pituitary gland. The ADH acts to make the filtration components of the kidney more able to accept water. When ADH concentration in the body is high, the kidney produces a smaller volume of highly concentrated urine. This would be advantageous when water intake is restricted due to lack of available drinking water.

The second vital hormone is known as aldosterone. The presence of the hormone affects the absorption of sodium by the kidneys. As a result, the urine that is produced contains less sodium and more potassium than when the aldosterone concentration is lower. This mechanism allows the body to adjust for changing internal needs for electrolytes such as sodium and potassium. If this mechanism is faulty, the changing sodium and potassium levels in the body promote a condition called hypertension.

The performance of the renal system can also be adversely affected by **diabetes mellitus**. The excess glucose in the blood may not be removed by the kidneys, leading to the appearance of glucose in the urine. Protein may also be present in the urine, a condition called proteinuria, which is indicative of a breakdown in kidney function. Medical attention is necessary for both conditions to prevent serious and permanent damage to the renal system, especially to the kidney.

See also Acid-base balance; Diuretics and antidiuretic hormones; Endocrine system and glands; Metabolic waste removal; Urogenital system; Urogenital system, embryonic development; Urology

RENAL SYSTEM, EMBRYOLOGICAL DEVELOPMENT

The **kidneys**, urinary tract, and the majority of the reproductive **organs** arise in the intermediate **mesoderm**. The kidney goes through three stages of development that mimic filogenesis (biological evolution) of the kidney; pronephros, mesonephros, and metanephros.

The pronephros arises at the C3-T1 vertebral levels by the dorsal proliferation of cords of cells. Such cords become pronephric tubules that grow caudally and get connections in order to form a common pronephritic duct that extends caudally toward the cloaca. Beyond the pronephros it is termed the Wolffian duct. The tubules of the pronephros are in continuity with the coelomic cavity. The pronephric kidney does not work in humans, but it seems to induce the normal formation of the further renal structures.

The mesonephros develops by the formation of menonephric tubules from the intermediate mesoderm of the C6-L3 vertebral levels. The tubules of the mesonephros do not communicate with the coelom, but the **blood** is filtrated directly into the mesonephric tubules through a capillary glomerulus structure from the aorta that is encapsulated by the proximal blind end of the tubule. The metanephros arises caudal to the mesonephros at five weeks of development. It derives from intermediate mesoderm (L4-S1 vertebral levels), the metanephrogenic blastema, lateral to the developing urogenital sinus and lateral to the mesonephric duct. The ureteric bud arises as a diverticulum from the mesonephric (Wolfian) duct close to the entrance to the cloaca. It grows towards and inside the metanephrogenic blastema, and invades the center of the metanephros. At this point, as the bud grows, it is named the metanephric bud. Its most caudal part has the name metanephric duct. This juxtaposition of the ureteric bud and the specialized mesoderm stimulates the metanephrogenic blastema to form glomeruli, and proximal and distal tubules. When the ureteric bud touches the metanephros, progressive branching of the ureteric bud occurs, creating the major and minor calyces, and the collecting tubules that will provide a conduit for urine drainage in the mature kidney. This process is known as the induction of the kidney.

The metanephrogenic blastema induces the bud to grow and to branch, so that it arborises to form a tree-like collecting duct system. The ureteric bud induces the metanephrogenic blastema to gain a stem- cell phenotype and to multiply. It then induces groups of **stem cells** to differentiate into nephrons, a complex developmental process. Such a reciprocal induction occurs as a result of sequential activation of a series of genes from different families. These encode growth factors, receptors, oncoproteins, transcription factors, **enzymes**, signal transducers, and extracellular matrix components.

There is a time gradient in the development of a fetal kidney. The most external cortex is composed of stem cells that are not yet committed to differentiation, the region just inside it contains cells undergoing the earliest phases of nephrogenic differentiation, and the most inside area that contains maturing nephrons and supporting stromal cells. During differentiation of the nephrons, condensation is the process by which disorganized mesenchymal cells (about 100) become a highly organized epithelial tubule to form a distinct mass. Then, during the transition to epithelium, condensed cells lose their mesenchymal characteristics and gain epithelial ones including the basement membrane, cell-cell junctions, and a

cellular apico-basal polarity. Condensation of the developing nephron then occurs as invaginations form comma-shape bodies and then an S-shaped body. At about this time, blood vessel progenitors begin construction of the vascular component of the glomerulus. The tubules mature as the peculiar transporting segments of the nephron differentiate, and the morphogenesis of convoluted tubules takes place. Finally, fusion with the collecting ducts leads to a continuous tubule system.

The cloaca is a common sinus that forms in the caudal region of the fetus. At the caudal end of the cloaca, **ectoderm** lies directly over **endoderm** forming the thin cloacal membrane. As development progresses, a mesoderm derivate septum named the urorectal septum forms, dividing the hindgut, the ventral (urogenital) sinus. This urorectal septum extends in a caudal direction. The mesonephric (Wolfian) duct descends from the mesonephros to meet the urogenital sinus. Once this connection is made, fetal urine drains into the urogenital sinus.

The urogenital sinus and a small portion of allantoidis enlarge to for the urinary bladder. Portion of the mesonephric and metanephric (ureter) ducts are incorporated into the urogenital sinus. The ureter enters the bladder and the mesonephric ducts enter more caudally in the less dilated portion of the urogenital sinus. This distal portion will become the urethra for both the genders, and the vestibule and part of the vagina for the female. The cranial portion of the bladder tapers to become the vescico-allantoic canal that successively closes, leaving the median umbilical ligament. The caudal end of the developing bladder thickens with **smooth muscle** in a triangle between the two ureteric orifices and the urethra.

See also Anatomical nomenclature; Embryonic development, early development, formation, and differentiation; Renal system

REPLICATION • *see* MOLECULAR BIOLOGY

REPRODUCTIVE SYSTEM AND ORGANS (FEMALE)

In both male and female embryos, primordial undifferentiated **gonads** form on the medial wall of the urogenital ridges. Two pair of ducts initially develop in the gonad, the mesonephric duct (Wolffian duct), and the paramesonephric (Mullerian) duct. The primordial undifferentiated **germ cells** then migrate from the yolk sac to the urogenital ridge during the fourth to eighth week of gestation, where they are required for development of the gonads (testes and ovaries). In the absence of a testicular differentiation factor from the Y chromosome, which directs the production of Mullerian inhibiting substance (MIS) by four to six weeks gestation in males, the germ cells differentiate into primitive oogonia that begin mitosis at about six weeks. The first meiotic division is initiated at about 15 weeks, signaling the transformation of oogonia to oocytes. This meiotic division is then arrested at the first prophase until primordial follicles are formed. Medullary structures infiltrate the ovarian cortex and surround the oocytes to invest each one

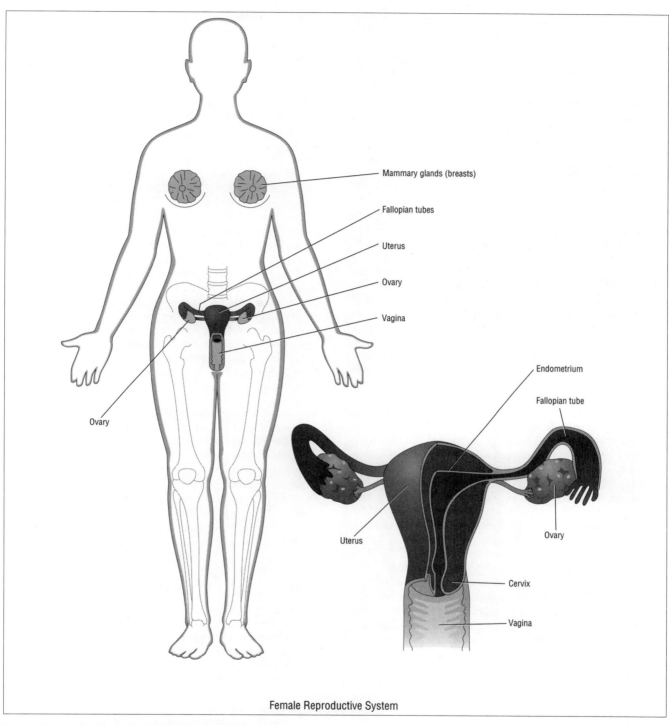

Mammary glands (breasts)

Fallopian tubes

Uterus

Ovary

Vagina

Ovary

Endometrium

Fallopian tube

Ovary

Uterus

Cervix

Vagina

Female Reproductive System

The female reproductive system. *Illustration by Argosy Publishing.*

with a single layer of primordial granulosa cells at about 20 weeks of gestation, thereby beginning the formation of primordial follicles. At birth, the human ovary is filled with approximately one million primordial follicles. In females, the Wolffian duct almost completely disappears.

In the female, the Mullerian duct starts to develop at six weeks of gestation as invaginations of the mesonephric kidney. The invagination forms a tube for each duct. The tubes on the cranial end open into a celomatic cavity (future peritoneal cavity), while the caudal segments fuse (8 weeks of gestation) into a Y-shaped uterovaginal primordium. From the cranial portion of each tube arises the fallopian tubes. The caudal (fused) portion becomes the uterus and part of the vagina.

At nine weeks of gestation, external sexual differentiation begins. In females, the genital tubercle develop into the clitoris, the urogenital folds form the labia minora, and the labioscrotal swellings become the labia majora.

The perineum is defined as the region of pelvic outlet below pelvic floor. A diamond shaped area, the perineum is separated into two triangles (the anterior urogenital triangle and the posterior anal triangle) by an imaginary line between the two ischeal tuberosities. The vulva rests on the urogenital triangle **diaphragm**. The vulva contains the mons pubis, the labia majora, the labia minora, the clitoris, and the vestibule.

The labia majora are folds of skin with underlying **adipose tissue**; they are fused anteriorly with the mons pubis and posteriorly with the perineum. The skin of the labia majora contains **hair** follicles and sebaceous and **sweat glands**. The labia minora merge anteriorly with the prepuce and frenulum of the clitoris and posteriorly with the labia majora and perineum. The labia minora contains sweat and sebaceous glands, but not hair follicles and adipose **tissue**. The clitoris is the embryologic equivalent of the penis. It consists in two crura (the equivalent of corpora cavernosa in the male). The vestibule lies between labia minora and is bounded anteriorly by the clitoris and posteriorly by perineum. The urethra and vagina open into the vestibule in the midline. The ducts of Skene(parauretral) glands and Bartolino's glands also empty into vestibule.

The vagina is a muscular tube that extends from above the inferior extent of the cervix of the uterus (that project into the upper position of the vagina) to its external opening in the vestibule. Its long axis is approximately parallel to the lower portion of sacrum. At its lower end, the vagina traverses the urogenital diaphragms and is the surrounded by two bulbocavernosus muscles that act as sphincters. The hymen, a fold of connective tissue, somewhat obscures the external vaginal orifice in childhood, and is fragmented into irregular remnants with sexual activity and in childbearing.

The vagina is related anteriorly to the base of the bladder and the urethra; posteriorly to the pouch of Douglas (posterior pelvic cul-de-sac), rectum and anal canal; and laterally to the levator ani muscle and the ureter,which passes near the lateral fornix. The major **blood** supply of vagina is the vaginal artery (branch of the hypogastric artery) and the **veins** that follow the path of the **arteries**. The vaginal wall consists of a mucous membrane, a submucosal layer of connective tissue, and an external muscular layer.

The uterus consists of the fundus, body, and cervix. The part of the body where the two uterine (fallopian) tubes enter is called the cornu. The part of the corpus above the cornu that projects above the cavity of the body is termed fundus. The body is the major portion of the uterus and the cervix is the inferior portion,part of which projects into vagina. They are separated by a narrowed isthmus. The cavity of uterus is continuous superolaterally with the narrow lumen of the uterine tubes and inferiorly with the cavity of vagina. The cavity of the uterus is the largest within the body. The lumen of the cervix (cervical canal) is narrow and ends inferiorly as the narrow

external os. The blood supply to the uterus comes primarily from the uterine arteries and also from the ovarian arteries.

The muscular layer of the uterus is continuous with the muscular wall of the vagina and the fallopian tubes. The terminal part of the fallopian tubes are fringed by fingerlike fimbriae. These fimbriae surround the ovary at the time of ovulation to capture the oocyte and to make **fertilization** possible (the union of the spermatozoan coming from the uterine cavity and the secondary oocyte). **Cilia** beating toward the uterus assist in oocyte transport to fertilization site.

The ovaries lie against the lateral pelvic wall just below the pelvic inlet and posteroinferior to the lateral aspect of uterine tubes. Each ovary is approximately 1.2–1.6 in. (3–4 cm), long, 0.79–1.2 in. (2–3 cm), wide, and 0.39–1.2 in. (1–3 cm) thick while fertile. The size of the ovaries decrease by two thirds after **menopause**.

The outer ovarian cortex consists of follicles embedded in a connective tissue stroma. Further follicular development requires stimulation of granulosa cells by **follicle-stimulating hormone** (**FSH**) to induce luteinizing hormone (LH) receptors, and therefore responsiveness to LH, leading to formation of preovulatory follicles and then to ovulation. After ovulation, the dominant follicle become corpus luteum,which secrets **progesterone** to prepare endometrium for the implantation of fertilized oocyte. If **pregnancy** does not occur, the corpus luteum undergoes involution, **menstruation** begins, and the cycle repeats.

See also Anatomical nomenclature; Embryonic development, early development, formation, and differentiation; Estrogen; Genitalia (female external); Gonads and gonadotropic hormone physiology; Implantation of the embryo; In vitro and in vivo fertilization; Infertility (female); Mammary glands and lactation; Ovarian cycle and hormonal regulation; Parturition; Pregnancy, maternal physiological and anatomical changes; Puberty; Pubic symphysis (gender differences); Reproductive system and organs (male); Secondary sexual characteristics; Semen and sperm; Sexual reproduction

REPRODUCTIVE SYSTEM AND ORGANS (MALE)

Human reproduction depends on the integrated action of **hormones**, the nervous system, and the reproductive system. Gametes are the cells involved in **sexual reproduction** produced by the **gonads**, and contain only one copy of each chromosome. A failure in the production of healtly **sperm** in the **semen** is termed **male infertility**.

Male gonads are the testes, which produce both sperm (the male mature gamete is spermatozoa) and male **sex hormones**. The testes are about 1.9 in. (5 cm) in length and 1.2 in. (3 cm) in diameter, and are enclosed in a tough, white fibrous capsule. The testes are complex **organs** containing a variety of different cell types that are structurally and functionally compartmentalized, thereby properly allowing the production of spermatozoa (**spermatogenesis**). The testes are microscopi-

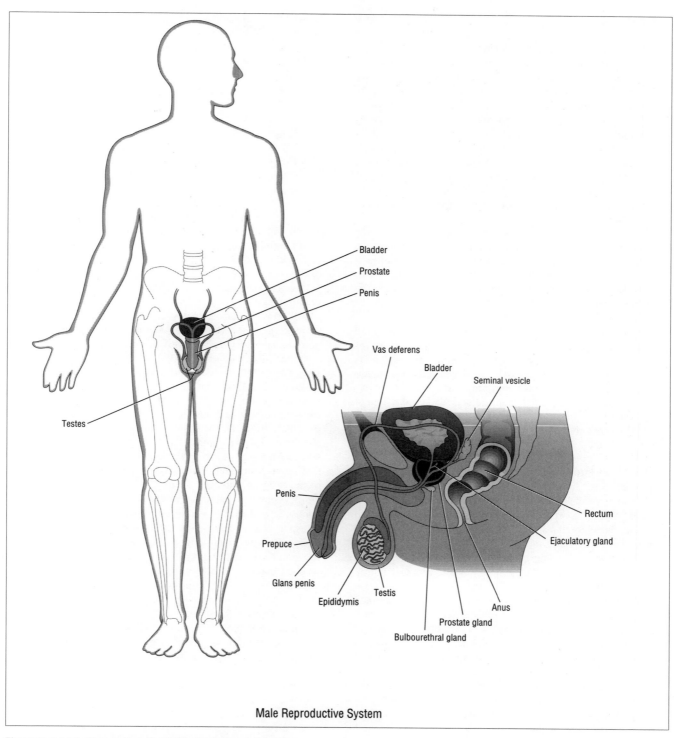

Labels in illustration: Bladder, Prostate, Penis, Testes

Vas deferens, Bladder, Seminal vesicle, Penis, Prepuce, Glans penis, Epididymis, Testis, Rectum, Ejaculatory gland, Anus, Prostate gland, Bulbourethral gland

Male Reproductive System

The male reproductive system. *Illustration by Argosy Publishing.*

cally organized into tubules named seminiferous tubules. About 273 yds. (250 m) of tubules are packed into each testis.

Spermatogenesis (sperm production) is controlled by a series of genes located mainly on the sex **chromosomes** and is

performed by the **germ cells**. Spermatogenesis occurs in the seminiferous tubules. At the outer edge of the tubules are the spermatogonia, the germ cells of the testis. They continue to divide throughout life and reproduce mitotically to maintain their numbers. Some spermatogonia begin meiosis after **puberty** to produce sperm for the whole reproductive life. A 2n spermatogonial cell replicates its **DNA** and becomes a 4n primary spermatocyte. Division (first meiotic division) of the primary spermatocyte produces two 2n secondary spermatocytes, and division (second meiotic division) of the secondary spermatocytes produces four 1n spermatids. Each spermatid becomes a sperm cell, with a residual body (excess cytoplasm) left behind. In human beings, spermatogenesis takes 74 days from spermatogonium to released sperm cells. Sperm capacitation (changes that render the sperm able to penetrate the egg) that occurs in the female tract (seven hours in human beings) before a sperm can fertilize an egg, is needed for the acrosome reaction (the release of egg-penetrating **enzymes** by sperm) to occur at **fertilization.**

Sertoli cells and Leydig cells are the most peculiar cells of the testicles. Sertoli cells are essential for spermatogenesis, assisting the germs cells in regulating hormonal conditions for their differentiation. Interstitial (Leydig) cells are located in stroma between seminiferous tubules, and represent the endocrine portion (primarily **testosterone**) of testis.

Male fertility depends on the proper balance of hormones and organs. The first step in fertility takes place in the **hypothalamus** pituitary axis, by means of the release of gonadotropin-releasing hormone (GnRH) in the hypothalamus that stimulates the pituitary gland to produce **follicle-stimulating hormone (FSH) and luteinizing hormone (LH).** FHS maintains spermatogenesis and LH stimulates the production of testosterone.

The other reproductive organs include the rete testis, a network of ducts leading to efferent ductules that transport spermatozoa to epididymis. Epididymis is single tube about 6.5 yd. (6 m) long (head, body, tail) that connects the ducts within the testes and is site of sperm maturation. Ductus deferens transport sperm from epididymis to ejaculatory ducts. They are mainly composed by **smooth muscle** and organized in a tube-shape about 18 in. (45 cm) long that passes upward along the medial side of the testis.

Ejaculatory Ducts are formed by the fusion of the vas deferens and the duct of the seminal vesicle. The ejaculatory duct ejects sperm into urethra. The accessory sex glands are the prostate that is located inferiorly to the urinary bladder and surrounds prostatic urethra seminal vesicles, and the bulbourethral gland (Cowper's) The bulbourethral gland (Cowper's gland) are two small structures about the size of peas, which are located below the **prostate gland** and enclosed by muscle fibers of the external urethral sphincter.

The prostate, located in the male pelvis, measuring 1.2–1.6 in. (3–4 cm) in length and 1.2–1.9 in. (3–5 cm) in width. The gland, surrounded by the pelvic muscles, is located behind the pubic bone, in front of the rectum, below and at the base of the bladder and surrounds the urethra. On average, the gland weighs about 0.71 oz. (20 g). There are 15–30 excretory ducts from the prostate entering the urethra as it passes through the prostate. Each of these excretory ducts receives prostatic secretions from 4–6 prostatic lobules that contain prostatic acini surrounded by tall columnar epithelium. The prostate has an essential function in human reproduction. Androgens control the growth of the prostate and formation of the prostatic secretions. The testosterone enters the prostatic cells in order to be then metabolized to a more potent metabolite, called dihydrotestosterone (DHT), which then binds to the androgen receptor within the cell. This binding participates in androgen-induced expression of genes such as prostatic specific antigen (PSA).

The seminal vesicles attach to the prostate and produce secrete that mixes with to form semen. They are a convoluted, saclike structure about 1.9 in. (5 cm) long that is attached to the vas deferens near the base of the urinary bladder. The growth of these glands and their secretory activity require androgen production from the testes as well as a functioning androgen receptor within the cells of the sex accessory tissues. They are involved in maintaining the viability and motility of the sperm.

A human male's ejaculate volume is about 3 ml of semen and ranges from 2 to 6 ml. The sperm constitutes less than 1% of the volume, with the seminal vesicles and prostate producing about 95% of the total volume. The largest portion of the semen, approximately 65%, is secreted from the seminal vesicles and appears in the latter portion of the ejaculate volume. 15–20% comes from the prostate gland and a very small proportion, originates from the bulbourethral gland (Cowper's gland). In the normal male, the ejaculate is rich in proteins and enzymes, as well as **prostaglandins**, citric acid, spermine and fructose. The significance of many components of prostate fluid is unknown, but is probably involved with reproduction. Thus, the prostate is essentially an organ for human reproduction. The proteins from the seminal vesicle cause the ejaculate to clot and form a coagulum after ejaculation. Subsequently, the PSA, a serine protease secreted from the prostate, lyses (breaks) the clot. Other proteins from the sex accessory tissues coat the sperm and are believed to protect sperm from environmental damage and agglutination, and to mask sperm **antigens** from the female's **immune system.** Other proteolytic enzymes in the secretions allow sperm traverse cervical **mucus**, while the prostaglandins stimulate the transport of the sperm toward the ovum.

The male external reproductive organs include the scrotum and the penis. The scrotum is a pouch of skin and subcutaneous **tissue** that hangs from the lower abdominal region behind the penis. Scrotum contains and suspends the testes outside the abdominal cavity and far from the body at an optimal temperature for sperm development. Rete testis and epididymis are also contained in the scrotum.

The penis is an erectile organ comprised primarily of two cylinders of sponge-like vascular tissue that fills with **blood** to create an erection. A series of valves keep the blood in the penis to maintain the erection. A third cylinder is the urethra, the tube that carries the urine and the ejaculate. The devopment of the genitial system includes an undifferentiated and a differentiated stage. This impicates a control of the the differentation toward the male system. Genetic sex is determined by the X and Y sex chromosomes at the time of fertilization, but phenotypic sex is determined by a number of factors as well as masculinizing hormones. Lack of masculinizing hormones results in a female phenotype. Testosterone and dihydrotestosterone promote masculinization of the mesonephric duct and external genitalia, respectively.

Primordial germ cells arise from the wall of the secondary yolk sac and migrate (via **peritoneum**) to the genital ridge by about six weeks of gestation. Genital ridges are a mesodermic proliferation of the ventromedial surface of the urogenital ridge. Germ cells must enter the genital ridge or the gonads will not develop. They proliferate yielding the medullary cords. Also epithelial cells contribute to form such cords. The primitive cords further develop to become testis cords and they generate a network of thin tubules, the rete testis, which become separated from the surface epithelium by a thickened layer connective tissue called the tunica albuginea. The cords also contain Sertoli cells derived from the surface epithelium of the testis. Leydig cells differentiate from the genital ridge mesenchyme between the testis cords.

Male duct system arises from the mesonephric (Wolffian) ducts, a proliferation from the primitive kidney that serves as excretory duct, and is under hormonal control. Testis determining factor (TDS) triggers the production of Mullerian Inhibiting Substance (MIS) and testosterone. MIS induces the degeneration of the paramesonephric (Muller) ducts in male. The androgens result in the development of the efferent ductules, the epididymis, and the ductus deferens. The seminal vesicles develop as outgrowths from the mesonephric duct. The remaining mesonephric duct becomes the ejaculatory duct. The prostate derives from **endoderm** proliferation from the developing urethra.

See also Infertility (male); Prostate gland; Reproductive system and organs (female); Sex determination; Sex hormones; Sexual reproduction

RESPIRATION CONTROL MECHANISMS

Respiration, the exchange of oxygen and carbon dioxide within the body's tissues, is primarily an involuntary process controlled by many mechanisms including nervous control, chemical control, some voluntary control, body temperature, drugs, **pain**, emotion, **sleep**, baroreceptors, and proprioceptors.

The medullary rhythmicity area is the respiratory control center of the **central nervous system** located in the **medulla oblongata**. The respiratory control center can be divided into the inspiratory center and expiratory center. The inspiratory center spontaneously controls the **diaphragm** and intercostal muscles responsible for inspiration. These inspiratory muscles are connected to the medulla oblongata via spinal nerves. Contraction of the diaphragm is controlled by two phrenic nerves that emanate from the third, fourth, and fifth cervical spinal nerves. Eleven pairs of intercostal nerves that commence from the first 11 thoracic spinal nerves regulate the intercostal muscles. Under normal, resting conditions, there is no activity within the expiratory center and expiration occurs passively by relaxation of the diaphragm and intercostal muscles and elastic recoil of the **lungs**. However, when **breathing** increases during exercise for example, the expiratory center stimulates forceful expiration by contraction of the abdominal muscles and the internal intercostals.

The medulla oblongata is able to change the rate, depth, and rhythm of respiration because of feedback mechanisms sent by sensory neural pathways. The first pathway is controlled by the vagus nerve and is called the Hering-Breuer reflex because it prevents the lungs from over inflating. The vagus nerve provides sensory information from the thoracic and abdominal viscera including the lungs. When the **bronchi** and bronchioles expand, the vagus nerve conveys this information to the medulla oblongata that in turn temporarily inhibits the inspiratory center so that expiration occurs. Once the bronchi and bronchioles return to normal size, the vagus nerve no longer sends inhibitory signals to the medulla oblongata and the inspiratory center becomes active and initiates inspiration.

The pneumotaxic center is another neural pathway that provides inhibitory impulses to the inspiratory center of the medulla oblongata. This center is located in the **pons** and is only active during increased breathing. In addition to the Hering-Breuer reflex, the pneumotaxic center ensures that the lungs do not overinflate. The apneustic center is also located in the pons and activates inspiration in the medulla oblongata under normal, resting conditions. However, the apneustic center is canceled out by the pneumotaxic center when the breathing rate increases.

Chemicals in the body such as oxygen, carbon dioxide, and hydrogen ion concentrations greatly influence respiration. Each chemical is required in certain amounts in the body and any deviation from the baseline level will change the rate, depth, or rhythm of respiration. Chemical changes are monitored by chemoreceptors located in the medulla oblongata as well as the **carotid arteries** and the **aortic arch**. For example, excess carbon dioxide in the **blood**, a condition called hypercapnia, is accompanied by a decrease in pH and can be sensed by chemoreceptive areas in the medulla oblongata. The medulla oblongata responds by increasing the rate and depth of respiration, called hyperventilation, to expel the excessive carbon dioxide during expiration and return the pH levels to normal. Conversely, if levels of carbon dioxide and hydrogen ions fall below the baseline level, hypocapnia may result. Hypocapnia occurs when carbon dioxide falls below its baseline level of 40 mm Hg in the blood and results in slow, shallow breathing called hypoventilation. Likewise, when oxygen levels are low and carbon dioxide and pH remain normal, ventilation will increase until oxygen levels return to normal. However, if oxygen levels fall lower than 50 mm Hg in the blood, tissues become starved for oxygen and impulses are not sent to the respiratory area. Thus, respiration is not increased and eventually the person may stop breathing.

To some extent, respiration can be controlled voluntarily because of neural pathways between the **cerebral cortex** and the respiratory control center. For example, people can chose to take slow, deep breaths in an attempt to relax themselves, or they can consciously increase respiration and self-induce hyperventilation. Additionally, people can voluntarily hold their breath. However, when there is too much carbon dioxide and hydrogen ions and not enough oxygen in the blood, loss of consciousness will occur and the respiratory center will return breathing to normal.

Respiration may increase or decrease in response to changes in body temperature. When body temperature is low, respiration will decrease in an effort to conserve heat that would otherwise be lost during expiration. Conversely, respiration increases when body temperature is high from fever or exercise for example, to dissipate some of the heat and return the body temperature to normal. Additionally, a sudden shock to cold may cause a temporary breathing cessation.

Certain drugs and medications can affect respiration. For example, narcotics such as Demerol and morphine can reduce the rate and depth of breathing while adrenaline, amphetamine, and cocaine typically have the reverse effect.

Pain and emotions often increase respiration. Similar to the shock of cold, a sudden pain such as being hit with a baseball may also cause temporary apnea (cessation of breathing). However, continual pain typically increases the rate and depth of breathing. Emotions such as crying are controlled by the **hypothalamus** and limbic system that, in turn, stimulate the respiratory center and increase respiration. When the pain or emotion subsides, respiration returns to normal.

Other influences that can affect respiration include baroreceptors and proprioceptors. Baroreceptors are pressure receptors located in the carotid and aortic sinuses that sense changes in blood pressure. Changes in respiration are inversely proportional to changes in blood pressure. For example, when blood pressure increases, respiration decreases. Proprioceptors are receptors located in muscles, **tendons**, and **joints** that sense movement. During exercise, these receptors transmit signals to the respiratory center that increase the rate and depth of respiration.

See also Acid-base balance; Alveoli; Anesthesia and anesthetic drug actions; Capillaries; Respiratory system embryological development; Yawn reflex

RESPIRATORY SYSTEM

The respiratory system is a series of **organs** designed to facilitate the exchange of gases, primarily oxygen and carbon dioxide, between red **blood** cells in the **circulatory system** and the outside environment. Human cellular **physiology** requires a constant supply of oxygen (O_2) and the ability to ventilate carbon dioxide (CO_2) produced as a byproduct of metabolic reactions.

The blood and circulatory system are responsible for the transportation of dissolved or bound gases to and from their principal site of exchange, the **lungs**. The morphology of the lungs is unique in its deviation from symmetrical bisymmetry. The displacement of the human **heart** to the left side of the body manifests itself in an unequal division of lobes or morphologically and functionally distinct compartments of lung **tissue** termed lobes. The heart-side left lung contains two lobes. The right lung normally contains three lobes.

In humans, during normal resting activity (not **sleep**) an average amount of half a liter (500 ml) of air is exchanged with each breath. This volume of air is termed the tidal volume of respiratory exchange. At maximum capacity, the vital

capacity, male athletes can exchange 5–6 L of air. The movement of the tidal volume of air is response to pressure differentials crated by the movement of the **diaphragm** and muscles associated with respiration.

The medulla normally exercises involuntary regulation or control of the diaphragm so that **breathing** is not a conscious effort. Specialized receptors monitor and respond to varying concentration of oxygen and carbon dioxide in the blood. The breathing rate involuntarily adjusts and fluctuates in response to varying O_2 and CO_2 levels. Of course, it is possible to exercise voluntary control over breathing and to speed up, slow down, or stop the breathing cycle for limited periods of time.

Specialized structures direct oxygen-containing air into the lungs and provide a pathway for the ventilation of carbon dioxide.

In humans, the normal air pathway begins at the nostril orifices and, when needed, at the mouth. In the nasal passages, air is filtered of large debris by nasal hairs. In addition to being filtered, air flowing over the nasal hairs is warmed and, in dry climates, humidified by the ambient moisture in the nasal passages. All of these changes are protective mechanisms designed to make the flow of air less traumatizing to lower passages. When increased volumes are air are required during physical exertion, or when the nasal passages are blocked by inflammation or physical obstructions, air may be inhaled through the mouth. Such mouth breathing, however, does not convey nearly the same level of filtering, warming, and moisturizing, as does the passing of air through the nasal passages. Regardless, air from both the mouth and **nose** ultimately mixes in the **pharynx** (a common pathway to both the gastrointestinal tract and the respiratory system).

After passing the pharynx, air enters the **larynx** (including the **vocal cords** and Adam's apple in males). The larynx provides a path to the trachea, a thickened cartilaginous tube with prominent rings of **cartilage** spaced along the tube. The rings serve a supportive function for the trachea that is often squeezed by food passing down the esophagus. Lining the trachea is a pseudo-stratified ciliated columnar epithelium that functions to further filter air passing trough the trachea.

At the lower terminus end of the trachea, it divides into left and right **bronchi**, or bronchial tubes that provide a pathway to the lungs. A mucous lining in the bronchi also serves to filter air passages. As the bronchi enter the lungs, they ramify or divide into many smaller branches termed bronchioles.

The lungs contain alveolar tissue that provide a termination for the smallest diameter bronchioles. An alveolus, with multiple lobed clusters of simple sqamous cell lined air sac is the site of **gaseous exchange** between the circulatory system and the respiratory system. Within an alveolus, thinned walled air passages are brought close to thin lined vascular arterial vessels. The barrier is so slight, and the communication so close, that gaseous exchange is possible. Differing concentrations of gases, establishing a concentration gradient, is the driving force guiding the exchange of gases from area of higher concentration to areas of lower concentration.

To keep the alveolar sacs moist and functional they are covered with a protective surfactant. The surfactant also prevents the total collapse of the air sacs during exhalation and

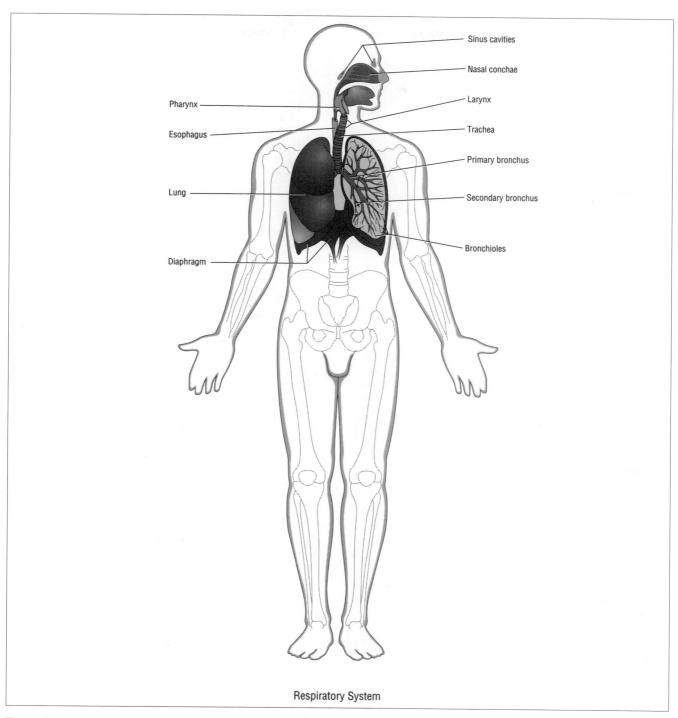

The respiratory system. *Illustration by Argosy Publishing.*

further prevents the linings of the sacs from sticking together during partial collapse. One of the major problems with premature birth involves the lack of surfactant production capacity in premature infants. The ability of alveolar tissue to produce surfactant does not occur until the last two months of gestation. It is also thought that certain physiological events such as yawning and coughing may be important in the redistribution of surfactants.

See also Acclimatization; Allergies; Alveoli; Anesthesia and anesthetic drug actions; Brachial system; Cough reflex; CPR (Cardiopulmonary resuscitation); Drowning; Ducts arteries;

Embryonic development: early development, formation, and differentiation; Epithelial tissues and epithelium; Gaseous exchange; Metabolic waste removal; Metabolism; Oxygen transport and exchange; Respiration control mechanisms; Respiratory system embryological development; Sense organs: olfactory (sense of smell) structures; Underwater physiology; Yawn reflex

RESPIRATORY SYSTEM, EMBRYOLOGICAL DEVELOPMENT

The **respiratory system**, and the **lungs** in particular, are the product of a set of complex developmental interactions between two distinct tissues, the endodermally derived epithelium and the **mesoderm**. Each **tissue** contributes to lung development by regulation of the spatial and temporal pattern of gene expression. Growth factors (GF) play a major role in the development of the lung. Five stages are described for the lung and respiratory system development.

In the embryonic stage (four to seven weeks), an endodermal respiratory diverticulum (laryngotracheal groove) develops from the ventral surface of the foregut, caudally to the last pharyngeal pouch. During the initial stages of lung laryngotracheal groove development, and the subsequent formation of the bronchial tree, fibroblast growth factor-10 (FGF-10) stimulates the proliferation and activation of cells that express FGF receptors. This is a critical step in development. The laryngotracheal groove grows, and is now called the primitive lung bud. The bud rapidly grows inferiorly and elongates into splanchnic mesoderm. It divides to form right and left bronchial buds (day 28) and successively branched structures that gives rise to the epithelial lining of the **larynx**, trachea, bronchious, and lower airways, including the **alveoli**. There is a transitory period when the cavitary structure is completed obliterated by the epithelial growth. All **cartilage**, muscle and connective tissue of the larynx, trachea, and lungs arise from splanchnic mesoderm that also form **blood** vessels, fibroblasts, lymphatics and **smooth muscle**. The bronchopulmonary segments appears at six weeks.

The pseudoglandular stage (6–16 weeks), is characterized by rapid growth and proliferation of the primitive airways caused by repeated branching of the distal ends of epithelial tubes. The conducting system undergoes 16 or more generations of branching. The formation of terminal bronchioles takes place. The airways are filled of fluid and are lined with tall columnar epithelium. These cells begin to differentiate into goblet-cells, mucous glands, and **ciliated epithelial cells**. The canalicular stage (16–26 weeks) is characterized by a widening and lengthening of the airways leading to a large increase of the future pulmonary airspace. Further terminations of the pulmonary tree develop, as well as the respiratory bronchioles, which also subdivide into terminal saccules. The surrounding mesenchyma thins and vascularisation (network of **capillaries**) associated with respiratory bronchioles and immature alveoli appears. Some growth factors as well as vascular endothelial growth factor (VEGF) participate in blood vessel formation. The cuboidal cells of the terminal sac epithelium differentiate into alveolar type-II cells, which secrete low levels of surfactant. Where cells with type-II phenotype juxtapose a capillary, they differentiate to type-I cells, which flatten and can provide a functional, though inefficient, blood/air barrier if the infant is born prematurely. At the end of this period, gas exchange is possible.

The terminal sac phase (24–36 weeks) is characterized by branching and growth of the terminal sacs or primitive alveolar ducts. Capillaries take proper apposition with the prospective alveoli. Functional type-II pneumonocytes differentiate via several intermediate stages from pluripotent epithelial cells in the prospective alveoli. Multilamellar bodies very specific to type II pneumocyte are indicative of the process of differentiation of type II pneumocyte. Pulmonary surfactant, produced by type II pneumocytes, is a lipoprotein complex produced exclusively by type II pneumonocytes of the lung alveolus, and acts to reduce surface tension at the air/liquid interface. Type-I pneumonocytes differentiate from cells with a type-II like phenotype. These cells continue to flatten and spread, increasing by dilation of the saccules, the surface area available for gas exchange. By 26 weeks, a primitive blood/gas barrier has formed. Maturation of the alveoli continues by further enlargement of the terminal sacs, deposition of elastin, and development of vascularized septae. The stroma continues to thin and the capillaries protrude into the alveolar spaces.

The mature alveolus is seen at 36 weeks. During alveolization, septation of the distal saccules to form the definitive alveoli requires FGF and platelet-derived growth factor (PDGF) signaling. New alveoli will continue to form for approximately three years. A decrease in the relative proportion of parenchyma to total lung volume still contributes significantly to growth for one to two years after birth; thereafter all components grow proportionately until adulthood.

During development of the laryx, the respiratory diverticulum forms a groove in the floor of the **pharynx** called the laryngotracheal groove. Cephalic to the laryngotracheal groove is the epiglottal swelling. On either side of this groove are the developing arytenoid swellings. The larynx epithelium develops form the **endoderm** of the laryngotracheal tube. Muscles and cartilage arise from the fourth and sixth arches. The cartilage develops from the neural crest cells. In the larynx, a temporary occlusion, followed by recanalization (reopening), occurs during the ninth and tenth week.

The **diaphragm**, which separates the abdominal cavity from the plerual cavities, is a composite structure. The embryonic structures from which it develops include the (1) septum transversum, the (2) dorsal esophageal mesentery, the (3) pleuroperitoneal folds, and (4) body wall mesoderm. Pleuroperitoneal folds arise from the posterior body wall and lie in a plane that is parallel to the septum transversum and perpendicular to the pleuropericardial folds. The diaphragm is innervated from the third through fifth cervical **spinal cord** segments via the phrenic nerve. The diaphragm develops initially in a cephalic region and during development (anteroposterior folding of the embryo) descends into a more caudal thoracic position. The phrenic nerves travel with the diaphragm and come to lie within the fibrous **pericardium**.

See also Embryology; Embryonic development, early development, formation, and differentiation

RETICULAR ACTIVIATING SYSTEM (RAS)

The reticular activating system (RAS) is a network of cells located in the reticular formation of the **medulla oblongata**. RAS cells receive information via collateral connections from **neurons** located in ascending sensory neural tracts and relay that information directly to higher cerebral structures. The function of the RAS is to regulate arousal reactions within the **central nervous system** (CNS). Proper RAS function is critical to **brain** or CNS alertness, as measured by an individual's responsiveness to external stimuli and electrical brain activity (as measured by an **electroencephalogram** or **EEG**).

Within the medullary brainstem, the reticular formation includes the areas of **gray matter** in the **pons** and mesencephalon, but does not include **cranial nerves** or cranial nerve nuclei. The RAS is continuous with the spinal reticular formation and with the thalamic system. The RAS also receives input from the auditory and visual tracts. In addition to stimulating **cerebral cortex** activity by stimulating the RAS it is possible to stimulate the RAS via the cerebral cortex.

The reticular activating system serves as a monitoring and switching system through which a wide number of signals, originating from stimuli in both the external and internal environment pass. RAS activity directly correlates with electrical activity of the cerebral cortex, and with electrical activity in the efferent nerves regulating both motor and vegetative processes.

When asleep, there are decreased levels of electrical activity in the RAS. Excitation of neural cells in the RAS, however, causes individuals to wake from **sleep** and corresponds to changes in both metabolic rate and general muscle tone. Electrical stimulation of the RAS in animal test subjects shows a corresponding increase in cortical activity corresponding to an arousal reaction as measured by an EEG. The input signals to the RAS to that initiate the arousal reaction may come from a variety of sensory inputs (e.g., touching sensations, light, noises, etc). The type and strength of the input signals to the RAS causes varying degrees of reaction. Both **pain** from external stimuli and protective (proprioceptive) signals from the body are the most potent signals in terms of eliciting RAS activity.

In addition to stimulation to the RAS that results in a generalized increase in alertness, specific stimulation of thalamic connected regions result in more specific stimulation of certain areas of the cerebral cortex and thus may provide a mechanism to increase or heighten certain types of thinking or awareness.

Recent research has indicated that there may be separate areas of the brain responsible for maintaining alertness from those responsible for initial arousal.

Barbiturates depress brain stem activity, especially electrical activity in the RAS, and contribute to a general **anesthesia** of drowsiness and sleep.

See also Biofeedback; Brain stem function and reflexes; Brain: intellectual functions; Nervous system overview; Neural plexuses; Neurology

RETICULO-ENDOTHELIAL SYSTEM

The reticulo-endothelial system (RES) is a diffuse system consisting of phagocytic cells. These are derived from bone marrow **stem cells**, which are associated with the connective **tissue** framework of the **liver**, spleen, and lymph nodes. It consists of the combination of mobile macrophages and fixed tissue macrophages. The macrophages in various tissues differ in appearance because of their environmental differences and they are known by various names: Kupfer cells in the liver, reticular cells in the lymph nodes, spleen and bone marrow, alveolar macrophages in the **alveoli** of the **lungs**, tissue histiocytes, clasmatocytes or fixed macrophages in the subcutaneous tissue, and microglia in the **brain**. The term "reticulo-endothelial system" was coined because it was formerly believed that a major part of the **blood** vessel endothelial cells could perform phagocytic functions similar to those performed by the macrophage system. More recent studies have disproved this, however, but the term is nevertheless widely used. It should be remembered that the reticulo-endothelial system is almost synonymous with the tissue macrophage system.

The phagocytic cells within this system comprise the mononuclear phagocyte system (MPS), and the macrophage is the major differentiated cell in the MPS. The MPS also consists of bone marrow monoblasts and pro-monocytes, peripheral blood monocytes and tissue macrophages. Cells of the RES and MPS, particularly the liver macrophage or Kupfer cells, are known to be important in the clearance of particles from the bloodstream. Negatively charged particles, in particular, are avidly scavenged by macrophages. The Kupfer cell is one of the most important cells in the MPS. These are highly phagocytic cells located in the sinusoid wall, usually on the endothelial surface, of the liver. They have a number of functions, but one of the most important is their ability to endocytose and remove from the blood potentially harmful materials and particulate matter such as bacterial endotoxins, microorganisms, immune-complexes and tumour cells. The recognition of these materials is mediate by an array of cell surface receptors that include receptors for IgG, IgA, complement components, galactose, mannose, CD4 and the carcinoembryonic antigen. Other receptors include the scavenger receptor, transferrin and DNA-binding receptors.

See also Antigens and antibodies; Cell differentiation; Hemopoiesis; Immune system; Immunity; Immunology

RETINA AND RETINAL IMAGING

The retina is an area at the back of the **eye** consisting of seven layers of specialized cells and **neurons** that convert the light signal, which has entered the eye through the **cornea** and has been focused by the lens, into an electrical signal. The electrical impulse is routed to the **brain** for interpretation and construction of an image.

The light to electrical signal transduction is accomplished by cells known as rods and cones, which are located in the outermost of the seven layers. The electrical signal then passes through the outer nuclear layer, outer plexiform layer, inner nuclear layer, inner plexiform layer, ganglion cell layer, and the optic fiber layer. In general, the nuclear layers contain cells while the plexiform, or paler appearing layers, contains axons and dendrites; specialized portions of neurons.

Incoming light contacts the receptor region of a rod or cone, which excites the bipolar cell lying just underneath in the inner nuclear layer, and enables the signal to pass across the outer plexiform layer. Also located in the inner nuclear layer are horizontal cells, named for their location along the inner periphery of the layer. These cells aid in the processing of the incoming electrical signals and determine how the bipolar cells will react to the signal. For example, diffuse light will excite a bipolar cell and horizontal cells. The excitation of the horizontal cells can inhibit the neighbouring bipolar cell. Thus, with diffuse light, few cells pass on the signal across the inner plexiform layer to ganglion cells located in the ganglion cell layer. An intense beam of light impinging on the retina will excite a bipolar cell, but not the neighbouring horizontal cells, so more of the signal will be passed on to the ganglion cells. The ganglion cells extend inward to form optic fibers, which are part of the optical fiber layer, and which collectively exit the eyeball at a single location, the optic disk.

The common exit site of the optic fibers creates a light-insensitive blind spot. Normally, each eye covers for the blind spot of the other, and the brain fills in the missing information with whatever informational pattern immediately surrounds the blind spot. Hence, the blind spot is usually not evident.

Rods and cones are also called photoreceptors. Rods are responsible for black and white vision and as such are sensitive to low intensities of light. Cones are responsible for color vision, and are not as sensitive to low intensities of light. Thus, at dusk an image is perceived as washed out or gray appearing.

The distribution of rods and cones over the light-receiving surface of the retina is not uniform. The majority of cones are positioned at the center of the retina, a region termed the fovea. For this reason, the fovea is the region of the greatest visual perception. When the eye is consciously directed at an object, it is the fovea that is receiving the bulk of the incoming visual information. Elsewhere, over the more peripheral light-receiving surface of the retina, rod cells predominate. These cells function more in peripheral vision, that is, in the information that is entering the eye from areas other than those on which the eye is directly focusing.

The different response of cells to diffuse and focused light, and the asymmetric arrangement of rods and cones over the retinal surface, enable the retina to image light as borders and contours. Humans perceive the visual world as a pattern of lines. The different responses of the retinal **imaging** system to diffuse and sharp light is the basis of the recognition of contours and borders.

Defects in the retina can decrease vision. Retinitis pigmentosa is an inherited, incurable condition in which the rod cells degenerate. This degeneration decreases the ability to see in dim light and can decrease peripheral vision. The defect is due to a mutation in a gene that codes for a protein called rhodopsin. As of 2002, over 50 different mutations have been discovered. The bursting of **blood** vessels that supply the retina can also cause retinal damage. Pooling of the blood can decrease vision. But testing of an infrared radiation, underway as of late 2001, may dissolve the pooled blood, restoring such decreased vision. In those people whose retinal damage is confined to the rod and cone cells, the use of an implanted microchip that substitutes functionally for the rods and cones may one day restore vision. The artificial retina research has achieved limited site in blind patients.

See also Aging processes; Eye and ocular fluids; Peripheral nervous system

RHEUMATOID ARTHRITIS • *see* AUTOIMMUNE DISORDERS

RIBONUCLEIC ACID (RNA)

Nucleic acids are complex molecules that contain a cell's genetic information and the instructions for carrying out cellular processes. In eukaryotic cells, the two nucleic acids, ribonucleic acid (RNA) and **deoxyribonucleic acid (DNA)**, work together to direct **protein synthesis**. Although it is DNA (deoxyribonucleic acid) that contains the instructions for directing the synthesis of specific structural and enzymatic proteins, several types of RNA actually carry out the processes required to produce these proteins. These include messenger RNA (mRNA), ribosomal RNA (rRNA), and transfer RNA (tRNA). Further processing of the various RNAs is carried out by another type of RNA called small nuclear RNA (snRNA). The structure of RNA is very similar to that of DNA, however, instead of the base thymine, RNA contains the base uracil. In addition, the pentose sugar ribose is missing an oxygen atom at position two in DNA, hence the name "deoxy-."

Nucleic acids are long chain molecules that link together individual nucleotides that are composed of a pentose sugar, a nitrogenous base, and one or more phosphate groups.

The nucleotides, the building blocks of nucleic acids, in ribonucleic acid are adenylic acid, cytidylic acid, guanylic acid, and uridylic acid. Each of the RNA subunit nucleotides carries a nitrogenous base: adenylic acid contains adenine (A), cytidylic acid contains cytosine (C), guanylic acid contains guanine (G), and uridylic acid contains uracil.

In humans, the DNA molecule is made of phosphate-base-sugar nucleotide chains, and its three-dimensional shape affects its genetic function. In humans and other higher organisms, DNA is shaped in a two-stranded spiral helix organized into structures called **chromosomes**. In contrast, most RNA molecules are single-stranded and take various shapes.

Nucleic acids were first identified by the Swiss biochemist Johann Miescher (1844–1895). Miescher isolated a cellular substance containing nitrogen and phosphorus. Thinking it was a phosphorus-rich nuclear protein, Miescher named it nuclein.

The substance identified by Miescher was actually a protein plus nucleic acid, as the German biochemist Albrecht Kossel discovered in the 1880s. Kossel also isolated nucleic acids' two purines (adenine and guanine) and three pyrimidines (thymine, cytosine, and uracil), as well as **carbohydrates**.

The American biochemist Phoebus Levene, who had once studied with Kossel, identified two nucleic acid sugars. Levene identified ribose in 1909 and deoxyribose (a molecule with less oxygen than ribose) in 1929. Levene also defined a nucleic acid's main unit as a phosphate-base-sugar nucleotide. The nucleotides' exact connection into a linear polymer chain was discovered in the 1940s by the British organic chemist Alexander Todd.

In 1951, American molecular biologist **James Watson** and the British molecular biologists **Francis Crick** and Maurice Wilkins developed a model of DNA that proposed its now accepted two-stranded helical shape in which adenine is always paired with thymine and guanine is always paired with the cytosine. In RNA, uracil replaces thymine.

During the 1960s, scientists discovered that three consecutive DNA or RNA bases (a codon) comprise the **genetic code** or instruction for production of a protein. A gene is transcribed into messenger RNA (mRNA), which moves from the nucleus to structures in the cytoplasm called ribosomes. Codons on the mRNA order the insertion of a specific amino acid into the chain of **amino acids** that are part of every protein. Codons can also order the translation process to stop. Transfer RNA (tRNA) molecules already in the cytoplasm read the codon instructions and bring the required amino acids to a ribosome for assembly.

Some proteins carry out cell functions while others control the operation of other genes. Until the 1970s cellular RNA was thought to be only a passive carrier of DNA instructions. It is now known to perform several enzymatic functions within cells, including transcribing DNA into messenger RNA and making protein. In certain **viruses** called retroviruses, RNA itself is the genetic information. This, and the increasing knowledge of RNA's dynamic role in DNA cells, has led some scientists to believe that RNA was the basis for the Earth's earliest life forms, an environment called the RNA World.

The first step in protein synthesis is the transcription of DNA into mRNA. The mRNA exits the nuclear membrane through special pores and enters the cytoplasm. It then delivers its coded message to tiny protein factories called ribosomes that consist of two unequal sized subunits. Some of these ribosomes are found floating free in the cytosol, but most of them are located on a structure called rough endoplasmic reticulum (rER). It is thought that the free-floating ribosomes manufacture proteins for use within the cell (cell proliferation), while those found on the rER produce proteins for export out of the cell or those that are associated with the cell membrane.

Genes transcribe their encoded sequences as a RNA template that plays the role of precursor for messenger RNA (mRNA), being thus termed pre-mRNA. Messenger RNA is formed through the splicing of exons from pre-mRNA into a sequence of codons, ready for protein translation. Therefore, mRNA is also termed mature mRNA, because it can be trans-ported to the cytoplasm, where protein translation will take place in the ribosomal complex.

Transcription occurs in the nucleus, through the following sequence of the events. The process of gene transcription into mRNA in the nucleus begins with the orginal DNA nitrogenous base sequence represented in the direction of transcription, e.g., from the 5' (five prime) end to the 3' (three prime end) as DNA 5'... AGG TCC TAG TAA...3' to the formation of pre-mRNA (for the exemplar DNA cited) with a sequence of 3'... TCC AGG ATC ATT...5' (exons transcribed to pre-mRNA template) then into a mRNA sequence of 5'... AGG UCC UAG UAA...3' (codons spliced into mature mRNA).

Messenger RNA is first synthesized by genes as nuclear heterogeneous RNA (hnRNA), being so called because hnRNAs varies enormously in their molecular weight as well as in their nucleotide sequences and lengths, which reflects the different proteins they are destined to code for translation. Most hnRNAs of eukaryotic cells are very big, up to 50,000 nucleotides, and display a poly-A tail that confers stability to the molecule. These molecules have a brief life, being processed during transcription into pre-mRNA and then in mRNA through splicing.

The molecular weight of mRNAs also varies in accordance with the protein size they encode for translation and they necessarily are much bigger than the protein itself, because three nucleotides are needed for the translation of each amino acid that will constitute the polypeptide chain during protein synthesis. Prokaryotic mRNA molecules usually have a short existence of about 2-3 minutes, but the fast bacterial mRNA turnover allows for a quick response to environmental changes by these unicellular organisms. In mammals, the average life span of mRNA goes from 10 minutes up to 2 days. Therefore, eukaryotic cell in mammals have different molecules of mRNA that show a wide range of different degradation rates. For instance, mRNA of regulatory proteins, involved either in cell **metabolism** or in the **cell cycle** control, generally has a short life of a few minutes, whereas mRNA for globin has a half-life of 10 hours.

The enzyme RNA-polymerase II is the transcriptional element in human eukaryotic cells that synthesizes messenger RNA. The general chemical structure of most eukaryotic mRNA molecules contain a 7-methylguanosine group linked through a triphosphate to the 5' extremity, forming a cap. At the other end (i.e., 3' end), there is usually a tail of up to 150 adenylils or poly-A. One exception is the histone mRNA that does not have a poly-A tail. It was also observed the existence of a correlation between the length of the poly-A tail and the half-life of a given mRNA molecule.

At the biochemical level, RNA molecules are linear polymers that share a common basic structure comprised of a backbone formed by an alternating polymer of phosphate groups and ribose (a sugar containing five carbon atoms). Organic nitrogenous bases i.e., the purines adenine and guanine, and the pyrimidines cytosine and uracil are linked together through phosphodiester bridges. These four nitrogenous bases are also termed heterocyclic bases and each of them combines with one of the riboses of the backbone to form a nucleoside, such as adenosine, guanosine, cytidine, and uridine. The combination of a ribose, a phosphate, and a given nitrogenous base by its turn results in a nucleotide, such as adenylate, guanylate, cytidylate, uridylate. Each phosphodiester bridge links the 3' carbon at the ribose of one nucleotide to the 5' carbon at the ribose of the subsequent nucleotide, and so on. RNA molecules fold on themselves and form structures termed hairpin loops, because they have extensive regions of complementary guanine-cytosine (G-C) or adenine-uracil (A-U) pairs. Nevertheless, they are single polynucleotide chains.

The mRNA molecules contain at the 5' end a leader sequence that is not translated, known as UTR (untranslated region) and an initiation codon (AUG), that precedes the coding region formed by the spliced exons, which are termed codons in the mature mRNA. At the end of the coding region, three termination codons (UAG, UAA, UGA) are present, being followed by a trailer sequence that constitutes another UTR, which is by its turn followed by the poly-A tail. The stability of the mRNA molecule is crucial to the proper translation of the transcript into protein. The poly-A tail is responsible by such stability because it prevents the precocious degradation of mRNA by a 3'to 5' exonuclease (a cytoplasmatic enzyme that digests mRNA starting from the extremity 3' when the molecule leaves the cell nucleus). The mRNA of histones, the nuclear proteins that form the nucleosomes, do not have poly-A tails, thus constituting an exception to this rule. The poly-A tail also protects the other extremity of the mRNA molecule by looping around and touching the 7-methylguanosine cap attached to the 5' extremity. This prevents the decapping of the mRNA molecule by another exonuclease. The removal of the 7-methylguanosine exposes the 5' end of the mRNA to **digestion** by the 5'to 3' exonuclease exonuclease (a cytoplasmatic enzyme that digests mRNA starting from the 5' end). When the translation of the protein is completed, the enzymatic process of deadenylation (i.e., enzymatic digestion of the poly-A tail) is activated, thus allowing the subsequent mRNA degradation by the two above mentioned exonucleases, each working at one of the ends of the molecule.

Transfer RNA (tRNA) is often referred to as the "Rosetta Stone" of **genetics**, as it translates the instructions encoded by DNA, by way of messenger RNA (mRNA), into specific sequences of amino acids that form proteins and polypeptides. This class of small globular RNA is only 75–90 nucleotides long, and there is at least one tRNA for every amino acid. The job of tRNA is to transport free amino acids within the cell and attach them to the growing polypeptide chain. First, an amino acid molecule is attached to its particular tRNA. This process is catalyzed by an enzyme called aminoacyl—tRNA synthetase that binds to the inside of the tRNA molecule. The molecule is now charged. The next step, joining the amino acid to the polypeptide chain, is carried out inside the ribosome. Each amino acid is specified by a particular sequence of three nucleotide bases called codons. There are four different kinds of nucleotides in mRNA. This makes possible 64 different codons (4^3). Two of these codons are called STOP codons; one of these is the START codon (AUG). With only 20 different amino acids, it is clear that some amino acids have more then one codon. This is referred to as the degeneracy of the genetic code. On the other end of the tRNA molecule are three special nucleotide bases called the anticodon. These interact with three complimentary codon bases in the mRNA by way of hydrogen bonds. These weak directional bonds are also the force that holds together the double strands of DNA.

In order to understand how this happens it was necessary to first understand the three dimensional structure (conformation) of the tRNA molecule. This was first attempted in 1965, where the two-dimensional folding pattern was deduced from the sequence of nucleotides found in yeast alanine tRNA. Later work (1974), using x-ray diffraction analysis, was able to reveal the conformation of yeast phenylalanine tRNA. The molecule is shaped like an upside-down L. The vertical portion is made up of the D stem and the anti-codon stem, and the horizontal arm of the L is made up of the acceptor stem and the T stem. Thus, the translation depends entirely upon the physical structure. At one end of each tRNA is a structure that recognizes the genetic code, and at the other end is the particular amino acid for that code. Amazingly, this unusual shape is conserved between **bacteria** plants and animals.

Another unusual thing about tRNA is that it contains some unusual bases. The other classes of nucleic acids can undergo the simple modification of adding a methyl (CH3-) group. However, tRNA is unique in that it undergoes a range of modifications from methylation to total restructuring of the purine ring. These modifications occur in all parts of the tRNA molecule, and increase its structural integrity and versatility.

Ribosomes are composed of ribosomal RNA (as much as 50%) and special proteins called ribonucleoproteins. In eukaryotes (an organism whose cells have chromosomes with nucleosomal structure and are separated from the cytoplasm by a two membrane nuclear envelope and whose functions are compartmentalized into distinct cytoplasmic organelles), there are actually four different types of rRNA. One of these molecules is called 18SrRNA and along with some 30–plus different proteins, it makes up the small subunit of the ribosome. The other three types of rRNA are called 28S, 5.8S, and 5S rRNA. One of each of these molecules, along with some 45 different proteins is used to make the large subunit of the ribosome. There are also two rRNAs exclusive to the mitochondrial (a circular molecule of some 16,569 base pairs in the human) genome. These are called 12S and 16S. A mutation in the 12SrRNA has been implicated in non-syndromic hearing loss. Ribosomal RNA's have these names because of their molecular weight. When rRNA is spun down by ultracentrifuge, these molecules sediment out at different rates because they have different weights. The larger the number, the larger the molecule.

The larger subunit appears to be mainly involved in such biochemical processes as catalyzing the reactions of polypeptide chain elongation and has two major binding sites. Binding sites are those parts of large molecule that actively participate in its specific combination with another molecule. One is called the aminoacyl site and the other is called the peptidyl site. Ribosomes attach their peptidyl sites to the membrane surface of the rER. The aminoacyl site has been associated with binding transfer RNA. The smaller subunit appears to be concerned with ribosomal recognition processes such as mRNA. It is involved with the binding of tRNA also. The smaller subunit combines with mRNA and the first "charged "tRNA to form the initiation complex for translation of the RNA sequence into the final polypeptide.

The precursor of the 28S, 18S and the 5.83S molecules are transcribed by RNA polymerase I (Pol I) and the 5S rRNA is transcribed by RNA polymerase III (PoIII). Pol I is the most active of all the RNA polynmerases, and is one indication of how important these structures are to cellular function.

Ribosomal RNAs fold in very complex ways. Their structure is an important clue to the evolutionary relationships found between different kinds of organisms. Sequence comparisons of the various rRNAs across various species show that even though their base sequences vary widely, **evolution** has conserved their secondary structures, therefore, organization must be important for their function.

Since the 1970s, nucleic acids' cellular processes have become the basis for genetic engineering, in which scientists add or remove genes in order to alter the characteristics or behavior of cells. Such techniques are used in agriculture, pharmaceutical and other chemical manufacturing, and medical treatments for **cancer** and other diseases.

See also Biochemistry; Genetic regulation of eukaryotic cells; Genetics and developmental genetics; Molecular biology

RIBS · *see* COSTOCHONDRAL CARTILAGE

RICHARDS, JR., DICKINSON W.
(1895-1973)
American physician

In refining the technique of cardiac catheterization, Dickinson Woodruff Richards made significant contributions to the study of cardiopulmonary function in human patients. In collaboration with his colleague André F. Cournand, Richards elaborated upon earlier research by German physician **Werner Forssmann**, leading ultimately to the discovery of how pulmonary efficiency could be measured. For their work, Richards, Cournand, and Forssmann shared the 1956 Nobel Prize in physiology or medicine.

Richards was born in Orange, New Jersey to Sally (Lambert) and Dickinson Woodruff Richards. Richards' maternal grandfather practiced general medicine in New York City, as did three of Richards' uncles; all either received their training or were otherwise affiliated with Bellevue Hospital or Columbia University's College of Physicians and Surgeons, where Richards himself would eventually study.

Richards received his A.B. from Yale University in 1917, and three months later enlisted in the United States Army, serving in France with the American Expeditionary Force. Upon his return to the United States, he entered the College of Physicians and Surgeons at Columbia; there he completed his M.A. in physiology in 1922, and his M.D. in 1923. Richards interned for two years at Presbyterian Hospital in New York, spending two additional years as a resident physician afterwards. In 1927, Columbia University granted

him a research fellowship to train at London's National Institute for Medical Research. From 1927 to 1928, he studied experimental physiology, working closely with Dr. Henry Hallett Dale, to whom Richards would later refer as one of his greatest influences. He then returned to Columbia University's Presbyterian Hospital to study pulmonary and circulatory physiology. In 1930, Richards became engaged to Constance Riley, a Wellesley College graduate who worked as a technician in his research lab at Presbyterian Hospital. They married in September, 1931.

Richards' collaboration with **André Cournand** began in 1931, at Bellevue Hospital. Basing their research on Richards' concept "that **lungs**, **heart**, and circulation should be thought of as one single apparatus for the transfer of respiratory gases between outside atmosphere and working tissues," these two physicians began a long and fruitful partnership. Their initial research involved the study of the physiological performance of the lungs and, in particular, a disorder known as chronic pulmonary insufficiency. Characterized by a malfunction in the heart's tricuspid and pulmonic valves, this defect causes **blood** to flow backward into the heart. Richards concluded, as had others before him, that it was necessary to be able to measure the amount of air in the lungs during different stages of pulmonary activity. Thus, he and Cournand unearthed studies done in 1929 by the German physician Werner Forssmann, wherein Forssmann had attempted to measure gases in the blood as it passed from the heart to the lungs.

Forssmann's technique was proven viable when he successfully inserted a narrow rubber catheter through a vein in his own arm and into the right atrium of his heart. This method gave access to blood as it entered the heart—blood that could then be examined in specific stages of pulmonary and cardiac activity and evaluated in terms of rate of flow, pressure relations, and gas contents. Catheterization would allow physicians to measure oxygen and carbon monoxide in blood returning from the right atrium, allowing for accurate measurement of blood flow through the lungs. Richards and Cournand sought to advance Forssmann's technique and to develop a safe procedure by first experimenting on animals. They began their research in 1936, and by 1941, they had successfully catheterized the right atrium of the human heart.

The measurements made possible through cardiac catheterization led Richards to other important assessments about circulation and functions of the heart. In 1941, he developed methods to measure the volume of blood pumped out of either ventricle (lower chamber) of the heart, and to measure blood pressure in the right atrium, the right ventricle, and the pulmonary artery, as well as total blood volume. More recent research has employed catheterization to diagnose abnormal exchange between the right and left sides of the heart, such as is present in some congenital cardiac defects. It has also contributed to the development of more sophisticated techniques such as angiocardiography (the x-ray examination of the heart after injection of dyes), which is used to determine whether normal circulation has resumed following a surgical procedure.

Richards and his colleagues also relied on their revolutionary research technique to study the effects of traumatic **shock** in heart failure and to identify congenital heart lesions.

During World War II, Richards and his colleagues were asked by the government to study the circulatory forces involved in shock, with Richards serving as chair of the National Research Council's subcommittee. The goal was to measure the effects of hemorrhage and trauma on the heart and cardiac circulation, and to evaluate various procedures for treatment. The most important result of this project was the discovery that whole blood, rather than just blood **plasma**, should be used in the treatment of shock to the cardiac system.

Richards was passionate about health issues in the social arena as well as in the laboratory. In 1957, he testified before the Joint Legislative Committee on Narcotics Study to suggest the construction of hospital clinics to legally distribute narcotics to recovering addicts; he lobbied for the building of a new hospital to replace the aging Bellevue; he spoke often about the need for constant reform within medical academia; and he supported the crusade to improve health care benefits for the elderly.

In 1945, Richards became the head of Columbia University's First Medical Division at Bellevue Hospital, and at the same time was promoted to the full-time Lambert Professorship of Medicine at the College of Physicians and Surgeons. He served as associate editor of *Circulation, The Journal of the American Heart Association,* and of the *American Review of Tuberculosis.* His articles have appeared in many publications, including *Physiological Review, Journal of Clinical Investigation,* and *Journal of Chronic Diseases.*

For their refinement of the catheterization procedure and the discoveries that followed, Richards, Cournand and Forssmann were awarded the Nobel Prize in physiology or medicine in 1956. In addition, Richards also received many individual honors and awards, including the John Phillips Memorial Award of the American College of Physicians (1960) and the Kober Medal of the Association of American Physicians (1970). He was made a chevalier of the Legion of Honor of France (1963), and was a fellow of the American College of Physicians, the American Medical Association, and the American Clinical and Climatological Association. He was offered numerous honorary degrees but accepted only two—Yale University, his alma mater, and Columbia University, where he did most of his work.

Richards was elected to the National Academy of Sciences in 1958, and retired from practice in 1961, though he continued to lecture and publish frequent articles for several years. He died at his home in Lakeville, Connecticut, after suffering a heart attack at the age of 78.

See also Circulatory system; Pulmonary circulation

RICHET, CHARLES ROBERT (1850-1935)
French physiologist

French physiologist Charles Robert Richet won the 1913 Nobel Prize for his discovery of a nonprotective, toxic **immune system** process that he called anaphylaxis, a process related to the **shock** and allergic reactions that occur when foreign substances are injected into the body. Richet also tried to

develop treatments for tuberculosis and to discover a serum to prevent tuberculosis.

Richet was born August 26, 1850, in Paris. His father, Alfred Richet, taught surgery at the University of Paris. His mother was Eugénie Rouard. After secondary school, Richet decided he wanted to practice medicine. He enrolled in the University of Paris medical school, but he soon found that he was more interested in research than in applied medicine. He also weighed the possibilities of a career in the humanities, and although he choose science instead, he maintained an active interest in literary, philosophical, and political subjects throughout his life.

In 1877, he received his medical degree. As a medical student, he did research on hypnotism, digestive tract fluids, and the function of the nerves and muscles in the presence of **pain**. He quickly went on to obtain his degree as a doctor of science in 1878. In his doctoral thesis, Richet showed that various forms of animal and marine life contain stomach hydrochloric acid. He also found the presence of a form of **lactic acid** in the human stomach. In that same year, he was appointed to the medical faculty of the University of Paris. With this appointment Richet focused his attention on the different ways the muscles contract.

After doing research in 1883 on heat maintenance in warm-blooded animals and the distribution of **bacteria** in the body fluids (an outgrowth of his work on the digestive system), in 1887 Richet began to work on the problem of creating a serum that could protect an animal against specific diseases. He followed the work of Louis Pasteur, who in 1880 found a way of protecting chickens from coming down with fowl cholera by injecting them with a weak form of the cholera microbe. The injection of the serum containing the weakened forms of the microbes created an antidote in the body that could then later fight off an invasion from a stronger force of the microbes. The injected serum contained the antigen, and the body receiving the injection produced the antibody.

Richet did extensive work in the development of techniques for immunization with his collaborator, Jules Hericourt. Over a ten-year period, Richet and Hericourt tried to develop a serum for tuberculosis, without any success. They were frustrated by the fact that **Emil Behring** had shown positive results for the development of an immunization serum for diphtheria during the same period of time.

In 1902, Richet was drawn to the problem of shock or allergic reactions in people after they received inoculations of disease-fighting serum. He noticed that some animals that had received a dosage of immunization serum would go into fatal shock when a second shot was administered. He found that the antibody produced by the first shot did not protect the animals against the second shot. The animal was now in a state of hypersensitivity caused by the production of too many antibodies against the foreign intruder. Richet called this condition anaphylaxis, a Greek word that means overprotection.

By 1906, the word allergy had been introduced to describe a wide range of adverse reactions to the use of antiserums and later antibiotics. The term also came into use to describe reactions to plants, animals, foods, chemicals, and many other substances. These substances fall into the category of **antigens**, meaning foreign substances that cause the immune system to produce antibodies, and they could therefore be understood in terms of Richet's concept of anaphylaxis. Richet was, therefore, a pioneer in the field of medicine dealing with the prevention and treatment of **allergies**.

For his development of the concept of anaphylaxis, Richet won the 1913 Nobel Prize. In his acceptance speech for the Nobel Prize, Richet acknowledged the difficulties anaphylaxis causes individuals, but he emphasized its biological significance in insuring the chemical integrity of the species. Such an argument was rooted in his philosophy of biological teleology, a view that maintains that there is a purpose in every biological process for the species concerned.

In 1926, Richet received the Cross of the Legion of Honor from France for his work during World War I studying the problems of **blood plasma** transfusion.

Richet married Amélie Aubry in 1877. The Richets had five sons and two daughters. Richet was also noted for his varied interests in non-scientific activities. He wrote poetry, novels, and plays, and for thirty years, he studied and wrote about hypnotism, parapsychology, telepathy, and extrasensory perception. In 1890, he participated in an early attempt to design an airplane. He was a pacifist who was outspoken on social and political issues, and he wrote on the subject of vivisection. He died in Paris on December 3, 1935.

See also Antigens and antibodies; Immune system; Immunity; Immunology; Shock

RIGOR MORTIS

Rigor mortis (from the Latin for "stiffness of death") is the rigidity that develops in a body after **death**. This rigidity may begin shortly after death—within 10 to 15 minutes—or may not begin until several hours later, depending on the condition of the body at the time of death and on environmental factors, such as moisture content of the air and particularly temperature. A colder temperature promotes a slower onset of rigor mortis.

Typically, rigor mortis affects facial muscles first. Spreading to other parts of the body follows. The body will remain fixed in the rigid position until decomposition of **tissue** begins, about 24 to 48 hours after death.

Rigor mortis occurs because **metabolism** continues in muscles for a short while after death. As part of the metabolic activity, **adenosine triphosphate** (**ATP**) is produced from the metabolism of a sugar compound called **glycogen**. ATP is a principal energy source for muscular activity. So, as long as it is present, muscles continue to maintain their tone. As the store of glycogen is exhausted, ATP can no longer be made and its concentration decreases. One of the consequences of ATP depletion is the formation of abnormal links between two components of muscle tissue, actin and myosin. The leakage of **calcium** into the muscle cells also contributes to the formation of abnormal actin-myosin links. The abnormality produces the stiffening of the muscle, which persists until the links are decomposed.

See also Death and dying

RNA • *see* RIBONUCLEIC ACID (RNA)

RODBELL, MARTIN (1925-1998)
American biochemist

Known for his part in the discovery of G-proteins, Rodbell performed groundbreaking work in cell biology, specifically advancing knowledge regarding how cells communicate. For his work in this area, Rodbell shared the Nobel Prize in physiology or medicine with scientist **Alfred Gilman**.

Rodbell was born on December 1, 1925 in Baltimore, Maryland. He attended a special high school in Baltimore that accepted boys from all over the city and prepared students to enter college as sophomores. The school emphasized languages, and Rodbell thought he might continue his language studies when he entered Johns Hopkins University in 1943. Rodbell, however, eventually became interested in chemistry.

Rodbell served in the Navy during World War II as a radio operator in the Philippine jungles until he contracted **malaria**. When he came back from the war, Rodbell continued his studies at Johns Hopkins, eventually concentrating his studies in **biochemistry**. Rodbell received a B.A. from Johns Hopkins University in 1949.

After graduation, Rodbell married and moved to Seattle to start graduate studies in biochemistry at the University of Seattle. He studied the chemistry of **lipids** (the fatty substances in cells), and his thesis was on the biosynthesis of lecithin (fats found in cell membranes) in the rat **liver**. Unfortunately, his thesis assertions were disproved by another scientist working on the same subject. This experience taught him not to assume that biological chemicals are pure, something that would help him later in his Nobel Prize-winning work.

Rodbell finished his Ph.D. in 1954 and then went to the University of Illinois for his post-doctoral fellowship. His research involved the biosynthesis of chloramphenicol, an antibiotic. After having taught a lecture course to freshman, only a few of whom passed his exams, Rodbell decided that teaching was not his calling. He accepted a position at the National Heart Institute in Bethesda, Maryland, and continued his research into fats, identifying important proteins that pertained to diseases concerning lipoproteins.

In the 1960s, Rodbell returned to his original interest in cell biology and was awarded a fellowship to work at the University of Brussels. There he learned new lab techniques and enjoyed European culture with his family. He returned to the United States and accepted a position at the NIH Institute of Arthritis and Metabolic Diseases in the Nutrition and Endocrinology lab. He developed a simple procedure that would separate and purify **fat** cells. He was also able to remove the fat from a cell, conserving most of the structure of the cell. He named these cells "ghosts."

In several groundbreaking experiments, Rodbell and his colleagues at the NIH showed that cell communication involves three different working devices: (1) a chemical signal; (2) a "second messenger" like a hormone; and (3) a transducer, something that converts energy from one form to

Martin Rodbell. *AP/Wide World Photos. Reproduced by permission.*

another. Rodbell's major contribution was in discovering that there was a transducer function. He and his colleagues also speculated that guanine nucleotides, components of **deoxyribonucleic acid (DNA)** and **ribonucleic acid (RNA)**, were somehow involved in cell communication, something that would later be confirmed by Alfred Goodman, the biochemist with whom he would share the Nobel Prize. Gilman searched for the chemicals involved with guanine nucleotides and discovered the G-proteins.

G-proteins are instrumental in the fundamental workings of a cell. They allow us to see and **smell** by changing light and odors to chemical messages that travel to the **brain**. Understanding how G-proteins malfunction could lead to a better understanding of serious diseases like cholera or **cancer**. Scientists have already linked improperly working G-proteins to diseases like alcoholism and diabetes. Pharmaceutical companies are developing drugs that would focus on G-proteins.

Rodbell served as director of the National Institute of Environmental Health Sciences in Chapel Hill, North Carolina, from 1985 until his retirement in 1994. Ironically, only a few months before receiving the Nobel Award, Rodbell opted for early retirement, because there were no funds to support the research he wanted to do. Upon receiving the Nobel Prize, Rodbell was vocal in his criticism of the government because of its unwillingness to provide adequate support for fundamental

research. He criticized them for favoring projects that yield obviously tangible and potentially profitable results, like drug treatments. Rodbell's other awards include the NIH Distinguished Service Award in 1973 and the Gairdner Award in 1984.

See also Endocrine system and glands

RODS • *see* VISION: HISTOPHYSIOLOGY OF THE EYE

ROSS, RONALD (1857-1932)
English physician and parasitologist

Ronald Ross is best known for his discovery of the method by which **malaria** is transmitted, research for which he was awarded the 1902 Nobel Prize in physiology or medicine. Ross's interest in bacteriology led him to study the causes of malaria, a disease that was widespread in India where he lived. His determination that the affliction was transmitted through a parasite common to mosquitoes led to more advanced treatments for the condition and more effective means of preventing it.

Ross was born in Almora, Nepal, on May 13, 1857. He was the first of ten children to be born to General Sir Campbell Claye Grant Ross, a British officer stationed in India, and the former Matilde Charlotte Elderton. In 1865, at the age of eight, Ross was sent to England for his schooling. When he returned to his family in India, he declared to his father that he wanted to pursue a career in the arts. Ross's true passion was the arts, and he became a doctor only because of his father's insistence. Ross returned to England in 1874 and began his medical education at St. Bartholomew's Hospital in London. He did poorly in his classes because he spent most of his time writing novels and reading. His father became so upset with his grades that he threatened to withdraw his son's financial support. In response, Ross took a job as a ship's doctor on Anchor Line ships plying the London-New York City route. DeKruif reports that Ross spent much of his time aboard ship "observing the emotions and frailties of human nature," which gave him more material for his novels and poems.

In 1879, Ross completed his course at St. Bartholomew's and was awarded his medical degree. He returned to India and held a series of posts in Madras, Bangalore, Burma, and the Andaman Islands. He soon became more interested in research than in the day-to-day responsibilities of medical practice and spent long hours working out new algebraic formulas.

An important turning point in Ross's life came with his first leave of absence in 1888. He returned to England and became interested in research on tropical diseases, many of which he had seen during his years in India. Ross took a course in bacteriology offered by E. Emanuel Klein and earned a diploma in public health. During this furlough, he also met his future wife. Married in 1889, the Rosses later had four children.

With his new found knowledge of bacteriology, Ross turned his attention to what was then the most serious health problem in India: malaria. In 1880 the French physician

Alphonse Laveran had discovered that malaria is caused by a one-celled organism called *Plasmodium*. Two decades of research had produced further data on the organism's characteristics, its means of reproduction, and its correlation with disease symptoms, but no one had determined how the disease was transmitted from one person to another.

Ross's original research led him to question Laveran's discovery, but for five years, he made little progress in his studies. Then, on a second leave of absence in England during 1894, he met Patrick Manson, an English physician particularly interested in malaria. During Ross's year in England, he studied with Manson and became convinced that Laveran's theory was correct and that the causative agent for malaria was transmitted by mosquitoes.

When Ross returned to India in March of 1895, he was prepared to take up an aggressive research program on the mosquito-transmission theory. However, he was frustrated by working conditions in India—especially the lack of support from his superiors and the primitive equipment available to him—but with Manson's constant letters of support and encouragement, he eventually succeeded.

The key discovery came on August 20, 1897, when Ross first observed in the stomach of an *Anopheles* mosquito Anopheles a cyst with black granules of the type described by Laveran. Ross worked out the life cycle of the disease-causing agent, including its reproduction within human **blood**, its transmission to a mosquito during the feeding process, its incubation within the mosquito, and then its transmission to a second human during a second feeding (a "bite") by the mosquito.

Ross's work, however, was complicated by several factors. For example, in the midst of his research he was transferred to Rajputana, a region in which human malaria did not exist. He spent his time there instead working on the transmission of another form of the disease that affects birds. In addition, Ross was continually distracted by his passion for writing, and he produced a number of poems when he could no longer work on his battle against malaria.

Adding to Ross's frustration was the news he received late in 1898 that an Italian research team led by Battista Grassi had published reports on malaria closely paralleling his own work. Although little doubt exists about the originality of the Italian studies, Ross called Grassi's team "cheats and pirates." The dispute was later described by DeKruif as similar to a spat between "two quarrelsome small boys."

To some extent, the dispute was resolved in 1902 when the Nobel Prize committee awarded Ross the year's prize in physiology or medicine. By that time, Ross had retired from the Indian Medical Service and returned to England as lecturer at the new School of Tropical Medicine in Liverpool. There he worked for the eradication of the conditions (such as poor sanitation) that were responsible for the spread of malaria. In 1917, after eighteen years at Liverpool, Ross was appointed physician of tropical diseases at King's College Hospital in London. In 1926, he became director of a new facility founded in his name, the Ross Institute and Hospital for Tropical Diseases near London. He remained in this post until his death on September 16, 1932. Among the honors granted to Ross were the 1895 Parke Gold Medal, the 1901 Cameron Prize,

and the 1909 Royal Medal of the Royal Society. He was knighted in 1911.

See also Bacteria and responses to bacterial infection; Malaria and the physiology of parasitic inflections

ROUS, PEYTON (1879-1970)
American physician

Francis Peyton Rous was a physician-scientist at the Rockefeller Institute for Medical Research (later the Rockefeller University) for over sixty years. In 1966, Rous won the Nobel Prize for his 1910 discovery that a virus can cause **cancer tumors**. His other contributions to scientific medicine include creating the first **blood** bank, determining major functions of the **liver** and gall bladder, and identifying factors that initiate and promote malignancy in normal cells.

Rous was born in Baltimore, Maryland, to Charles Rous, a grain exporter, and Frances Wood, the daughter of a Texas judge. His father died when Rous was eleven, and his mother chose to stay in Baltimore to ensure that her three children would have the best possible education. His sisters were professionally successful, one a musician, the other a painter.

Rous, whose interest in natural science was apparent at an early age, wrote a "flower of the month" column for the *Baltimore Sun*. He pursued his biological interests at Johns Hopkins University, receiving a B.A. in 1900 and an M.D. in 1905. After a medical internship at Johns Hopkins, however, he decided (as recorded in *Les Prix Nobel en 1966*) that he was "unfit to be a real doctor" and chose instead to concentrate on research and the natural history of disease. This led to a full year of studying lymphocytes with Aldred Warthin at the University of Michigan and a summer in Germany learning morbid **anatomy** at a Dresden hospital.

After Rous returned to the United States, he developed pulmonary tuberculosis and spent a year recovering in an Adirondacks sanatorium. In 1909, Simon Flexner, director of the newly founded Rockefeller Institute in New York City, asked Rous to take over cancer research in his laboratory. A few months later, a poultry breeder brought a Plymouth Rock chicken with a large breast tumor to the Institute and Rous, after conducting numerous experiments, determined that the tumor was a spindle-cell sarcoma. When Rous transferred a cell-free filtrate from the tumor into healthy chickens of the same flock, they developed identical tumors. Moreover, after injecting a filtrate from the new tumors into other chickens, a malignancy exactly likes the original formed. Further studies revealed that this filterable agent was a virus, although Rous carefully avoided this word. Now called the Rous sarcoma virus (RSV) and classed as an **RNA** retrovirus, it remains a prototype of animal tumor **viruses** and a favorite laboratory model for studying the role of genes in cancer.

Rous's discovery was received with considerable disbelief, both in the United States and in the rest of the world. His viral theory of cancer challenged all assumptions, going back to **Hippocrates**, that cancer was not infectious but rather a spontaneous, uncontrolled growth of cells and many scientists dismissed his finding as a disease peculiar to chickens. Discouraged by his failed attempts to cultivate viruses from mammal cancers, Rous abandoned work on the sarcoma in 1915. Nearly two decades passed before he returned to cancer research.

After the onset of World War I, Rous, J. R. Turner, and O. H. Robertson began a search for emergency blood transfusion fluids. Nothing could be found that worked without red blood corpuscles so they developed a citrate-sugar solution that preserved blood for weeks as well as a method to transfuse the suspended cells. Later, behind the front lines in Belgium and France, they created the world's first blood bank from donations by army personnel. This solution was used again in World War II, when half a million Rous-Turner blood units were shipped by air to London during the Blitz.

During the 1920s, Rous made several contributions to **physiology**. With P. D. McMaster, Rous demonstrated the concentrating activity of **bile** in the gall bladder, the acid-alkaline balance in living tissues, the increasing permeability along **capillaries** in muscle and skin, and the nature of gallstone formation. In conducting these studies, Rous devised culture techniques that have become standard for studying living tissues in the laboratory. He originated the method for growing viruses on chicken embryos, now used on a mass scale for producing viral vaccines, and found a way to isolate single cells from solid tissues by using the enzyme trypsin. Moreover, Rous developed an ingenious method for obtaining pure cultures of Kupffer cells by taking advantage of their phagocytic ability; he injected iron particles in animals and then used a magnet to separate these iron-laden liver cells from suspensions.

In 1933, a Rockefeller colleague's report stimulated Rous to renew his work on cancer. Richard Shope discovered a virus that caused warts on the skin of wild rabbits. Within a year, Rous established that this papilloma had characteristics of a true tumor. His work on mammalian cancer kept his viral theory of cancer alive. However, another twenty years passed before scientists identified viruses that cause human cancers and learned that viruses act by invading genes of normal cells. These findings finally advanced Rous's 1910 discovery to a dominant place in cancer research.

Meanwhile, Rous and his colleagues spent three decades studying the Shope papilloma in an effort to understand the role of viruses in causing cancer in mammals.

Careful observations, over long periods of time, of the changing shapes, colors, and sizes of cells revealed that normal cells become malignant in progressive steps. Cell changes in tumors were observed as always evolving in a single direction toward malignancy.

The researchers demonstrated how viruses collaborate with carcinogens such as tar, radiation, or chemicals to elicit and enhance tumors. In a report co-authored by W. F. Friedewald, Rous proposed a two-stage mechanism of carcinogenesis. He further explained that a virus can be induced by carcinogens, or it can hasten the growth and transform benign tumors into cancerous ones. For tumors having no apparent trace of virus, Rous cautiously postulated that these spontaneous growths might contain a virus that persists in a masked or latent state, causing no harm until its cellular environment is disturbed.

Rous maintained a rigorous workday schedule at Rockefeller. His meticulous editing and writing, both scientific and literary, took place during several hours of solitude at the beginning and end of each day. At midday, he spent two intense hours discussing science with colleagues in the Institute's dining room. Rous then returned to work in his laboratory on experiments that often lasted into the early evening.

Rous was appointed a full member of the Rockefeller Institute in 1920 and member emeritus in 1945. Though officially retired, he remained active at his lab bench until the age of ninety, adding sixty papers to the nearly three hundred he published. He was elected to the National Academy of Sciences in 1927, the American Philosophical Society in 1939, and the Royal Society in 1940. In addition to the 1966 Nobel Prize in physiology or medicine, Rous received many honorary degrees and awards for his work in viral oncology, including the 1956 Kovalenko Medal of the National Academy of Sciences, the 1958 Lasker Award of the American Public Health Association, and the 1966 National Medal of Science.

As editor of the *Journal of Experimental Medicine,* a periodical renowned for its precise language and scientific excellence, Rous dominated the recording of forty-eight years of American medical research. He died in New York City, just six weeks after he retired as editor.

See also Viruses and responses to viral infection

S

S-A NODE

The S-A node or sino-atrial node (also spelled unhyphenated as the sinoatrial node) is a specialized area of **tissue** located along the upper margin of the right atrium of the **heart** that acts to regulate the contractions of the heart. The S-A node is often referred to as the pacemaker of the **cardiac cycle**.

It is important to note that the signals from the S-A node are not required in order for cardiac muscle to contract. Cardiac muscle is inherently contractile (can contract on its own). S-A node contractions serve only to stimulate and coordinate cardiac muscular contractions. In effect, the S-A node is the cardiac rhythm regulator.

The S-A node is composed of Purkinje fibers that spontaneously contract. This contraction generates a nerve impulse that then propagates (travels) throughout the cardiac tissue of the heart and results in the contraction of both atria. The contraction of the right atria forces **blood** from the right atrium into the right ventricle where it is then pumped into the pulmonary **circulatory system**. The contraction of the left atrium forces blood into the left ventricle that then pumps the blood into the **systemic circulation**.

Along with the atrioventricular node (A-V node), the S-A node is an area of specialized tissue (nodal tissue) located in the cardiac subendothelium that has the capacity to functionally act as both muscular and neural tissue. The contraction of nodal tissue generates nerve impulses. The A-V nodal tissue, also a part of the **Purkinje system**, serves to dampen or delay the signal to contract from the S-A node so that the atria have time to fully contract and empty their contents into their respective ventricles.

See also Coronary circulation; Heart defects; Heart, embryonic development and changes at birth; Heart, rhythm control and impulse conduction; Nerve impulses and conduction of impulses

SABIN, FLORENCE (1871-1953)
American physician

Florence Sabin's studies of the **central nervous system** of newborn infants, the origin of the lymphatic system, and the body's responses to infections—especially by the bacterium that causes tuberculosis—carved an important niche for her in the annals of science. In addition to her research at Johns Hopkins School of Medicine and Rockefeller University, she taught new generations of scientists and thus, extended her intellectual reach far beyond her own life. In addition, Sabin's later work as a public health administrator left a permanent imprint upon the communities in which she served. Some of the firsts achieved by Sabin include becoming the first woman faculty member at Johns Hopkins School of Medicine, as well as its first female full professor, and the first woman to be elected president of the American Association of Anatomists.

Sabin was born Florence Rena Sabin in Central City, Colorado, to George Kimball Sabin, a mining engineer and son of a country doctor, and Serena Miner, a teacher. Her early life, like that of many in that era, was spare: the house where she lived with her parents and older sister Mary had no plumbing, no gas and no electricity. When Sabin was four, the family moved to Denver; three years later her mother died.

After attending Wolfe Hall boarding school for a year, the Sabin daughters moved with their father to Lake Forest, Illinois, where they lived with their father's brother, Albert Sabin. There the girls attended a private school for two years and spent their summer vacations at their grandfather Sabin's farm near Saxtons River, Vermont.

Sabin graduated from Vermont Academy boarding school in Saxtons River and joined her older sister at Smith College in Massachusetts, where they lived in a private house near the school. As a college student, Sabin was particularly interested in mathematics and science, and earned a bachelor of science in 1893. During her college years she tutored other students in mathematics, thus beginning her long career in teaching.

A course in zoology during her junior year at Smith ignited a passion for biology, which she made her specialty. Determined to demonstrate that, despite widespread opinion to the contrary, an educated woman was as competent as an educated man, Sabin proceeded to chose medicine as her career. This decision may have been influenced by events occurring in Baltimore at the time.

The opening of Johns Hopkins Medical School in Baltimore was delayed for lack of funds until a group of prominent local women raised enough money to support the institution. In return for their efforts, they insisted that women be admitted to the school—a radical idea at a time when women who wanted to be physicians generally had to attend women's medical colleges.

In 1893, the Johns Hopkins School of Medicine welcomed its first class of medical students; but Sabin, lacking tuition for four years of medical school, moved to Denver to teach mathematics at Wolfe Hall, her old school. Two years later she became an assistant in the biology department at Smith College, and in the summer of 1896 she worked in the Marine Biological Laboratories at Woods Hole. In October of 1896, she was finally able to begin her first year at Johns Hopkins.

While at Johns Hopkins, Sabin began a long professional relationship with Dr. Franklin P. Mall, the school's professor of **anatomy**. During the four years she was a student there and the fifteen years she was on his staff, Mall exerted an enormous influence over her intellectual **growth and development** into prominent scientist and teacher. Years after Mall's death, Sabin paid tribute to her mentor by writing his biography, *Franklin Paine Mall: The Story of a Mind.*

Sabin thrived under Mall's tutelage, and while still a student she constructed models of the medulla and mid-brain from serial microscopic sections of a newborn baby's nervous system. For many years, several medical schools used reproductions of these models to instruct their students. A year after her graduation from medical school in 1900, Sabin published her first book based on this work, *An Atlas of the Medulla and Midbrain,* which became one of her major contributions to medical literature, according to many of her colleagues.

After medical school, Sabin was accepted as an intern at Johns Hopkins Hospital, a rare occurrence for a woman at that time. Nevertheless, she concluded during her internship that she preferred research and teaching to practicing medicine. However, her teaching ambitions were nearly foiled by the lack of available staff positions for women at Johns Hopkins. Fortunately, with the help of Mall and the women of Baltimore who had raised money to open the school, a fellowship was created in the department of anatomy for her. Thus began a long fruitful period of work in a new field of research, the embryologic development of the human lymphatic system.

Sabin began her studies of the lymphatic system to settle controversy over how it developed. Some researchers believed the vessels that made up the lymphatics formed independently from the vessels of the **circulatory system**, specifically the **veins**. However, a minority of scientists believed that the lymphatic vessels arose from the veins themselves, budding outward as continuous channels. The studies that supported this latter view were done on pig embryos that were already so large (about 3.54 in [90mm] in length) that many researchers—Sabin included—pointed out that the embryos were already old enough to be considered an adult form, thus the results were inconclusive.

The young Johns Hopkins researcher set out to settle the lymphatic argument by studying pig embryos as small as 0.91 in (23 mm) in length. Combining the painstaking techniques of injecting the microscopic vessels with dye or ink and reconstructing the three-dimensional system from two-dimensional cross sections, Sabin demonstrated that lymphatics did in fact arise from veins by sprouts of endothelium (the layer of cells lining the vessels). Furthermore, these sprouts connected with each other as they grew outward, so the lymphatic system eventually developed entirely from existing vessels. In addition, she demonstrated that the peripheral ends (those ends furthest away from the center of the body) of the lymphatic vessels were closed and, contrary to the prevailing opinion, were neither open to **tissue** spaces nor derived from them. Even after her results were confirmed by others they remained controversial. Nevertheless, Sabin firmly defended her work in her book *The Origin and Development of the Lymphatic System.*

Sabin's first papers on the lymphatics won the 1903 prize of the Naples Table Association, an organization that maintained a research position for women at the Zoological Station in Naples, Italy. The prize was awarded to women who produced the best scientific thesis based on independent laboratory research.

Back at Hopkins from her year abroad, she continued her work in anatomy and became an associate professor of anatomy in 1905. Her work on lymphatics led her to studies of the development of **blood** vessels and blood cells. In 1917, she was appointed professor of **histology**, the first woman to be awarded full professorship at the medical school. During this period of her life, she enjoyed frequent trips to Europe to conduct research in major German university laboratories.

After returning to the United States from one of her trips abroad, she developed methods of staining living cells, enabling her to differentiate between various cells that had previously been indistinguishable. She also used the newly devised "hanging drop" technique to observe living cells in liquid preparations under the microscope. With these techniques she studied the development of blood vessels and blood cells in developing organisms—once she stayed up all night to watch the "birth" of the bloodstream in a developing chick embryo. Her diligent observation enabled her to witness the formation of blood vessels as well as the formation of **stem cells** from which all other red and white blood cells arose. During these observations, she also witnessed the **heart** make its first beat.

Sabin's technical expertise in the laboratory permitted her to distinguish between various blood cell types. She was particularly interested in white blood cells called monocytes, which attacked infectious **bacteria**, such as *Mycobacterium tuberculosis,* the organism that causes tuberculosis. Although this organism was discovered by the German microbiologist **Robert Koch** during the previous century, the disease was still a dreaded health menace in the early twentieth century. The National Tuberculosis Association acknowledged the importance of Sabin's research of the body's immune response to the tuberculosis organism by awarding her a grant to support her work in 1924.

In that same year, she was elected president of the American Association of Anatomists, and the following year Sabin became the first woman elected to membership in the National Academy of Sciences. These honors followed her 1921 speech to American women scientists at Carnegie Hall during a reception for Nobel Prize-winning physicist Marie Curie, an event that signified Sabin's recognized importance in the world of science.

Although her research garnered many honors, Sabin continued to relish her role as a professor at Johns Hopkins. The classes she taught in the department of anatomy enabled her to influence many first-year students—a significant number of whom participated in her research over the years. She also encouraged close teacher-student relationships and frequently hosted gatherings at her home for them.

One of her most cherished causes was the advancement of equal rights for women in education, employment, and society in general. Sabin considered herself equal to her male colleagues and frequently voiced her support for educational opportunities for women in the speeches she made upon receiving awards and honorary degrees. Her civic-mindedness extended to the political arena where she was an active suffragist and contributor to the Maryland *Suffrage News* in the 1920s.

Sabin's career at Johns Hopkins drew to a close in 1925, eight years after the death of her close friend and mentor Franklin Mall. She had been passed over for the position of professor of anatomy and head of the department, which was given to one of her former students. Thus, she stepped down from her position as professor of histology and left Baltimore.

In her next position, Sabin continued her study of the role of monocytes in the body's defense against the tubercle bacterium that causes tuberculosis. In the fall of 1925, Sabin assumed a position as full member of the scientific staff at the Rockefeller Institute for Medical Research (now Rockefeller University) in New York City at the invitation of the institute's director, Simon Flexner. At Rockefeller Sabin continued to study the role of monocytes and other white blood cells in the body's immune response to infections. She became a member of the Research Committee of the National Tuberculosis Association and aspired to popularize tuberculosis research throughout Rockefeller, various pharmaceutical companies, and other universities and research institutes. The discoveries that she and her colleagues made concerning the ways in which the responded to tuberculosis led her to her final research project: the study of antibody formation.

Meanwhile, she continued to accrue honors. She received fourteen honorary doctorates of science from various universities, as well as a doctor of laws. *Good Housekeeping* magazine announced in 1931 that Sabin had been selected in their nationwide poll as one of the twelve most eminent

women in the country. In 1935, she received the M. Carey Thomas prize in science, an award of $5,000 presented at the fiftieth anniversary of Bryn Mawr College. Among her many other awards was the Trudeau Medal of the National Tuberculosis Association (1945), the Lasker Award of the American Public Health Association (1951), and the dedication of the Florence R. Sabin Building for Research in Cellular Biology, at the University of Colorado Medical Center.

In 1938, Sabin retired from Rockefeller and moved to Denver to live with her older sister, Mary, a retired high school mathematics teacher. She returned to New York at least once a year to fulfill her duties as a member of both the advisory board of the John Simon Guggenheim Memorial Foundation and the advisory committee of United China Relief.

Sabin quickly became active in public health issues in Denver and was appointed to the board of directors of the Children's Hospital in 1942, where she later served as vice president. During this time she became aware of the lack of proper enforcement of Colorado's primitive public health laws and began advocating for improved conditions. Governor John Vivian appointed her to his Post-War Planning Committee in 1945, and she assumed the chair of a subcommittee on public health called the Sabin Committee. In this capacity she fought for improved public health laws and construction of more health care facilities.

Two years later she was appointed manager of the Denver Department of Health and Welfare, donating her salary of $4,000 to the University of Colorado Medical School for Research. She became chair of Denver's newly formed Board of Health and Hospitals in 1951 and served for two years in that position. Her unflagging enthusiasm for public health issues bore significant fruit. A *Rocky Mountain News* reporter stated that "Dr. Sabin... was the force and spirit behind the Tri-County chest x-ray campaign" that contributed to cutting the death rate from tuberculosis by fifty percent in Denver in just two years.

But Sabin's enormous reserve of energy flagged under the strain of caring for her ailing sister. While recovering from her own illness, Sabin sat down to watch a World Series game on October 3, 1953, in which her favorite team, the Brooklyn Dodgers, was playing. She died of a heart attack before the game was over.

The state of Colorado gave Sabin a final posthumous honor by installing a bronze statue of her in the National Statuary Hall in the Capitol in Washington, D.C., where each state is permitted to honor two of its most revered citizens. Upon her death, as quoted in *Biographical Memoirs,* the Denver *Post* called her the "First Lady of American Science."

See also Antigens and antibodies; Immune system; Immunity, cell mediated; Immunity, humoral regulation; Immunology

SACRUM · *see* VERTEBRAL COLUMN

SAGITTAL PLANE · *see* ANATOMICAL NOMENCLATURE

SAKMANN, BERT (1942-)
German physician and cell physiologist

Bert Sakmann, along with physicist **Erwin Neher**, was awarded the 1991 Nobel Prize in physiology or medicine for inventing the patch clamp technique. The technique made it possible to realize a goal that had eluded scientists since the 1950s: to be able to examine individual ion channels—pore-forming proteins found in the outer membranes of virtually all cells that serve as conduits for electrical signals. Introduced in 1976, the patch clamp technique opened new paths in the study of membrane physiology. Since then, researchers throughout the world have adapted and refined patch clamping, contributing significantly to research on problems in medicine and neuroscience. The Nobel Committee credited Sakmann and Neher with having revolutionized modern biology.

Sakmann was born in Stuttgart, Germany, on June 12, 1942. His later education involved much time around the laboratory. From 1969 to 1970, he was a research assistant in the department of neurophysiology at the Max Planck Institute for Psychiatry in Munich. Between 1971 and 1973, Sakmann studied biophysics with Nobel Laureate **Bernard Katz** at University College in London as a British Council scholar. In 1974 he received his medical degree from the University of Göttingen. From that year until 1979 he was a research associate in the department of neurobiology at the Max Planck Institute for Biophysical Chemistry in Göttingen.

In the 1950s and 1960s, the existence of ion channels that allow for the transmission of electrical charges from one cell to another was inferred from research since no one had been able to actually locate the sites of these channels. Cell physiologists were being drawn to the question of how electrically charged ions control such biological functions as the transmission of nerve impulses, the contraction of muscles, vision, and the process of conception. Sakmann's early interest in ion channels was stimulated by two papers published in 1969 and 1970 that gave strong evidence for the existence of ion channels. As stronger evidence began to accumulate for their existence, it became clear to Sakmann and Neher, who were sharing laboratory space at the Max Planck Institute, that they would have to develop a fine instrument to be able to locate the actual sites of the ion channels on the cell membrane.

Bedeviling efforts of researchers to that point was the electrical "noise" generated by the cell's membrane, which made it impossible to detect signals coming from individual channels. Sakmann and Neher set about to reduce the noise by shutting out most of the membrane. They applied a glass micropipette one micron wide and fitted with a recording electrode to a cell membrane and were able to measure the flow of current through a single channel. "It worked the first time," Sakmann recalled in *Science* magazine. The biophysical community was exultant.

Over the next few years, Sakmann and Neher refined their "patch clamp" technique, solving a residual noise problem caused by leaks in the seals between pipette and cell by applying suction with freshly made and fire-polished pipettes. The refinements made it possible to measure even very small

currents, and established the patch clamp as a tremendously versatile tool in the field of cell biology. Patch clamping has been instrumental in studies of cystic fibrosis, hormone regulation, and insulin production in diabetes. The technique has also made possible the development of new drugs in the treatment of **heart** disease, epilepsy, and disorders affecting the nervous and muscle systems. In 1991, Sakmann and Neher won the 1991 Nobel Prize in physiology or medicine for their work on ion channels. The Nobel Awards citation congratulated the researchers for conclusively establishing the existence and function of the channels, and contributing immeasurably to the understanding of disease mechanisms.

Sakmann has continued to work with other research teams, altering the genes for identified ion channels in order to trace the molecules in the channel responsible for opening and closing the ion pore. Even though Sakmann expressed surprise at receiving the Noble Prize, given all the other important work going on in cell physiology, the opinion of many of his colleagues was that the award was long overdue.

In 1989, Sakmann moved from the Max Planck Institute in Göttingen to the University of Heidelberg as a professor on the medical faculty. Among his other awards are the Spencer and Louisa Gross-Horwitz Awards from Columbia University in 1983 and 1986, respectively.

Sakmann is married to an ophthalmologist; they have three children.

See also Action potential; Cell membrane transport; Electrolytes and electrolyte balance; Nerve impulses and conduction of impulses

SALIVA

Saliva is the principal secretion of the mouth and contains **enzymes** that play an important role in **digestion**. Saliva also lubricates the mouth and upper digestive tract. Contact with saliva assures that food is softer and moister, and therefore, more able to be swallowed and less irritating to the esophageal mucosa (the lining of the esophagus). The secretion of saliva is under the control of the **autonomic nervous system**. Salivation controlled by parasympathetic stimulation from the **brain**, was demonstrated first in dogs by Russian physiologist **Ivan Petrovich Pavlov**.

In contrast to the acidic contents of the stomach, saliva is alkaline, and provides a protective coating from acid reflux from the stomach. Because saliva contains an antibacterial lysozyme that lyses **bacteria** (ruptures bacterial cells), adequate amounts of saliva in the mouth also reduce the amount of bacteria in the oral cavity.

Saliva is secreted by a number of glands including the salivary glands that include mucous glands, parotid, submaxillary (mandibular), and sublingual glands. More specifically, saliva is secreted from specialized clusters of cells termed acini.

The formation of saliva is a multi-step process. Initially formed of an aqueous solution (water based solution) of **electrolytes**, proteins (mostly enzymes), and **mucus**, saliva undergoes several chemical changes before it is release from the glandular collecting ducts into the oral cavity. The sodium content is reduced and potassium levels increase along with the addition of bicarbonate ions that make the saliva alkaline.

Depending on their particular histophysiology, the paired salivary glands each produce subtle variations on the compositional mixture of the components of saliva. The differences depend on the amount of serous or mucosal cells present in each gland.

Adequate amounts of saliva are also needed to facilitate **taste** sensations because moist substances provide greater amounts of soluble molecules that can bind to taste receptors.

See also Acid-base balance; Mammary glands and lactation; Mastication; Taste, physiology of gustatory structures

SAMUELSSON, BENGT INGEMAR (1934-   )
Swedish biochemist

Bengt Ingemar Samuelsson shared the 1982 Nobel Prize in physiology or medicine with his compatriot Sune K. Bergström and British biochemist John R. Vane "for their discoveries concerning **prostaglandins** and related biologically active substances." Because prostaglandins are involved in a diverse range of biochemical functions and processes, the research of Bergström, Samuelsson, and Vane opened up a new arena of medical research and pharmaceutical applications.

Samuelsson was born on May 21, 1934, in Halmstad, Sweden, to Anders and Kristina Nilsson Samuelsson. Samuelsson entered medical school at the University of Lund, where he came under the mentorship of Sune K. Bergström. Renowned for his work on prostaglandin chemistry, Bergström was on the university faculty as professor of physiological chemistry. In 1958, Samuelsson followed Bergström to the prestigious Karolinska Institute in Stockholm. There, Samuelsson received his doctorate in medical science in 1960 and his medical degree in 1961, and he was subsequently appointed as an assistant professor of medical chemistry. In 1961, he served as a research fellow at Harvard University, and then in 1962 he rejoined Bergström at the Karolinska Institute, where he remained until 1966.

At the Karolinska Institute, Samuelsson worked with a group of researchers who were trying to characterize the structures of prostaglandins. Prostaglandins are hormone-like substances found throughout the body, which were so named in the 1930s on the erroneous assumption that they originated in the prostate. They play an important role in the **circulatory system**, and they help protect the body against sickness, **infection**, **pain**, and stress. Expanding on their earlier research, Bergström, Samuelsson, and other researchers discovered the role that arachidonic acid, an unsaturated fatty acid found in meats and vegetable oils, plays in the formation of prostaglandins. By developing synthetic methods of producing prostaglandins in the laboratory, this group made prostaglandins accessible for scientific research worldwide. It was Samuelsson who discovered the process through which arachidonic acid is converted

into compounds he named endoperoxides, which are in turn converted into prostaglandins.

Prostaglandins have many veterinary and livestock breeding applications, and Samuelsson joined the faculty of the Royal Veterinary College in Stockholm in 1967. He returned to the Karolinska Institute as professor of medicine and physiological chemistry in 1972. Samuelsson served as the chair of the department of physiological chemistry from 1973 to 1983, and as dean of the medical faculty from 1978 to 1983, combining administrative duties with a rigorous research schedule. During 1976 and 1977, Samuelsson also served as a visiting professor at Harvard University and the Massachusetts Institute of Technology.

During these years, Samuelsson continued his investigation of prostaglandins and related compounds. In 1973, he discovered the prostaglandins that are involved in the clotting of the **blood**; he called these thromboxanes. Samuelsson subsequently discovered the compounds he called leukotrienes, which are found in white blood cells (or **leukocytes**). Leukotrienes are involved in asthma and in anaphylaxis, the **shock** or hypersensitivity that follows exposure to certain foreign substances, such as the toxins in an insect sting.

In the wake of such research, prostaglandins have been used to treat fertility problems, circulatory problems, asthma, arthritis, menstrual cramps, and ulcers. Prostaglandins have also been used medically to induce abortions. As noted by *New Scientist* magazine, the 1982 Nobel Prize shared by Bergström, Samuelsson, and Vane acknowledged that they had "carried prostaglandins from the backwaters of biochemical research to the frontier of medical applications." In 1983, succeeding Bergström, Samuelsson was appointed as president of the Karolinska Institute.

The importance of Samuelsson's research has been recognized by numerous awards and honors in addition to the Nobel Prize. Such acknowledgments include the A. Jahres Award in medicine from Oslo University in 1970; the Louisa Gross Horwitz Prize from Columbia University in 1975; the Albert Lasker Medical Research Award in 1977; the Ciba-Geigy Drew Award for biomedical research in 1980; the Gairdner Foundation Award in 1981; the Bror Holberg Medal of the Swedish Chemical Society in 1982; and the Abraham White Distinguished Scientist Award in 1991. Samuelsson has published widely on the **biochemistry** of prostaglandins, thromboxanes, and leukotrienes.

See also Hormones and hormone action; Inflammation of tissues

SCHALLY, ANDREW V. (1926-   )
Polish-born American biochemist

Andrew V. Schally conducted pioneering research concerning **hormones**, identifying three **brain** hormones and greatly advancing scientists' understanding of the function and interaction of the brain with the rest of the body. His findings have proved useful in the treatment of diabetes and peptic ulcers, and in the diagnosis and treatment of hormone-deficiency diseases. Schally shared the 1977 Nobel Prize with French-born American endocrinologist **Roger Guillemin** and Rosalyn Yalow (an American scientist whose work in the discovery and development of radioimmunoassay—the use of radioactive substances to find and measure minute substances, especially hormones in **blood** and tissue—helped Schally and Guillemin isolate and analyze peptide hormones).

Schally was born on November 30, 1926, in Wilno, Poland, to Casimir Peter Schally and Maria Lacka Schally. His father served in the military on the side of the Allies during World War II, and Schally grew up during Nazi occupation of his homeland. The family later left Poland and immigrated to Scotland, where Schally entered the Bridge Allen School in Scotland. He studied chemistry at the University of London and obtained his first research position at London's highly regarded National Institute for Medical Research. Leaving London for Montreal, Canada, in 1952, Schally entered McGill University, where he studied endocrinology and conducted research on the adrenal and pituitary glands. He obtained his doctorate in **biochemistry** from McGill in 1957. Also, in 1957, Schally became an assistant professor of **physiology** at Baylor University School of Medicine in Houston, Texas. There he was able to pursue his interest in the hormones produced by the **hypothalamus.**

Scientists had long thought that the hypothalamus, a part of the brain located just above the pituitary gland, regulated the **endocrine system**, which includes the pituitary, thyroid and adrenal glands, the **pancreas**, and the ovaries and testicles. They were, however, unsure of the way in which hypothalamic hormonal regulation occurred. In the 1930s British anatomist Geoffrey W. Harris theorized that hypothalamic regulation occurred by means of hormones, chemical substances secreted by glands and transported by the blood. Harris was able to support his hypothesis by conducting experiments that demonstrated altered pituitary function when the blood vessels between the hypothalamus and the pituitary were cut. Harris and others were unable to isolate or identify the hormones from the hypothalamus.

Schally devoted his work to identifying these hormones. He and Roger Guillemin, who also worked at Baylor University's School of Medicine, were engaged in research to unmask the chemical structure of corticotropin-releasing hormone (CRH). Their efforts, however, were unsuccessful—the structure was not determined until 1981. The two then focused their work, independently, on other hormones of the hypothalamus. Schally left Baylor in 1962, when he became director of the Endocrine and Polypeptide Laboratory at the Veterans Administration (VA) Hospital in New Orleans, Louisiana. Also that year, Schally became a United States citizen and took on the post of assistant professor of medicine at Tulane University Medical School.

Schally's first breakthrough came in 1966 when he and his research group isolated TRH, or thyrotropin-releasing hormone. In 1969 Schally and his VA team demonstrated that TRH is a peptide containing three **amino acids**. It was Guillemin, though, who first determined TRH's chemical structure. The success of this research made it possible to decipher the func-

tion of a second hormone, called luteinizing-hormone releasing factor (LHRH). Identified in 1971, LHRH is a decapeptide and controls reproductive functions in both males and females. The chemical makeup of the growth-releasing hormone (GRH) was also discovered by Schally's team in 1971. Schally was able to show that GRH, a peptide consisting of ten amino acids, causes the release of gonadotropins from the pituitary gland. These gonadotropins, in turn, cause male and female **sex hormones** to be released from the testicles and ovaries. In conjunction with this, Schally was able to identify a factor that inhibits the release of GRH in 1976. Guillemin, however, had determined its structure earlier and named it somatostatin. Subsequent studies by Schally showed that somatostatin serves multiple roles, some of which relate to insulin production and growth disorders. This led to speculation that the hormone could be useful for treating diabetes and acromegaly, a growth-disorder disease.

The hormone research done by Schally and his colleagues was tedious and expensive. Thousands of sheep and pig hypothalami were required to extract the smallest amount of hormone. These **organs** were solicited from many area slaughterhouses and required immediate dissection to prevent the hormones from degrading. Their accomplishment of isolating the first milligram of pure thyrotropin-releasing hormone, Guillemin stated, cost many times more than the NASA space mission that brought a kilogram of moon rock back to earth.

Schally's intense years of hard work and accomplishment were capped by the Nobel Prize, but he has also received many other awards and honors. In 1974, Schally received the Charles Mickle Award of the University of Toronto, and the Gairdner Foundation International Award. He received the Borden Award in the Medical Sciences of the Association of American Medical Colleges in 1975 and, that same year, the Lasker Award, and the Laude Award. He has held memberships in the National Academy of Sciences, the American Society of Biological Chemists, the American Physiology Society, the American Association for the Advancement of Science, and the Endocrine Society. In the years prior to receiving the Nobel Prize, Schally and his colleagues published more than 850 papers. Married to Brazilian endocrinologist, Ana Maria de Medeiros-Comaru, Schally often lectures in Latin America and Spain.

See also Adrenal glands and hormones; Endocrine system and glands; Hormones and hormone action; Ovarian cycle and hormonal regulation; Parathyroid glands and hormones; Pituitary gland and hormones; Thyroid histophysiology and hormones

SCHLEIDEN, MATTHIAS JACOB (1804-1881)

German botanist

Matthias Schleiden was first to recognize the importance of cells as fundamental units of life. Schleiden made other accurate observations about plant cells and cell activity and his conclusions marked one of the important landmarks in the rise of modern cytology. In 1839, **Theodor Schwann** would expand Schleiden's **cell theory** to include the animal world, establishing cell theory as the fundamental concept in biology. Schwann (first to articulate that cells—one type of which are now known as Schwann cells—comprise the nerve sheath) and Schleiden published an 1839 text, Microscopical Researches, that proved a pivotal and influential argument for the advancement of cell theory.

Schleiden described Robert Brown's 1832 discovery of the cell nucleus (which he renamed cytoblast). Schleiden argued that the cell nucleus must somehow be connected with **cell division**, but he mistakenly asserted that new cells erupted from the nuclear surface like blisters.

Schleiden did not originally pursue his interest in botany; instead, he studied law at Heidelberg University from 1824 to 1827. After graduation, Schleiden became a barrister in Hamburg, Germany, but he soon grew dissatisfied with his legal practice.. He abandoned the profession altogether in 1831 and returned to college to pursue his real interests—botany and medicine. After graduation, Schleiden became professor of botany at Jena University. Instead of spending his time classifying plants, however, he preferred to observe their development using the microscope because he argued that was the only way plants could be studied. By 1838, his methods led him to propose the cell theory for plants.

Schleiden's approach to educating students was very different and his social, political, and philosophical views often put him at odds with other scientists. However, his great abilities and his introduction of improved techniques earned him the title "reformer of scientific botany."

See also Cell cycle and cell division; Cell differentiation; Cell membrane transport

SCHWANN, THEODOR AMBROSE HUBERT (1810-1882)
German biologist

Theodor Schwann was a pioneer of **cell theory** and the first scientists to articulate that cells—one type of which are now known as Schwann cells—comprise the nerve sheath.

Schwann was a shy child, preferring to absorb himself in his studies, his family, and religion rather than deal with the outside world. He left his hometown to attend the Jesuit College in Cologne, Germany, in 1826. Schwann eventually gave up theology to study medicine at the University of Bonn, where he enrolled in premedical studies and graduated in 1831.

While in Berlin, Germany, Schwann attended clinical demonstrations of **Johannes Müller** and began preparing a dissertation under Müller's guidance. Schwann's dissertation, *De necessitate aeris atmosph'rici ad evolutionem pulli in ovo incubato*, involved a study of the **breathing** of the embryo in a hen's egg. After his graduation, Schwann immediately became Müller's assistant and devoted his time to research. Around 1834, Schwann became interested in digestive processes. Within two years, he had isolated a chemical (called pepsin) from the stomach lining that he thought was responsible for the **digestion**

of protein. This discovery was also significant because it was the first time an enzyme had ever been isolated from animal **tissue**.

Schwann then turned to researching fermentation in an attempt to disprove the theory of spontaneous generation, which at that time was enjoying resurgence in the German scientific community. In the course of his experiments, Schwann proved that yeast consists of tiny plant-like organisms and that the fermentation of sugar is a result of the physiological processes of these living yeast cells. Schwann later coined the term **metabolism** to describe the chemical changes that occur in living tissue. Schwann's contemporaries, however, ridiculed his findings, which were not confirmed until nearly thirty years later by Louis Pasteur. Their harsh criticisms eventually prompted Schwann to leave Germany for Belgium, where he became a professor at the universities of Liège and Louvain.

In 1839, one year after **Matthias Schleiden** presented his cell theory regarding plants, Schwann, who was familiar with Schleiden's work, published a more precise version of the theory that he extended to animals. He had realized that the small units containing a nucleus that he had observed while assisting Müller were the animal equivalents of the plant cells Schleiden had observed. Schwann noted that the fertilized egg from which an animal grows is a single cell that contains a nucleus and is surrounded by a membrane—much like the nucleus and membrane of animal tissues. Previously, no one thought an organism was composed solely of cells, although cells had been observed in plants and animals under the microscope before Schwann's explanation.

Schwann's 1839 publication on cell theory, *Mikroskopische Untersuchung* (*Microscopical Research*), is divided into three parts. First, Schwann presented his microscope study of frog larvae. His observations convinced him that cells of the frog's **notochord** (supporting structure) and of **cartilage** were from the same kind of structures as plant cells—all had a nucleus, membrane and vacuoles. The second part of his publication asserted that all elementary parts (such as teeth, bone, muscle, cartilage, nerve tissue) are products of **cell differentiation**. The third part, philosophical in nature, centered on Schwann's attempt to replace theological explanations for natural phenomena with physical explanations. He believed that living things could not be seen as shaped by a divine plan, but must be understood as a product of the general characteristics of all matter and the random physical forces of nature. At the center of his philosophy was the cell—the basic unit of all life and the smallest entity capable of independent reproduction. Following the publication of his treatise, Schwann gradually abandoned science for religious and mystical studies. Cell theory, however, soon became widely accepted and today is recognized as one of the basic concepts of biology.

See also Cell cycle and cell division; Cell differentiation; Cell membrane transport; Cell structure; Cell theory; Gray matter and white matter; Nerve impulses and conduction of impulses; Nervous system overview; Neurons

SCROTUM · *see* GENITALIA (MALE EXTERNAL)

SECONDARY SEXUAL CHARACTERISTICS

Secondary sexual characteristics are the set of anatomical structures and features unique to males and females that are not directly related (and therefore secondary) to the production of sex cells (gametes).

Although in humans, the differences in male and female secondary sexual characteristics are more easily defined and determined, in many animals the differences are subtle. Regardless, the role of secondary sexual characteristics, especially with regard to the evolutionarily essential processes of mate identification and selection, can be extremely important.

Secondary sexual characteristics are generally distinguishing of gender in the human male and female. Accordingly, characteristics such as male facial **hair** or the enlarged **breasts** of a female may also carry importance in a particular social or religious culture.

The development of easily identifiable secondary sexual characteristics begins with the onset of **puberty** and continues throughout the teenage years into young adulthood.

In females, secondary sexual characteristics include the development of breast and mammary **tissue** needed to suckle infants. During puberty, the female hips take on a more broad and rounded appearance as a result of a general widening in pelvic structure that will ultimately allow for easier passage of the fetus through the birth canal. There is also an increased amount of fatty tissue deposited throughout the body that results in a normally higher percentage of body **fat** for females.

Maturation of the uterus, ovaries, and the **endocrine system** results in the onset of **menstruation**, a normally 28-day (with variation) cyclic growth and shedding of the uterine mucosal lining. Changes in the endocrine system also result in the more outwardly visible secondary sexual characteristics associated with increased hair growth in the pubic region and under the arms.

In males, the outward secondary sexual characteristics include the growth of the **larynx** and the appearance of the "Adam's apple." Changes in the larynx and elongation of the **vocal cords** result in a deepening voice associated with males. The release of androgens also promote growth and the generalized development of body mass and muscle. As a result, males are, on average, heavier, taller, and stronger than females.

In males, hormonal changes, especially those associated with the release of androgens, associated with puberty also stimulate the growth of course facial hair. Across the body, hair begins to thicken and grow, especially in the axillary (underarm), chest, back, and pubic regions of the body.

Odor can also be considered a secondary sexual characteristic. The prevalence of the use of perfume in many cultures provides evidence of the strong link between olfactory sensations and sexual selection. The changes in the endocrine system during puberty greatly increase the output of apocrine glands that are located in the axillary, anal, peritoneal, genital, and breast regions. The enhanced apocrine gland production of sebum leads to sometimes distinguishing smells for males and females. Especially in males, increased sebaceous production of oils can produce acne.

At puberty, there is a general enlargement of the penis and testicles in males and an enlargement of the vulva, labia, and clitoris in the female.

The development of secondary sexual characteristics in females is generally complete by the age of 16–18. In males, most secondary sexual characteristics are well established by age 20.

See also Adipose tissue; Adolescent growth and development; Endocrine system and glands; Sex determination; Sexual reproduction; Urogenital system

SEMEN AND SPERM

In the male, semen is the fluid expelled during ejaculation. In addition to **plasma**, the semen ejaculate contains secretions from the seminal vesicles and other glands to support and nourish the living sperm cells (spermatozoa) contained within the semen. Sperm cells are haploid sex cells of the male. Unlike eggs (oocytes and the mature ovum) that are large, non-motile, and generally ovulated one at a time, sperm are tiny, motile, and produced in the millions. While the human sperm contain a relatively long tail (flagella), the volume of an entire sperm, tail and all, is only 1/85,000 of the mature ovum.

Reproduction in humans may occur when semen—containing sufficient numbers of living and healthy (viable) sperm cells—is deposited in the vagina of a female near the cervical opening of the uterus. The haploid sperm move through the cervix, into the uterus and then migrate into the Fallopian (uterine) tube where, if a mature ovum is encountered, **fertilization** may occur. With fertilization, the egg finishes its second meiotic division to become haploid. The haploid sperm and mature egg together form a diploid zygote (single-celled embryo).

The male gonad (testis singular, testes plural; testicle is derived from the diminutive of testis and perhaps is best used to describe the **gonads** of a sexually immature boy) produces the hormone **testosterone** and sex cells. Early in embryonic development, primordial **germ cells**, which are diploid, migrate to the embryonic gonad. The primordial germ cells give rise to the diploid **stem cells** of the testis, known as spermatogonia. Each of the many spermatogonia, after the first meiotic division, form two primary spermatocytes that in turn, after the second meiotic division, form four haploid spermatids. In the process of forming mature sperm the spermatids lose much of their cytoplasm and develop a long, propulsive tail.

Motility of the sperm is due to the long tail which is a modified flagellum. **Cilia** and flagella, from protozoa through humans, all have a similar structure that has been intensively investigated since first described in early electron microscope studies. Microtubules that run the length of the sperm tail are arranged in a ring of nine pairs surrounding a pair in the center. Ciliary dynein is associated with each of the nine microtubule pairs. It is the interaction of the dynein with the microtubules which causes flagellar bending and thus propulsion.

It is estimated that a quarter of a billion sperm are released in a single ejaculate of semen in a healthy male human. Tin addition to a nutrient function the semen plays an important

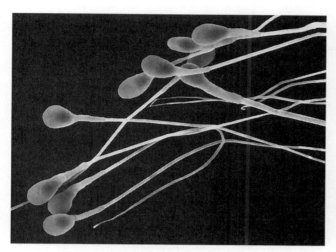

Scanning electron micrograph (SEM) of human spermatozoa, magnified 760 times. © Dr. Dennis Kunkel/Phototake. Reproduced by permission.

role in thermal and hydration regulation that promotes viable sperm cells. The semen also provides initial protection against the acidic gradient of the vagina and cervical region.

See also Adolescent growth and development; Cell cycle and cell division; Embryonic development: early development, formation, and differentiation; Genitalia (female external); Genitalia (male external); Gonads and gonadotropic hormone physiology; Implantation of the embryo; In vitro and in vivo fertilization; Puberty; Sexual reproduction; Urogenital system (female); Urogenital system (male); Urogenital system: Embryonic development

SEMINIFEROUS TUBULES • *see* REPRODUCTIVE SYSTEM AND ORGANS (MALE)

SENSE ORGANS: BALANCE AND ORIENTATION

Embedded in the temporal bone inside the inner ear lays the vestibular system, which contains fluid-filled sacs and cavities that monitor the position and movement of the head and transmit that information to the **brain**. This system contains three semicircular canals, each oriented at right angles to the other two. The canals are connected to a saclike utricle, below which lies the sacule, another hollow structure.

The utricle and sacule contain receptors consisting of groups of hair-like cells, **cilia**, that are embedded in a gelatinous material. The gelatinous material contains many small particles of **calcium** carbonate called otoliths. These increase the sensitivity of the cilia. At the base of each receptor is a nerve fiber. The nerve fibers collectively carry information to the brain via cranial nerve VIII, the auditory vestibular nerve. The receptors

Balance is a coordination of muscular, skeletal, and neurological actions. © 1996 P. Stocklein. Custom Medical Stock Photo. Reproduced by permission.

of the saccule and utricle respond to static positions of the head. In other words, they tell the brain which way is up.

The semicircular canals also contain gelatin-embedded receptors bearing cilia. These cilia detect changes in the rate of rotation or angular movements of the head. When the head moves, the fluid in the semicircular canals presses against the **hair** cells, causing them to bend. Bending of the cilia triggers action potentials in the nerve fibers connected to the receptors. More rotation causes more bending, causing more action potentials. The specific direction of head movement stimulates different receptors in the canals in each of the three planes. A three-dimensional message about direction of head movement is therefore compiled and sent to the brain along the eighth cranial nerve.

Low frequency movements that a person can't control often lead to motion sickness. Most often, a person may experience motion sickness as a passenger, but not as the driver. Motion sickness is probably the result of the brain's receiving contradictory information from the eyes and the vestibular system. The eyes, fixed on the interior of the vehicle, report "no motion" to the brain, while the stimulation of the hair cells reports "motion." Destruction of the semicircular canals by antibiotics or other drugs eliminates motion sickness.

Some of the **neurons** of the auditory vestibular nerves **synapse** in the vestibular nuclei of the lower brain stem. From here, neurons synapse on the motor neurons that control the muscles that move the eyes and on nuclei in the **thalamus**, **cerebral cortex**, and other locations. Some of the neurons, however, carry information from the inner ear directly to the **cerebellum**, a structure that coordinates motor control. The cerebellum uses information about the position and movement of the head to regulate to regulate output from the motor cortex, helping to maintain balance.

See also Ear (external, middle and internal)

SENSE ORGANS: EMBRYONIC DEVELOPMENT

The human senses, which include sight, **smell**, hearing, and balance, develop primarily from the outermost **tissue** layer (called the **ectoderm**) during embryogenesis as highly specialized **organs**. These organs (the eyes, **nose**, and ear) appear as regions on the surface of the developing embryo called placodes, which are connected to the **central nervous system**. Placodes can be subdivided into two major groups. Each group has a pre-determined fate in terms of the specific tissues it gives rise to. For example, one group develops into organs related to the inner ear, the lens of the **eye**, and the olfactory sensory epithelium (tissues related to the sense of smell). The other group develops into a complex array of nerve cells and tissues that are wired to the **brain**.

The eye is one of the most complex structures of the sensory organs. Various signaling molecules or cues stimulate cells to differentiate into various cell types that lead to different structural components of the eye. A coordinated, natural developmental progression must take place to ensure proper development. For example, the **cornea** must be properly aligned with respect to the lens to allow light to reach the **retina**. Similarly, the retina must also develop properly in terms of structure and location to allow the reception of visual images and transmission of these signals through the optic nerve to specific parts of the brain.

Primitive eye structures are called diverticula and begin developing by 22 days of gestation. The retina, iris, and optic nerve develop from an ectodermal outgrowth of the developing brain. The optic grooves appear as the neural tube (the progenitor to the central nervous system) closes and these grooves enlarge to form optic vesicles. The optic vesicles are connected to the ectoderm. The connection is necessary in order for inductive processes that result in thickening of the surface of the ectoderm, which forms the lens of the eye. Failure for these inductive processes to take place results in loss of proper eye development.

By six weeks, the optic vesicles become the optic cup, an area that invaginates to encompass the developing eyes. A **blood** vessel in the groove of the optic cup (called the choroid fissure) develops. The choroid fissure closes a week later, and the vessel becomes the major artery of the retina. The retina is the receptive area of the eye and forms light-sensitive elements within specialized cells from the optic cup. The eyelid and iris develop after eight weeks. The choroid and sclera (other components that make up the eye) develop from tissue surrounding the optic cup.

These structures along with the cornea and retina provide additional inductive signals for development of the lens.

Shortly after development of the eye, the **ear** begins to develop from an area of ectodermal tissue called the auditory plate. The ear can be divided into three anatomical parts: the external ear, the middle ear or tympanic cavity, and the internal ear or labyrinth. The inner ear is the first of the three to develop and serves as a sensory apparatus that controls both hearing and balance through structures called the cochlea and the vestibular apparatus, respectively. During the fourth week of embryogenesis, the otic placode (a thickened region on the surface of the ectoderm) appears on each side of an area of the brain called the hindbrain. An otic pit is formed as the otic placode invaginates and it sinks into the ectodermal tissue. The auditory plate becomes the auditory vesicle, from which an epithelial membrane of the labyrinth develops. The end of the membranous labyrinth elongates and forms a coiled tube called the cochlear duct. The vestibular apparatus develops from the central portion of the labyrinth.

The middle ear primarily functions to convert sound pressure waves into mechanical waves in the middle ear ossicles that connect to the **tympanic membrane**. It continues to grow through **puberty**. The auditory tube (also called the **Eustachian tube**), important for the equalization of pressure, and the tympanic cavity together make up the middle ear. The typmpanic cavity is derived from the first pharyngeal pouch, a region near the wall of the hindbrain, which becomes the tubotympanic recess as the first sign of **cartilage** develops. The tympanic cavity expands and encloses on the ossicles, **tendons**, ligaments, and nerves.

The external ear begins developing during the fifth week from the opposite end of the first pharyngeal groove that the middle ear develops from. The structures that develop include the auricle (pinna), the external auditory meatus (ear canal), and the tympanic membrane (eardrum). The external auditory meatus, which guides sound to the eardrum, develops from ectodermal tissue when a funnel-like tube called the meatal plug forms into a cavity. It is small at birth increasing the risk of damage to the tympanic membrane causing permanent hearing loss in the event of an injury until it fully elongates.

The sense of smell relies on the olfactory nerves, which develop between 11 and 15 weeks. They project into the upper part of the nose and are distributed in the mucous membranes of the nasal cavity. The nose, however, begins to develop during the fourth week of embryogenesis, from a pair of thickened areas of the ectoderm called olfactory placodes. The placodes form nasal pits that are deepened by rapid growth of the surrounding tissue and converge toward the midline of the face. They progressively billow into the oral cavity, separated by tissue that thins into a membrane called the osonasal membrane, which eventually breaks. Olfactory receptors project through the tissue surfaces of the nasal cavity and form nerve-like connections through an area called the olfactory bulb. The olfactory bulb consists of mitral cells, which relays messages to the olfactory centers of the brain, producing a sense of smell. The number and location of these sensory receptors determines the level of sensory perception. Bloodhounds, for example, develop a keen sense of smell due to an extremely large number of these receptors in the nose compared to humans.

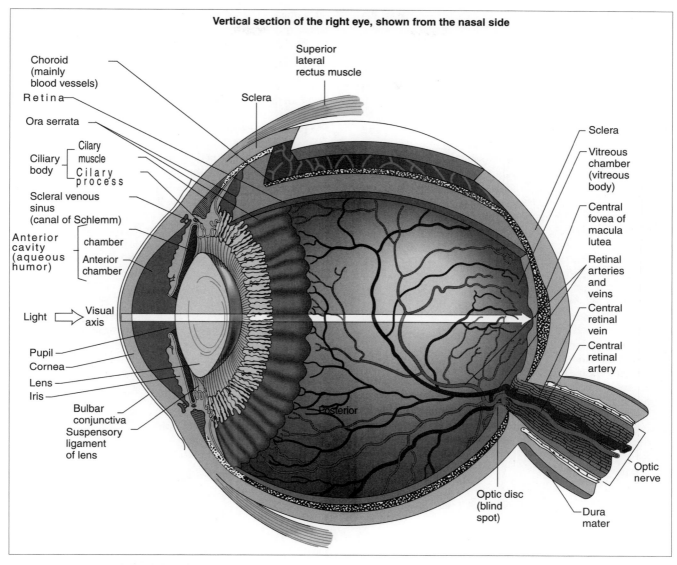

Vertical section of the right eye, shown from the nasal side

The human eye. *Illustration by Hans & Cassidy.*

See also Embryonic development: early development, formation, and differentiation; Hearing (physiology of sound transmission); Integumentary system: embryonic development; Nervous system: embryological development; Sense organs: balance and orientation; Sense organs: ocular (visual) structures; Sense organs: olfactory (sense of smell) structures; Sense organs: otic (hearing) structures

SENSE ORGANS: OCULAR (VISUAL) STRUCTURES

The **eye** and its constituent structures are capable of taking in light, **imaging** the light, and converting the visual signal into an electrical signal. The latter information can be manipulated by the **brain** to produce the multi-colored image that we perceive.

Light first enters a transparent and bulging structure called the **cornea**. The cornea is the covering structure of the eye. In addition to this protective function, the cornea is the primary light focusing structure of the eye. The cornea is made up of four layers of connective **tissue** with a layer of epithelium (the same tissue that covers the skin) on the surface. The four layers, from the front of the cornea to the back, are: Bowman's (anterior limiting) membrane, stroma (also called substantia propria), Descemet's (posterior limiting) membrane, and the endothelium. Even with multiple layers, the cornea remains transparent as it is almost devoid of cells and does not contain **blood** vessels.

The cornea is bathed in tear fluid, which keeps the cornea moist and restricts the growth of **bacteria** and other microorganisms on the surface of the eye, because of the presence of a destructive enzyme called lysozyme.

The colored disc inside the eye, visible through the cornea, is called the iris. The iris is a diaphragm—a structure designed to open and close. Another example of an iris, which in fact operates similarly to the iris of the eye, is that of a camera lens. The iris of the eye is composed of connective tissue and **smooth muscle**. The various colors of the iris in human and other mammals is genetically determined.

At the center of the iris lies a circular hole. This is the pupil. As the iris opens or contracts, the diameter of the pupil can increase, which allows more light to pass into the eye, or can decrease, restricting the amount of light entering the eye. This response is regulated by the light intensity. In bright light, the pupil will be smaller than in a dark room.

Behind the pupil lies the crystalline lens, which focuses the incoming image. The lens is made up of four layers of cells that are suspended and held firm by ligaments. This allows the lens to change shape as needed to focus on nearby objects and objects further away. The image, which is actually now inverted, passes through a syrupy fluid within the eyeball called the vitreous humor. The image is projected onto the **retina** at the back of the eyeball. Cells within the retina collect the incoming light and convert the information to electrical signals, which are conveyed to the brain via the optic nerve.

All the optic structures are set in a protective cavity in the **skull** called the orbit or the socket. Soft, fatty tissue surrounds the socket, as protection for the eye and to aid the eye in turning. Turning is possible because of the action of three pairs of muscles: the medial/external rectus muscles, superior/inferior rectus muscles and the superior/inferior oblique muscles.

See also Cornea and corneal transplantation; Retina and retinal imaging

SENSE ORGANS: OLFACTORY (SENSE OF SMELL) STRUCTURES

The olfactory system (**smell** system) allows the body to detect chemicals in the environment. Along with the gustatory system, the olfactory system also allows the body to distinguish flavors. Most of the olfactory system, however, is devoted to alerting the body to dangers in the environment. Humans can smell several hundred thousand substances, but only about 20% of them are interpreted as pleasant.

In the dorsal part of the nasal cavity is a thin bony structure called the cribiform plate. On the ventral surface of the plate is the olfactory epithelium, which contains three types of cells. Olfactory receptor cells, the dendrites of olfactory **neurons**, transduce chemical messages into nerve impulses. Supporting cells, which are somewhat like the glia of the **central nervous system**, produce **mucus**. Basal cells produce new olfactory receptor cells, which wear out and need replacement every four to eight weeks.

The mucus that covers the olfactory epithelium dissolves chemicals, or odorants, that enter the dorsal region of the nasal cavity. This watery substance contains, among other substances, odorant-binding proteins. The area occupied by the human olfactory epithelium is several square centimeters. Not surprisingly, the olfactory epithelium of a dog is much larger and has more densely packed receptors.

Each of the six million olfactory receptor neurons has one knob-like dendrite that ends in the olfactory epithelium. Several **cilia** (tiny, hair-like structures) extend from the dendritic knob into the mucus. The odorant-binding proteins in the mucus bring dissolved odorants into contact with the receptors in the dendritic knobs and moving cilia. The binding of the odorant to the receptor initiates transduction of the chemical signal into a nerve impulse. There are 500–1,000 different genes that encode odor receptors. Each olfactory receptor cell probably expresses only one of the genes.

Each receptor cell has on the other side of its cell body a thin, unmyelinated axon. These axons merge into the olfactory nerve (cranial nerve I). About 25,000 of these axons on each side pass through the cribiform plate into one of the two olfactory bulbs, where they form synapses with about 100 secondary olfactory neurons. These neurons leave the olfactory bulbs through the olfactory tracts, which project to the olfactory cortex, amygdala, and **hypothalamus**. From the olfactory cortex, the information is sent to the **thalamus** and then to other **brain** areas. The olfactory system is the only sensory system in which information reaches the cortex before it reaches the thalamus. This fact and the presence of unmyelinated axons suggest that olfaction (smell) is the oldest of the senses.

If the olfactory receptor cells are damaged due to trauma, anosmia, or the inability to smell, may result.

See also Smell, physiology of olfactory senses

SENSE ORGANS: OTIC (HEARING) STRUCTURES

In humans, hearing structures are also commonly known as the **ear**. The human ear is divided into three functional areas; the outer, middle and inner ear.

The outer ear includes the visible portion of the ear. This portion of the ear acts to collect and focus sound. Without the dish shaped portion of the ear, low intensity sound might escape detection. The visible portion of the ear also includes a structure known as the pinna (or auricle). This is visible as a hole. It leads to a passageway, typically about 0.98 in (2.5 cm) in length that leads deeper into the ear. The passageway is commonly called the ear canal (or auditory canal). The ear canal is typically not a uniform diameter all down its length. Rather, it narrows about three-quarters of the way inward toward the eardrum to a diameter of typically a bit less than 0.098 in (0.8 mm), and then widens near a taut membrane stretched across the end of the canal. The membrane is known as the eardrum. The ear drum is the boundary between the outer and middle portions of the ear.

The construction of the ear canal—open at one end and closed off at the other end by a membrane—is what enables sound vibrations to be amplified. An example of a similar effect is the resonance that can be achieved by the pipes of an organ.

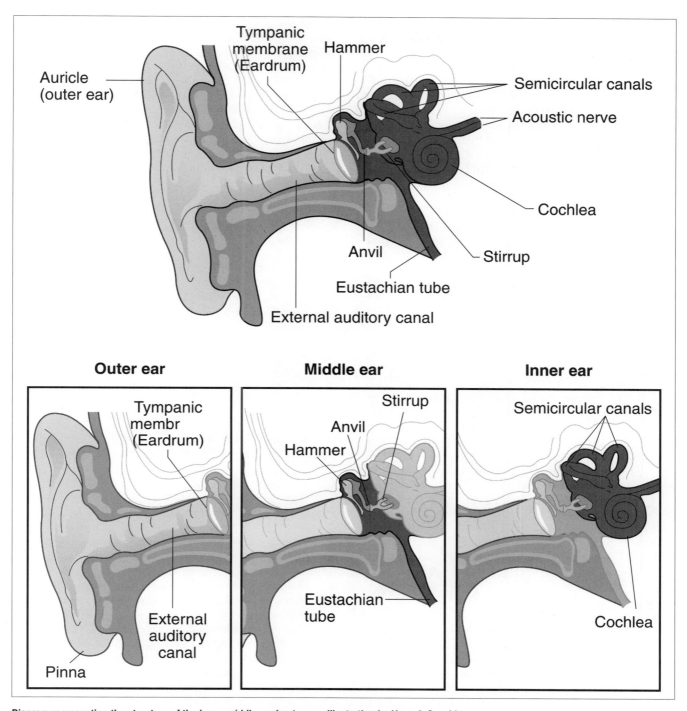

Diagram representing the structure of the inner, middle, and outer ear. *Illustration by Hans & Cassidy.*

The middle ear houses three small bones: the malleus (or hammer), incus (or anvil), and the stapes (or stirrup). The malleus is attached at one end to the inner surface of the eardrum and at the other end to the Incus. The latter is in turn attached at its other end to the stapes. The three bones form a "U" shape. They function by acting as levers, driven by the movement of the eardrum, with malleus pushing incus and incus pushing stapes. These three bones are located in a cavity. At the bottom of the cavity lies the Eustachian tube. This tube connects the middle ear with the nasopharynx and functions to equalize the pressure between the external air and the middle ear. An example of the Eustachian tube in action occurs when traveling in an airplane. The relief of the ear full-ness upon **swallowing** or chewing gum is due to the equalization of air via the Eustachian tube.

The stapes bone of the middle ear contacts a structure termed the oval window. The oval window is the boundary between the middle and inner portions of the ear. It is another membrane, similar to the eardrum, but is approximately one-third the size of the eardrum. As with the eardrum, the oval window is designed to vibrate, and so to pass on and amplify sound waves.

Beyond the oval window lies the region of the inner ear. This region consists of a structure that is termed the cochlea. The cochlea looks something like a snail shell and is only the size of an average-size fingernail. The interior of the cochlea contains a looping channel, which is further divided by thin membranes into a triangular arrangement of two canals, the vestibular canal and the tympanic canal, and a duct called the cochlea duct. The membrane demarcating the vestibular canal and the cochlea duct is Reissner's membrane. The membrane demarcating the tympanic canal and the cochlear duct is the basilar membrane. The intersection of the triangular division houses a body called the organ of Corti.

The cochlea converts the sound energy to hydraulic energy by virtue of the fluid that circulates through it. The vestibular and tympanic channels contain perilymph, which is similar to spinal fluid. The cochlea duct contains endolymph, which is very similar to the fluid found within the cells of the body. These fluids and the Organ of Corti accomplish the sound-to-hydraulic energy conversion. In other words, sound energy is converted to waves in the cochlear fluids. The wave height provides the information for the nerve cells positioned around the cochlea, which telegraph the information to the **brain**.

See also Hearing (physiology of sound transmission)

SERATONIN **•** *see* NEUROTRANSMITTERS

SEROLOGY

The basis of serology is the recognition of an antigen by immune mechanisms, with the subsequent production of an antibody. Serology is the studies of antigen-antibody reactions outside of a living organism (i.e., *in vitro*, in a laboratory setting).

In medical terminology, serology refers to a **blood** test to detect the presence of antibodies against a microorganism. The detection of antibodies can be qualitative (i.e., determining whether the antibodies are present) or quantitative (i.e., determining the quantity of an antibody produced). Some microorganisms can stimulate the production of antibodies that persist in a person's blood for a long time. Thus, in a qualitative assay the detection of a particular antibody does not mean that the person has a current **infection**. However, it does mean that it is likely that at some time that person was infected with the particular microbial pathogen. Serology assays can be

performed at various times and the level of antibody determined. If the antibody level rises, it usually is indicative of a response to an infection. The body produces elevated amounts of the antibody to help fight the challenging antigen.

Serology as a science began in 1901. Austrian American immunologist **Karl Landsteiner** identified groups of red blood cells as A, B, and O. From that discovery came the recognition that cells of all types, including blood cells, cells of the body, and microorganisms carry proteins and other molecules on their surface that are recognized by cells of the **immune system**. There can be many different **antigens** on the surface of a microorganism, with many different antibodies being produced.

When the antigen and the antibody are in suspension together, they react together. The reaction can be a visible one, such as the formation of a precipitate made up of a complex of the antigen and the antibody. Other serology techniques are agglutination, complement-fixation and the detection of an antigen by the use of antibodies that have been complexed with a fluorescent compound.

Serological techniques are used in basic research to, for example, decipher the response of immune systems and to detect the presence of a specific target molecule. In the clinical setting, serology is used to confirm infections and to type the blood from a patient. Serology has also proven to be very useful in the area of forensics, where blood typing can be vital to establishing the guilt or innocence of a suspect. or the identity of a victim.

See also Immunity; Infection and resistance; Autoimmune disorders; Immunity, cell mediated; Immunity, humoral regulation

SEX DETERMINATION

The sex of an individual is determined by the genetic information present in the individual's sex **chromosomes**. Each diploid cell has one pair of sex chromosomes that differs from the other homologous pairs of chromosomes. For example, in humans, one of the 23 pairs of homologous chromosomes makes up the sex chromosomes. However, unlike all of the other homologous pairs, the sex chromosomes may not be the same as each other in size. There are two types of sex chromosomes, the X chromosome and the Y chromosome. The X chromosome is larger than the Y chromosome, and therefore has spaces for genes that are not present on the Y chromosome.

In humans and many other species, females have two X chromosomes (symbolized as XX). However, the chromosomes that make up the pair of sex chromosomes in males are different. Males have one X chromosome and one Y chromosome (symbolized as XY). In some species, such as birds, moths, and butterflies, the sex chromosomes are reversed; males are XX and females are XY. In some insects the Y chromosome is absent altogether. Thus, a female would be XX, but a male would be XO (the "O" indicates that the Y chromosome is absent).

The sex of an individual is based on its sex chromosomes. This is determined at the time of **fertilization**, when the male's **sperm** joins with the female's egg. For example, because all human female cells have two X chromosomes, when the egg forms by meiosis, it must contain an X chromosome. However, since all human male cells have both an X and a Y chromosome, when the sperm form by meiosis, they can contain either an X chromosome or a Y chromosome. Thus, during fertilization, the egg always contributes an X chromosome and the sperm can contribute either an X or a Y chromosome. If the sperm has an X chromosome, the resulting offspring will have two X chromosomes and will be a female (XX). If, on the other hand, the sperm contributes a Y chromosome, the zygote that forms as a result of fertilization will have one X and one Y chromosome and will be a male (XY). In humans and many other species, it is the sex chromosome contributed by the male that is responsible for determining the sex of the offspring. Since the sperm have a 50% chance of receiving an X chromosome and a 50% chance of receiving a Y chromosome, there are approximately equal numbers of X and Y sperm. As a result, there is an almost equal chance of having a male or female offspring.

See also Sexual reproduction

SEX HORMONES

Gender in humans, whether male or female, is primarily determined by the sex **chromosomes** inherited by the individual. However, the normal development and maturation of sex glands (i.e., **gonads**) depends on the proper production of some **hormones** by the anterior pituitary gland, such as the growth hormone, the adrenocorticotrophic hormone (ACTH), prolactin, and the gonadotropic hormones known as **follicle-stimulating hormone (FSH) and luteinizing hormone (LH)**. The anterior pituitary, in turn, is controlled by the **hypothalamus**, which secretes six different regulatory hormones. Gonads, or sex glands in the male and the female synthesize specific hormones in response to gonadotropic hormonal stimulation that confer each gender the particular characteristics that distinguish one sex from the other.

Male hormones, or androgens, are steroid hormones soluble in **lipids**, and synthesized either from cholesterol or from acetyl coenzyme A. Some androgens are secreted by the testes and others by the adrenal glands. **Testosterone**, the most abundant hormone found in males, is synthesized in the testes by the interstitial cells of Leydig. This cell type is found in great numbers in the testes after **puberty**, but it is scarce in children. Other androgens secreted by the testes are dihydrotestosterone and androstenedione. The adrenal glands secrete five other androgens; but testosterone is the main hormone responsible for adult male characteristics. Testosterone is first produced in the male embryo in about the seventh week of development and its production continues during the first ten weeks after birth, when many interstitial cells of Leydig are present in the infant testes. After that period, the number of these cells is dramatically reduced and almost no testosterone is synthesized

during childhood, until the beginning of puberty, when the anterior pituitary starts producing the luteinizing hormone that stimulates testosterone production. In late puberty or early adolescence, the follicle-stimulating hormone induces **sperm** maturation in the Sertoli cells of the testes. From puberty on, testosterone promotes the gradual development of the male characteristics, such as, pubic **hair**, change in voice tonality, body hair, and beard. During adolescence, testosterone promotes the complete development of the scrotum, the penis, and the testes, and promotes an increase in muscular body mass of up to fifty per cent more than that of females.

Female ovary development and maturation are also induced by gonadotropic hormones synthesized by the anterior pituitary and controlled by the hypothalamic hormones. The female sex hormones are steroid hormones, which are mostly synthesized from cholesterol with a discrete synthesis also occurring from acetyl coenzyme A. The ovaries produce two types of sex hormones: **estrogen** and **progesterone**. From puberty on, estrogen promotes the gradual transformation and development of the female body, such as breast growth, maturation of the ovaries, the fallopian tubes, the uterus, and the vagina. Estrogen causes the selective deposition of subcutaneous **fat**, especially in the breast, thighs, and buttocks, during puberty and adolescence as well. Among the many physiological roles of estrogen, it interferes with the electrolyte balance, causing in females a tendency to retain water, due to its molecular similarity with aldosterone and other adrenocortical hormones. Water retention usually occurs for one or two weeks during the monthly menstrual cycle and is especially noticeable during **pregnancy**.

Hair growth and distribution on the pubic region and axillae during adolescence is not due to estrogen, but to the production by the female adrenal glands of discrete levels of androgens. During the sexual monthly cycle, the levels of synthesis of estrogen oscillate and the same is true of progesterone. The hypothalamic hormones and the gonadotropic hormones control the monthly cycle that prepares the female body for conception, by causing changes in progesterone and estrogen synthesis, and affect the ovary, the fallopian tubes, and the uterus. The follicle-stimulating hormone (FSH), for instance, induces the growth and maturation of the ovarian follicles and the formation of a mature ovum. The luteinizing hormone (LH) induces progesterone synthesis, stimulates ovulation, and the corpus luteum formation. Progesterone causes the uterine endometrium (i.e., the internal mucosa lining the uterine walls) to change its secretory pattern, preparing a layer of nutrients to receive a fertilized egg. It also increases the secretion in the internal mucosa of the fallopian tubes in order to supply adequate **nutrition** to the fertilized egg and induces the swelling of the **breasts**, while estrogen promotes mammary cell proliferation in the weeks before **menstruation**. Once pregnancy does not occur, the excess of proliferated cells is induced to programmed cell death or apoptosis during and after menstruation. In the case of pregnancy, the **mammary glands** and related structures will continue to be developed in preparation for **lactation** (milk production), near the end of pregnancy. Lactation however is induced by another anterior pituitary hormone known as prolactin.

See also Adolescent growth and development; Adrenal glands and hormones; Diuretics and antidiuretic hormones; Embryonic development: early development, formation, and differentiation; Human growth hormone; Hypothalamus; Mammary glands and lactation; Menopause; Ovarian cycle and hormonal regulation; Pituitary glands and hormones; Prenatal growth and development; Prostate gland; Sex determination; Sexual reproduction

SEXUAL ORGANS (FEMALE) • *see* REPRODUCTIVE SYSTEM AND ORGANS (FEMALE)

SEXUAL ORGANS (MALE) • *see* REPRODUCTIVE SYSTEM AND ORGANS (MALE)

SEXUAL REPRODUCTION

Sexual reproduction is the process through which two parents produce offspring that are genetically different from themselves and have new combinations of their characteristics. During sexual reproduction, each parent contributes one haploid gamete (a sex cell with half the normal number of **chromosomes**). The two sex cells fuse during **fertilization** and form a diploid zygote containing the normal number of chromosomes.

Mutation and recombination (the production of variations in gene combinations), also bring together new combinations of alleles (a form of a gene located on a chromosome). In addition, crossing-over, the exchange of pieces of chromosomes by two homologous chromosomes, brings about genetic variation during the formation of gametes.

Sexual reproduction is advantageous because it generates variations in characters that can adapt a species over time and improve its chances of survival. There are also a number of strong cultural and religious influences on human behaviors associated with sexual reproduction, especially strong are influences on mate selection.

Physiologists have long studied the hormonal changes that facilitate and influence mating behavior and reproductive readiness. These influences, the coordination and control of reproduction by **hormones** are, in humans, also heavily influenced by societal factors so that **sperm** and egg are brought together at the appropriate time. During human sexual reproduction, a haploid sperm deposited by the male in the vagina of the female makes its way through the cervix, uterus, and the fallopian tube carrying the unfertilized ovum (egg). Sperm are provided with a fluid (**semen**) that provides an aquatic medium for the sperm to swim when inside the male's body. Of the millions of sperm cells deposited, only a small percentage make it to the unfertilized ovum.

Under normal circumstances, only one haploid sperm cell may ultimately penetrate the haploid ovum to produce fertilization and to form a diploid zygote. If development is to proceed, the fertilized egg implants in the uterus, where the growth and differentiation of the embryo occur. The zygote divides mitotically and differentiates into an embryo. Embryonic **nutrition** and respiration occur by diffusion from the maternal bloodstream through the **placenta**. The embryo grows and matures. When development is complete, the birth process takes place. After birth, the normal human matures and develops into an adult capable of producing haploid sex cells (gametes) and repeating the sexual reproduction process.

See also Human genetics; Implantation of the embryo; Infertility (female); Infertility (male); Parturition; Reproductive system and organs (female); Reproductive system and organs (male)

SHERRINGTON, CHARLES SCOTT (1857-1952)
English neurophysiologist

Charles Scott Sherrington became one of the founders of the discipline of neurophysiology through his research on how nerve impulses are transmitted between the **central nervous system** and muscles. Sherrington focused much of his career on understanding the structure and the function of the nervous system. Drawing on the research of Spanish neuroanatomist Santiago Ramón Cajal, Sherrington proposed viewing nervous activity as part of an integrated and complex system. For his work on how the central nervous system elicits motor activity from muscles, Sherrington shared the 1932 Nobel Prize in physiology or medicine with Edgar Douglas Adrian.

Born November 27, 1857, in London, England, Sherrington was the son of James Norton and Anne (Brookes) Sherrington. James Sherrington died while his son was still very young, and later Sherrington's mother married Caleb Rose, Jr., a physician in Ipswich, England. Rose was broadly and classically educated, and his home served as a gathering place for artists, writers, and scholars. Exposure to these diverse arts influenced Sherrington and was reflected in his own broad interests in the humanities and the sciences. After attending Ipswich Grammar School, Sherrington began medical training in 1875 at St. Thomas's Hospital in London. In 1879, he enrolled in Caius College at Cambridge University. Two years later, Sherrington began work in the laboratory of Michael Foster, England's foremost physiologist. In Foster's laboratory, Sherrington also met John Newport Langley, Newell Martin, Walter Gaskell, and Sheridan Lea, individuals who would become important physiologists in their own right.

After earning a bachelor's degree in medicine in 1884, Sherrington left Cambridge to pursue graduate studies in German laboratories. He remained abroad for three years, receiving training and conducting research in physiology, **histology**, and **pathology**, and working in the laboratories of **Rudolf Virchow**, **Robert Koch**, and Friedrich Goltz, with whom he studied the central nervous system. Upon returning to England, Sherrington assumed a post teaching systematic physiology to medical students at his training site, St. Thomas's Hospital in London. He left this position in 1891 to become professor and superintendent of the Brown Institute for Advanced Physiological and Pathological Research. A year later, Sherrington married and the couple later had one son.

Sherrington accepted the physiology chair at the University of Liverpool in 1895. Seeking to understand the structures and the mechanisms that operated the nervous system, Sherrington began to draw on the work of **Santiago Ramón y Cajal**. Prior to the latter scientist's work in the late 1880s, neurophysiologists believed that nerve fibers formed a continuous network or system through the body. This proposition was known as the reticular theory. Ramón y Cajal refuted the reticular theory by using a silver-based dye developed by the Italian anatomist **Camillo Golgi**. Golgi's preparation stained individual nerve cells a black color and demonstrated to neuroanatomists that nerve cells were discrete entities and not part of a nexus as was previously thought. The new theory that saw nerve cells as independent units was called the neuron theory, or popularly, the neuron doctrine. Although nerve cells were discrete units, **neurons** in a series could form pathways through which information can be transmitted. Nerves—consisting of a bundle of fibers—relay sensations (like touch and **smell**) and instructions on motor activity (like moving an arm or a leg) by electrical impulses. Sherrington became interested in understanding how nerves formed integrative pathways between the central nervous system and muscles. He considered some simple reflexive behavior, such as the knee-jerk, and attempted to explain the neurophysiology of the phenomena. Finding that he had an insufficient knowledge of neural **anatomy** to conduct the research, Sherrington stoically devoted the next decade to mapping the pathways between the central nervous system and muscle groups and to identifying the sensory nerves that innervated muscle **tissue**.

Sherrington's commitment to understanding the neural pathways proved to have an important impact. He came to realize that a particular reflexive behavior was not controlled by a single pathway or an isolated response to a single stimulus. Rather, a simple reflex was the product of a complex process that involved the inhibition and excitation of many nerve cells in many different pathways. Sherrington concluded that the central nervous system was an integrated whole that coordinated multiple pathways to produce any single action. His contributions on this point were not only theoretical but also experimental. He introduced seminal research strategies for studying questions of the central nervous system. For example, the spinal animal, an animal with a transected **spinal cord**, and the decerebrate rigid animal, an animal partially paralyzed by the excision of the **cerebral cortex**, were introduced as important approaches to exploring the activity of the nervous system. Sherrington's analysis of the hind limb scratch of a dog helped to elucidate neuronal action.

Sherrington's study of the scratch reflex in dogs elucidated other important principles of how the central nervous system is organized. He concluded that **reflexes** can have "reciprocal innervation" so that inhibitory and excitatory **reflexes** are coordinated simultaneously. Sherrington also concluded that there are two levels on which actions are controlled—higher level control by the **brain** and lower level control by the muscle nerves. His most important idea perhaps

reflected in the integrative scheme is that there is a break between one nerve cell and another, between brain and muscles, between inhibitory and excitatory processes. To describe this break, Sherrington coined the term "synapse." The idea of a **synapse** became important for two reasons. First, it acknowledged that nerve cells were not organized in the reticular fashion as it was previously argued. Second, understanding how synapses were transcended became the next challenge for twentieth-century neurophysiologists. Sherrington lucidly offered these ideas about the nervous system in his seminal work, *The Integrative Action of the Nervous System,* published in 1906.

In 1913, Sherrington left the University of Liverpool after eighteen years of service to assume the Waynflete Professorship of Physiology at Oxford University. The post offered Sherrington the opportunity to continue his research on the central nervous system, but the entry of Great Britain into World War I in August, 1914, meant Sherrington had to postpone his studies for some time. He joined the war effort, serving as chair of the Industrial Fatigue Board. Not satisfied with merely reading about the conditions of war-time industrial workers, in 1915 Sherrington worked incognito in a shell factory to experience first-hand the hardships and long shifts faced by workers. Although he managed to complete a textbook of physiology during the war period, Sherrington did not return to his normal research work until the mid–1920s. He successfully recruited a number of promising assistants, including E. G. T. Liddell and John Carew Eccles. Eccles would go on to win the 1963 Nobel Prize in physiology or medicine for research that had its roots in his stint in Sherrington's Oxford laboratory. Eccles, Liddell, and Sherrington's other students grew in reputation as the "Sherrington school," and their assistance allowed Sherrington to complete a minimum of an experiment a week.

Sherrington's research at Oxford after the 1920s differed from the work that he had been doing prior to World War I. Rather than studying the nervous system as a whole, Sherrington focused his attention on specific mechanisms in the central nervous system. He developed with Eccles the idea of a "motor unit"—a nerve cell that coordinates many muscle fibers. He also concluded that neuronal excitation and inhibition were separate and distinct processes; one was not merely the absence of the other.

Although Sherrington retired in 1936, four years after being named a Nobel Prize winner, he maintained an active life after his formal retirement. He cultivated many of the interests that he had as child in the eclectic home of his stepfather, including poetry, history, and philosophy. In 1925 Sherrington wrote and published a book of poems titled *The Assaying of Brabantius*. His deep interests in philosophy and history were reflected in two post-retirement publications, 1941's *Man on His Nature* and 1946's *The Endeavor of Jean Fernel*. In addition to being a popular and sought-after speaker, Sherrington was a trustee of the British Museum in London and served as governor of the Ipswich School from which he had graduated.

In addition to the Nobel Prize, Sherrington garnered virtually every honor that could be given to a British scientist. At the time of his death in 1952, he held memberships in more than forty scholarly societies and had been given honorary degrees from twenty-two universities. Most notably, Sherrington was a past president of the Royal Society of London (1920–1925), and recipient of the Knight Grand Cross of the British Empire in 1922 and the Order of Merit in 1924. He died on March 4, 1952, from heart failure.

See also Action potential; Muscular innervation; Nerve impulses and conduction of impulses

SHOCK

Shock is a medical emergency in which the **organs** and tissues of the body are not receiving an adequate flow of **blood**. This deprives the organs and tissues of oxygen (carried in the blood) and allows the buildup of waste products. Shock can result in serious damage or even **death**.

There are three stages of shock: Stage I (also called compensated, or nonprogressive), Stage II (also called decompensated or progressive), and Stage III (also called irreversible).

In Stage I of shock, when low blood flow (perfusion) is first detected, a number of systems are activated in order to maintain/restore perfusion. The result is that the **heart** beats faster, the blood vessels throughout the body become slightly smaller in diameter, and the kidney works to retain fluid in the **circulatory system**. All this serves to maximize blood flow to the most important organs and systems in the body. The person in this stage of shock has very few symptoms, and treatment can completely halt any progression.

In Stage II of shock, these methods of compensation begin to fail. The systems of the body are unable to improve perfusion any longer, and the patient's symptoms reflect that fact. Oxygen deprivation in the **brain** causes the patient to become confused and disoriented, while oxygen deprivation in the heart may cause chest **pain**. With quick and appropriate treatment, this stage of shock can be reversed.

In Stage III of shock, the length of time that poor perfusion has existed begins to take a permanent toll on the body's organs and tissues. The heart's functioning continues to spiral downward, and the **kidneys** usually shut down completely. Cells in organs and tissues throughout the body are injured and dying. The endpoint of Stage III shock is the patient's death.

Shock is caused by three major categories of problems: cardiogenic (meaning problems associated with the heart's functioning); hypovolemic (meaning that the total volume of blood available to circulate is low); and septic shock (caused by overwhelming **infection**, usually by **bacteria**).

Cardiogenic shock can be caused by any disease, or event, which prevents the heart muscle from pumping strongly and consistently enough to circulate the blood normally. Heart attack, conditions which cause inflammation of the heart muscle (myocarditis), disturbances of the electrical rhythm of the heart, any kind of mass or fluid accumulation and/or blood clot which interferes with flow out of the heart can all significantly affect the heart's ability to adequately pump a normal quantity of blood.

Hypovolemic shock occurs when the total volume of blood in the body falls well below normal. This can occur when there is excess fluid loss, as in dehydration due to severe vomiting or **diarrhea**, diseases which cause excess urination (diabetes insipidus, **diabetes mellitus**, and kidney failure), extensive **burns**, blockage in the intestine, inflammation of the **pancreas** (pancreatitis), or severe bleeding of any kind.

Septic shock can occur when an untreated or inadequately treated infection (usually bacterial) is allowed to progress. Bacteria often produce poisonous chemicals (toxins),which can cause injury throughout the body. When large quantities of these bacteria, and their toxins, begin circulating in the bloodstream, every organ and **tissue** in the body is at risk of their damaging effects. The most damaging consequences of these bacteria and toxins include poor functioning of the heart muscle; widening of the diameter of the blood vessels; a drop in blood pressure; activation of the blood clotting system, causing blood clots, followed by a risk of uncontrollable bleeding; damage to the **lungs**, causing acute respiratory distress syndrome, **liver** failure, kidney failure, and **coma**.

Initial symptoms of shock include cold, clammy hands and feet; pale or blue-tinged skin tone; weak, fast **pulse** rate; fast rate of **breathing**; low blood pressure. A variety of other symptoms may be present, but they are dependent on the underlying cause of shock.

The most important goals in the treatment of shock include: quickly diagnosing the patient's state of shock; quickly intervening to halt the underlying condition (stopping bleeding, re-starting the heart, giving antibiotics to combat an infection, etc.); treating the effects of shock (low oxygen, increased acid in the blood, activation of the blood clotting system); and supporting vital functions (blood pressure, urine flow, heart function).

See also Bacteria and bacterial infection; Blood pressure and hormonal control mechanisms; Circulatory system; Oxygen transport and exchange

SICKLE CELL ANEMIA

Sickle cell anemia is an inherited **blood** disorder that arises from a single amino acid substitution in one of the component proteins of **hemoglobin**. The component protein, or globin, that contains the substitution is defective. Hemoglobin molecules constructed with such proteins have a tendency to stick to one another, forming strands of hemoglobin within the red blood cells. The cells that contain these strands become stiff and elongated—that is, sickle shaped.

A child who inherits the sickle cell trait from both parents—a 25% possibility if both parents are carriers—will develop sickle cell anemia. Sickle cell anemia is characterized by the formation of stiff and elongated red blood cells, called sickle cells. These cells have a decreased life span in comparison to normal red blood cells. Normal red blood cells survive for approximately 120 days in the bloodstream; sickle cells last only 10-12 days. As a result, the bloodstream is chronically short of red blood cells and the affected individual develops anemia.

However, the severity of the symptoms cannot be predicted based solely on the genetic inheritance. Some individuals with sickle cell anemia develop health or life-threatening problems in infancy, but others may have only mild symptoms throughout their lives. For example, genetic factors, such as the continued production of fetal hemoglobin after birth, can modify the course of the disease. Fetal hemoglobin contains gamma-globin in place of beta-globin; if enough of it is produced, the potential interactions between hemoglobin S molecules are reduced.

Worldwide, millions of people carry the sickle cell trait. Individuals whose ancestors lived in sub-Saharan Africa, the Middle East, India, or the Mediterranean region are the most likely to have the trait. The areas of the world associated with the sickle cell trait are also strongly affected by **malaria**, a disease caused by blood-borne parasites transmitted through mosquito bites. According to a widely accepted theory, the genetic mutation associated with the sickle cell trait occurred thousands of years ago. Coincidentally, this mutation increased the likelihood that carriers would survive malaria outbreaks. Survivors then passed the mutation on to their offspring, and the trait became established throughout areas where malaria was common.

Although modern medicine offers drug therapies for malaria, the sickle cell trait endures. Approximately two million Americans are carriers of the sickle cell trait. Individuals who have African ancestry are particularly affected; one in 12 African Americans are carriers. An additional 72,000 Americans have sickle cell anemia, meaning they have inherited the trait from both parents. Among African Americans, approximately one in every 500 babies is diagnosed with sickle cell anemia. Hispanic Americans are also heavily affected; sickle cell anemia occurs in one of every 1,000-1,400 births. Worldwide, it has been estimated that 250,000 children are born each year with sickle cell anemia.

Symptoms typically appear during the first year or two of life, if the diagnosis has not been made at or before birth. However, some individuals do not develop symptoms until adulthood and may not be aware that they have the genetic inheritance for sickle cell anemia.

The gel electrophoresis test is also used as a screening method for identifying the sickle cell trait in newborns. More than 40 states screen newborns in order to identify carriers and individuals who have inherited the trait from both parents.

Early identification of sickle cell anemia can prevent many problems. The highest death rates occur during the first year of life due to **infection**, aplastic anemia, and acute chest syndrome. If anticipated, steps can be taken to avert these crises. With regard to long-term treatment, prevention of complications remains a main goal.

Screening at birth offers the opportunity for early intervention; more than 40 states include sickle cell screening as part of the usual battery of blood tests done for newborns. Screening is recommended for individuals in high-risk populations; in the United States, African Americans and Hispanic Americans have the highest risk of being carriers. Pregnant

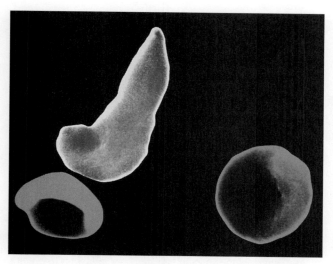

A scanning electron micrograph comparing healthy erythrocytes (red blood cells) with a sickle-shaped erythrocyte. *Photograph by Dr. Gopal Murti. National Audubon Society Collection/Photo Researchers, Inc. Reproduced by permission.*

women and couples planning to have children may also wish to be screened to determine their carrier status. Carriers have a 50% chance of passing the trait to their offspring. Children born to two carriers have a 25% chance of inheriting the trait from both parents and having sickle cell anemia. Carriers may consider genetic counseling to assess any risks to their offspring. The sickle cell trait can also be identified through prenatal testing; specifically through use of amniotic fluid testing or chorionic villus sampling.

See also Amniocentesis; Human genetics

SIGHT • *see* VISION: HISTOPHYSIOLOGY OF THE EYE

SKELETAL MUSCLE

Skeletal muscle is also known as striated muscle because of its striped appearance under the microscope and as voluntary muscle because it can be controlled at will. These muscles are attached to, and move, the bones, and are arranged in antagonistic pairs that enable movement in opposite directions. For example, the biceps flexes the lower arm while the triceps extends it.

A muscle is like a set of Russian nesting dolls. Each level of structure has a smaller element within it. A muscle such as the biceps is composed of many long fibers running in parallel. Each muscle fiber contains many myofibrils, and each myofibril contains many copies of two types of myofilaments, actin, or thin filaments; and myosin, or thick filaments. These, in turn are composed of aggregates of actin and myosin molecules, respectively.

The actin and myosin filaments are arranged in repeating units called sarcomeres. Each sarcomere is bound by a Z-

•

line on each end. The actin filaments are attached to the Z-lines on each end of the sarcomere and extend inward toward the center of the sarcomere. Since they do not quite contact each other, they leave a space between them in the center of the sarcomere. There are several rows of these actin filaments parallel to each other in each sarcomere. The pairs of actin filaments alternate with single myosin filaments, which do occupy the center of the sarcomere and extend outward toward, but do not touch, the Z-lines.

A nerve impulse from a motor neuron causes a **muscle contraction**, in which the myosin filaments "grab" the actin filaments, pulling them closer together and shortening each sarcomere, and therefore the entire muscle. Each myosin molecule has a "head" and a "tail." The tails combine to make the myosin filament, and the heads stick out to the sides of the filament. The myosin heads are capable of hydrolyzing **adenosine triphosphate (ATP)**, and when they do, some of the energy released changes the shape of the head and allows it to bind to a specific site on the actin filament. When it returns to its normal shape, it pulls the actin filament toward the center of the sarcomere and then releases the actin. This attachment-pull-release process can be repeated five times per second. The myosin heads act like the oars of a crew team, but instead of pulling a boat through the water, they pull actin filaments closer together.

Rigor mortis, the stiffness of the body that occurs after **death**, is a result of lack of ATP and the inability of myosin to release the actin to which it is bound.

See also Muscular innervation; Muscular system overview; Skeletal and muscular systems, embryonic development; Skeletal system overview (morphology)

SKELETAL MUSCLE SYSTEM, EMBRYONIC DEVELOPMENT

The skeletal and muscular systems are taught as individual subjects in traditional **anatomy** courses. However, they are intimately linked in both function and development. Early in development, the skeletal system is formed by **cartilage**. This embryonic structure is often labeled as a template from which bone cells deposit and maintain the health of growing bones. In turn, muscle cells have a completely different mode of development. Eventually, it is the action and growth of both systems that contribute to the formation of each other.

At first glance, bone may appear to be anything but a **tissue**. It is, however, a connective tissue complete with cells, fibers, and ground substance (minerals that make crystalline structures). Early in the growth process, osteogenic (osteoprogenitor) cells form from embryonic mesenchyme tissue. They are found in the endosteum and lie within the Haversian (central) canals. Many osteogenic cells are also located on the inner surface of the periosteum, which is logical since bone growth continues to occur from the inside to the outside of the bone. These cells are the only bone cells capable of mitosis. As a result, they are the only cells to produce new bone cell types.

One of the cells to develop from the osteogenic cells is the osteoblast. These cells are the actual bone-forming cells. The osteogenic cells lie on the surface of the cartilaginous template forming a small depression or lacuna. When they divide, a permanent resting place for future bone cells, the lacuna, becomes the site of bone formation. The osteoblasts synthesize collagen and glycosaminoglycans (GAGs), which are part of the bone matrix (supporting tissue). Osetoblasts cannot undergo mitosis, but the osteogenic cells rapidly multiply under stress or fractures and produce many more osteoblasts. This process of bone formation under stress will become very important when the role of muscle formation is discussed.

As the osteoblasts continue to deposit bone matrix by depositing the two molecules just discussed in addition to proteoglycans and plycoproteins. The mineral portion of the bone, hydroxyapatite is also deposited and is about 85% of the individual bone's mineral composition. Other depositional minerals are magnesium, sodium, potassium, fluoride, sulfate, carbonate, and hydroxyl ions.

Osteoblasts eventually become trapped in the lacunae as the bone grows around them. At this point the cells are renamed osteocytes. The osteocytes no longer produce bone matrix. Instead, they remain as the active and living part of bone. They monitor **calcium** and phosphate levels and work to maintain the correct balance between the two. Nutrients for proper bone growth and maintenance are passed from one osteocyte to another through small connections (canaliculi) between one osteocyte and another.

Physical stress is one of the most important factors in bone growth. Most of the physical stress put on bones during early development comes from the simultaneous growth of muscles. Skeletal muscles of the trunk arise from cells located along the neural crest. The **somites** first appear as wedge-shaped segments along either side of the **notochord**. They arise from the **mesoderm** and at first only one-three pairs are visible. At various stages during growth, more somite pairs appear and may reach up to 40 pairs. The first four are occipital and eventually migrate to form muscles of the head and neck region. The next eight somites form the cervical section. There are twelve thoracic and five lumbar somites. The first few sacral and the remaining caudal somites will develop into the muscles of the lower region and limbs.

The cells of the somites follow two different paths. The first group (the myotome) becomes the muscles of the axial skeleton. The second give rise to the muscles of the limbs and body wall. Cells for these latter structures migrate from their source and become the precursor cells for later development.

As the somites grow, they begin their origination on certain bones. Their insertion is on other target bones. With continued growth, the muscles begin to contract and gain strength. This puts stress on the individual bones and stimulates the deposition of osteoblasts and formation of newer bone. In the case of the **spinal cord**, each embryonic somite pulls on a region of the notochord. This biomechanical stress causes the bone to fortify. Differentiation of the notochord occurs and the individual sections become the vertebrae. This is the inseparable relationship between bone and muscle growth. As muscle grows so does bone. As bones become larger and stronger they provide a greater surface area for additional muscle fibers of the somites to grow upon. This then causes the muscles to enlarge and become stronger. While the two systems are quite different in growth mechanics and tissue structure, they cannot function and grow without each other.

See also Bone histophysiology; Bone injury, breakage, repair, and healing; Bone reabsorbtion; Muscle tissue damage, repair, and regeneration; Muscles of the thorax and abdomen; Muscular innervation; Muscular system overview; Osteology; Osteoporosis; Posture and locomotion; Sense organs: balance and orientation; Skeletal system overview (morphology)

SKELETAL SYSTEM OVERVIEW (MORPHOLOGY)

The skeletal system is composed of a network of hard and soft tissues including bone, **cartilage**, ligaments, and **tendons**. Together, these tissues form an overall framework, thus giving the body its characteristic shape and providing structural support. Additionally, the skeletal system protects internal **organs** and facilitates movement by providing attachment sites for muscles. Other functions of the skeletal system include a storage facility for minerals such as **calcium**, formation of **blood** cells, and an energy reserve of **lipids**.

The bones of the skeletal system can be classified according to their shape and location. The types of bones categorized by shape—long, short, flat, and irregular—also provide evidence of their function. Long bones consist of an elongated shaft called the diaphysis. Each end of the diaphysis is an expanded portion of the shaft and is called an epiphysis. Examples of long bones include the femur in the thigh and the humerus in the arm. These bones function as levers when muscles contract, thus providing support to enable movement. Short bones often have equal dimensions, like those of a cube. Compared to long bones, short bones have a limited range of motion but are able to withstand force. Examples of short bones include the carpals in the wrist and the tarsals in the ankle. Flat bones are thin, flat bones that protect internal organs and provide sites for muscle attachment. The ribs, cranial bones, and scapula are all examples of flat bones. Irregular bones are not shaped like any of the three aforementioned bones and therefore form their own category. The vertebrae and facial bones are categorized as irregular bones.

Location rather than shape classifies other types of bones, such as sesamoid and sutural bones. Sesamoid bones bear pressure as the result of being buried in tendons. The kneecap, or patella, is the best-known example of a sesamoid bone. Sutural bones are tiny bones located between the **joints**, or sutures, of the cranial bones.

The adult skeleton consists of 206 bones. A baby is born with 270 bones, many of which fuse together during adolescence and adulthood. The bones of males and females differ in that male bones tend to be larger and heavier than female bones.

The skeletal system can be divided into the axial skeleton and the **appendicular skeleton**. The axial skeleton is composed of the bones that surround the midline or axis of the body, forming the head and trunk. These bones include the **skull** bones, auditory ossicles, hyoid bone, **vertebral column**, **sternum**, and ribs.

The skull can be subdivided into eight cranial bones and 14 facial bones. The cranial bones include the frontal bone, two parietal bones, two temporal bones, occipital bone, sphenoid bone, and ethmoid bone. The facial bones include two lacrimal bones, two nasal bones, two inferior nasal conchae, vomer, two zygomatic bones, two maxillae, two palatine bones, and mandible. Within the middle **ear** are three auditory ossicles: the maleus, incus, and stapes. These tiny bones transmit vibrations from the eardrum to the inner ear. The hyoid bone is located in the superior part of the neck and attaches the muscles of the tongue.

The vertebral column typically consists of 26 vertebrae that protect the **spinal cord** and provide attachment sites for ribs and back muscles. The seven most superior vertebrae are the cervical vertebrae. The first **vertebra** is called the atlas and enables the head to move forward and backward. The second vertebra, the axis, is unique in that it is the only vertebra that has a process called the dens or odontoid process. The axis enables the head to rotate from side to side. The vertebrae immediately inferior to the cervical vertebrae are the 12 thoracic vertebrae. These vertebrae are larger than the cervical vertebrae and, except for the eleventh and twelfth thoracic vertebrae, have facets that articulate with the ribs. The point where two bones meet forms a joint and the bones are said to articulate with one another. Just below the thoracic vertebrae are the five lumbar vertebrae. The lumbar vertebrae are the largest of the vertebrae because they support a tremendous amount of the body's weight. The five sacral vertebrae are actually fused together in adults to form the sacrum. Inferior to the lumbar vertebrae, the sacrum articulates with the pelvic girdle to form the pelvis. The four remaining bones of the vertebral column constitute the coccyx. These individual bones also become fused together in adults.

The sternum, also known as the breastbone, consists of three parts. The manubrium and the body are the superior and middle parts of the sternum that articulate with the ribs. Additionally, the manubrium articulates with the clavicles. The xiphoid process is the inferior part of the sternum that provides attachment for abdominal muscles.

There are 12 pairs of ribs that make up the rib cage. The first seven pairs are true ribs because they are attached directly to the sternum by cartilage. The next three pairs of ribs are false ribs because they are indirectly attached to the sternum by the cartilage of the seventh pair. The two remaining ribs are known as floating ribs because they do not connect to the sternum at all.

The appendicular skeleton is comprised of two pectoral girdles, two pelvic girdles, and the bones of the upper and lower extremities. Each pectoral girdle, or shoulder girdle, includes the clavicle and scapula responsible for attaching the upper extremities to the axial skeleton. The clavicle, or collarbone, is the anterior component of the shoulder that articulates

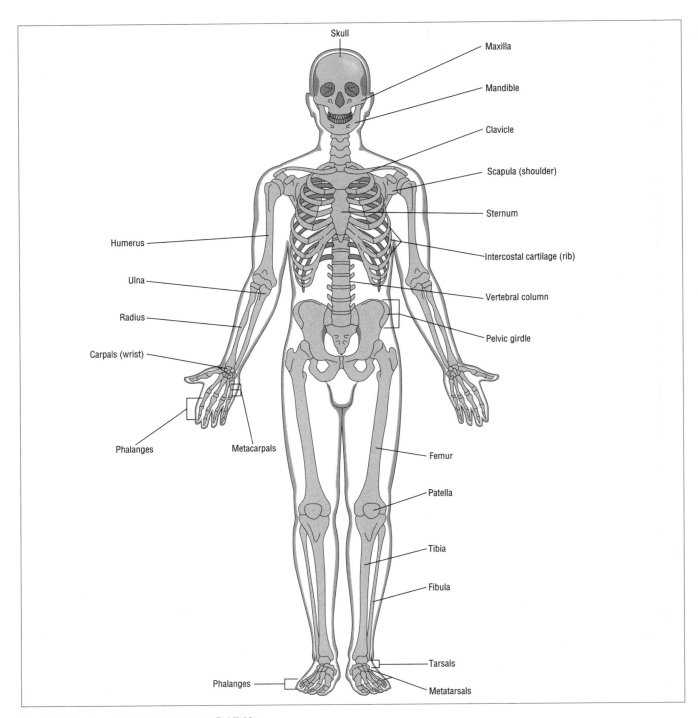

The skeletal system. *Illustration by Argosy Publishing.*

with the scapula and manubrium of the sternum. The scapula, or shoulder blade, is positioned posterior to the clavicle and articulates with the humerus. The humerus constitutes the upper arm and articulates with the two bones of the forearm. The radius is the lateral bone and the ulna is the medial bone of the forearm. The distal end of the radius articulates with the carpals, the first row of bones in the hand. The proximal row of carpals located from lateral to medial includes the scaphoid, lunate, triquetrum, and pisiform. The distal row of carpals that articulates with the metacarpals are the trapezium, trapezoid, capitate, and hamate. The metacarpals are numbered one through five beginning on the lateral palm of the hand extending medially. The 14 bones of the fingers, named **phalanges**, articulate with the metacarpals. Each finger has a proximal, middle, and distal phalanx except for the thumb, which only has two phalanges.

Each pelvic girdle, or hipbone, in an adult is made of three fused bones. Also known as the coxal bones, the hipbones consist of the ilium, ischium, and pubis. The ilium articulates posteriorly with the sacrum. The ischium connects the ilium and pubis. The two pubis bones meet anteriorly to form the pubis symphysis. Together, the hipbones, sacrum, and coccyx constitute the pelvis. One major difference between the male and female skeleton is the bones of the pelvis. In the female, the pelvic bones form a wide, round opening called the pelvic inlet to accommodate for childbirth. The pelvic inlet of males is **heart** shaped and much narrower than in women. Additionally, the sacrum is wider and shorter in women than in men.

The bones of the lower extremities include the femur, patella, tibia, fibula, tarsals, metatarsals, and phalanges. The femur is the leg bone that articulates with the pelvic girdle. The distal end of the femur articulates with the foreleg to form the knee. Anterior to the knee lies the patella, or kneecap. Each foreleg consists of two bones: tibia and fibula. The tibia is the larger of the two bones and forms the shin. The fibula is the lateral bone in the foreleg. At the distal end of the forelegs are the proximal bones of the foot called the tarsals. The tarsals include the calcaneus, talus, navicular, cuboid, and three cuneiforms. The metatarsals form the sole of the foot and are labeled one through five beginning on the medial side of the foot. Each toe consists of three phalanges, the proximal, middle, and distal phalanx. The exception is the big toe that contains only two phalanges.

See also Bone histophysiology; Bone injury, breakage, repair, and healing Bone injury, breakage, repair, and healing; Arthrology (joints and movement)

SKIN • *see* INTEGUMENT

SKODA, JOSEF (1805-1881)
Czech physician

With his detailed classification and interpretation of chest sounds, Josef Skoda improved the science of diagnosing thoracic diseases. His *Abhandlung über Perkussion und Auskultation* (Treatise on Percussion and Auscultation, 1839) is a classic of diagnostics and has been translated into many languages.

Skoda was born the son of a locksmith in Pilsen, Bohemia. After a solid secondary education at the gymnasium in Pilsen, he entered the University of Vienna in 1825 and received his M.D. there in 1831, just when the worldwide cholera epidemic of 1831–1832 was beginning to sweep Europe. Skoda returned to his native land to help fight the cholera, but became so frustrated by the inefficacy of his med-

ical knowledge that he decided to get further training in Vienna. At first he worked in the pathological **anatomy** laboratory of Karl von Rokitansky (1804–1878) and took an unpaid position as assistant physician at the General Hospital of Vienna, doing mostly charity cases and barely supporting himself with a small supplemental clinical practice.

Around 1836, influenced by the pioneer work of **René-Théophile-Hyacinthe Laënnec**, Jean Nicolas Corvisart des Marets (1755–1821), Pierre Adolphe Piorry (1794–1879), William Stokes (1804–1878), and other stethoscopists, Skoda began studying chest diseases and their diagnoses in depth. His investigations in this area resulted in many discoveries and reinterpretations, so that he achieved instant prominence as a diagnostician when he published his results in the 1839 *Abhandlung*. Several chest sounds are named after him, notably "Skoda's resonance," characteristic of pneumonia, pericardial effusion, and a few other conditions. But even though Skoda's book had immediate sway on the European continent, it was not translated into English until 1853.

Because of the recent fame of his book and through Rokitansky's intercession on his behalf, Skoda received several prestigious appointments in Vienna culminating with his being named professor of internal medicine at the University of Vienna in 1846. He found his niche as a teacher. He was the first medical professor in Vienna to lecture in German rather than Latin. Throughout the 1840s he actively encouraged the efforts of Ferdinand von Hebra (1816–1880) to establish dermatology as a separate discipline. After 1848, perhaps because of his debilitating heart condition but more probably because of his dedication to his teaching duties and his clinical practice, he published very little. When he retired from teaching in 1871, his students arranged for a torchlight parade in his honor. He lived simply, almost monastically, never married, and died in Vienna.

Rokitansky, Skoda, and other leading nineteenth-century Viennese physicians and medical educators prescribed drugs sparingly and often just advised their patients to change their lifestyles or diets rather than undergo rigorous medical interventions. Some Viennese doctors even advocated withholding treatment entirely in some cases. Instead, led in part by Skoda, they emphasized public health and sanitation. Because of this minimalist therapeutic philosophy, the Viennese medical community had a bad reputation, especially in France, as "nihilistic." Yet the fact remains that, in some cases, the simple Viennese therapies proved more effective against disease than the more complicated French therapies.

See also Cardiac disease; Dermis; Fever and febrile seizures; Heart defects; Lungs; Pathology; Respiratory system

SKULL

The skull is the ossified, bony structure that encloses and protects the **brain**, internal extensions of sensory **organs**, and some facial structures. The skull is usually considered to consist of a cranial section (the **cranium**) and a facial region.

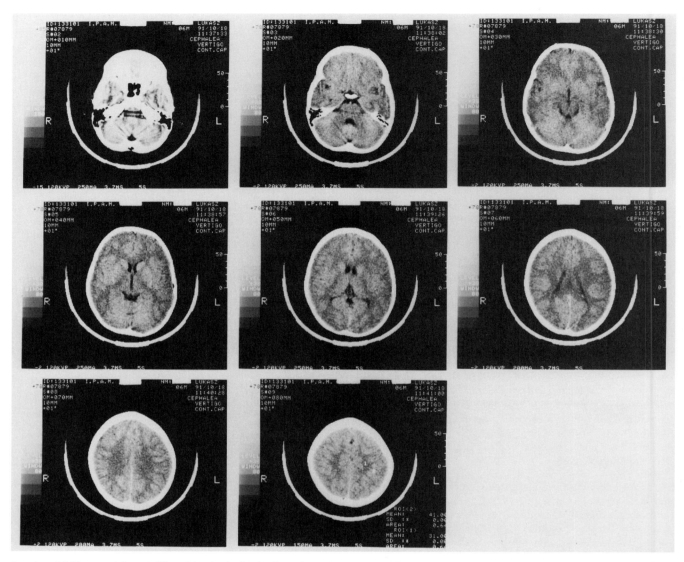

A series of CAT scans, taken at different levels, depict the dense bone tissue composition of the human skull. *Photograph by Mauritus GMBH. Phototake NYC. Reproduced by permission.*

The cranium is a large rounded, dome-shaped region of the skull that is composed of paired left and right frontal bones, parietal bones, temporal bones, and an unpaired occipital bone that forms the posterior base of the skull.

The bones of the cranium are fused by sutures—joints that run jaggedly along the interface between the bones. At birth, the sutures are soft, broad, and cartilaginous. This flexibility allows the skull to grow as the child matures. The sutures eventually fuse and become rigid and ossified near the end of **puberty** or early in adulthood. The coronal suture unites the frontal bone with the parietal bones. In **anatomical nomenclature**, the primary coronal plane is, of course, the plane that runs through the length of the coronal suture. At right angles to the coronal suture, the metopic suture separates the frontal bones in the midline region. The area formed by the fusion of

the four bones near the top of the skull is termed the anterior fontanel or bregmatic fontanel (also commonly known as the topmost "soft spot" in a baby's skull). As with the sutures, the fontanels are soft at birth to permit growth. The fontanels shirk and close during childhood and are usually fully closed and hardened by young adulthood.

The sagittal suture unties the two large domed-shaped parietal bones along the midline of the body. The suture is used as an anatomical landmark in anatomical nomenclature to establish what are termed sagittal planes of the body. The primary sagittal plane is the sagittal plane that runs through the length of the sagittal suture. Sagittal planes run anteriorly and posteriorly, are always at right angles to the coronal planes. The lambdoidal suture unites the left and right parietal bones with occipital bone. The area where the two parietals and the unpaired occipital bone meet is termed the posterior fontanel, lamdoidal fontanel, or lambda point (also commonly called the rear "soft spot" on a baby's skull). Like the anterior fontanel, the posterior fontanel closes and hardens with age, but is an important feature that allows growth of the skull during embryological and childhood development.

Along the sides of the cranium, the squamosal suture unties the temporal bone lying above (superior to) the ear and ear canal with the parietal bone. The anterior region of the temporal bones is united with the great wing of the sphenoid bone by continuation of the squamosal suture. The junction of the temporal, parietal, frontal and great wing of the sphenoid takes place at the sphenoid fontanel. The posterior border of the temporal bone on each side unties with the corresponding mastoid bone

A mastoid fontanel lies at the posterior region of the side of the skull where the parietal, occipital, and mastoid bones unite. A mastoid process extends anteriorly toward the ear canal. A bony finger-like styloid process protrudes the area interior to the external auditory opening (external auditory meatus).

The facial area of the skull is composed of the left and right zygomatic arches that extend from the lowest, most anterior margins of the temporal bone where the temporal bones articulate with the mandible (the temporomandibular joint) into the zygomatic bone itself. The zygomatic arches and zygomatic bones thicken to become prominent facial landmarks forming the lower and side orbits of the eyes. The orbits are separated by a number of smaller bones in the nasal region including the ethmoid, lacrimal, and nasal bones. The maxilla and upper teeth form the most inferior region of the facial portion of the skull and are fused to the zygomatic bones.

The mandible is not considered a formal portion of the skull. In decayed bodies, the mandible becomes detached from the skull as the temporomandibular joint and supporting ligaments deteriorate.

A number of small openings allow nerves and **blood** vessels to penetrate the skull. These openings are termed foramen and are generally named for the bone they penetrate. For example, opening in the parietal bones are termed parietal foramen. A large foramen magnum at the rear and base of the skull allows the **spinal cord** to exit the skull into the **vertebral column**. Rounded, smooth, bony protuberances termed the occipital condyles lie on the anterior sides of the foramen magnum and help articulate the skull with the vertebral column.

The external occipital crest marks the posterior midline of the occipital bone. The crest runs from the foramen magnum upward (superiorly) to a bony knot-like external occipital protuberance.

See also Bone histophysiology; Child growth and development; Meninges; Skeletal and muscular systems, embryonic development; Skeletal system overview (morphology)

SLEEP

Sleep is a natural state of rest that is as essential to a person's well being as food and water. Without enough sleep, the ability to perform even simple tasks declines dramatically. Sleep is easily distinguished from other sleeplike states, such as a **coma**, because it is easily interrupted by external stimulation.

The sleep process is governed by a complex architecture and characterized by five stages. During the first stage, muscles relax and a person's **brain** waves become irregular and rapid. But the official onset of sleep doesn't begin until stage two, when **eye** movements cease and brain activity intensifies. Eventually, brain activity subsides and the person drifts into a profound slumber. The next stages, three and four, are marked by large, slow brain waves. A person who awakens during these stages of deep sleep will usually feel groggy and disoriented. About 75% of sleep is spent in these first four stages.

Stage five is called REM (rapid-eye-movement) sleep. This is the period when dreams occur and the brain reviews and discards the day's memories. **Blood** flow in the brain increases and brain waves appear as if a person were awake. The entire body is immobile with the exception of the eyes and **respiratory system**. Scientists contend that this is a natural defense mechanism that prevents a person from acting out his or her dreams.

A complete sleep cycle takes about 90 to 110 minutes and then repeats. With each new cycle, a person falls into progressively deeper stages of non-REM sleep and each successive REM period is longer than the last.

A person's sleep pattern is regulated by their body's circadian rhythms and an internal biological 'clock' located in the brain. For most adults, sleepy peaks occur every 12 hours, as the body's temperature naturally decreases. Normal sleepers have a relatively predictable pattern of REM (rapid-eye-movement) and non-REM sleep.

How long a person sleeps varies significantly with age and each individual. For most adults, doctors advise 7–8 hours of sleep every day. Newborns, which sleep between 17 and 18 hours daily, spend more time in deep sleep than adults. Older adults spend more time in the lighter sleep stages.

Difficulty falling or staying asleep is a common problem. About half of Americans have trouble sleeping at least occasionally; 40 million suffer from chronic, long-term sleep disorders. Sleeplessness can take a serious toll on a person's health and livelihood. Lack of sleep can result in poor concentration, mood swings, and an inability to perform even simple tasks.

See also Brain: Intellectual functions

SMALL INTESTINE • *see* GASTROINTESTINAL TRACT

SMELL, PHYSIOLOGY OF OLFACTORY SENSES

The **physiology** of smell in humans begins in the nasal cavity. There, a huge number of receptors (over 40 million) are located in the upper roof of the cavity. The receptors have **cilia** projections that stick out into the cavity space. These increase the surface area and the sensitivity of the receptors.

One reason for the receptor sensitivity concerns the mechanics of airflow in the nasal cavity. The air rushes in quickly (at about 250 milliliters per second) and is turbulent. Thus, not all of a particular odor will have a chance to contact a receptor. So, a receptor must be able to swiftly detect a low concentration of a molecule.

The olfactory receptor cells are replaced every three to four weeks.

The receptors are responsible for detecting a large number of odours (about 2,000, depending on the individual). A group of genes is known to encode proteins associated with the receptors that may function in the specific detection of an odor. There may be upwards of 1,000 very specific odor receptors.

Odors reach a receptor by diffusing through the air and physically contacting the receptor. Surrounding the cilia is a mucous membrane. It is into this membrane that an odor dissolves. The binding of an odor molecule to a receptor stimulates the activation of a protein called the G-protein and the release of **calcium** from the receptor membrane. These events begin the process whereby an electrical potential is generated. The potential constitutes the signal that is sent off to the **brain**.

A signal is relayed to the anterior olfactory nucleus, which is essentially a collection point for the receptor signals. The signals are then routed to a region of the brain responsible for the processing of the information. This region is known as the primary olfactory cortex.

Following the stimulation of a receptor, the odor molecule is rapidly destroyed and the stimulation ended. This frees the receptor for stimulation by another odor molecule. In this way the sensitivity of the smell sensory system is maintained.

Dogs have a much greater sense of smell than humans. Their receptors have almost 20 times the surface area as human receptors, and there are 100 times more receptors per square centimeter in a dog's nasal cavity than in a humans.

See also Action potential; Breathing; Cilia and ciliated epithelial cells

SMOOTH MUSCLE

Smooth muscle gets its name because there are no striations visible in the muscle **tissue**. In other muscles, dark and light bands are seen under microscopic examination. Nevertheless, smooth muscles do contain different types of filaments. Thick filaments are called myosin and thin filaments are called actin. These filaments are present in other types of muscles, such as striated muscle, in addition to smooth muscle.

The cells of a smooth muscle are long and spindle-shaped. In hollow **organs** they tend to be arranged in bundles in an outer layer that is oriented with the long axis of the organ, and in a circular pattern around the inner surface.

Smooth muscles are found in internal organs. Examples include the uterus, the **bronchi** of the **lungs**, the bladder, and the walls of **blood** vessels.

The myosin and actin filaments are designed to slide against each other. In smooth and other muscles this is known as the sliding filament model of contraction. The movement of the actin and myosin filaments against one another is an energy-requiring process. This back-and-forth ability allows the muscles to contract. The individual muscle filaments do not become shorter. The process is analogous to the operation of a bicycle air pump, where a central rod slides in and out of the surrounding housing.

The contraction of smooth muscle can be stimulated by the release of chemicals known as **neurotransmitters** by nearby motor **neurons** of the **autonomic nervous system**. Two examples of neurotransmitters are noradrenaline and nitric oxide. Smooth **muscle contraction** can occur in the absence of motor neurons. For example, if a compound called histamine is released, the smooth muscles lining the air passages will contract. This is the basis of an asthma attack, where the muscle contraction restricts airflow through air passages. **Hormones** present in the blood also stimulate the contraction of smooth muscles. A well-known example is the contractions of the uterus during childbirth by the hormone **oxytocin**.

The contraction of smooth muscles is typically slower than the contraction of striated muscle. As well, smooth muscle contraction tends to be maintained for a longer time. Smooth muscle contraction is not intended to provide the explosive power that contraction of a striated muscle provides.

Smooth muscles are not under conscious control. As such, they are also known as involuntary muscles. As an example, the smooth muscles of the bladder contract involuntarily to expel urine. In fact, trying to suppress the contractions requires great concentration and is often uncomfortable.

See also Adenosine triphosphate (ATP)

SNEEZE REFLEX

The sneeze reflex is a coordinated neural and muscular response to the irritation of the upper **respiratory system**, especially the nasal orifice (opening) and nasal passages. As with the **cough reflex**, sneezing is a reflex action that does not require conscious direction or control. Sneezing is a nociceptive reflex, designed to protect the body from injury and maintain respiratory integrity.

Sneezing is initiated by irritation of the afferent sensory lining of the respiratory passages. These afferent neural impulses travel via the fifth cranial nerve to the medulla of the **brain** and result in the appropriate muscular excitations to produce a violent expulsion of air designed to clear the respiratory passages. Irritation of the respiratory passages may result from debris, dust, mechanical obstruction of the airway, or by an excessive buildup of fluid that obstructs the nasal passages.

The coordinate sneeze reflex involves the depression of the uvula so that air is forced out of the **nose** and mouth. The forceful contraction of abdominal muscles, **diaphragm**, and intercostal muscles (muscles between the ribs) produce the high pressure needed to generate the air velocity required to expel the source of irritation.

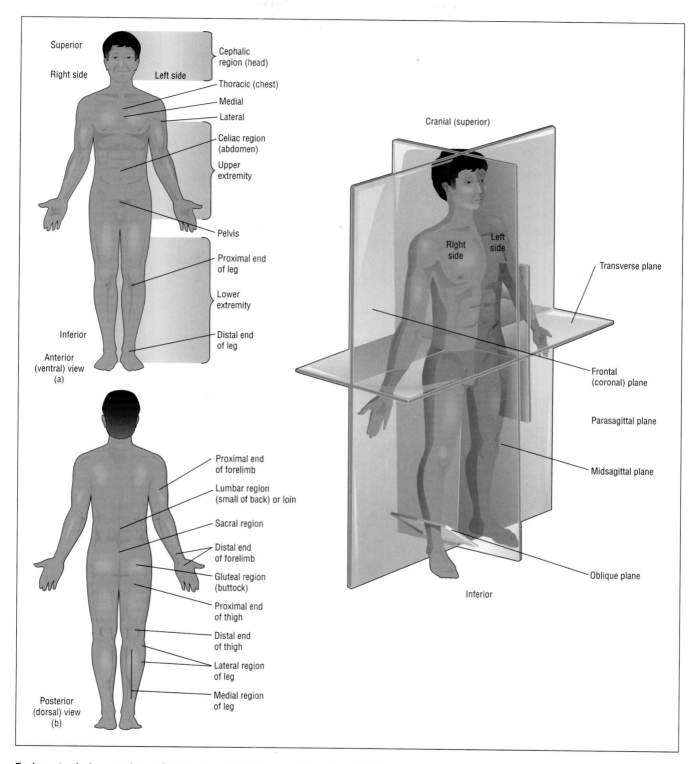

Basic anatomical nomenclature. See entry, "Anatomical nomenclature," page 20. *Illustration by Argosy Publishing.*

A comparison of hearts during transplant surgery shows the diseased heart (held in the hand) beside the implanted healthy heart. See entry, "Cardiac disease," page 83. *Photograph by Alexander Tsiaras. National Audubon Society Collection/Photo Researchers, Inc. Reproduced by permission.*

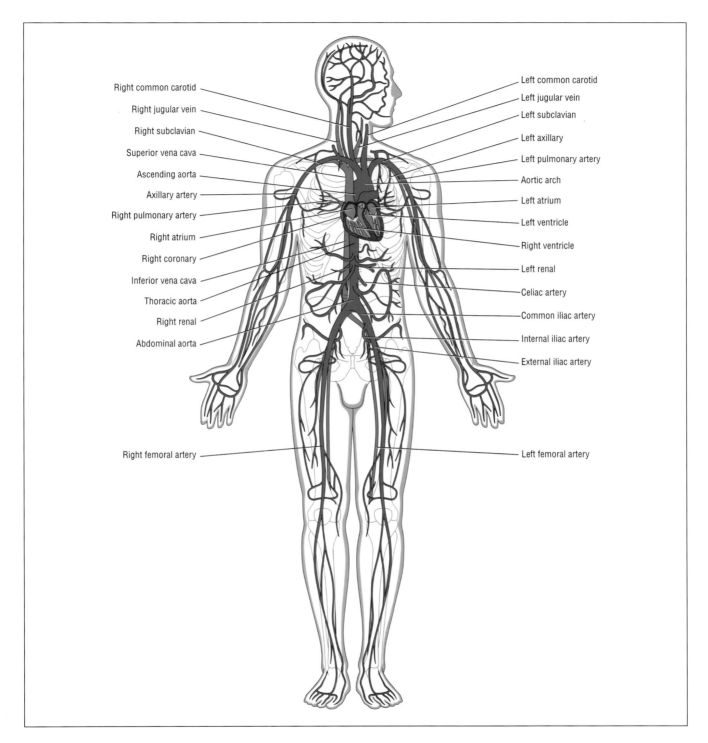

Right common carotid

Right jugular vein

Right subclavian

Superior vena cava

Ascending aorta

Axillary artery

Right pulmonary artery

Right atrium

Right coronary

Inferior vena cava

Thoracic aorta

Right renal

Abdominal aorta

Right femoral artery

Left common carotid

Left jugular vein

Left subclavian

Left axillary

Left pulmonary artery

Aortic arch

Left atrium

Left ventricle

Right ventricle

Left renal

Celiac artery

Common iliac artery

Internal iliac artery

External iliac artery

Left femoral artery

The circulatory system. See entry, "Circulatory system," page 108. *Illustration by Argosy Publishing.*

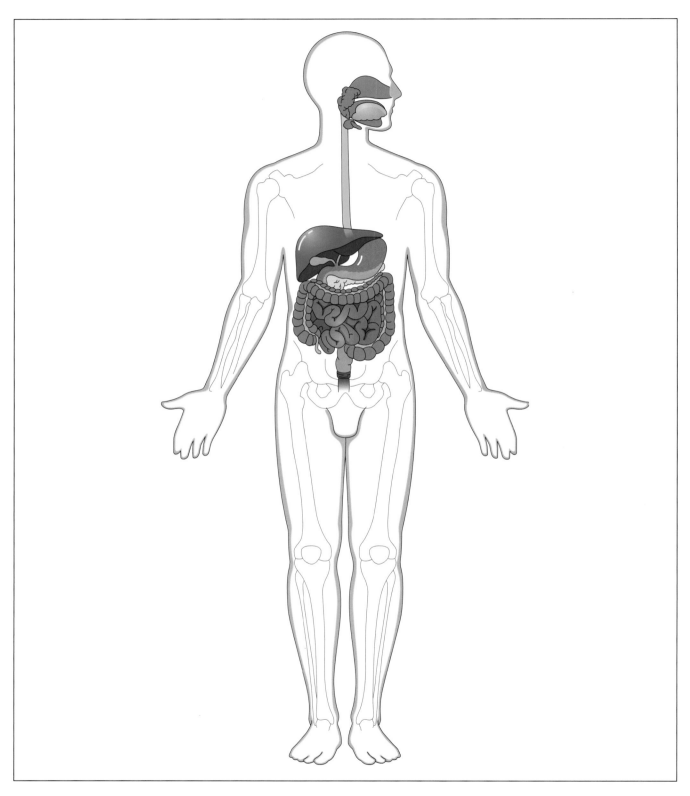

The digestive system: salivary glands (turquoise); esophagus (bright yellow); liver (bright red); stomach (pale gray-blue); gall bladder (bright orange against the red liver); colon (green); small intestine (purple); rectum (shown in pink, continuing the colon); anus (dark blue). See entry, "Digestion," page 143. *Illustration by Argosy Publishing.*

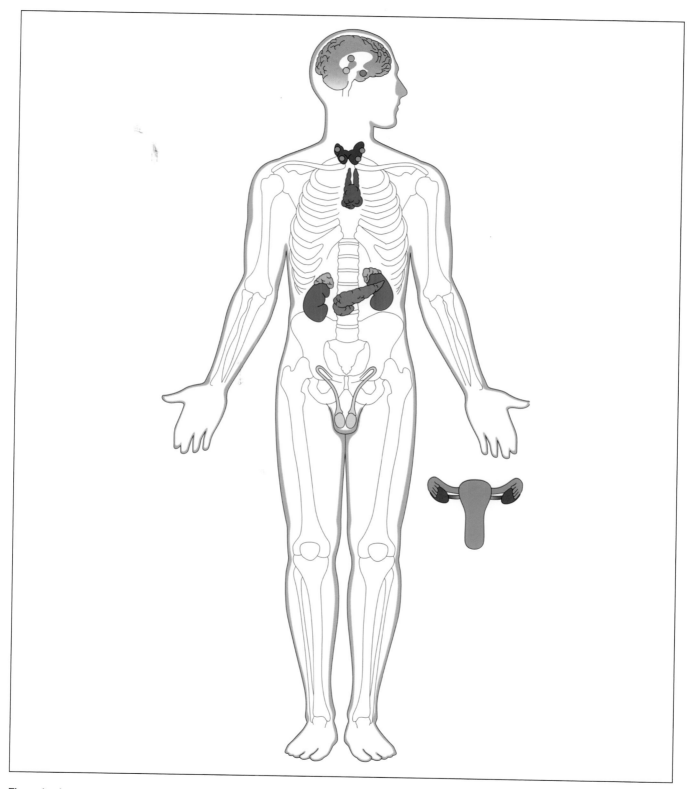

The endocrine system. In the brain: pituitary gland (blue); hypothalamus (pale green); pineal gland (bright yellow). Throughout the rest of the body: thyroid (dark blue); parathyroid glands (four, adjacent to the thyroid); thymus (green); pancreas (turquoise); adrenal glands (in apricot, above kidneys); testes (in males, shown in yellow); ovaries (in females, shown in dark blue in inset image). See entry, "Endocrine system and glands," page 176. *Illustration by Argosy Publishing.*

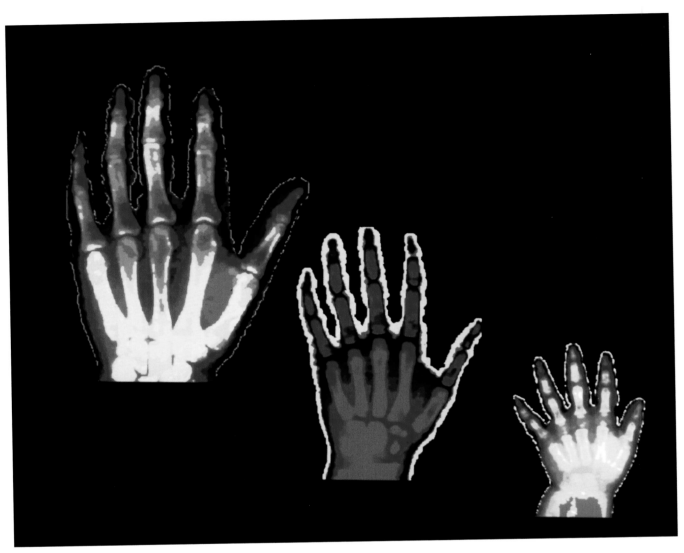

A computer-enhanced x-ray image depicting the hand development of a male at two (far right), six, and nineteen years (far left) of age. See entry, "Growth and development," page 246. *X-ray by Scott Camazine.* © *Scott Camazine, Photo Researchers, Inc. Reproduced by Permission.*

Healthy lung tissue (left) contrast with the diseased lung tissue (right) damaged by cigarette smoking. See entry, "Lungs," page 364. *Photograph by A. Glauberman. National Audubon Society Collection/Photo Researchers, Inc. Reproduced by permission.*

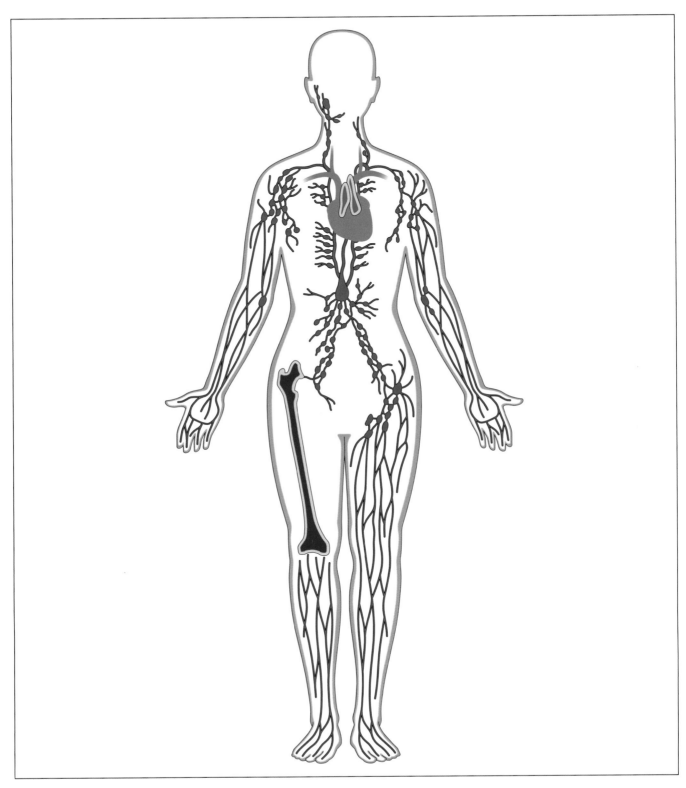

The lymphatic system: lymph nodes (pale green); thymus (deep blue); and one of the bones rich in bone marrow (essential in the circulatory and lymphatic processes)—the femur—shown in purple. See entry, "Lymphatic system," page 366. *Illustration by Argosy Publishing.*

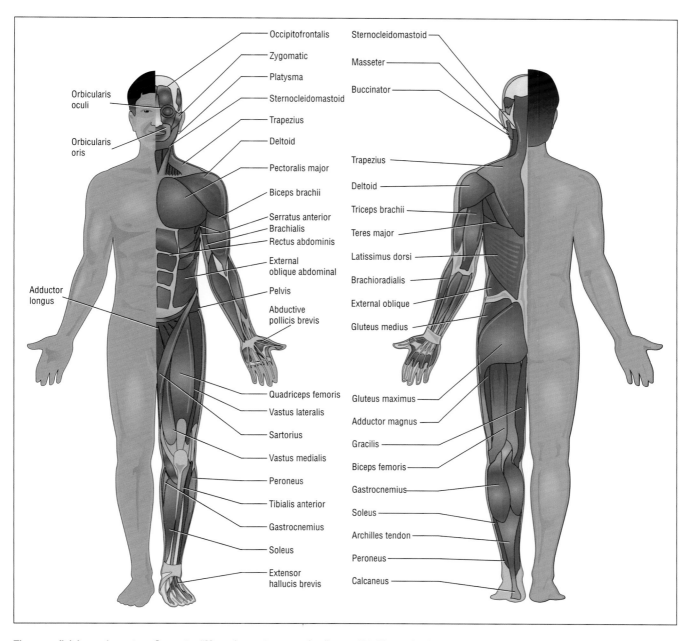

Occipitofrontalis
Zygomatic
Platysma
Orbicularis oculi
Sternocleidomastoid
Orbicularis oris
Trapezius
Deltoid
Pectoralis major
Biceps brachii
Serratus anterior
Brachialis
Rectus abdominis
External oblique abdominal
Adductor longus
Pelvis
Abductive pollicis brevis
Quadriceps femoris
Vastus lateralis
Sartorius
Vastus medialis
Peroneus
Tibialis anterior
Gastrocnemius
Soleus
Extensor hallucis brevis

Sternocleidomastoid
Masseter
Buccinator
Trapezius
Deltoid
Triceps brachii
Teres major
Latissimus dorsi
Brachioradialis
External oblique
Gluteus medius
Gluteus maximus
Adductor magnus
Gracilis
Biceps femoris
Gastrocnemius
Soleus
Archilles tendon
Peroneus
Calcaneus

The superficial muscle system. See entry, "Muscular system overview," page 406. *Illustration by Argosy Publishing.*

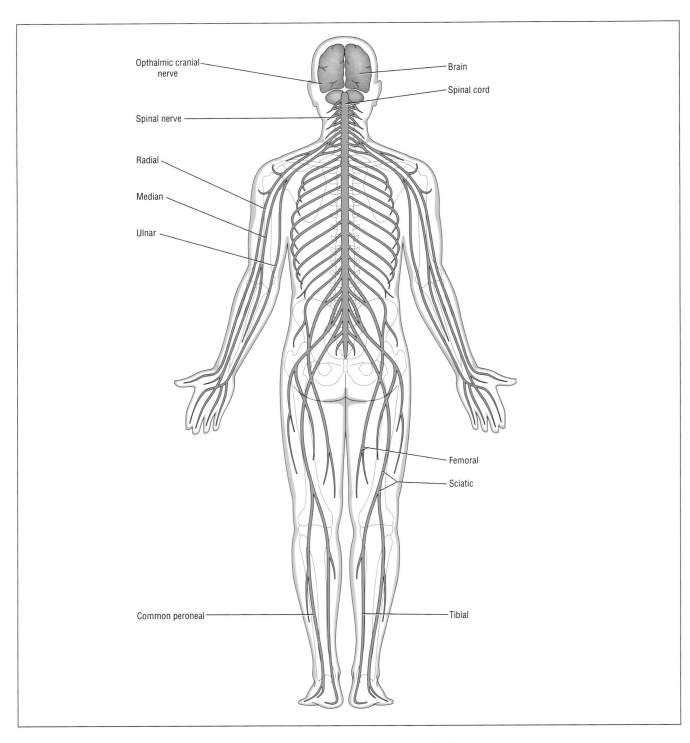

Opthalmic cranial nerve

Brain

Spinal cord

Spinal nerve

Radial

Median

Ulnar

Femoral

Sciatic

Common peroneal

Tibial

The nervous system. See entry, "Nervous system overview," page 415. *Illustration by Argosy Publishing.*

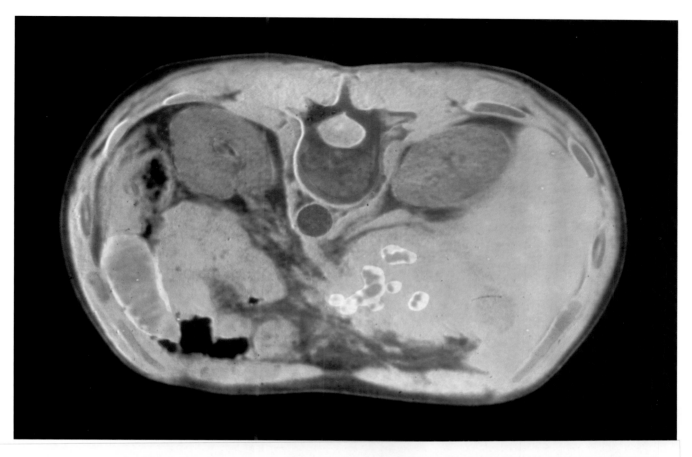

Computed tomography (CT) x-ray scan through the human abdomen. Visible structures include the liver (right), spleen (left), abdominal aorta (center), vertebral column, spinal cord, and kidneys (upper left and upper right). See entry, "Organs and organ systems," page 429. *Photo Researchers, Inc. Reproduced by permission.*

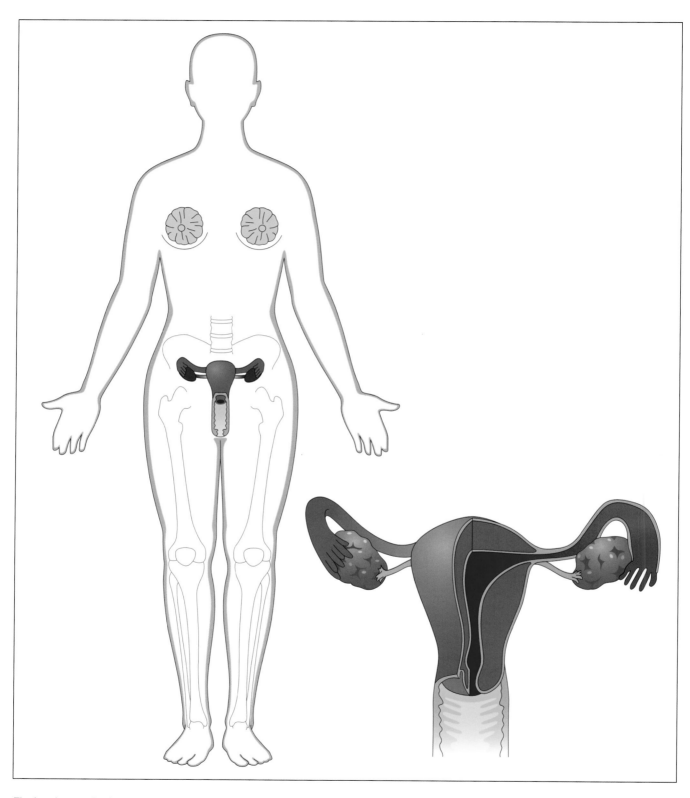

The female reproductive system: uterus (shown in red with the Fallopian tubes); ovaries (blue); vagina (pink with a yellow lining); breasts (apricot). Inset shows detail of ovaries, uterus, Fallopian tubes, and cervix (turquoise). See entry, "Reproductive system (female)," page 480. *Illustration by Argosy Publishing.*

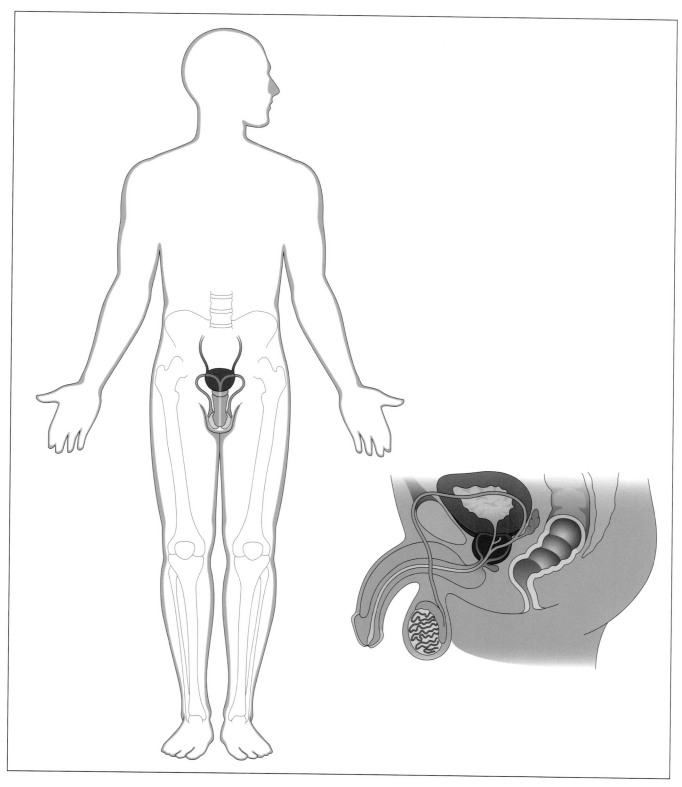

The male reproductive system: penis (pink); testes (yellow); prostate gland (in full-body illustration, shown in peach/apricot, and in the inset as dark blue gland between the bladder and the penis); vas deferens (apricot, in insert); epididymis (yellow-green). See entry, "Reproductive system (male)," page 482. *Illustration by Argosy Publishing.*

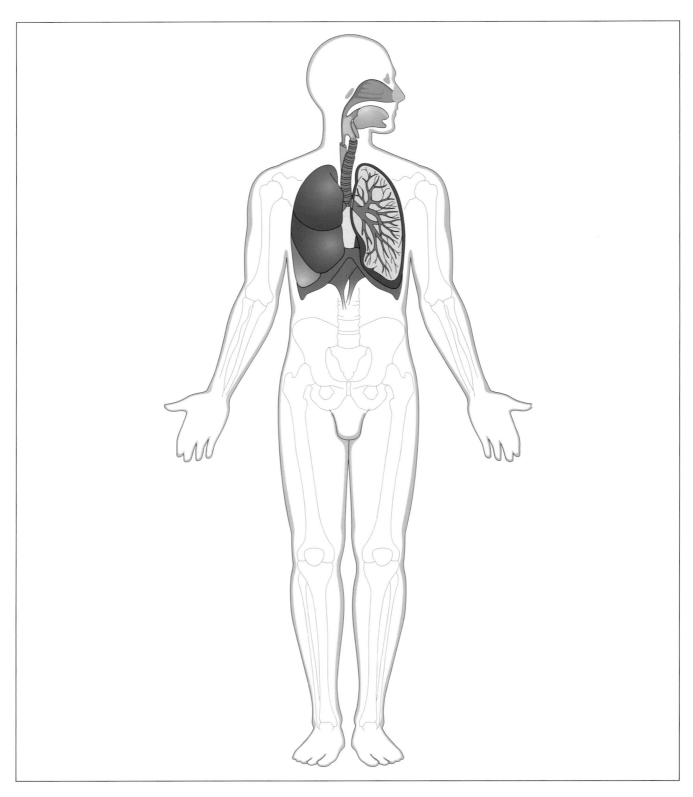

The respiratory system: pharynx (orange); larynx (green ridged tube); esophagus (involved in digestion, not breathing, is shown as smooth green tube); trachea (purple); lungs (deep blue). See entry, "Respiratory system," page 486. *Illustration by Argosy Publishing.*

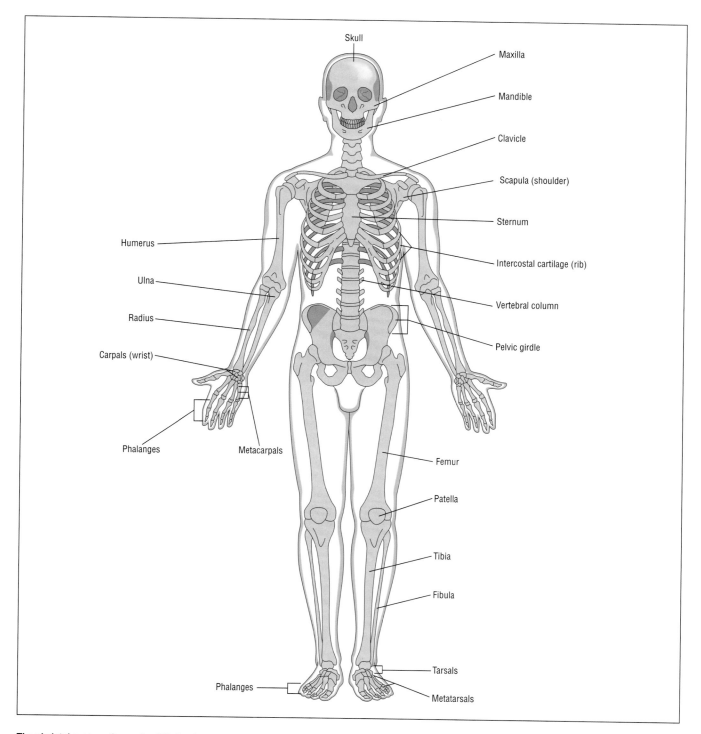

The skeletal system. See entry, "Skeletal system overview (morphology)," page 519. *Illustration by Argosy Publishing.*

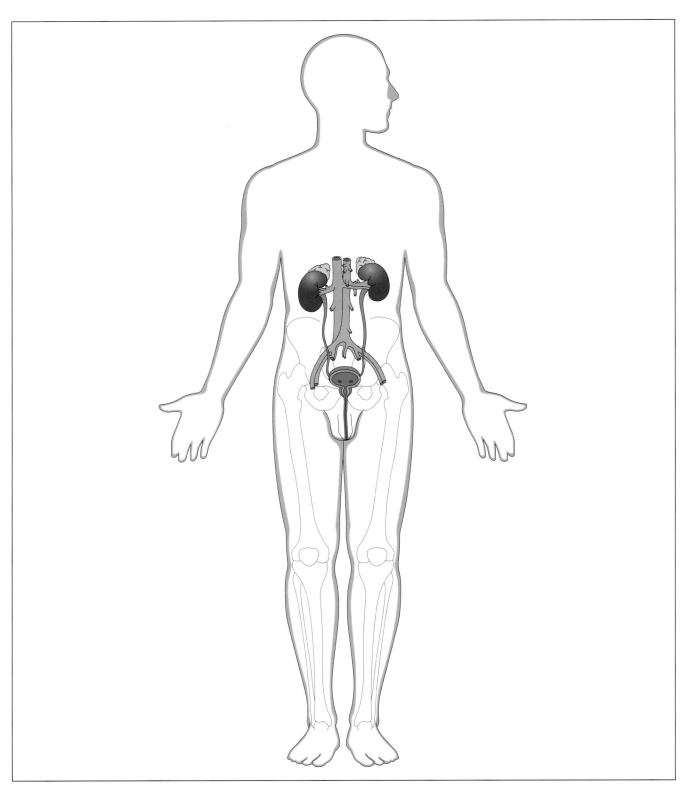

The urogenital system: kidneys (purple); ureters (green); bladder (blue-green). See entry, "Urogenital system," page 570. *Illustration by Argosy Publishing.*

Just as with other forms of neural transmission, the transmission of sneeze producing neural **reflexes** occurs via well established mechanisms and pathways of alternating electrical and chemical neural transmission. Within the neuron cell body (axon), neural impulses travel electrically via an **action potential** that moves down the axon to the presynaptic terminal end of the presynaptic neuron. At the **synapse** (the intercellular gap or space between **neurons**), the neural signal is transmitted through the release of **neurotransmitters** from the presynaptic neuron, diffusion of neurotransmitters across the synaptic gap, and the site-specific binding of neurotransmitters to the dendrites of the next neuron in the neural chain. Reflex impulses are limited to pathway dependent physical limitations (e.g., the physical attributes of the nerve fibers) and are subject to reflex rebound and reflex fatigue.

The sneeze reflex works to maintain a relatively free flow of air. In addition to foreign bodies and specific allergens, irritation by corrosive gases and chemicals may induce a sneeze reflex designed to expel the irritating particles or gas that initiate the reflex.

See also Brain stem function and reflexes; Nerve impulses and conduction of impulses; Nervous system overview; Neurology; Yawn reflex

SNELL, GEORGE DAVID (1903-1996)

American immunogeneticist

Geneticist George David Snell's pioneering research on the **immune system** in the 1930s and 1940s enabled medical science to develop the process of organ transplantation. Through skin grafts performed on mice at the Jackson Hole Laboratory, Snell discovered the factor (known as histocompatibility) that enables doctors to determine whether **organs** and tissues can be successfully transplanted from one body to another. Snell's research earned him the 1980 Nobel Prize in physiology or medicine.

One of three children, Snell was born in Bradford, Massachusetts, to Cullen Snell and the former Kathleen Davis. Snell's father developed and manufactured many inventions, including a mechanism for starting motorboat engines. In Snell's fifth year, the family moved to nearby Brookline. Snell's interests while growing up were varied, and included science, math, sports, and music.

After enrolling at New Hampshire's Dartmouth College in 1922, Snell was influenced to major in biology after taking a **genetics** course taught by Professor John Gerould. He obtained a B.S. degree in that subject in 1926 and enrolled at Harvard that same year to study genetics under the renowned biologist William Castle, who was among the first American scientists to delve into the biological laws of inheritance regarding mammals. Snell received a Ph.D. in 1930 after completing his dissertation on linkage, the means by which two or more genes on a chromosome are interrelated. That same year he became an instructor of zoology at Rhode Island's Brown University, only to leave in 1931 to work at the University of Texas at Austin following receipt of a National Research Council Fellowship.

Snell's decision to accept the fellowship turned out to be a momentous one, as he began work for the famed geneticist Hermann Joseph Muller, whose research with fruit flies led to the discovery that x rays could produce mutations in genes. At the university, Snell experimented with mice, showing that x rays could produce mutations in rodents as well. Although Snell left the University of Texas in 1933 to serve as assistant professor at the University of Washington, he ventured to the Jackson Laboratory in Bar Harbor, Maine, in 1935 to return to research work. The laboratory, specialized in mammalian genetics, and was well-known for its work in spite of its small size.

After continuing his work with x rays and mice, Snell decided to embark on a new study. Snell's project was concerned with the notion of transplants. Earlier scientific research had indicated that certain genes are responsible for whether a body would accept or reject a transplant. The precise genes responsible had not, however, then been identified.

Snell began his experiments by performing transplants between mice with certain physical characteristics. He quickly discovered those mice with certain identical characteristics—in particular a twisted tail—tended to accept each other's skin grafts. In 1948, Peter Gorer came to Jackson Laboratory from London, England. Gorer, who had also conducted experiments on mice, developed an antiserum. He had discovered the existence of a certain antigen (foreign protein) in the **blood** of mice that induced an immune reaction when injected into other mice. Gorer had called this type of substance "Antigen II."

In collaboration, Snell and Gorer proved that Antigen II was present in mice with twisted tails, indicating that the genetics code for Gorer's antigen and the code found by Snell to be vital for **tissue** acceptance were identical. They called their discovery of this factor "H–2," for "Histocompatibility Two" (a term invented by Snell to describe whether a transplant would be accepted or rejected).

Later research revealed that instead of only a single gene being responsible for this factor, a number of closely related genes controlled histocompatibility. As a result, this was subsequently designated as the Major Histocompatibility Complex (MHC). The discovery of the MHC, and subsequent research by other scientists in the 1950s, proving it also existed in humans, made widespread organ transplantation possible. Donors and recipients could be matched (as had been done with blood types) to see if they were compatible.

Eventually Snell was able to produce what he called "congenic mice"—animals that are genetically identical except for one particular genetic characteristic. Unfortunately, the first strains of these mice were destroyed in a 1947 forest fire, which burned down the laboratory. However, Snell's tenacity and dedication enabled him to rebound from this setback. Within three years, Snell created three strains of mice which differed genetically only in their ability to accept tissue grafts. The development of congenic strains of mice opened up a new field for experimental research, with Jackson Laboratory eventually being able to supply annually tens of thousands of these mice to other laboratories.

In 1952, Snell became staff scientific director and, in 1957, staff scientist at Jackson Laboratories. In those capacities, Snell continued his research, particularly on the role that MHC plays in relation to **cancer**. Experiments he conducted with congenic mice found that on some occasions the mice rejected **tumors**, which had been transplanted from their genetic twins. This "hybrid resistance" indicated that some tumors provoke an immune response, causing the body to produce antibodies to fight the tumor. This discovery could eventually be of great importance in developing weapons to fight cancer.

Although he retired in 1968, Snell continued to visit the lab, discuss scientific and medical matters with colleagues, and write articles and books. Elected to the American Academy of Arts and Sciences in 1952 and to the National Academy of Science in 1970, he was also a member of international scientific societies, including the French Academy of Science and the British Transplantation Society. Snell won numerous awards during the 1960s and 1970s, such as the Hectoen Silver Medal from the American Medical Association, the Gregor Mendal Award for genetic research, and a career award from the National Cancer Institute. This culminated in his winning the 1980 Nobel Prize in physiology or medicine for his work on histocompatibility. He shared this with two other immunogeneticists, **Jean Dausset** and **Baruj Benacerraf**. After being told of the Nobel committee's decision, Snell said there should have been a fourth recipient—his colleague Peter Gorer who died in 1962, and was thus ineligible to receive the prize.

See also Immune system; Transplantation of organs

SODIUM-POTASSIUM PUMP

Found in the membranes of all animal cells, the sodium-potassium pump is a protein complex used to transport potassium ions (K^+) into the cell and sodium ions (Na^+) out of the cell. For every three Na^+ the pump sends out of the cell, two K^+ are pushed in. This activity results in the net loss of positive charges within the cell and sets up a concentration gradient of both ions.

Because the ions must be transported against their natural tendencies to equalize the concentration inside and outside the cell, the pump is an active process, requiring the hydrolysis of one molecule of **adenosine triphosphate (ATP)** for every cycle. This hydrolysis releases free energy, which is then used to transport the ions.

The pump is composed of two proteins: large and small subunits. The large subunit spans the cell membrane and is the catalytic subunit. This protein harbors binding sites for Na^+ and ATP on its cytoplasmic surface and for K^+ on the extracellular surface. The function of the smaller subunit is unknown. Binding of three Na^+ ions and hydrolysis of ATP causes shape changes (conformational changes) in the large subunit, which results in the opening of the pump to the outside. Release of the Na^+ to the outside and binding of two K^+ ions follows, causing another conformational change that opens the protein to the inside. The K^+ is released, and the cycle repeats.

The action of the pump drives a net current across the membrane, creating an electrical potential. The unequal concentration of K^+ also causes an electrical force. The pumps are especially important in muscle and nerve cells, which require a **membrane potential** to function correctly. During nerve transmission and **muscle contraction**, K^+ exits the cell and Na^+ enters, resulting in an electrical charge change that causes a nerve impulse or muscle contraction. The pump resets the electrical charge within the cell. The pump also brings K^+ into the cell to offset the large negative charge from organic molecules (e.g., **DNA** and proteins) found in the cell.

The gradient resulting from unequal Na^+ concentrations provides a source of free energy used to drive conformational changes in other membrane-bound proteins, resulting in the import of other molecules, such as glucose, against their concentration gradient. The Na^+ gradient is also used to prevent cellular swelling: the high concentration of solutes found inside the cell creates an osmotic gradient that tends to draw water into the cell. This force is counteracted in animal cells by a high concentration of Na^+ in the **extracellular fluid**.

See also Action potential; Cell membrane transport; Nerve impulses and conduction of impulses; Osmotic equilibria between intercellular and extracellular fluids

SOMATIC AND VISCERAL

Somatic and visceral are terms often applied to anatomical processes. In the strictest sense, somatic processes are those that pertain to body general body structure. The large interior **organs** located in the major body cavities (e.g., **heart, liver, lungs**) are referred to as the viscous and the organs described as visceral organs.

For example, there are visceral and somatic sensory tracts in nerve pathways, and sensory **neurons** can be divided into somatic or visceral receptors.

Pain is often divided into somatic or visceral pain (there is also a third type of pain termed neuropathic pain). Somatic pain results from stimulation of pain receptors in musculoskeletal or cutaneous **tissue** (skin and associated integumentary system structures). In contrast, visceral pain resulted from the stimulation of pain receptors in the visceral organs situated in the thoracic or abdominal cavities.

The types of pain differ not only in origin, but also in quality. Musculoskeletal somatic pain is often described as a dull or aching pain that can be localized as emanating from a particular structure. Cutaneous somatic pain is often a burning or pricking sensation localized to a particular surface area of the body. In contrast, visceral pain is often described as deep, pressure or squeezing related sensation that seems to emanate form several areas or as a pain diffuse (spread) over the thoracic, abdominal, or pelvic areas.

Physiologists often study how processes differ between somatic and visceral structures. For example, physiologists may study the conduction of action potentials and differences in the contractile mechanisms of **skeletal muscle** versus visceral **smooth muscle**.

See also Ectoderm; Endoderm; Nervous system overview

SOMITES

Somites are aggregations of cells that lie in pairs along the transient (temporary) **notochord** in developing human embryos. Somites are also found in the midline paraxial (near the axis) mesodermal **tissue** of all vertebrates in early embryonic stages. Somites are formed from mesodermal tissue that thickens and then divides transversely into blocks. Somites ultimately develop into **vertebra**, ribs, muscles, and dermal structures.

Segmentation of the **mesoderm**, starting about the beginning of the fourth week of embryonic development, proceeds in a cranial-caudal direction (from head to tail) as the embryo develops. Ultimately, there are four occipital somites (some researchers assert that there are as many as nine somatic divisions of the same tissue) that contribute to the development of the **skull**. As development proceeds, eight cervical, twelve 12 thoracic, 5 lumbar, 5 sacral and approximately 8 coccygeal somites come to lie along the notochord and developing **spinal cord**.

Somites are comprised of densely packed epitheloid cells. Individual somites are further divided into a ventromedial sclerotome. Cells from this region ultimately form the vertebrae and ribs of the axial skeleton. The dorsolateral portion of the somite cells comprises the dermatome (also called the dermomyotome) that further divides mytotomes and dermatomes.

Cells from the myotomes that ultimately produce muscle tissue striated **skeletal muscle** tissue. Cells from the dermatome undergo a number of additional changes during **cell differentiation**, losing their epitheloid characteristics and joining with other cells to form the **dermis**.

See also Embryology; Embryonic development: early development, formation, and differentiation; Human development (timetables and developmental horizons)

SONOGRAM · *see* IMAGING

SPACE PHYSIOLOGY

Space **physiology** is concerned with the structure and functioning of the body under the conditions encountered by space travelers. To date, these conditions have been confined to the environment of the spacecraft that houses the astronauts. In the future, however, as travel to other bodies in the solar system is undertaken, space physiology will include the atmospheric and gravitational conditions found on these planets, moons, or other stellar bodies.

Aside from the lunar missions of the 1960s, man's extraterrestrial voyages have been confined to orbital forays aboard space capsules or space stations. But even orbiting around

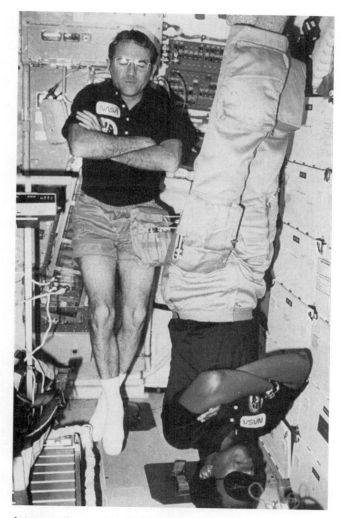

Astronauts living and working in a weightless environment for extended periods of time face special physiological challenges upon their return to Earth. *National Aeronautics and Space Administration. Reproduced by permission.*

Earth poses difficulties for the astronauts. The reduced gravity of a spacecraft makes it difficult for the body to distinguish "up" from "down." On Earth, such distinction by the vestibular organ of the inner ear is easy, because of the orienting power of gravity. In the space shuttle and the developing International Space Station, all writing on the walls is oriented in the same direction, to provide the **brain** with a reference point.

Low gravity (also known as microgravity) affects other body systems besides the vestibular system. The propioceptive system—the system of nerves in the **joints** and muscles that tell us where the arms and legs are without any visual inspection—can also be affected. Low gravity reduces or eliminates the tensions impinging on the joints and muscles, which can make the appendages appear invisible to the brain.

Such confusion between what the **eye** sees and the brain perceives can result in what has been termed space sickness. This is somewhat analogous to the feeling of **nausea** experienced by someone trying to read in a moving car. The inner ear detects the motion of the car, or the spacecraft, but the eyes staring at the page of the book or the space outside the spacecraft does not detect motion. Space sickness is usually transient, and astronauts acclimate soon after going into orbit.

Microgravity also affects the skeletal structure of astronauts. The absence of stress-bearing activity and the loss of components of the bones, particularly **calcium**, have produced shortening and weakening of bones (essentially the development of **osteoporosis**) and the atrophy (wasting away) of muscles in astronauts who have been orbiting the Earth for just several months. Even the **heart** becomes smaller. So far these conditions have reversed upon return to Earth. The extended missions of the future will need to incorporate more Earth-like gravitation conditions, or an exercise regimen, or both.

Space flight also affects the cardiovascular system of astronauts. The weakened muscles and bones cannot support the maintenance of the same rate of flow of **blood** as on Earth. Also, the diminished downward pull of gravity affects the ability of the body to pump blood to extremities like the legs. Fluid flow to the upper regions of the body is not affected, however. As a result, faces of astronauts often appear puffy. In a very real sense, astronauts become out of shape. So much so that Russian cosmonauts who spend months in orbit around the Earth are sometimes carried away from the spacecraft on a stretcher upon their return.

A physiological parameter that will become important when manned travel to other parts of the solar system begins is exposure to higher levels of radiation that will be encountered on planets where atmospheric constituents do not absorb the harmful energies. Genetic material can be damaged by high-energy (ionizing) cosmic radiation and high-energy particles, with adverse effects on the functioning of the body. Thus far, the relatively short-term voyages into space have not proven to be harmful. But the hazards posed by extended voyages of years or even decades are as yet unknown.

See also Acclimatization

SPALLANZANI, LAZZARO (1729-1799)

Italian biologist and physiologist

Spallanzani, the son of a lawyer from Scandiano, Italy attended the University of Bologna and began his studies in law. However, his cousin, Laura Bassi, a professor of physics and mathematics, introduced him to a broad range of scientific studies. Spallanzani altered his educational course and, in 1754, he earned a Ph.D. in philosophy. He joined the priesthood to support himself while he studied natural phenomena, hoping to determine explanations for such events as a stone skipping on water, the regeneration of decapitated snail heads, and the electric discharge of torpedo fish. Over the course of his career, Spallanzani would examine the pits of spitting volcanoes, the waters of eels, the dark depths of the bat's home, and the intricacies of the reproductive and vascular systems.

Yet, Spallanzani's greatest contribution was in the area of what was assumed to spontaneous generation of microorganisms. The theory of spontaneous generation held that living creatures could develop from lifeless matter, especially from decaying matter. For instance, **Aristotle** believed that animal life generated spontaneously from mud, dung, or decaying timber. Other scientists believed alligators arose from Nile River mud, worms came from Thames River mud, and mites came from cheese.

Francesco Redi, (1626-1697) an Italian physician and naturalist, conducted experiments in the seventeenth century that first dispelled the myths of spontaneous generation. Using the theory that decaying products only served as a nesting site for maggots to lay eggs, Redi showed that in hot weather maggots would appear on exposed meat or dead animals. If the fresh meat was placed in a jar covered with a fine gauze, no maggots appeared.

Spallanzani, meanwhile, set out in 1765 to prove that microorganisms existed because they were already present in some form in the solution, the container, or the air. He took solutions that he knew would "breed" organisms and boiled them for up to an hour. The flasks were hermetically sealed to keep out contaminated air. Under such conditions, no new organisms appeared.

But proponents of the spontaneous generation theory dismissed Spallanzani's experiments, saying only that the boiling process had destroyed elements vital to the propagation of the organisms. It was not until Louis Pasteur's experiments on **bacteria** a century later, that Spallanzani was proved right. Spallanzani's work regarding spontaneous generation eventually led to means of food preservation through heat sterilization and canning.

Spallanzani also turned his attention to the **circulatory system**. Viewing the system of **blood** vessels within a hen's egg in 1771, he was able to determine that an arteriovenous network existed in a warm-blooded animal. With further study of the circulatory system, in which Spallanzani studied the changes that occur upon impending **death** as well as the effects of wounds on various parts of the system, he eventually developed a theory of blood pressure. He determined that the arterial **pulse** was not due simply to displacement of the cardiac muscle, but to an intentional and forceful push of blood against the vascular walls.

One of his next inquiries involved the **fertilization** of eggs. He began with the mating practices of frogs and toads. By 1785, when he was working with dogs, he induced the first case of artificial insemination.

Spallanzani's curiosity surrounding natural phenomena took him on an expedition to the volcanoes of Vesuvius, Stromboli, Vulcano, and Etna. Spallanzani's volcanic studies earned him status as a pioneer in the volcanology.

Spallanzani was fascinated by their ability to maneuver without light. Even blinded, the bats could travel and eat without interruption or hesitation. Spallanzani went through the senses, one-by-one, in an attempt to discover which governed the habits of the bat. Through process of elimination, he found that obstructing the bats' ears rendered them directionless. While Spallanzani accepted the theory of echolocation, this theory wasn't explained until 1941 when Donald R. Griffin described the bat's sensitivity to sound waves.

See also Embryology; History of anatomy and physiology: The Classical and Medieval periods; History of anatomy and physiology: The Renaissance and Age of Enlightenment; Reproductive system and organs (female); Reproductive system and organs (male)

SPEMANN, HANS (1869-1941)

German embryologist

Hans Spemann was recognized for his research into the development of embryos, and in particular for his studies into the causes behind the specialization and differentiation of embryonic cells. In the mid–1930s, Spemann discovered "organizers," regions within developing embryos that cause undifferentiated **tissue** to evolve in a specific way. For this finding, Spemann was awarded the 1935 Nobel Prize in physiology or medicine. In addition, Spemann is credited with founding the early techniques of microsurgery, the minute manipulations of tissue or living structure.

The son of a well-known book publisher, Spemann was born in Stuttgart, Germany. He was the eldest of four children of Johann Wilhelm Spemann and the former Lisinka Hoffman. The family was intellectually, socially and culturally active, and lived in a large home that was well stocked with books. Upon entering the Eberhard Ludwig Gymnasium, Spemann first wished to study the classics. Although he later turned to embryology—the branch of biology that focuses on embryos and their development—Spemann never relinquished his love of artistic endeavors. Throughout his lifetime, Spemann organized evening gatherings of friends to discuss art, literature, and philosophy.

Before entering the University of Heidelberg in 1891 to study medicine, Spemann worked at his father's business and served a tour of duty in the Kassel hussars. His strict interest in medicine lasted only until he met German biologist and psychologist Gustav Wolff at the University of Heidelberg. Only a few years older than Spemann, Wolff had begun experiments on the embryological developments of newts and had shown how, if the lens of an embryological newt's **eye** is removed, it regenerates. Spemann remained interested and intrigued by both Wolff's finding, and also in the newt, on which he based much of his future work. Spemann was interested in more than the regeneration phenomenon, especially how the eye develops from the start. Spemann proceeded to devote his scientific career to the study of how embryological cells become specialized and differentiated in the process of forming a complete organism.

Spemann left Heidelberg in the mid–1890s to continue his studies at the University of Munich. He then transferred to the University of Würzberg's Zoological Institute to study under the well-known embryologist **Theodor Boveri**. Spemann quickly became Boveri's prize student, and completed his doctorate in botany, zoology, and physics in 1895. Spemann

stayed at Würzburg until 1908, when he accepted a post as professor at the University of Rostock. During World War I, he served as director of the Kaiser Wilhelm Institute of Biology (now the Max Planck Institute) in Berlin-Dahlem, and following the war, in 1919, Spemann took a professorship at the University of Freiburg.

By the time Spemann began research at the Zoological Institute in Würzburg, he had already developed a keen facility and reputation for conducting well-designed experiments that centered on highly focused questions. His early research followed Wolff's closely. The eye of a newt is formed when an outgrowth of the **brain**, called the optic cup, reaches the surface layer of embryonic tissue (the **ectoderm**). The cells of the ectoderm then form into an eye. In removing the tissue over where the eye would form and replacing it with tissue from an entirely different region, Spemann found that the embryo still formed a normal eye, leading him to believe that the optic cup exerted an influence on the cells of the ectoderm, inducing them to form into an eye. To complete this experiment, as well as others, Spemann had to develop a precise experimental technique for operating on objects often less than two millimeters in diameter. In doing so, he is credited with founding the techniques of modern microsurgery, considered one of his greatest contributions in biology. Some of his methods and instruments are still used by embryologists and neurobiologists today.

Spemann's contributions to **embryology**, for which he won the 1935 Nobel Prize, was his discovery of what he termed the "organizer" effect. In experimenting with transplanting tissue, Spemann found that when an area containing an organizer is transplanted into an undifferentiated host embryo, this transplanted area induces the host embryo to develop in a certain way, or into an entirely new embryo. Spemann called these transplanted cells organizers, and they include the precursors to the **central nervous system**. In vertebrates, they are the first cells in a long series of differentiations of which the end product is a fully formed fetus.

In an earlier series of experiments conducted in the 1920s, Spemann used a method less technically demanding to make another discovery. By tying a thin **hair** around the jelly-like egg of a newt early in embryogenesis (embryo development), he could split the egg entirely, or squeeze it into a dumbbell shape. When the egg halves matured, Spemann found that the split egg would produce either a whole larva and an undifferentiated mass of cells, or two whole larva (although smaller than normal size). The split egg never produced half an embryo. In the case of the egg squeezed into a dumbbell shape, the egg formed into an embryo with a single tail and two heads. Spemann's primary finding in these experiments was that if an egg is split early in embryogenesis, the two halves do not form into two halves of an embryo; they either become two whole embryos, or an embryo and a mass of cells.

This led Spemann to the conclusion that at a certain stage of development, the future roles of the different parts of the embryo have not been fixed, which supported his experiments with the newt's eye. In an experiment conducted on older eggs, however, Spemann found that the future role of some parts of the embryo had been decided, meaning that somewhere in between, a process he called "determination" must have taken place to fix the "developmental fate" of the cells.

Spemann remained at the University of Freiburg until his retirement in the mid–1930s. When not busy with his scientific endeavors, he cultivated his love of the liberal arts. He died at his home near Freiburg at age 72.

See also Cell differentiation; Embryonic development: Early development, formation, and differentiation; Organizer experiment

SPERM · *see* SEMEN AND SPERM

SPERMATOGENESIS

Spermatogenesis is male gametogenesis, the process in males by which diploid cells are reduced to haploid gametes. In a male fetus, the primordial **germ cells** migrate to the developing testes. Here, they undergo 20–30 mitotic divisions to generate a population of spermatogonia in the seminiferous tubules. At **puberty**, some of the spermatogonia begin the final differentiation into male gametes or **sperm**. The first step in this sequence is development into a primary spermatocyte. These are diploid cells with the normal number of 46 **chromosomes** that replicate their total **DNA** and enter meiosis. After the first meiotic division, the reduction division, a group of secondary spermatocytes containing only half of the normal human chromosome complement, or 23 chromosomes, is produced. Each secondary spermatocyte then undergoes a second **cell division**, that of meiosis II, resulting in two haploid spermatids. In this way, a total of four haploid spermatids are generated from each original diploid primary spermatocyte. Further differentiation gives rise to the functional sperm.

This entire process takes between 42 and 64 days, and produces approximately 200 million sperm per ejaculate. Spermatogenesis is continuous throughout the male adult life. Over an average lifetime, it has been estimated that a normal male with produce 10^{12} sperm.

Because of the large number of cell divisions that occur in male gametogenesis, it has been suggested that every sperm has one or more new mutations. Most of these are benign, but based on current data, it is estimated that one in ten sperm will have a deleterious mutation. Indeed, a number of autosomal dominant disorders including neurofibromatosis I, achondroplasia, and hemophilia B have been associated with new paternally derived mutations. This appears to be an important mechanism for the introduction of new dominant mutations into the population.

Although male and female gametogenesis have the same endpoint, the production of haploid gametes, the actual processes are slightly different. Spermatogenesis doesn't begin until the individual enters puberty, but then the cell divisions giving rise to the spermatids proceed continuously

throughout that person's life. Each primary spermatocyte gives rise to four similar sperm cells that differ primarily in their genetic content, but, because of the large number of cell divisions that occur, about 10% of the sperm produced carry potentially deleterious new mutations. **Oogenesis**, on the other hand, begins before birth in the developing female fetus. The process advances to a point and then stops, is started again at puberty, but ceases entirely around 50 years of age. Only one functional gamete that includes the majority of the cytoplasm is produced from each starting primary oocyte, and because there are relatively few cell divisions required for each egg, few DNA level mutations are thought to occur in oogenesis. However, cell division errors resulting in gametes with too many or too few chromosomes become increasingly common as a female ages, resulting in an increased risk of spontaneous fetal loss or an abnormal liveborn infant.

The culmination of gametogenesis is the fusion of a sperm with an egg to produce a zygote that will develop into a new human being. This is not the end of the story, but is a new beginning to the process of the fetal development.

See also Cell cycle and cell division; Cell differentiation; Embryonic development, early development, formation, and differentiation; Gonads and gonadotropic hormone physiology; Semen and sperm

SPERMATOZOA · *see* SPERMATOGENESIS

SPERRY, ROGER W. (1913-1994)
American biologist

Roger W. Sperry, a major contributor to at least three scientific fields—developmental neurobiology, experimental psychobiology, and human split-brain studies—conducted pioneering research in the functions of the left and right hemispheres of the **brain**. He was awarded the Nobel Prize in physiology or medicine in 1981. The system of split-brain research that Sperry enabled scientists to better understand the workings of the human brain.

Sperry was born on August 20, 1913, in Hartford, Connecticut, to Francis Bushnell Sperry, a banker, and Florence Kramer Sperry. When Sperry was 11 years old, his father died and his mother returned to school and took work as an assistant to a high school principal. Sperry attended local public schools through high school and then went to Oberlin College in Ohio on a scholarship. There, he competed on the track team and was captain of the basketball squad. Although he majored in English, Sperry was especially interested in his undergraduate psychology courses with R. H. Stetson, an expert on the physiology of speech. Sperry earned his B.A. in English in 1935 and then worked as a graduate assistant to Stetson for two years. In 1937 he received an M.A. in psychology.

Thoroughly committed to research in the field of psychobiology by that time, Sperry went to the University of

Chicago to conduct research on the organization of the **central nervous system** under the renowned biologist Paul Weiss. Before Weiss's research, scientists believed that the connections of the nervous system had to be very exact to work properly. Weiss disproved this theory by surgically crossing a subject's nerve connections. After the surgery was performed, the subject's behavior did not change. From this, Weiss concluded that the connections of the central nervous system were not predetermined, so that a nerve need not connect to any particular location to function correctly.

Sperry tested Weiss's research by surgically crossing the nerves that controlled the hind leg muscles of a rat. Under Weiss's theory, each nerve should eventually "learn" to control the leg muscle to which it was now connected. This did not happen. When the left hind foot was stimulated, the right foot responded instead. Sperry's experiments disproved Weiss's research and became the basis of his doctoral dissertation, "Functional results of crossing nerves and transposing muscles in the fore and hind limbs of the rat." He received a Ph.D. in Zoology from the University of Chicago in 1941.

Sperry did other related experiments that confirmed his findings and further contradicted Weiss's theory that "function precedes form" (that is, the brain and nervous system learn, through experience, to function properly). In one experiment, Sperry rotated a frog's eyeball and cut its optic nerve. If Weiss's theory proved correct, the frog would reeducate itself, adjust to seeing the world upside down, and change its behavior accordingly. This did not happen. In fact, the nerve fibers became tangled in the scar **tissue** during healing. When the nerve regenerated, it ignored the repositioning of the eyeball and reattached itself correctly, albeit upside down. From this and other experiments, Sperry deduced that genetic mechanisms determine some basic behavioral patterns. According to his theory, nerves have highly specific functions based on genetically predetermined differences in the concentration of chemicals inside the nerve cells.

In 1941, Sperry moved to the laboratory of the renowned psychologist Karl S. Lashley at Harvard to work as a National Research Council postdoctoral fellow. A year later, Lashley became director of the Yerkes Laboratories of Primate Biology in Orange Park, Florida. Sperry joined him there on a Harvard biology research fellowship. While there, he disproved some Gestalt psychology theories about brain mechanisms, as well as some theories of Lashley's.

During World War II, Sperry fulfilled his military service duty by working for three years in an Office of Scientific Research and Development (OSRD) medical research project run by the University of Chicago and the Yerkes laboratory. His work involved research on repairing nerve injuries by surgery. In 1946, Sperry returned to the University of Chicago to accept a position as assistant professor in the school's **anatomy** department. He became associate professor of psychology during the 1952–53 school year and also worked during that year as section chief in the Neurological Diseases and **Blindness** division of the National Institutes of Health.

In 1954, he transferred to the California Institute of Technology (Caltech) to take a position as the Hixon Professor of Psychobiology. At Caltech, Sperry conducted research on split-brain functions that he had first investigated when he worked at the Yerkes Laboratory. It had long been known that the cerebrum of the brain consists of two hemispheres. In most people, the left hemisphere controls the right side of the body and vice versa. The two halves are connected by a bundle of millions of nerve fibers called the corpus callosum, or the great cerebral commissure.

Neurosurgeons had discovered that this connection could be cut into with little or no noticeable change in the patient's mental abilities. After experiments on animals proved the procedure harmless, surgeons began cutting completely through the commissure of epileptic patients in an attempt to prevent the spread of epileptic seizures from one hemisphere to the other. The procedure was generally successful, and beginning in the late 1930s, cutting through the forebrain commissure became an accepted treatment method for severe epilepsy. Observations of the split-brain patients indicated no loss of communication between the two hemispheres of the brain.

From these observations, scientists assumed that the corpus callosum had no function other than as a prop to prevent the two hemispheres from sagging. Scientists also believed that the left hemisphere was dominant and performed higher cognitive functions such as speech. This theory developed from observations of patients whose left cerebral hemisphere had been injured; these patients suffered impairment of various cognitive functions, including speech. Since these functions were not transferred over to the uninjured right hemisphere, scientists assumed that the right hemisphere was less developed.

Sperry's work shattered these views. He and his colleagues at Caltech discovered that the corpus callosum is more than a physical prop; it provides a means of communication between the two halves of the brain and integrates the knowledge acquired by each of them. They also learned that in many ways, the right hemisphere is superior to the left. Although the left half of the brain is superior in analytic, logical thought, the right half excels in intuitive processing of information. The right hemisphere also specializes in non-verbal functions, such as understanding music, interpreting visual patterns (such as recognizing faces), and sorting sizes and shapes.

Sperry discovered these different capacities of the two **cerebral hemispheres** through a series of experiments performed over a period of several decades. In one such experiment, Sperry and a graduate student, Ronald Myers, cut the nerve connections between the two hemispheres of a cat's brain. They discovered that behavioral responses learned by the left side of the brain were not transferred to the right, and vice versa. In an article published in *Scientific American* in 1964, Sperry observed that "it was as though each hemisphere were a separate mental domain operating with complete disregard—indeed, with a complete lack of awareness—of what went on in the other. The split-brain animal behaved in the test situation as if it had two entirely separate brains." It was evident from this experiment that the severed nerves had been responsible for communication between the two halves of the brain.

In another experiment on a human subject, he showed a commissurotomy patient (one whose corpus callosum had been surgically severed) a picture of a pair of scissors. Only the patient's left visual field, which is governed by the nonverbal right hemisphere, could see the scissors. The patient could not verbally describe what he had seen because the left hemisphere, which controls language functions, had not received the necessary information. However, when the patient reached behind a screen, he sorted through a pile of various items and picked out the scissors. When asked how he knew the correct item, the patient insisted it was purely luck.

Sperry started published technical papers on his split-brain findings in the late 1960s. The importance of Sperry's research was recognized relatively quickly, and in 1979 he was awarded the prestigious Albert Lasker Basic Medical Research Award in recognition of the potential medical benefits of his research, including possible treatments for mental or psychosomatic illnesses.

Two years later, Sperry was honored with the 1981 Nobel Prize in physiology or medicine. He shared it with two other scientists, Torsten N. Wiesel and David H. Hubel, for research on the central nervous system and the brain. In describing Sperry's work, the Nobel Prize selection committee praised the researcher for demonstrating the difference between the two hemispheres of the brain and for outlining some of the specialized functions of the right brain. The committee, as quoted in the *New York Times,* stated that Sperry's work illuminated the fact that the right brain "is clearly superior to the left in many respects, especially regarding the capacity for concrete thinking, spatial consciousness and comprehension of complex relationships."

In his acceptance speech, as quoted in *Science* in 1982, Sperry talked about the significance of his discovery of the previously unrecognized skills of the nonverbal right half-brain. He commented that an important gain from his work is increased attention to "the important role of the nonverbal components and forms of the intellect." Because split-brain research increased appreciation of the individuality of each brain and its functions, Sperry believed that his work helped to point out the need for educational policies that took into consideration varying types of intelligence and potential.

Sperry rejected conventional scientific thinking that viewed human consciousness solely as a function of physical and chemical activity within the brain. In his view, which he discussed in his Nobel Prize lecture, "cognitive introspective psychology and related cognitive science can no longer be ignored experimentally.... The whole world of inner experience (the world of the humanities) long rejected by twentieth-century scientific materialism, thus becomes recognized and included within the domain of science."

Known as a private, reserved person, Sperry was camping with his wife in a remote area when the news of his Nobel Prize award was announced. Sperry and his wife had two children. In addition to camping, Sperry's personal interests included sculpture, drawing, ceramics, folk dancing, and fossil hunting. He retired from Caltech in 1984 as Professor Emeritus. In 1989, Sperry was awarded a National Medal of Science. He died in Pasadena.

In addition to the Nobel prize, Sperry received many awards and honorary doctorates. He was member of many scientific societies, including the Pontifical Academy of Sciences and the National Academy of Sciences. Sperry was always held in high regard by his students. One of them, Michael Gazzaniga, described him in *Science* as "exceedingly generous" to many students at Caltech. Gazzaniga also defined Sperry as a teacher "constitutionally only able to be interested in critical issues," who drove his "herd of young scientists to consider nothing but the big questions."

See also Brain: Intellectual functions; Central nervous system (CNS); Cerebral cortex; Cerebral hemispheres

SPINAL CORD

The spinal cord is the principle route for the passage of sensory information to and from the **brain**. The spinal cord is a long column of nerve **tissue** that extends from the base of the brain, (the **medulla oblongata**) downward through a canal created by the spinal vertebral foramina, usually terminating around the first lumbar **vertebra**.

The spinal cord is enveloped and protected by the vertebra of the spinal column. There are four regions of vertebrae. Beginning at the **skull** and moving downward, these are the eight cervical vertebrae, twelve thoracic vertebrae, five lumbar vertebrae, five sacral vertebrae, and one coccygeal vertebra. The sacral and coccygeal vertebrae are also known as the tailbone. In cross-section, the vertebrae appear as a flat plate-like region (body) with antler-like projections (processes) perched above much like a hat. Between the body and the processes there is an open space called the vertebral foramen, where the spinal cord is located.

The spinal cord is between 17–18 in. (43–45 cm) long in the average woman and man, respectively. The spinal column is quite bit longer. This is because the spinal cord nerves begin to branch off to various regions of the body below its termination site. The more diffuse appearance of these nerves, relative to the tighter bundle of the cord above, gives the region its name (cauda equina, meaning "horse tail").

The spinal cord is one component of the **central nervous system**. Information flows to the central nervous system from the **peripheral nervous system**, which senses signals from the environment outside the body (sensory-somatic nervous system) and from the internal environment (**autonomic nervous system**). The brain's receives, processes and responds to the information it receives. If the response is an action, the signal for that action flows thought the spinal cord nerve network to connect with nerves that o to whatever regions the responsive action is to take place.

Both the spinal cord and the brain are made up of hair-like nerve cells called axons. The axons are coated with a material called myelin. Also, there are bundles of cell bodies, out of which the axons grow, and branched regions of nerve cells (dendrites). There is a space (**synapse**) between the axon of one cell body and the dendrite of another nerve cell. In the spinal cord of humans, the myelin-coated axons are on the surface and the axon-dendrite network is on the inside. In cross-section, the pattern of contrasting color of these regions produces an axon-dendrite shape that is reminiscent of a butterfly.

The nerves of the spinal cord correspond to the arrangement of the vertebrae. There are 31 pairs of nerves, grouped as eight cervical pairs, twelve thoracic pairs, five lumbar pairs, five sacral pairs, and one coccygeal pair. The nerves toward the top of the cord are oriented almost horizontally. Those further down are oriented on a progressively upward slanted angle. Towards the bottom of the cord, this produces an appearance similar to that of ropes twined together in a bundle. Also towards the bottom of the spinal cord, the spinal nerves connect with cells of the **sympathetic nervous system**. These cells are called pre-ganglionic and ganglionic cells. One branch of these cells is called the gray ramus communicans and the other branch is the white ramus communicans. Together they are referred to as the rami. Other rami connections led to the pelvic area.

The spinal cord is also protected by layers of connective tissue known as the **meninges. Cerebrospinal fluid** flows between two of the layers of the meninges. This arrangement provides cushioning for the spinal cord. Both the connective tissue and the cerebrospinal fluid are susceptible to microbial contamination. Infections of the meninges are called meningitis. All such infections are serious and require prompt medical attention.

The two-way communication network of the spinal cord allows the reflex response to occur. This type of rapid response occurs when a message from one type of nerve fiber, the sensory fiber, stimulates a muscle response directly, rather than the impulse traveling to the brain for interpretation. For example, if a hot stove burner is touched with a finger, the information travels from the finger to the spinal cord and then a response to move muscles away from the burner is sent rapidly and directly back. This response is initiated when speed is important.

See also Autonomic nervous system; Nerve impulses and conduction of impulses; Nervous system overview; Spinal nerves and rami

SPLANCHNOLOGY

Splanchnology is that branch of **anatomy** devoted to the study of the viscera. The internal body structure of vertebrates, including that of humans, is divided by anatomists into two cavities, the dorsal cavity (containing only the **brain** and **spinal cord**) and the ventral cavity. The ventral cavity is further subdivided by the **diaphragm** muscle into the thoracic (chest) cavity and abdominopelvic cavity. The **organs** contained by the dorsal and ventral body cavities—that is, in the chest and abdomen—are the viscera.

The term splanchnology is at the same level of medical generality as **osteology** (study of the bones), syndesmology (study of the **joints**), or **neurology** (study of the nervous system). However, since the viscera include the **heart**, digestive system, **respiratory system**, and urogenital organs, splanch-

nology is too broad to be an actual field of study. Textbooks are not written about splanchnology as such. It is, rather, a term of convenience—a quick way of indicating the study of all that part of the body that does not include the skin, skeleton, nervous system, and other non-visceral parts.

Some definitions of the visceral organs studied by splanchnology, such as that used by the classic *Gray's Anatomy of the Human Body*, exclude the heart.

See also Abdominal aorta; Liver

SPLEEN HISTOPHYSIOLOGY

The spleen is not only a lymphoid organ, but is also the largest one in which lymphocytes and **plasma** cells are formed. The spleen belongs to the **reticulo-endothelial system** that removes from the circulation old red **blood** cells and other impurities in the blood. The spleen is located below the **diaphragm** on the left side of the **abdomen**, almost completed surrounded by the stomach and the **peritoneum**. The splenetic structure is similar to that of the lymph nodes, and is covered by a serous membrane containing a fibromuscular capsule with central elongations known as pulp trabeculae that radiate from the hilus through the organ, forming a mesh of fibrous bands. The splenic pulp is constituted by the white and red pulps. Between the trabeculae, there are islands of the white pulp, composed of white blood cells and the ramifications of the splenic artery that supply the spleen with blood. The red pulp consists of the sustaining reticulum and a dense web of venous **capillaries** surrounded by venous sinuses.

The spleen functions as a blood reservoir that admits more blood into circulation by contraction under adrenergetic stimulation, and stores blood in two different locations: the venous sinuses and the red pulp. The venous sinuses receive blood directly from the capillaries, swelling as they are filled up while the red pulp is supplied via the trabeculae that trap the blood that oozes out the highly permeable capillary walls. In the red pulp, the red blood cells (**erythrocytes**) are retained as the plasma flows into the venous sinuses, and are returned to circulation. Phagocytic cells of the reticulo-endothelial system line the venous sinuses and are also present inside the pulp, where they remove from the blood damaged red blood cells, **bacteria**, and other pathogens, as well as cellular debris.

Like the lymph glands, the spleen can also enlarge in order to gain function in the presence of **infection**. The spleen is one of the lymphogenous **organs** where lymphocytes and plasma cells are formed. The other lymphogenous organs are: the thymus, the lymph glands, the **tonsils**, as well as the many islands of lymphoid **tissue** present in the bone marrow and under the intestinal epithelial walls.

See also Bacteria and responses to bacterial infection; Circulatory system; Hemopoiesis; Immune system; Lymphatic system; Platelets

SPORTS PHYSIOLOGY

Exercise induces changes in the body of the athlete. These changes are varied and depend upon the nature of the activity. Endurance athletes, such as marathon runners and long distance cyclists, must supply their muscles with oxygen, swiftly remove the waste products of metabolic activity from their tissues, and quickly propel their bodies along for mile after mile. In contrast, weight lifters and sprinters need to generate sudden explosive power in their arms and legs in order to propel massive weights vertically upwards or move down the track at high speed.

While the athletic activities may demand different responses from the body, the underlying **physiology** is similar. Sports physiology is concerned with the response of muscles. Other functions of the body change to support the demand for increased muscular activity. For example, the **heart** will beat faster in order to supply more **blood** to the muscles. Also areas of the body such as the stomach will shut down during exercise, so as not to divert blood from where it is needed most. Indeed, one of the beneficial aspects of exercise, especially where weight loss is a goal, is the diminished appetite that results from a work-out.

The fuel for muscular work is a molecule called **adenosine triphosphate** (**ATP**). ATP production requires oxygen. Hence, during exercise, the oxygen demand of the body, and of muscles in particular, rises. Waste products of the fuel production, such as **lactic acid** and carbon dioxide, need to be removed, or else the muscles will loss their ability to perform work. The heavy feeling and burning sensation felt in muscles during intense activity is due to the buildup of waste products. Lactic acid is produced when ATP is manufactured from a compound called **glycogen** in an anaerobic ("without oxygen") metabolic pathway. Competitive athletes train to build up their capacity to perform their athletic activity while still keeping their muscles supplied with oxygen. In this state, known as aerobic **metabolism**, a sugar called glucose is the energy source.

Exercising muscles are able to extract oxygen from the blood three times as fast as resting muscles can. This is accomplished by increased blood flow (that involves the increased heart beat and the enlarging of the diameter of the blood vessels in the muscles), increased rate and depth of **breathing** (also known as the VO$_2$MAX), and increasing the rate by which oxygen molecules are released from **hemoglobin** (the **oxygen transport** molecule in the blood).

Exercise produces heat in the muscles. This heat must be dissipated to sustain the level of work. An important means of dissipating heat is through sweat. Blood vessels in the skin become larger in diameter. More blood is able to flow through them to the skin, where the large surface area of the skin helps the heat in the blood pass of to the air. The body's response to heat is controlled by an area of the **brain** called the **hypothalamus**. In essence, the hypothalamus is the body's thermostat.

Sweat removes water and ions like sodium and potassium from the body. These need to be replenished for the level of exercise to continue. Hydration, replenishment of the body with water, is vitally important. Severe dehydration can be fatal.

The changes in the body that accompany exercise can be developed through training. Training at higher altitudes can stimulate the production of more hemoglobin molecules, which has the effect of increasing the oxygen carrying capacity of the blood. For this reason, endurance athletes often train at higher altitudes for a prescribed period of time prior to an event. Weight training can increase muscle mass. Even the heart can be made fitter through exercise.

But not all aspects of the body can be changed by training. For example, muscles are a mixture of fast twitch and slow twitch fibers. Fast twitch fibers contact more quickly than slow twitch. The content of the fiber types is determined genetically. So, someone with more fast twitch fibers is likely to achieve more success as a sprinter than as a marathon runner.

See also Muscle contraction; Muscle tissue damage, repair, and regeneration; Muscular innervation; Muscular system overview; Oxygen transport and exchange; Respiratory system

STEM CELLS

Stem cells are undifferentiated cells that have the capability of self replication as well as being able to give rise to diverse types of differentiated or specialized cell lines. They are subclassified as embryonic stem cells, embryonic **germ cells**, or adult stem cells. Embryonic stem cells are cultured cells that were originally collected from the inner cell mass of an embryo at the blastocyst stage of development (four days post **fertilization**). Embryonic germ cells are derived from the fetal **gonads** that arise later in fetal development. Both of these stem cell types are pluripotent, that is, they are capable of producing daughter cells that can differentiate into all of the various tissues and **organs** of the body that are derived from the **endoderm**, **ectoderm** and **mesoderm**. Adult stem cells, found in both children and adults, are somewhat more limited, or multipotent, since they are associated with a single **tissue** or organ and function primarily in cell renewal for that tissue.

Because they are undifferentiated, stem cells have unique properties that may make them useful for new clinical applications. Initially, stem cells were considered as a potential source of tissue for transplantation. The current standard of care for many diseases that result in total tissue and/or organ destruction is transplantation of donor tissues, but the number of available organs is limited. Data on bone marrow transplantation supported the idea of using stem cells in transplantation. Initial studies showed that collection of **blood** enriched with hematopoetic stem cells had a higher engraftment rate than an equivalent bone marrow sample. Expanding on that idea, it was hypothesized that if adult stem cells from a specific organ could be collected and multiplied, it might be possible to use the resultant cells to replace a diseased organ or tissue. One drawback to this is that adult stem cells are very rare and although they have been isolated from bone marrow, **brain**, eyes, muscle, skin, **liver**, **pancreas**, and the digestive system, there are many tissues and organs for which it is not known if stem cells exist. Adult stem cells are also difficult to identify and isolate, and even when successfully collected, the cells often fail to survive outside of the body. However, despite the obstacles, the theory appears to be sound, so research is continuing.

Approaching the problem from another direction, researchers realized that embryonic stem cells and embryonic germ cells were available in cell culture in several laboratories, and that, under the right conditions, these stem cells might be induced to produce a broad range of different tissues that could be utilized for transplantation. Research on Parkinson disease, a neurodegenerative disorder that results in loss of brain function following the death of dopamine producing cells, underscored the potential of this approach. In the 1980s, studies on monkeys and rats showed that when fetal brain tissue rich in stem cells was implanted into the brains of diseased animals, there was a regeneration of functional brain cells and a reduction or elimination the symptoms of the disease. One disadvantage to this as a clinical procedure is that random pieces of undefined tissue are used resulting in the significant possibility of variability from one patient to the next. A better solution would be to isolate the embryonic stem cells, induce these cells to differentiate, and generate a population of dopamine producing cells. Theoretically, if these cells were transplanted back into the brains of Parkinson patients, they would replace the defective cells and reverse the course of the disease. However, the mechanisms that trigger differentiation of embryonic stem cells into various specialized tissue types are not yet well understood, so it will require additional research before transplantable tissues derived from embryonic stem cells will be a reality.

In addition to possible applications in transplantation, embryonic stem cells may be useful tools in other clinical disciplines. These cells represent a stage of development about which relatively little is known. Close observation in the laboratory could provide a better understanding of normal development versus abnormal development and what triggers fetal demise. Studies on the causes and control of childhood **tumors** may also be possible. Embryonic stem cell lines could aid in testing the effect of new drugs and investigating appropriate drug dosages, eliminating the need for human subjects. Similarly, such cell lines may be utilized to investigate the biological effects of toxins on human cells.

It has also been suggested that embryonic stem cells might be used in **gene therapy**. If a population of engineered embryonic stem cells containing a known, functional gene can be produced, these cells might function as vectors to transfer the gene into target tissues. Once in place, the cells would hopefully become part of the unit, begin to replicate, and restore lost function. Initial studies in mice confirmed the idea was feasible. Investigators in Spain incorporated an insulin gene into mouse embryonic stem cells. After demonstrating the production of insulin *in vitro*, the cells were injected into the spleens of diabetic mice that subsequently showed evidence of disease reversal.

There are many different diseases, ranging from **heart** disease to **spinal cord** injury and **autoimmune disorders**, that could benefit from a better understanding of and the use of

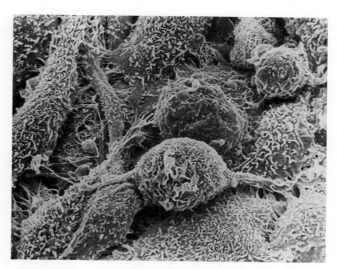

Scanning electron micrograph (SEM) of mammalian embryonic stem cells. *Illustration by Yorgos Nikas. Photo Researchers. Reproduced by permission.*

stem cells as therapeutic agents. Although work is ongoing, research on embryonic stem/germ cells is limited due to an ethical dilemma regarding the source of the cells. For research purposes, embryonic stem cells are primarily derived from leftover products of *in vitro* fertilization procedures. Embryonic germ cells from later gestational age fetuses have been obtained from elective termination of **pregnancy** or spontaneous fetal demise with appropriate parental consent. However, because it is feared that an increase in research using these cell types would encourage the "buying and selling" of embryos for profit, researchers have been asked to use currently existing cell lines rather than establishing new cell cultures.

Though still in its infancy, stem cell research holds great promise for providing important new medical treatments in the future. In addition, study of these cells will impart new knowledge about human cells and early fetal development.

See also Cell differentiation; Ectoderm; Embryonic development, early development, formation, and differentiation; Endoderm; Ethical issues in embryological research; Ethical Issues in genetics research; Fertilization; Human development; *In vitro* and *in vivo* fertilization; Mesoderm

STEPTOE, PATRICK (1913-1988)
English physician

Patrick Steptoe, an English gynecologist and medical researcher, helped develop the technique of *in vitro* **fertilization**. In this process, a mature egg is removed from the female ovary and is fertilized in a test tube. After a short incubation period, the fertilized egg is implanted in the uterus, where it develops as in a typical **pregnancy**. This procedure gave women whose fallopian tubes were damaged or missing, and were thus unable to become pregnant, the hope that they could conceive children. Steptoe and his colleague, English physiologist Robert G. Edwards, received international recognition—both positive and negative—when the first so-called test tube baby was born in 1978.

Patrick Christopher Steptoe was born in Oxfordshire, England. His father was a church organist, while his mother served as a social worker. Steptoe studied medicine at the University of London's St. George Hospital Medical School and, after being licensed in 1939, became a member of the Royal College of Surgeons. World War II interrupted his medical career for a time, when Steptoe volunteered as a naval surgeon. Following the war, Steptoe completed additional studies in obstetrics and gynecology. In 1948, he became a member of the Royal College of Obstetricians and Gynecologists and moved to Manchester to set up a private practice. In 1951, Steptoe began working at Oldham General and District Hospital in northeast England.

While at Oldham General and District Hospital, Steptoe pursued his interest in fertility problems. He developed a method of procuring human eggs from the ovaries by using a laparoscope, a long thin telescope replete with fiber optics light. After inserting the device—through a small incision near the navel—into the inflated abdominal cavity, Steptoe was able to observe the reproductive tract. Eventually, the laparoscope would become widely used in various types of surgery, including those associated with infertility. But, at first, Steptoe had trouble convincing others in the medical profession of the merits of laparoscopy; observers from the Royal College of Obstetricians and Gynecologists considered the technique fraught with difficulties. Five years passed before Steptoe published his first paper on laparoscopic surgery.

In 1966, Steptoe teamed with Cambridge University physiologist Robert G. Edwards to propel his work with fertility problems. Utilizing ovaries removed for medical reasons, Edwards had pioneered the fertilization of eggs outside of the body. With his laparoscope, Steptoe added the dimension of being able to secure mature eggs at the appropriate moment in the monthly cycle when fertilization would normally occur. A breakthrough for the duo came in 1968, when Edwards successfully fertilized an egg that Steptoe had extracted. Not until 1970, however, was an egg able to reach the stage of cell division—into about 100 cells—when it generally moves to the uterus. In 1972, the pair attempted the first implantation, but the embryo failed to lodge in the uterus. Indeed, none of the women initially implanted with embryos carried them for a full trimester.

As their work progressed and word of it leaked out, the researchers faced criticism from scientific and religious circles concerning the ethical and moral issues relating to tampering with the creation of human life. Some opponents considered the duo's work akin to the scenario in Aldous Huxley's 1932 work, *Brave New World,* in which babies were conceived in the laboratory, cloned, and manipulated for society's use. Members of Parliament demanded an investigation and sources of funds were withdrawn. A *Time* reporter quoted Steptoe as saying, "All I am interested in is how to help women who are denied a baby because their tubes are inca-

pable of doing their small part." Undaunted, Steptoe and Edwards continued their work at Kershaw's Cottage Hospital in Oldham. Disturbed with the criticism, Steptoe and Edwards became more secretive, which made the speculation and criticism more intense.

In 1976, Steptoe met thirty-year-old Leslie Brown, who experienced problems with her fallopian tubes. Steptoe removed a mature egg from her ovary, and Edwards fertilized the egg using her husband Gilbert's **sperm**. The fertilized egg—implanted after two days—thrived, and on July 25, 1978, Joy Louise Brown, a healthy 5–lb. 12 oz. (2.6 kg) girl was born in Oldham District and General Hospital. Even before the birth, reporters and cameramen congregated outside of the four story brick hospital, hoping for a glimpse of the expectant mother.

Steptoe and Edwards were reluctant to discuss the procedures in press conferences and did not immediately publish their findings in a medical journal. The procedures were fully presented at the January 26, 1979 meeting of the Royal College of Obstetricians and Gynecologists and at the conference of the American Fertility Society in San Francisco. Steptoe reported that with modified techniques, ten percent of the *in vitro* fertilization attempts could succeed. He further predicted that there could one day be a fifty percent success rate for the procedure.

In the aftermath of the first successful test tube baby, Steptoe received thousands of letters from couples seeking help in conception. He retired from the British National Health Service and constructed a new clinic near Cambridge. For their efforts, Steptoe and Edwards were both named Commanders of the British Empire, and in 1987, Steptoe was honored with fellowship in the Royal Society. Steptoe and his wife, a former actress, had one son and one daughter. His interests outside of medicine included piano and organ, cricket, plays, and opera. Steptoe died in Canterbury at age 74. Since the birth of baby Brown and the pioneering techniques of Steptoe, thousands of couples throughout the world with certain infertility problems have been able to have children through *in vitro* fertilization.

See also Embryology; *In vitro* and *in vivo* fertilization

STERNBERG, ESTHER MAY (1951-)
Canadian-born American physician

Esther M. Sternberg has made many contributions to the study of rheumatology, neuroendocrinology, stress and neurological disorders, and the relationship between emotions and disease. Her research was fundamental to elucidating the etiology of a puzzling 1989 epidemic of eosinophilia-myalgia syndrome. Sternberg's work on the new field of research known as "mind-body interactions" has been instrumental in explaining how the **immune system** and the nervous system communicate with each other.

Sternberg was born in Montreal in 1951. In 1991, she became a citizen of the United States. Her choice of a career was greatly influenced by her family background. Joseph Sternberg, her father, was a physician-scientist and a pioneer in the fields of radiation biology and nuclear medicine.

In 1972, Sternberg received a B.Sc., with Great Distinction, from McGill University, located in Montreal. Two years later, she was awarded an M.D. from McGill. During her post-graduate medical training at the Royal Victoria Hospital, she selected rheumatology as her area of clinical and research specialization. Between 1981 and 1986, she was a research associate in the Division of Allergy and Clinical Immunology at the Washington University School of Medicine in St. Louis, Missouri. In 1987, she accepted a research position in clinical neurosciences at the National Institute of Mental Health (NIMH) in Bethesda, Maryland. This job has led to a series of increasingly prestigious positions at the NIMH, including the directorship of the Integrative Neural Immune Program. Her current position there is Chief of the Section on Neuroendocrine Immunology and Behavior. She is also a research full professor at the American University in Washington, D.C.

Sternberg has published over 60 articles and books, including *The Balance Within: The Science Connecting Health and Emotions*, in 2000. These writings explore many medical problems, including rheumatoid arthritis, sclero-derma, fibromyalgia, multiple sclerosis, the effect of serotonin on the immune response, the design of drugs to inhibit HIV receptor binding, the relationship between L-tryptophan and human eosinophilia-myalgia syndrome (EMS), neuroen-docrinology and the immune response, and the stress response and regulation of inflammatory disease. Other works by Sternberg consider hyperimmune fatigue syndrome, tamox-ifen, lymphokines, the relationship between exercise and the immune system, neuroendocrine aspects of autoimmunity, neuroimmune stress interactions, neuroendocrine factors in susceptibility to inflammatory disease, emotions and disease, and other aspects of mind-body interactions. Sternberg has received many awards and honors for her research on rheuma-tology, EMS, and the relationship between emotions and dis-ease. Her expertise has been sought in areas as diverse as scleroderma, asthma, mind-body interactions, influenza and pneumoccocal vaccines, stress in neurological disorders, the health of deployed United States military forces, and military **nutrition** research.

In 1989, a previously unknown disorder that came to be called eosinophilia-myalgia syndrome (EMS) appeared in an epidemic pattern in the United States. About 1,500 cases and 40 deaths were reported. Case studies, epidemiological data, and laboratory research on animals linked the syndrome to the amino acid L-tryptophan, sold by a particular manufacturer as a dietary supplement. Sternberg's research was fundamental in elucidating the etiology (origin) of EMS, and in drawing atten-tion to the pressure that the manufacturer of the suspected L-tryptophan was bringing against researchers trying to investigate the disorder.

For much of the twentieth century, medical science and even medical practice had become, according to many critics, increasingly specialized and mechanistic. The field of neu-roimmune interactions, in contract, is an intensely interdisci-

plinary field that encompasses immunology, neurobiology, neuroendocrinology, and the behavioral sciences. Beliefs about the relationship between emotions and disease that go back to the writings of **Hippocrates** are now being investigated in terms of molecules, cells, and nerve signals. Biomedical scientists and physicians involved in this challenging area have been able to integrate research results from **molecular biology** with clinical observations of behaviors, emotions, and disease. As a leader in the study of mind-body interactions, Sternberg has called for research that is precise, focused, and integrative. For example, vague references to "stress" are giv-ing way to precise definitions and measurements of both external stressors and internal responses (e.g., neural, neu-roendocrine, and immune factors). The task of future research, according to Sternberg, will be to provide rigorous evidence that neuroimmune interactions play a role in susceptibility and resistance to inflammatory and infectious diseases and to find ways to apply these new scientific insights to human health and healing.

See also Arthrology (joints and movement); Autoimmune dis-orders; Immunity

STERNUM AND THORACIC ARTICULA-TIONS

The sternum is a long bony plate located in the anterior (front) region of the chest. Three distinct bones can be seen. The superior or uppermost plate is the manubrium, followed by the body or gladiolus. At the end of this series is a tiny triangular shaped bone named the xiphoid. Both muscles of the chest and costal **cartilage** articulate with these bones. They help anchor the ribcage for protection of the **lungs** and upper body **organs**.

The manubrium is triangular and thicker at its top than bottom. It is somewhat convex. On the anterior side, the pec-toralis and sterno portion of the sterno-cleido-mastoid muscle originate. The site of attachment can be identified by the ster-nal notch. On the posterior surface the heads of the sterno-hyoid and sterno-thyroid muscles find their origin.

The thickest portion of the manubrium, the superior border, the pre-sternal notch, indicates the position where the sternal end of the clavicle (collar bone) articulates. On either side of this flat bone there are facets for the costal cartilage of the first pair of ribs. These lie just under the clavicle.

The long body of the gladiolus joins the manubrium at the sternal angle. It is possible to feel this ridge with the hands, which makes it useful as a locator for **cardio-pulmonary resus-citation** (**CPR**). In some people, however, this region is con-cave and very distinctive. Facets occur along the lateral ridges of this flat bone for articulation with costal cartilage and ribs two, three, four, and five. The inferior margin of the gladiolus supports two costal cartilage heads that then differentiate to support ribs 6-10. Ribs 11 and 12 do not articulate with the sternum.

The xiphoid process terminates the sternum. It is small and somewhat elongate. In adolescence it is cartilaginous, but ossifies in adulthood. Although tiny in size, the bone offers an attachment site for part of the seventh rib, the chondro-xiphoid ligament, and the linea alba. This bone has a highly diversified appearance from one body to another.

See also Skeletal system overview (morphology)

STOMACH HISTOPHYSIOLOGY

The enlarged segment of the digestive tube between the esophagus and the small intestines is termed the stomach, and is located in the abdominal cavity below the **diaphragm**, occupying the left hypochondriac, the epigastric, and part of the right hypochondriac space. The stomach contains two orifices, the cardia that communicates with the esophagus, and the pylorus that gives access to the duodenum. The main portion of the stomach is known as the body and the part proximal to the pyloric canal, or pyloric region, is termed antrum; the superior pole is termed fundie or fundus. The body is subdivided in two physiological portions: the first two thirds is termed orad, and the last third of the body is the caudad. The stomach walls contain four layers of tissues; the internal face is coated with the mucous or mucosa, followed by the submucous, the muscular, and the serous (external layer) epithelia. The mucous contains the gastric glands, which secrete digestive juices and a gel-like alkaline **mucus** that protects the mucous epithelium against the hydrochloric acid that is secreted during **digestion**. The two major types of glands are the gastric glands, distributed in most of the body and fundus, and the pyloric glands, found in the antrum. The products of gastric (or oxyntic) glands are hydrochloric acid, mucus, pepsinogens, and intrinsic factor, whereas the pyloric glands secrete mainly mucus, the hormone gastrin, and discrete amounts of pepsinogens. The mucus secreted by the pyloric glands is thin, and helps to lubricate the stomach walls, easing food movement. The hydrochloric acid in the stomach breaks the pepsinogen molecules into pepsins, a family of **enzymes** that digests proteins. The intrinsic factor is secreted together with hydrochloric acid and promotes the absorption of vitamin B_{12}, an essential nutrient for red **blood** cells maturation in the bone marrow.

When food enters the stomach, gastric and pyloric glands are stimulated to secrete digestive juices, while the mid-portion of the stomach wall starts a discrete pattern of peristaltic contractions. The contractions are termed mixing waves because they actually mix the food with the gastric juice, promoting digestion and the formation of an emulsified semi-fluid paste, termed chyme. Antral contractions promote the emptying of the stomach as the chime gradually passes through the pylorus into the duodenum, where the final process of digestion continues with the processing of fats and starches through either hydrolysis or enzymatic digestion.

See also Bile, bile ducts, and the biliary system; Enzymes and coenzymes; Gastrointestinal tract; Intestinal histophysiology

STRIATED MUSCLE • *see* SKELETAL MUSCLE

STROKE

A stroke is the sudden death of **brain** cells in a localized area due to inadequate **blood** flow.

A stroke occurs when blood flow is interrupted to part of the brain. Without blood to supply oxygen and nutrients and to remove waste products, brain cells quickly begin to die. Depending on the region of the brain affected, a stroke may cause paralysis, speech impairment, loss of memory and reasoning ability, **coma**, or death. A stroke is also sometimes called a cerebrovascular accident (CVA).

There are four main types of stroke. Cerebral thrombosis and cerebral **embolism** are caused by blood clots that block an artery supplying the brain, either in the brain itself or in the neck. These account for 70–80% of all strokes. Subarachnoid hemorrhage and intracerebral hemorrhage occur when a blood vessel bursts around or in the brain.

Cerebral thrombosis occurs when a blood clot, or thrombus, forms within the brain itself, blocking the flow of blood through the affected vessel. Clots most often form due to "hardening" (atherosclerosis) of brain **arteries**. Cerebral thrombosis occurs most often at night or early in the morning. Cerebral thrombosis is often preceded by a transient ischemic attack, or TIA, sometimes called a "mini-stroke." In a TIA, blood flow is temporarily interrupted, causing short-lived stroke-like symptoms. Recognizing the occurrence of a TIA, and seeking immediate treatment, is an important step in stroke prevention.

Cerebral embolism occurs when a blood clot from elsewhere in the **circulatory system** breaks free. If it becomes lodged in an artery supplying the brain, either in the brain or in the neck, it can cause a stroke. The most common cause of cerebral embolism is atrial fibrillation, a disorder of the **heart** rhythm. In atrial fibrillation, the upper chambers (atria) of the heart beat weakly and rapidly, instead of slowly and steadily. Blood within the atria is not completely emptied. This stagnant blood may form clots within the atria, which can then break off and enter the circulation. Atrial fibrillation is a factor in about 15% of all strokes. The risk of a stroke from atrial fibrillation can be dramatically reduced with daily use of anticoagulant medication.

Hemorrhage, or bleeding, occurs when a blood vessel breaks, either from trauma or excess internal pressure. The vessels most likely to break are those with preexisting defects such as an aneurysm. An aneurysm is a "pouching out" of a blood vessel caused by a weak arterial wall. Brain aneurysms are surprisingly common. According to autopsy studies, about 6% of all Americans have them. Aneurysms rarely cause symptoms until they burst. Aneurysms are most likely to burst when blood pressure is highest, and controlling blood pressure is an important preventive strategy.

Intracerebral hemorrhage affects vessels within the brain itself, while subarachnoid hemorrhage affects arteries at the brain's surface, just below the protective arachnoid membrane.

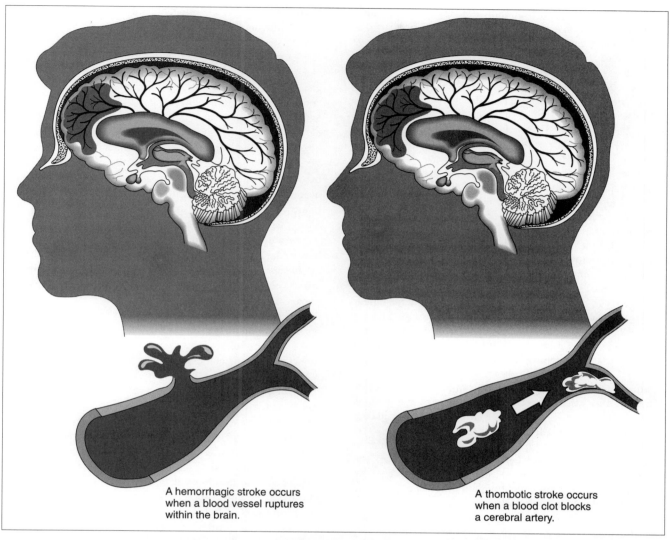

A hemorrhagic stroke occurs when a blood vessel ruptures within the brain.

A thombotic stroke occurs when a blood clot blocks a cerebral artery.

Diagram representing damage to the brain by either a hemorrhagic (blood loss) or thrombolic (clot-caused) stroke. *Illustration by Hans & Cassidy.*

Intracerebral hemorrhages represent about 10% of all strokes, while subarachnoid hemorrhages account for about 7%.

In addition to depriving affected tissues of blood supply, the accumulation of fluid within the inflexible **skull** creates excess pressure on brain **tissue**, which can quickly become fatal. Nonetheless, recovery may be more complete for a person who survives hemorrhage than for one who survives a clot, because the blood deprivation effects are usually not as severe.

Death of brain cells triggers a chain reaction in which toxic chemicals created by cell death affect other nearby cells. This is one reason why prompt treatment can have such a dramatic effect on final recovery.

Symptoms of an embolic stroke usually come on quite suddenly and are at their most intense right from the start, while symptoms of a thrombotic stroke come on more gradually. Symptoms may include blurring or decreased vision in one or both eyes, severe **headache**, weakness, numbness, or paralysis of the face, arm, or leg, usually confined to one side of the body, and dizziness, loss of balance or coordination, especially when combined with other symptoms.

See also Angiology; Blood coagulation and blood coagulation tests; Blood pressure and hormonal control mechanisms; Brain stem function and reflexes; Brain, intellectual functions

STRUCTURE AND FUNCTION

One of the easiest ways to learn **anatomy** is to relate function with structure. Quite often, the anatomy is intimately connected to its function. **Evolution** has produced the most efficient structure for the maximum performance of the body. However, bodies are under constant change and as researchers look back at the history of the human, many structures have disappeared or become more prominent. Some vestigial (obsolete and undergoing reduction) structures remain in the body. The **appendix** is one such structure.

The skeleton is a good place to observe structure and function. As humans began to walk upright, the stress on the sacral vertebrae and pelvic girdles became greater. As a result, the sacral vertebrae fused to make a strong support for the pelvic bones. These bones fused to provide additional strength. The articulation with the femur gained importance since the torso rested on the pelvic region. The greater trochanter of the femur became more robust and lateral. Its deep insertion into the acetabulum of the pelvis provided increased support for bipedal **locomotion**.

The structure of the intestine is a great example of increasing complexity for better digestive functioning. The increased length, infolding along the intestinal lining, and addition of microvilli provide increased surface for greater absorption of **nutrition**. For animals, such as humans, with high metabolisms this increase allows higher metabolic rates.

Today one of the problems in dentistry is that the mouth of many humans is becoming smaller. This is a result of eating softer food that is cooked. The large masseter and surface grinding are not as robust as they once were. However, the **genetics** of tooth production has not kept pace with the decrease in buccal (mouth) volume. As a consequence, more and more people have crowded teeth that may require straightening. Even the wisdom teeth have no room in some mouths and must be removed. It may be that eventually the human teeth may be much smaller and have less surface area.

The **lungs** are not equal in size. The left is smaller in volume and has only two lobes while the left has three. The left side of the lungs accommodates the **heart**, which lies to the left of the thoracic cavity. This structural difference allows the heart to function unencumbered in its chamber.

It is almost impossible to separate structure from function. The more human anatomy is understood, the more easily recognizable the structures that become based on their function.

See also Anatomical nomenclature

SUBCLAVIAN ARTERIES

Branching from the **aortic arch** of the ascending aorta, the subclavian **arteries**, as their name specifies, run below the clavicle. The right subclavian artery normally splits from the brachiocephalic trunk off the aortic arch of the ascending aorta. The left subclavian artery, however, usually splits directly from the aortic arch of the ascending aorta.

The subclavian arteries ultimately provide the origin of the thyrocervical and costocervical trunks of the internal **thoracic arteries**, and origin of the vertebral arteries. Accordingly, the left and right subclavian arteries ultimately supply oxygenated **blood** to the internal thoracic region (including the thoracic wall and internal thoracic region), parts of the upper arm (upper limb), neck, **spinal cord**, **meninges**, and **brain**. The

internal thoracic arteries supply blood to the **diaphragm**, mediastinal structures, and the anterior thoracic wall.

After branching into the internal thoracic and vertebral arteries, the subclavian arteries run to the edge of the first rib where they then continue as the axillary arteries that supply oxygenated blood to the axilla, thorax, shoulder region and upper arm (upper limb). Although the left and right subclavian arteries have different origins and initial anatomical relations, both subclavian arteries run over the top region (apex) of their respective **lungs** before continuing as the left and right axillary arteries.

After following a path below the clavicle that starts below the sternoclavicular joint, the subclavian arteries come nearest the surface of the skin (i.e., become superficial) as they branch into the axillary arteries in a region described by anatomists as the supraclavicular triangle (a feature of surface **anatomy** defined by the omohyoid, clavicle and sternocleidomastoid). In this region the respective (right or left) subclavian artery runs close, and in some areas next to, the vulnerable and critical networking junction of nerves known as the **brachial plexus**. Because the subclavian arteries are superficial, it is usually possible to feel the pulsations of the subclavian arteries in this region.

As with all arteries, there are variations in the normal branching, course and anatomical relations of the subclavian arteries

In the embryo, the initial (proximal) portion of the right subclavian artery develops from the fourth arch artery of the embryonic pharyngeal branch arteries. Other portions of the right subclavian artery and the left subclavian artery develop from the sixth or seventh intersegmental arteries that supply blood to the **limb buds**. The fusions of embryonic arteries that accompany the development of the pharyngeal arch system into the ascending aorta, aortic arch region of the ascending aorta, and the continuing descending aorta found in adults explain the often tortuous (twisting and turning) course around other **organs**, especially the esophagus and trachea, followed by the aortae and the major arteries that branch off from them.

The subclavians are major arteries. As such, the pressures found in the subclavian arteries are at, or near, the extremes of blood pressure—the highest pressures in the arteries are normally at or near the systolic high pressure (normally, 90-140 mm Hg). As with all arteries, due to the cumulative effects of vascular resistance (i.e., the drag exerted on the flow of blood by blood vessels), there is a generalized decline in pressure with increasing distance from the **heart**.

See also Angiology; Blood pressure and hormonal control mechanisms; Systemic circulation; Vascular system (embryonic development)

SUPERIOR · *see* ANATOMICAL NOMENCLATURE

SUPINATION · *see* ANATOMICAL NOMENCLATURE

SUTHERLAND, EARL W. (1915-1974)

American biochemist

Earl Sutherland was a biochemist who extended research and knowledge into the mechanisms by which **hormones** regulate body functions. His early work showed how the hormone adrenaline regulates the breakdown of sugar in the **liver** to release a surge of energy when the body is under stress. Later, Sutherland discovered a chemical within cells called cyclic adenosine 3'5'-monophosphate, or cyclic AMP. This chemical provided a universal link between hormones and the regulation of **metabolism** within cells. For this work, Sutherland was awarded the Nobel Prize in physiology and medicine in 1971.

Earl Wilbur Sutherland, Jr., the fifth of six children in his family, was born on November 19, 1915, in Burlingame, Kansas, a small farming community. His father, Earl Wilbur Sutherland, a Wisconsin native, had attended Grinnell College for two years and his mother attended college and received some nursing training. In 1933, Sutherland entered Washburn College in Topeka, Kansas. Supporting his studies by working as an orderly in a hospital, Sutherland graduated with a B.S. in 1937. He married the same year. Sutherland then entered Washington University Medical School in St. Louis, Missouri. There he enrolled in a **pharmacology** class taught by **Carl Ferdinand Cori**, who would share the 1947 Nobel Prize in medicine and physiology with his wife **Gerty Cori**. Impressed by Sutherland's abilities, Cori offered him a job as a student assistant. This was Sutherland's first experience with research. The research on the sugar glucose that Sutherland undertook in Cori's laboratory started him on a line of inquiry that led to his later groundbreaking studies.

Sutherland received his M.D. in 1942. He then worked for one year as an intern at Barnes Hospital while continuing to do research in Cori's laboratory. Sutherland was called into service during World War II as a battalion surgeon under General George S. Patton. Later in the war he served in Germany as a staff physician in a military hospital.

In 1945, Sutherland returned to Washington University in St. Louis. He was unsure whether to continue practicing medicine or to commit himself to a career in research. Sutherland later attributed his decision to stay in the laboratory to the example of his mentor Carl F. Cori. By 1953, Sutherland had advanced to the rank of associate professor at Washington University. During these years he came into contact with many leading figures in **biochemistry**, including **Arthur Kornberg**, Edwin G. Krebs, T. Z. Posternak, and others now recognized as among the founders of modern **molecular biology**. But Sutherland preferred, for the most part, to do his research independently. While at Washington University, Sutherland began a project to understand how an enzyme known as phosphorylase breaks down **glycogen**, a form of the sugar stored in the liver. He also studied the roles of the hormone adrenaline, also known as **epinephrine**, and glucagon, secreted by the **pancreas**, in stimulating the release of energy-producing glucose from glycogen.

Sutherland was offered the chairmanship of the Department of Pharmacology at Western Reserve (now Case Western) University in Cleveland in 1953. It was during the ten years he spent in Cleveland that Sutherland clarified an important mechanism by which hormones produce their effects. Scientists had previously thought that hormones acted on whole **organs**. Sutherland, however, showed that hormones stimulate individual cells in a process that takes place in two steps. First, a hormone attaches to specific receptors on the outside of the cell membrane. Sutherland called the hormone a "first messenger." The binding of the hormone to the membrane triggers release of a molecule known as cyclic AMP within the cell. Cyclic AMP then goes on to play many roles in the cell's metabolism, and Sutherland referred to the molecule as the "second messenger" in the mechanism of hormone action. In particular, Sutherland studied the effects of the hormone adrenaline, also called epinephrine, on liver cells. When adrenaline binds to liver cells, cyclic AMP is released and directs the conversion of sugar from a stored form into a form the cell can use.

Sutherland made two more important discoveries while at Western Reserve. He found that other hormones also spur the release of cyclic AMP when they bind to cells, in particular, the adrenocorticotropic hormone and the thyroid-stimulating hormone. This implied that cyclic AMP was a sort of universal intermediary in this process, and it explained why different hormones might induce similar effects. In addition, cyclic AMP was found to play an important role in the metabolism of one-celled organisms, such as the amoeba and the bacterium *Escherichia coli,* which do not have hormones. That cyclic AMP is found in both simple and complex organisms implies that it is a very basic and important biological molecule and that it arose early in **evolution** and has been conserved throughout millennia.

In 1963, Sutherland became professor of physiology at Vanderbilt University in Nashville, Tennessee, a move which relieved him of his teaching duties and enabled him to devote more of his time to research. Sutherland remarried and the couple later had two girls and two boys.

At Vanderbilt Sutherland continued his work on cyclic AMP, supported by a Career Investigatorship awarded by the American Heart Association. Sutherland studied the role of cyclic AMP in the contraction of heart muscle. He and other researchers continued to discover physiological processes in different tissues and various animal species that are influenced by cyclic AMP, for example in **brain** cells and **cancer** cells. Sutherland also did research on a similar molecule known as cyclic GMP (guanosine 3',5'-cyclic monophosphate). In the meantime, his pioneering studies had opened up a new field of research. By 1971, as many as two thousand scientists were studying cyclic AMP.

For most of his career, Sutherland was well-known mainly to his scientific colleagues. In the early 1970s, however, a rush of awards gained him more widespread public recognition. In 1970, he received the prestigious Albert Lasker Basic Medical Research Award. In 1971, he was awarded the Nobel Prize for "his long study of hormones, the chemical substances that regulate virtually every body function," as well as the American Heart Association Research Achievement Award. In 1973, he was bestowed with the National Medal of Science of the United States. During his career, Sutherland was also elected to membership in the National Academy of Sciences, and he belonged to the American Society of Biological Chemists, the American Chemical Society, the American Society for Pharmacology and Experimental Therapeutics, and the American Association for the Advancement of Science. He received honorary degrees from Yale University and Washington University. In 1973, Sutherland moved to the University of Miami. Shortly thereafter, he suffered a massive esophageal hemorrhage, and he died on March 9, 1974, after surgery for internal bleeding, at the age of 58.

See also Endocrine system and glands; Hormones and hormone action

SUTURES • *see* ANATOMICAL NOMENCLATURE

SWALLOWING AND DYSPHAGIA

Swallowing is a series of events where food or liquid is passed from the mouth into esophagus. The process is largely involuntary and is highly coordinated. Approximately 50 pairs of muscles and many nerves are operative in each swallowing event.

When food is ingested, it is prepared for swallowing by chewing (**mastication**). During mastication, there is an introduction and immersion of the food bolus into **saliva**. Serving as a predigestive fluid that breaks food down into biochemically useful components, saliva also acts as a lubricant for the swallowing process.

The swallow response begins with a pushing of the food or liquid to the back of the mouth by the tongue. This is known as the oral stage. The next stage, called the pharyngeal stage, starts as the food or liquid moves past the **pharynx** (a canal that connects the mouth with the esophagus) into the esophagus. The final, or esophageal, stage is the passage of the food or liquid down the esophageal tube to the stomach. Contractions of the esophagus help propel the bolus of food or liquid downward.

Dysphagia is a condition in which swallowing food or liquid becomes difficult or impossible. Severe dysphagia is almost always associated with a severe underlying **pathology** (illness, disease, or dysfunction). A common manifestation of mild dysphagia is the feeling of dry mouth and throat and near painful swallowing that accompanies nervousness. Another cause of dysphagia is the upward migration of stomach acid into the esophagus. This is called gastroesophageal reflux. The burning discomfort is what makes swallowing difficult. Conditions such these are usually temporary and abate by themselves. Other causes, such as reflux of stomach acid, may be successfully treated using conventional antiacid tablets. Severe dysphagia can sometimes, albeit rarely, be due to a tumor, nervous system disorder, or a neurological malady such as **stroke**. Disorders that weaken the muscles of the mouth or throat or which cause inefficient functioning of facial or throat nerves can lead to weight loss because of the inability to get much food to the digestive system. The more serious cases of dysphagia may require physiotherapy, adoption of new ways of eating and drinking, alteration of the diet to avoid problematic food or drink, or corrective surgery.

Symptoms of dysphagia include drooling, a sensation of sticking or of a lump in the throat, and coughing or choking due to improperly-swallowed food being sucked into the **lungs**. The latter can also cause a lung **infection** called aspiration pneumonia.

See also Gustatory structures

SWEAT GLANDS

Sweat glands are coiled tubes of cuboidal epithelial cells embedded in the skin. The human body contains approximately three million sweat glands. Also known as sudoriferous glands, sweat glands produce and secrete a fluid that is composed mostly of water but also includes salts, urea, **uric acid**, **amino acids**, ammonia, sugar, **lactic acid**, and ascorbic acid. The fluid produced is called sweat and is secreted during a process called perspiration.

The function of perspiration is to regulate body temperature and eliminate wastes from the body. Sweat glands are able to regulate body temperature because of the cooling effect produced when sweat evaporates from the skin. Regulated by the **hypothalamus**, the sweat glands are under control of the **sympathetic nervous system**. During a 24-hour period, an adult secretes approximately 100 ml of sweat under resting conditions. Exposure to heat and/or physical activity can cause sweat loss to exceed 100 times that of normal conditions in a day.

The two types of sweat glands found in human skin include eccrine (merocrine) and apocrine glands. The most common type is the eccrine glands; these are most abundant on the palms of the hands and soles of the feet. However, eccrine glands are located on all skin surfaces except eardrums, lips, fingers, toes, **nails**, penis, clitoris, and labia minora. Eccrine glands consist of a secretory duct, located in the subcutaneous **tissue** below the **dermis**, and an excretory duct embedded in the dermis that opens onto the surface of the **epidermis**.

Much less common than the eccrine glands are the apocrine glands. Although similar in structure to eccrine glands, apocrine glands open into **hair** follicles. The secretory portion of the apocrine gland originates in either the dermis or subcutaneous tissue. During secretion, the apical region of the cells is eliminated. Apocrine glands are located in the axilla, pubic region, and areolae of the **breasts** and do not become active until **puberty**. Additionally, **mammary glands** are essentially modified apocrine glands. The fluid produced by apocrine glands is thicker than that of eccrine glands. Initially, sweat secreted from apocrine glands is odorless. However, **bacteria** from the skin's surface invade the region and nourish themselves with **carbohydrates** and proteins lost during secretion, thus producing the characteristic body odor.

See also Homeostatic mechanisms; Sympathetic nervous system; Temperature regulation

Sylvius, Franciscus dele Bo
(1614-1672)
German physician

Franciscus dele Bo Sylvius, the first physician in the Netherlands to defend that the **blood** circulated in the vessels, was a descendent from an aristocratic protestant family (Calvinist) originally from Flanders with a successful business tradition. His grandfather immigrated to Germany because of religious persecution. Franciscus was born in Hanau, Germany, but spent most of his life in the Netherlands, where he received his education. He attended Medical School at the University of Leiden from 1633 to 1635, but finished in Basel, Switzerland, in 1637. After obtaining his M.D., he practiced medicine for a year and a half in Hanau, but decided to return to Leiden to work as an unpaid lecturer of **anatomy** at the University of Leiden, while hoping for a paid academic position.

Sylvius had a special interest in **brain** anatomy, and his studies led to the discovery of the lateral cerebral fissure (cleft), also known as the Sylvian fissure, and the Sylvian angle, that is formed by the lateral cerebral fissure with a line perpendicular to the superior border of the brain hemisphere. The Sylvian fissure, also known as the Aqueduct of Sylvius, is a groove at the external face of the brain, which separates the frontal lobe from the parietal temporal-occipital lobe. This structure is a narrow, elongated cavity of the midbrain that connects the third and fourth ventricles.

Besides anatomy, Sylvius had a special interest in iatrochemistry (from the Greek *iatros* for physician, i.e., physician's chemistry), a new theory of the seventeenth century that claimed that the cure of diseases through the administration of medicinal preparations, as well as all physiologic and pathologic processes, could be explained on a chemical basis. The theory also defended the use of chemistry as a tool to both the understanding of the ways drug compounds acted and the development of new treatments. Sylvius made a synthesis of the chemical notions of Paracelce (Theophrastus Bombastus von Hohenheim) and Van Helmont, and is considered the founder of the Iatrochemical School. From the iatrochemical perspective, Sylvius studied the process of **digestion**. One experiment attributed to Sylvius was the use of an inexpensive diuretic (fluid eliminating) compound made from the oil of Juniper berries mixed with grain alcohol.

Without peospects of being appointed to a paid position at the University of Leiden, Sylvius moved to Amsterdam in 1641, where his medical practice became very successful. He was already famous in Amsterdam as an anatomist, especially because of his studies on brain anatomy, what granted him a membership in the prestigious Amsterdam College of Physicians. However, it was only in 1658 that he was finally invited to become professor of medicine at the University of Leiden. From 1669 until 1670, Sylvius occupied the position of vice-chancellor of the university. In Amsterdam, he began writing a treatise entitled *New Idea In Medical Practice (Praxeos Medica Idea Nova)*, but died at the age of 58, after completing only the first volume. Franciscus Sylvius also wrote a descriptive treatise on the natural history of pulmonary tuberculosis that became a classic in late Renaissance Europe.

See also Cerebral hemispheres; Circulatory system; History of anatomy and physiology: The Renaissance and Age of Enlightenment

Sympathetic nervous system

The sympathetic nervous system and **parasympathetic nervous system** comprise two divisions of the **autonomic nervous system** (ANS). Sympathetic fibers innervate **smooth muscle**, cardiac muscle, and glandular **tissue**. In general, stimulation via sympathetic fibers increases activity and metabolic rate. Sympathetic system stimulation is a critical component of the fight or flight response.

Most target **organs** and tissues are innervated by neural fibers from both the sympathetic and parasympathetic systems. The systems can act to stimulate organs and tissues in opposite ways (antagonistically). For example, sympathetic stimulation acts to increase **heart** rate. In contrast, parasympathetic stimulation results in decreased heart rate. The systems can also act in concert to stimulate activity (e.g., both increase the production of **saliva** by salivary glands, but sympathetic stimulation results in viscous or thick saliva as opposed to a watery saliva). Although they share a number of common features, the classification of the parasympathetic and the sympathetic systems of the ANS is based both on anatomical and physiological differences between the two subsystems.

Both the sympathetic and parasympathetic systems contain myelinated preganglionic nerve fibers that usually connect with (**synapse** to) unmyelinated postganglionic fibers, via a cluster of neural cells termed ganglia. The cell bodies of sympathetic fibers traveling toward the ganglia (preganglionic fibers) are located in the thoracic and lumbar spinal nerves. These thoraco-lumbar fibers then travel only a short distance within the spinal nerve (composed of an independent mixture of fiber types) before leaving the nerve as myelinated white fibers that synapse with the sympathetic ganglia that lie close to the side of the **vertebral column**. The sympathetic ganglia lie in chains that line both the right and left sides of the vertebral column, from the cervical to the sacral region. Portions of the sympathetic preganglionic fibers do not travel to the vertebral ganglionic chains, but travel instead to specialized cervical or abdominal ganglia. Other variations are also possible. For example, preganglionic fibers can synapse directly with cells in the adrenal medulla.

In contrast to the parasympathetic system, the preganglionic fibers of the sympathetic nervous system are usually short and the sympathetic postganglionic fibers are long fibers that must travel to the target tissue. The sympathetic postganglionic fibers usually travel back to the spinal nerve via unmyelineted or gray rami before continuing to the target effector organs.

With regard to specific target organs and tissues, sympathetic stimulation of the pupil dilates the pupil. The dilation allows more light to enter the **eye** and acts to increase acuity in depth and peripheral perception.

Sympathetic stimulation acts to increase heart rate and increase the force of atrial and ventricular contractions. Sympathetic stimulation also increases the conduction velocity of cardiac muscle fibers. Sympathetic stimulation also causes a dilation of systemic arterial **blood** vessels, resulting in greater oxygen delivery.

Sympathetic stimulation of the **lungs** and smooth muscle surrounding the **bronchi** results in bronchial muscle relaxation. The relaxation allows the bronchi to expand to their full volumetric capacity and thereby allow greater volumes of air passage during respiration. The increased availability of oxygen, and increased venting of carbon dioxide are necessary to sustain vigorous muscular activity. Sympathetic stimulation can also results in increased activity by glands that control bronchial secretions.

Sympathetic stimulation of the **liver** increases glycogenolysis and lipolysis to make energy more available to metabolic processes. Constriction of gastro-intestinal sphincters (smooth muscle valves or constrictions), and a general decrease in gastrointestinal motility assure that blood and oxygen needed for more urgent needs (i.e., fight or flight) is not wasted on digestive system processes that can be deferred for short periods.

Sympathetic stimulation results in renin secretion by the **kidneys** and causes a relaxation of the bladder. Accompanied by a constriction of the bladder sphincter, sympathetic stimulation tends to decrease urination and promote fluid retention.

Acetylcholine is the neurotransmitter most often found in the sympathetic preganglionic synapse. Although there are exceptions (e.g., **sweat glands** utilize acetylcholine), **epinephrine** (noradrenaline) is the most common neurotransmitter found in postganglionic synapses.

See also Muscular innervation; Nerve impulses and conduction of impulses; Nervous system overview; Nervous system, embryological development; Neural plexi; Neurology; Neurons; Neurotransmitters

Synapse

A synapse is a functional gap or intercellular space between neural cells (**neurons**). The neural synapse is bound by the presynaptic terminal end of one neuron, and the dendrite of the postsynaptic neuron. Neuromuscular synapses are created when neurons terminate on a muscle. Neuroglandular synapses occur when neurons terminate on a gland. The major types of neural synapses include axodendritic synapses, axosomatic synapses, and axoaxonic synapses—each corresponding to the termination point of the presynaptic neuron.

The synapse is more properly described in structural terms as a synaptic cleft. The cleft is filled with extra cellular fluid and free **neurotransmitters**.

Nerve impulses are transmitted through the synaptic gap via chemical messengers—a special group of chemicals termed neurotransmitters. The arrival of an **action potential** (a

moving wave of electrical changes resulting from rapid exchanges of ions across the neural cell membrane) at the presynaptic terminus of a neuron, expels synaptic vesicles into the synaptic gap.

The four major neurotransmitters found in synaptic vesicles are noradrenaline, actylcholine, dopamine, and serotoin. Acetylchomine is derived from acetic acid and is found in both the **central nervous system** and the **peripheral nervous system**. Dopamine, **epinephrine**, and norepinephrine are catecholamines derived from tyrosine. Dopamine, epinephrine, and norepinephrine are also found in both the central nervous system and the peripheral nervous systems. Serotonin and histamine neurotransmitters are indolamines that primarily function in the central nervous system. Other **amino acids**, including gama-aminobutyric acid (GABA), aspartate, glutamate, and glycine along with neuropeptides containing bound amino acids also serve as neurotransmitters. Specialized neuropeptides include tachykinins and **endorphins** (including enkephalins) that function as natural painkillers.

Neurotransmitters diffuse across the synaptic gap and bind to neurotransmitter specific receptor sites on the dendrites of the postsynaptic neurons. When neurotransmitters bind to the dendrites of neurons across the synaptic gap they can, depending on the specific neurotransmitter, type of neuron, and timing of binding, excite or inhibit postsynaptic neurons.

After binding, the neurotransmitter may be degraded by **enzymes** or be released back into the synaptic cleft where in some cases it is subject to reuptake by a presynaptic neuron.

A number of neurons may contribute neurotransmitter molecules to a synaptic space. Neural transmission across the synapse is rarely a one-to-one direct diffusion across a synapse that separates individual presynaptic-postsynaptic neurons. Many neurons can converge on a postsynaptic neuron and, accordingly, presynaptic neurons are often able to affect the many other postsynaptic neurons. In some cases, one neuron may be able to communicate with hundreds of thousands of postsynaptic neurons through the synaptic gap.

Excitatory neurotransmitters work by causing ion shifts across the postsynaptic neural cell membrane. If sufficient excitatory neurotransmitter binds to dendrite receptors and the postsynaptic neuron is not in a refractory period, the postsynaptic neuron reaches threshold potential and fires off an electrical action potential that sweeps down the post synaptic neuron.

A summation of chemical neurotransmitters released from several presynaptic neurons can also excite or inhibit a particular postsynaptic neuron. Because neurotransmitters remain bound to their receptors for a time, excitation or inhibition can also result from an increased rate of release of neurotransmitter from the presynaptic neuron or delayed reuptake of neurotransmitter by the presynaptic neuron.

Bridge junctions composed of tubular proteins capable of carrying the action potential are found in the early embryo. During development, the bridges degrade and the synapses become the traditional chemical synapse.

See also Nerve impulses and conduction of impulses; Nervous system: embryological development; Nervous system overview

SYNOVIAL JOINTS

Synovial joints are lubricated by a synovial fluid and have a coating of **cartilage** to reduce the friction between the adjacent bones and to provide shock absorption. These joints hold the skeletal together while permitting a range of motions. As such, they are important for proper locomotion of the body.

The name synovial derives from synovium, meaning "like the white of a egg," referring to its visual appearance. A so-called synovial membrane, a very thin and delicate cell layer, lines the cavity containing the synovial fluid. Some of the cells in this membrane produce the synovial fluid.

Synovial joints vary in mobility (freedom of movement) and stability. Very mobile joints, such as the shoulder, are not stable and are easily dislocated. The elbow is not nearly as mobile, having a restricted range of movement, but it is much more stable and is seldom dislocated.

There are six ways that a synovial joint can move. The first, exemplified by the shoulder and the fit between the hip-bone and the femur, is a ball-and-socket arrangement. Back-and-forth and rotational movements are possible with this type of joint.

A condyloid joint allows for a pivoting motion. The joint accommodates convex and concave shapes of adjacent bones, allows them to fit together and to pivot against each other. Examples are the fingers and the jaw.

Synovial joints can glide over each other and rotate against each other in one plane. This can be best visualized as two blocks moving against each other in a back-and-forth or a twisting fashion. An example of this joint is in the carpus region of the lower hand.

A hinge joint offers movement in one plane, with no twisting, sliding or side-to-side motion. The elbow and the knee contain hinged synovial joints.

A pivot joint allows one bone is able to spin around on the spindle of another bone. The elbow is also a pivot joint. Another example is the first two vertebrae in the spinal column, whose back-and-forth capability allows for the "no" movement of the head.

The final type of synovial joint is called the saddle joint. This is similar to a condyloid joint, except that the fit between adjacent bones is not concave-to-convex. In a saddle joint the shapes can be distinctive, but they are still complimentary. The name derives from the saddle shape of the fit between the trapezium and first metacarpal bones of the hand.

See also Arthrology (joints and movement); Fibrous joints

SYNOVIAL MEMBRANES · *see* JOINTS AND SYN-
OVIAL MEMBRANES

SYSTEMIC CIRCULATION

The systemic circulation describes the extensive collection of **arteries** that distribute oxygenated **blood** to the **tissue** of the body and return deoxygenated blood to the **heart**.

Once returned to the right atrium of the heart, deoxygenated venous blood is pumped from the right ventricle into the **pulmonary circulation** where it is oxygenated in the **alveoli** of the **lungs** before being returned to the left atrium of the heart. This freshly oxygenated blood then moves into the right ventricle of the heart where a forceful muscular contraction of the heart's cardiac muscle expels the oxygenated blood into the aorta.

Because the **coronary circulation** derives from coronary arteries that almost immediately branch off the aorta, the coronary circulation can be described as a specialized part of the systemic circulation.

Blood in the aorta blood passes upward (superiorly) through the artic arch. From this arch, the aorta gives rives to a number of important arterial groups, including the carotid system of arteries (internal and external carotids) that ultimately branch to form smaller arteries that supply oxygenated blood to the head and neck, and the **subclavian arteries** that supply blood to the shoulder and upper limbs. After turning downward, the arch of the aorta continues as the **thoracic aorta** and then the **abdominal aorta**, before ultimately dividing into the iliac arterial system that serves the lower limb. All along its course, the aorta gives off branches of arteries that supply particular structures, **organs**, and tissues.

Each branch of the aorta gives off increasing smaller arteries or arterioles before multiplying (ramifying) into a capillary bed that most intimately services the target tissue or organ. It is within the **capillaries**, with membranes as thin as one cell, that gas and nutrient exchange takes place between blood, tissue, and interstitial fluids.

Ultimately, deoxygenated blood and metabolic waste products are removed from target tissues through a venous network. Smaller venules fuse together to create larger vessels. Small valves help keep the blood flowing toward the heart and venous pools help assure constant blood supply to the heart. The larger **veins** are classified by the region they drain, and the particular vessel through which they return blood to the heart.

Veins that drain blood from the head, neck, thorax, and upper limbs ultimately direct that blood into the right atrium of the heart through the superior vena cava. Blood from the **abdomen**, pelvis, and lower limbs returns to the right atrium of the heart through the inferior vena cava. A coronary sinus collects blood from cardiac veins and enters the right atrium near a point where the inferior vena cava enters the chamber.

The flow of blood within the systemic **circulatory system** flows a rhythmic pattern of higher and lower pressure associated with systole and diastole in the **cardiac cycle**. Accordingly, the flow of blood in the arterial system is pulsating rather than smooth and continuous.

At any given time, the venous portion of the systemic circulatory systems contains approximately 65% of the total volume of blood in the body. Blood in the systemic arterial system accounts for about 15 to 12.5% of the total blood volume.

Regulation of the systemic circulatory system is mostly achieved through the **autonomic nervous system**. Accordingly, sympathetic and parasympathetic stimulation can greater change vessel size and the flow of blood to particular organs, muscles groups, and tissues. Circulating **hormones** also play a regulatory role in altering systemic circulation and blood pressure. The proper maintenance of blood pressure within physiological limits is important. Low pressures reduce the availability of blood to tissues (tissue perfusion) and pressures that are too high can lead to **stroke** or **embolism**.

See also Abdominal veins; Angiology; Blood pressure and hormonal control mechanisms; Brachial system; Carotid arteries; Circle of Willis; Collateral circulation; Drug treatment of cardiovascular and vascular disorders; Fetal circulation; Lymphatic system; Metabolic waste removal; Nervous system overview; Oxygen transport and exchange; Pulse and pulse pressure; Thoracic aorta and arteries; Thoracic veins; Vascular exchange; Vascular system (embryonic development); Vascular system overview

SYSTOLE AND SYSTOLIC PRESSURE • *see*
BLOOD PRESSURE AND HORMONAL CONTROL MECHANISMS

SZENT-GYÖRGYI, ALBERT (1893-1986)
Hungarian-born American biochemist, molecular biologist, and physiologist

Albert Szent-Györgyi was awarded the Nobel Prize in physiology or medicine for his work in isolating vitamin C and his advances in the study of intercellular respiration; in 1954 he received the Albert and Mary Lasker Award from the American Heart Association for his contribution to the understanding of heart disease through his research in muscle physiology. In later years, Szent-Györgyi studied matter smaller than molecules, seeking the substances that would define the basic building blocks of life. In his late seventies, he founded the National Foundation for Cancer Research.

Albert Szent-Györgyi von Nagyrapolt was born in Budapest, Hungary, on September 16, 1893, to Miklos and Josephine Szent-Györgyi von Nagyrapolt. He was the second of three sons. His father, whose family claimed a title and was said to have traced their ancestry back to the seventeenth century, was a prosperous businessman who owned a two-thousand-acre farm located outside Budapest. His mother came from a long line of notable Hungarian scientists.

As a student, Szent-Györgyi did not begin to develop his potential until his last two years in high school, when he decided to become a medical researcher. In 1911, he entered Budapest Medical School. His education was interrupted by World War I, when he was drafted into the Hungarian Army. In the wake of the Austrian defeat, with Budapest under Communist rule, Szent-Györgyi decided to leave and accepted a research position at Pozony, Hungary, one hundred miles away. It was there, at the Pharmacological Institute of the Hungarian Elizabeth University, that Szent-Györgyi gained experience as a pharmacologist. In 1919, war broke out between Hungary and the Republic of Czechoslovakia. The Czechs seized Pozony, renaming it Bratislava. In order to con-

tinue his scientific training, Szent-Györgyi joined the millions of intellectuals who left Hungary during this time.

In 1921, he accepted a position at the Pharmaco-Therapeutical Institute of the University of Leiden in The Netherlands. This began a period of intense productivity for Szent-Györgyi: by the time he was twenty-nine years old, he had written nineteen research papers, and his research spanned the disciplines of physiology, **pharmacology**, bacteriology, and **biochemistry**. Szent-Györgyi is quoted by Ralph W. Moss, author of his biography, *Free Radical: Albert Szent-Györgyi and the Battle over Vitamin C,* as saying: "My problem was: was the hypothetical Creator an anatomist, physiologist, chemist or mathematician? My conclusion was that he had to be all of these, and so if I wanted to follow his trail, I had to have a grasp on all sides of nature." The scientist added that he "had a rather individual method. I did not try to acquire a theoretical knowledge before starting to work. I went straight to the laboratory, cooked up some senseless theory, and started to disprove it."

It was while in The Netherlands, as assistant to the professor of physiology at Groningen, that he presented the first of a series of papers on cellular respiration (the process by which organic molecules in the cell are converted to carbon dioxide and water, releasing energy), a question whose answer was considered central to biochemistry. Competing theories put forth on this question (one citing the priority of oxygen's role in the process; the other championing hydrogen as having the primary role) had caused biochemists to take one side or the other. Szent-Györgyi's contribution was that both theories were correct: active oxygen oxidized active hydrogen. Szent-Györgyi's research into cellular respiration laid the groundwork for the entire concept of the respiratory cycle. The paper discussing his theory is considered a milestone in biochemistry. Here, also, was the beginning of the work for which he was eventually given the Nobel Prize.

While still at Groningen, Szent-Györgyi began studying the role of the adrenal glands (responsible for secreting adrenaline and other important **hormones**), hoping to isolate a reducing agent (electron donor) and explain its role in the onset of Addison's disease. This work was to occupy him for almost a decade, produce unexpected results, and bring him worldwide attention as a scientist. He was sure he had made a breakthrough when silver nitrate added to a preparation of minced adrenal glands turned black. That indicated a reducing agent was present, and he set out to explain its function in oxidative **metabolism**. He thought the reducing agent might be a hormone equivalent to adrenalin. Frustrated because scientists in Groningen seemed unconvinced of the importance of his discovery, he wrote to **Henry Hallett Dale**, a prominent British physiologist. As a result of their correspondence, Szent-Györgyi was invited to England for three months to continue his work.

Unfortunately, his testing proved a failure—the color change of the silver nitrate turned out to be a reaction of adrenaline with the iron in the mincer in which he ground the adrenal glands. Szent-Györgyi returned to The Netherlands, where he continued his work on cellular respiration in plants, writing a paper on respiration in the potato. But increasing friction with the head of the laboratory caused him to resign his position. Unable to support his wife and daughter, he sent them home to Budapest. In August 1926, he attended a congress of the International Physiological Society in Stockholm, Sweden. It was there that his luck turned. The chairman of the event was Sir Frederick Gowland Hopkins, considered the greatest living biochemist of his day. Much to Szent-Györgyi's surprise, Hopkins referred to Szent-Györgyi's paper on potato respiration in his address to the congress. After the address, Szent-Györgyi introduced himself to Hopkins, who invited Szent-Györgyi's to Cambridge, where he was to remain until he returned to Hungary in 1932, eventually becoming president of the University of Szeged.

With the assurance of a fellowship from the Rockefeller Foundation (the foundation was to be a source of much financial support throughout his career), Szent-Györgyi sent for his family, rented a house, and set to work. Hopkins became his mentor—and the man Szent-Györgyi regarded as having the most influence on him as a scientist. While at Cambridge, he was awarded a Ph.D. for the isolation of hexuronic acid, the name given to the substance he had isolated from adrenal glands. One of the puzzling things about this substance was its similarity to one also found in citrus fruits and cabbage. Szent-Györgyi set out to analyze the substance, but the main obstacle to doing this was obtaining a sufficient supply of fresh adrenal glands. He finally was able to isolate a small quantity of a similar substance from orange juice and cabbage, learning that it was a carbohydrate and a sugar acid.

In 1929 Szent-Györgyi made his first visit to the United States. It was at this time that he visited the scientific community at Woods Hole, Massachusetts. He then went on to the Mayo Clinic in Rochester, Minnesota, where he had been invited to use the research facilities to continue his work isolating the adrenal substance. He managed to purify an ounce of the substance, and sent ten grams of it back to England for analysis. Nothing came of this, however, as the amount sent was too small. After almost ten years, the research appeared to be at a dead end. Szent-Györgyi took what remained of the purified crystals and returned to Cambridge.

In 1928 Szent-Györgyi had been offered a top academic post at the University of Szeged in Hungary. He accepted, but did not take up his duties there until 1931 because of delays in completing the Szeged laboratory. At Szeged, in addition to his duties as teacher, Szent-Györgyi continued his research, still trying to solve the puzzle of the adrenal substance, hexuronic acid.

It had been known since the sixteenth century that certain foods, especially citrus fruits, prevented scurvy, a disease characterized by swollen gums and loosened teeth. Although scurvy could be prevented by including citrus fruit in the diet, isolation of the antiscurvy element from citrus eluded researchers. It was not until after World War I that drug companies began a concentrated search for the antiscorbutic element (now called vitamin C). Scientists in Europe and the United States began competing to be the first to isolate this element. Vitamin C was not unfamiliar to Szent-Györgyi, and he had written of its possible connection with hexuronic acid. Now he was able to positively identify hexuronic acid as vita-

min C and not an adrenal hormone, as he had previously thought. He suggested the compound be called ascorbic acid, and continued his study of its function in the body, using vitamin C-rich Hungarian paprika as the source material.

Although Charles Glen King had also isolated Vitamin C and made the connection between it and hexuronic acid—and announced his findings just two weeks before Szent-Györgyi made his report, in 1937 Szent-Györgyi was given the Nobel Prize. His acceptance speech, "Oxidation, Energy Transfer, and Vitamins," gave details of the extraordinary circumstances under which his discoveries were made.

In 1941 Szent-Györgyi remarried. Bitterly opposed to Nazi rule in Hungary, he became an active member of the Hungarian underground. It was during the war years that he made some of his most important discoveries. His work during this time still concentrated on cellular respiration. His research in this area proved to be the basis for one of the fundamental breakthroughs in biology: the citric acid cycle. This cycle explains how almost all cells extract energy from food. It was during the war years that he also studied the chemical mechanisms of **muscle contraction**. His discoveries about how muscles move and function were fundamental to twentieth-century physiology, and made him a pioneer in **molecular biology**.

By 1944 Szent-Györgyi's outspoken opposition to Hitler's regime had put his life in danger. He and his wife went into hiding for the remainder of the war, surfacing in Budapest when the Russians liberated Hungary from the Nazis in 1945. Disillusioned with Soviet rule, he emigrated to the United States in 1947, and became an American citizen in 1954.

Szent-Györgyi settled in Woods Hole, Massachusetts. Although research facilities were provided for him at the Marine Biological Laboratories, he struggled to find backing to continue his work. With the help of five wealthy businessmen, he set up the Szent-Györgyi Foundation (later called the Institute for Muscle Research), whose purpose was to raise money for muscle research and bring a group of Hungarian scientists to America to assist him. In 1948, Szent-Györgyi took a position with the National Institutes of Health (NIH). He left there in 1950 for a short assignment at Princeton University's Institute for Advanced Studies. Then grants began to come in for his muscle research.

During these years, Szent-Györgyi and his team of researchers continued to make strides in the analysis of muscle protein. He also published three books: *Chemistry of Muscular Contraction, The Nature of Life,* and *Chemical Physiology of Contraction in Body and Heart,* and 120 scientific papers. These writings brought him to the attention of the American scientific community and had great influence on scientists worldwide.

Szent-Györgyi remarried twice after his second wife's death. In addition to his scientific writings, he wrote books that characterized his personal philosophy: *Science, Ethics, and Politics* and *The Crazy Ape* (which included his poem series, "Psalmus Humanus and Six Prayers"). He spoke out against the Vietnam war on numerous occasions, both in public lectures and through letters to newspapers and periodicals.

Szent-Györgyi was almost eighty years old when he founded the National Foundation for Cancer Research.

Funding from the NFCR supported his research until the end of his life. For more than forty years, his research had been concerned with the development of a basic theory about the nature of life. Szent-Györgyi called this new field of endeavor "submolecular biology." It was not just a cure for cancer that he was looking for, but a new way of looking at biology. He was convinced that his study of the structure of life at the level of electrons would not only make possible a cure for cancer but would also provide the knowledge to ensure the human body's optimum health.

Ralph Moss, the author of *Free Radical,* asked Szent-Györgyi for his philosophy of life shortly before the scientist's death of kidney failure on October 22, 1986. He scrawled on a piece of paper: "Think boldly. Don't be afraid of making mistakes. Don't miss small details, keep your eyes open and be modest in everything except your aims."

See also Adrenal glands and hormones; Antioxidents; Cell membrane transport; Cell structure; Metabolism; Myology; Respiration control mechanisms; Vitamins

T

T LYMPHOCYTES

When a vertebrate encounters substances that are capable of causing it harm, a protective system known as the **immune system** comes into play. This system is a network of many different **organs** that work together to recognize foreign substances and destroy them. The immune system can respond to the presence of a disease-causing agent (pathogen) in two ways. Immune cells called the B cells can produce soluble proteins (antibodies) that can accurately target and kill the pathogen. This branch of immunity is called "humoral immunity." In cell-mediated immunity, immune cells known as the T cells produce special chemicals that can specifically isolate the pathogen and destroy it.

The T cells and the B cells together are called the lymphocytes. The precursors of both types of cells are produced in the bone marrow. While the B cells mature in the bone marrow, the precursors to the T cells leave the bone marrow and mature in the thymus. Hence the name, "T cells" for thymus-derived cells.

The role of the T cells in the immune response is to specifically recognize the pathogens that enter the body and to destroy them. They do this either by directly killing the cells that have been invaded by the pathogen, or by releasing soluble chemicals called "cytokines," which can stimulate other killer cells specifically capable of destroying the pathogen.

During the process of maturation in the thymus, the T cells are taught to discriminate between "self" (an individual's own body cells) and "non-self" (foreign cells or pathogens). The immature T cells, while developing and differentiating in the thymus, are exposed to the different thymic cells. Only those T cells that are "self-tolerant," that is to say, they will not interact with the molecules normally expressed on the different body cells are allowed to leave the thymus. Cells that react with the body's own proteins are eliminated by a process known as "clonal deletion." The process of clonal deletion ensures that the mature T cells, which circulate in the **blood**, will not interact with or destroy an individual's own tissues and organs. The mature T cells can be divided into two subsets, the T-4 cells (that have the accessory molecule CD4), or the T-8 (that have CD8 as the accessory molecule).

There are millions of T cells in the body. Each T cell has a unique protein structure on its surface known as the T cell receptor (TCR), which is made before the cells ever encounter an antigen. The TCR can recognize and bind only to a molecule that has a complementary structure. It is kind of like a lock-and key arrangement. Each TCR has a unique binding site that can attach to a specific portion of the antigen called the epitope. As stated before, the binding depends on the complementarity of the surface of the receptor and the surface of the epitope. If the binding surfaces are complementary, and the T cells can effectively bind to the antigen, then it can set into motion the immunological cascade which eventually results in the destruction of the pathogen.

The first step in the destruction of the pathogen is the activation of the T cells. Once the T lymphocytes are activated, they are stimulated to multiply. Special cytokines called interleukins that are produced by the T-4 lymphocytes mediate this proliferation. It results in the production of thousands of identical cells, all of which are specific for the original antigen. This process of clonal proliferation ensures that enough cells are produced to mount a successful immune response. The large clone of identical lymphocytes then differentiates into different cells that can destroy the original antigen.

The T-8 lymphocytes differentiate into cytotoxic T-lymphocytes (CTLs) that can destroy the body cells that have the original antigenic epitope on its surface, e.g., bacterial infected cells, viral infected cells, and tumor cells. Some of the T lymphocytes become memory cells. These cells are capable of remembering the original antigen. If the individual is exposed to the same **bacteria** or virus again, these memory cells will initiate a rapid and strong immune response against

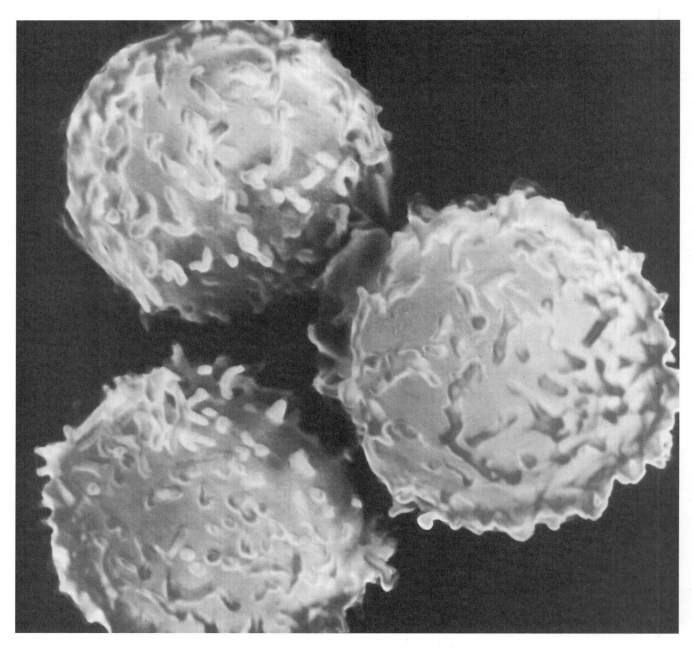

Scanning electron micrograph (SEM) image of three T lymphocytes. © Microworks/Phototake. Reproduced by permission.

it. This is the reason why the body develops a permanent immunity after an infectious disease.

Certain other cells known as the T-8 suppressor cells play a role in turning off the immune response once the antigen has been removed. This is one of the ways by which the immune response is regulated.

See also Bacteria and bacterial infection; Antigens and antibodies; Immunity, cell mediated; Immunity, humoral regulation; Viruses and responses to viral infection

TASTE, PHYSIOLOGY OF GUSTATORY STRUCTURES

The ability to taste depends on the presence and function of the 2,000 to 5,000 taste buds in the oral cavity. These are located on the tongue, where they are associated with papillae, and on the **soft palate**, the **pharynx**, **larynx**, and epiglottis.

Each taste bud is made up of cells arranged together much like the slices of an orange. At the end of the bud that

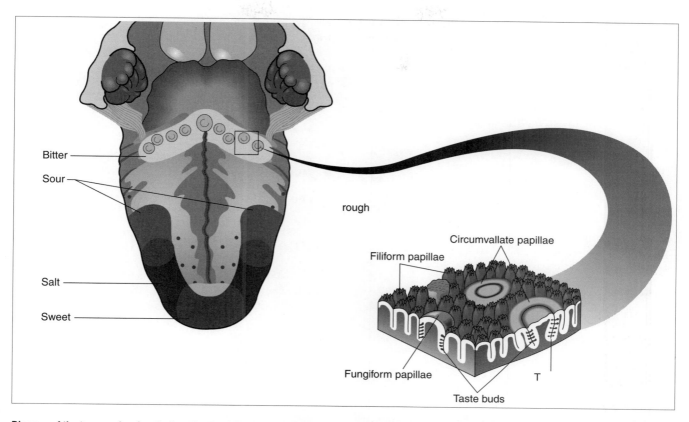

Diagram of the tongue showing the locations and basic structure of taste buds. *Illustration by Hans & Cassidy.*

connects with the surface is a taste pore, which houses a recep-tor for the various chemicals that confer taste. For example, sweet molecules are sugars and alcohols, salty chemicals are dominated by sodium, and sour chemicals are dominated by hydrogen ions.

The interaction of the various molecules with the taste bud receptor stimulates a depolarization of the membrane at the receptor site. In depolarization, the membrane becomes more permeable to the passage of ions such as sodium and hydrogen (this is dependent on the particular receptor). The depolarization causes the activation of a protein called the G-protein, which acts to alter the countermovement of potas-sium out of the cell. The resulting ionic imbalance between the inside of the receptor cell and the outside generates an electrical potential. The potential can be harnessed to drive the movement of calicium ions into the cell. The excess **cal-cium** and the release of a neurotransmitter converts the recep-tor signal to an electrical signal, which is conveyed to the **brain**.

Depolarization and re-polarization of the membranes of the cells are rapid events. This allows a taste cell to respond to a stimulus and quickly regain its potential for another response.

The taste receptors have a very low threshold, or the concentration at which a substance can be distinguished from water. The intensity of a taste sensation sems to be related to the number of taste cells that respond to the presence of the

particular chemical. For example, if a food is exceptionally salty, the great number of sodium molecules in the food will stimulate the activity of a large number of taste buds. In more complex foods, which have a blended variety of tastes, the response by the population of taste buds will become diverse.

See also Sodium-potassium pump

TEETH · *see* TOOTH DEVELOPMENT, LOSS, REPLACEMENT, AND DECAY

TEMPERATURE REGULATION

The human body is able to actively regulate its internal tem-perature somewhere between 98°F and 100°F (36.6°C and 37.7°C) in spite of various forces that act to cool off the body.

These mechanisms operate through a control region of the **brain** called **hypothalamus**, which contains the mecha-nisms to control the internal temperature and sensors to mon-itor the body.

The sensors of temperature regulation are on the skin surface and within the body. These all "report" to the hypo-thalamus.

The body is monitored to gauge the ebb and flow that can occur by various means. Evaporation of liquid draws heat away, as does loss of heat by moving air (convection) and by the radiation of heat from the body surface to the cooler surrounding air. The body can also be heated by a source of radiation. For example, sitting out in the sun on a summer day will warm up the body. Another form of heat loss is via the transfer from the body directly to a solid surface, such as a tile floor. This is heat loss by conduction.

If the internal temperature increases out of the normal range the hypothalamus acts to maximize heat loss by triggering the production of sweat. The evaporation of the sweat cools the body. If this is not sufficient to bring the temperature back to normal, the degree of sweating can be increased. On the other hand, if body temperature drops below the normal range, a number of responses can be triggered by the hypothalamus to minimize heat loss and maximize heat production. Vasoconstriction (shrinkage of the diameter of **blood** vessels) can occur. This decreases the flow of blood (and so of heat) to the skin. Sweating can be ended. The body can begin to shiver, which heats up the muscles. The secretion of compounds such as norepinephrine, **epinephrine**, and thryoxine can be stimulated. These compounds also increase heat production.

Temperature regulation also accounts for the temperature of the surrounding air. In cold winter conditions, the body will monitor so as to maintain the proper internal temperature in the core of the body (head and truck). The limbs will be allowed to get cooler before a regulatory alteration is made. In the warm summer months, the zone of the internal temperature that is maintained covers a larger area of the body, excluding only the hands and feet. Body **fat** provides further temperature regulation via insulation.

See also Adipose tissue; Homeostatic mechanisms

TEMPOROMANDIBULAR JOINT

One of the most multifunctional articulations in the body is the temporomandibular joint (TMJ). It is the point of connection between the jaw and the **cranium**. The large mandibular condyle on top of the ascending ramus of the mandible fits neatly into the mandibular fossa of the temporal bone. Both the fossa and condyle are large and covered with a serous membrane for easy movement. The synovial membrane of both elements allows for gliding and hingelike movements. The jaw can be both depressed (moved down) opening the mouth and rotated from side to side. This type of movement allows for chewing of food and is not seen in most vertebrates. For example, crocodiles have only an up and down movement of the jaw. Their temporomandibular joint is much more restricted than those in humans.

This high degree of motion is accomplished by the size of the fossa and condyle in addition to some important ligaments. The synovial cavity of the TMJ has both superior and inferior chambers that are covered by the articular disc. The temporomandibular ligament lies deep or under the parotid salivary gland on the lateral side. Its function is to keep the jaw from moving from side to side. This ligament is especially helpful in preventing the jaw, when receiving a hard blow, from pushing upwards into the **skull**. The companion ligament on the medial side is the sphenomandibular ligament. It extends from the sphenoid bone to the ramus of the mandible.

Two additional ligaments are not considered part of the TMJ, but are nonetheless important. The stylomandibular ligament connects the styloid process to the inferior ramus of the mandible. It prevents too much extension of the jaw. Covering the articular capsule itself is the capsular ligament. This **tissue** keeps the joint from popping out to the external side.

See also Joints and synovial membranes; Mastication

TENDONS AND TENDON REFLEX

Tendons form the attachment between muscles and bones. They are strong, whitish inelastic cords consisting of bundles of collagen fibers, held together by a "wrapping" of membrane called the epitendineum. At their point of attachment, some of the tendon fibers merge with the periosteum, the bone's fibrous covering. Others, called Sharpey's perforating fibers, penetrate into the bone. The penetrating fibers are engulfed by the bone as it grows.

When a muscle, contracting to move a joint, is overloaded, the tendon may tear. The damage can range from a tear affecting only some of the fibers, a strain that heals quickly, to a full rupture. If the tendon is torn completely off the bone, muscle movement usually pulls it away from its original position. In such a case, it cannot heal properly on its own, and surgical repair is required.

Tendons have very little **blood** supply. The nerves supplying the tendons have special receptors called organs of Golgi at their attachment points. The organs of Golgi consist of millimeter-long capsule of muscle and collagen fibers in which the nerve endings are intertwined.

The nerves in the organs of Golgi detect changes in muscle tension when they are squeezed by the tendon fibers. If the tension is excessive, the organs of Golgi prevent damage to the muscle and tendon by signaling the motor **neurons** in the **spinal cord**, which inhibit the muscle from contracting further. This feedback phenomenon is called the tendon reflex.

Aponeuroses are flattened ribbons of connective **tissue**. They are generally consider a type of tendon and are similar in structure. In addition to attaching muscle to bone, aponeuroses sometimes connect two muscles or other tissues. For example, the lingual aponeurosis connects the tongue with the lingual muscles.

See also Arthrology; Muscle contraction; Muscular innervation; Muscle tissue damage, repair and regeneration

TESTIS · *see* REPRODUCTIVE SYSTEM AND ORGANS (MALE)

TESTOSTERONE

Testosterone is a male sex hormone (androgen) that is principally produced by the testes. Testosterone is an important component the normal development of male sexual **organs** and prostate. Elevated levels of testosterone during **puberty** cause the development of **secondary sexual characteristics** (broadening of shoulders, beard development, etc.). Testosterone plays an important role in the **growth and development** of bone and muscle, and in the maintenance of muscle strength.

It is important to note that although testosterone is an androgen, it is also manufactured (secreted) and found in much lower levels in females. Elevated levels of testosterone in females may lead to the development of male-like secondary sexual characteristics (masculinization).

Teststosterone and other steroid **hormones** (e.g., **estrogen**, cortisone, etc) are derived from cholesterol. At the molecular level, steroid hormones such as testosterone consist of four interlocking rings of carbon atoms (three of the rings contain six carbon atoms and the fourth ring contains five carbon atoms) that with hydrogen atoms bonded to the ringed carbon atoms, form a hydrogenated cyclopentophenanthrenering. Testosterone is produced in the interstitial cells of Leydig that lie between the seminiferous tubules.

After secretion by the testes, testosterone enters the **blood** stream to circulate throughout the body. As the testosterone circulates, it binds to specific receptors on target tissues. Within the cell, testosterone is converted into dihydrotestosterone—and it is in this form that testosterone enters the cellular **metabolism**. Dihydrotestosterone binds with other proteins to form a complex that is capable of entering the nucleus and stimulating **DNA** transcription.

Testosterone that fails to bind to tissues degrades into inactive substances that are removed by excretion. Most of these biochemical breakdowns occur in the **liver**.

Although there are other male **sex hormones** produced by the testes, testosterone is by far the most abundant hormone produced. Elevated testosterone levels occur at two very different and distinct times in normal male development. Increased testosterone secretion occurs in newborn male infants and with the onset of puberty.

In the embryo, testicular **tissue**, stimulated by placental gonadotropin hormones, produced the higher levels of testosterone needed for normal male development during organogenisis (formation of organs). The synthesis of testosterone begins during the second month of gestation and acts principally to influence the differential development of the genital ridge in males. Physiological tests on other animals demonstrate that the introduction of testosterone during this stage of development can lead to the formation of male genitalia even if the developing embryo is genetically female. Testosterone stimulates the genital ridge form a penis, scrotum, seminal vesicles, and **prostate gland**, and suppresses the development of a female clitoris and vagina.

In the newborn male, testosterone also induces the descent of the testis into the scrotum through the inguinal canals.

Luteinizing hormone (LH) produced by the pituitary gland stimulates testosterone production in the pubescent and adult male. LH acts in coordination with **follicle stimulating hormone** (FSH) to promote the production of **sperm** cells (**spermatogenesis**). In the adult, increased testosterone secretion causes enlargement of the penis, scrotum, and testes and induce the development of secondary sexual characteristics, including the development of body **hair**. Interestingly, testosterone inhibits the growth of hair on the top of the head and—along with a genetic predisposition to baldness—may eventually cause baldness. Testosterone also causes a thickening of skin and of the laryngeal mucosa and **larynx** that produce the deeper male voice.

In addition to promoting a higher metabolic rate, testosterone cases increase nitrogen retention and **protein synthesis** associated with muscle growth. In addition, testosterone acts to promote **calcium** retention and bone matrix development that produces the thickened male skeletal structure.

See also Adolescent growth and development; Adrenal glands and hormones; Embryonic development: early development, formation, and differentiation; Homologous structures; Hormones and hormone action; Human development (timetables and developmental horizons); Larynx and vocal cords; Muscular system overview; Urogenital system (male)

THALAMUS

The thalamus is a part of the **brain** that is found in the third ventricle. It is egg-shaped and along with the **hypothalamus** comprises the area known as the diencephalon.

The thalamus acts as a central relay station for all sensory impulses travelling from other parts of the **spinal cord** and brain to the **cerebral cortex** (outer portion of the brain). It receives all sensory impulses (except those associated with **smell**) and directs them to the appropriate regions of the cerebrum for appropriate interpretation and integration. The thalamus acts as the last portion of the brain for sensory input before the cerebrum.

The thalamus connects to various parts of the brain via specialized nerve fibers that originate in the upper part of the reticular activating system (RAS). The RAS (which extends from the **medulla oblongata** to the thalamus) sorts out stimuli from the sense **organs** and passes on only those impulses that require immediate attention. We are not aware of most of the sensory impulses received by the **central nervous system**. Because the thalamus channels only certain sensory impulses to the cerebrum, our brains do not become overloaded with stimuli. For this reason, the thalamus is sometimes referred to as the "gatekeeper of the cerebrum." The thalamus also receives input from the cerebrum and from other parts of the brain that regulate emotion and arousal.

Encephalopathy or brain diseases of the thalamus have been associated with Parkinson's disease, epilepsy, bipolar disorder and chronic solvent abuse.

Removal of the ventral thalamus (thalamotomy) has been reported to significantly reduce or completely stop tremor in 80–90% of Parkinson's patients who undergo the procedure. Neurostimulation of the thalamus also reduces tremors without removal of thalamic **tissue**. In this procedure, a tiny electrical **pulse** generator is surgically implanted near the patient's collarbone. The generator is connected to four electrodes that have been implanted in the thalamus and can deliver programmed pulses which the patient can turn on or off using a magnet held over the skin. When the pulse is turned on, the tremor is suppressed. Neurostimulation of the thalamus is also being tested on epileptics to suppress seizures.

It has been found that patients suffering from bipolar disorder (maniac-depression) have elevated levels of VMAT2 (a protein that regulates the transport of important **neurotransmitters**) in the thalamus and brain stem. Since the thalamus plays a role in the **anatomy** of mood and emotion, this finding may be related to the extreme emotional swings evident in this disorder.

The thalamus of chronic solvent abusers who inhale volatile fumes in order to attain a high have been found by MRI (magnet resonance **imaging**) pictures to have a combination of diffuse white matter changes and low signal intensity. This may be related to the partitioning of solvents into the lipid membranes of the thalamus.

It is also estimated that 10–15% of strokes (**blood** clots in the brain) that involve loss of consciousness are thalamic in origin.

See also Brain stem function and reflexes; Brain, intellectual functions; Nerve impulses and conduction of impulses

THEILER, MAX (1899-1972)
South African virologist

Max Theiler (pronounced Tyler) was a leading scientist in the development of the yellow-fever vaccine. His early research proved that yellow-fever virus could be transmitted to mice. He later extended this research to show that mice that were given serum from humans or animals that had been previously infected with yellow fever developed immunity to this disease. From this research, he developed two different vaccines in the 1930s, which were used to control this incurable tropical disease. For his work on the yellow-fever vaccine, Theiler was awarded the Nobel Prize in physiology or medicine in 1951.

Theiler was born on a farm near Pretoria, South Africa, on January 30, 1899, the youngest of four children of Emma (Jegge) and Sir Arnold Theiler, both of whom had emigrated from Switzerland. His father, director of South Africa's veterinary services, pushed him toward a career in medicine. In part to satisfy his father, he enrolled in a two-year premedical program at the University of Cape Town in 1916. In 1919, soon after the conclusion of World War I, he sailed for England, where he pursued further medical training at St. Thomas's Hospital Medical School and the London School of Hygiene and Tropical Medicine, two branches of the University of London. Despite this rigorous training, Theiler never received the M.D. degree because the University of London refused to recognize his two years of training at the University of Cape Town.

Theiler was not enthralled with medicine and had no intention of becoming a general practitioner. He was frustrated by the ineffectiveness of most medical procedures and the lack of cures for serious illnesses. After finishing his medical training in 1922, the 23-year-old Theiler obtained a position as an assistant in the Department of Tropical Medicine at Harvard Medical School. His early research, highly influenced by the example and writings of American bacteriologist Hans Zinsser, focused on amoebic dysentery and rat-bite fever. From there, he developed an interest in the yellow-fever virus.

Yellow fever is a tropical viral disease that causes severe fever, slow **pulse**, bleeding in the stomach, jaundice, and the notorious symptom, black vomit. The disease is fatal in 10% to 15% of cases, the cause of **death** being complete shutdown of the **liver** or **kidneys**. Most people recover completely, after a painful, extended illness, with complete immunity to reinfection. The first known outbreak of yellow fever devastated Mexico in 1648. The last major breakout in the continental United States claimed 435 lives in New Orleans in 1905. Despite the medical advances of the twentieth century, this tropical disease remains incurable. As early as the eighteenth century, mosquitoes were thought to have some relation to yellow fever. Cuban physician Carlos Finlay speculated that mosquitoes were the carriers of this disease in 1881, but his writings were largely ignored by the medical community. Roughly 20 years later, members of America's Yellow Fever Commission, led by Walter Reed, the famous U.S. Army surgeon, concluded that mosquitoes were the medium that spread the disease. In 1901, Reed's group, using humans as research subjects, discovered that yellow fever was caused by a blood-borne virus. Encouraged by these findings, the Rockefeller Foundation launched a world-wide program in 1916 designed to control and eventually eradicate yellow fever.

By the 1920s, yellow-fever research shifted away from an all-out war on mosquitoes to attempts to find a vaccine to prevent the spread of the disease. In 1928, researchers discovered that the Rhesus monkey, unlike most other monkeys, could contract yellow fever and could be used for experimentation. Theiler's first big breakthrough was his discovery that mice could be used experimentally in place of the monkey and that they had several practical research advantages. When yellow-fever virus was injected into their brains, the mice didn't develop human symptoms. Instead, "when you give a mouse yellow fever, he gets not jaundice but encephalitis, not a fatal bellyache but a fatal headache," Theiler stated, according to Greer Williams author of *Virus Hunters*.

One unintended research discovery kept Theiler out of his lab and in bed for nearly a week. In the course of his experiments, he accidentally contracted yellow fever from one of his mice, which caused a slight fever and weakness. Theiler was much luckier than some other yellow-fever researchers. Many had succumbed to the disease in the course of their investigations. However, this small bout of yellow fever simply gave Theiler immunity to the disease. In effect, he was the first recipient of a yellow-fever vaccine.

In 1930, Theiler reported his findings on the effectiveness of using mice for yellow fever research in the respected journal *Science*. The initial response was overwhelmingly negative; the Harvard faculty, including Theiler's immediate supervisor, seemed particularly unimpressed. Undaunted, Theiler continued his work, moving from Harvard University, where he was considered an upstart, to the Rockefeller Foundation in New York City. Eventually, yellow-fever researchers began to see the logic behind Theiler's use of the mouse and followed his lead. His continued experiments made the mouse the research animal of choice. By passing the yellow-fever virus from mouse to mouse, he was able to shorten the incubation time and increase the virulence of the disease, which enabled research data to be generated more quickly and cheaply. He was now certain that an attenuated live vaccine, one weak enough to cause no harm yet strong enough to generate immunity, could be developed.

In 1931, Theiler developed the mouse-protection test, which involved mixing yellow-fever virus with human **blood** and injecting the mixture into a mouse. If the mouse survived, then the blood had obviously neutralized the virus, proving that the blood donor was immune to yellow fever (and had most likely developed an immunity by previously contracting the disease). This test was used to conduct the first worldwide survey of the distribution of yellow fever.

A colleague at the Rockefeller Foundation, Dr. Wilbur A. Sawyer, used Theiler's mouse strain, a combination of yellow fever virus and immune serum, to develop a human vaccine. Sawyer is often wrongly credited with inventing the first human yellow-fever vaccine. He simply transferred Theiler's work from the mouse to humans. Ten workers in the Rockefeller labs were inoculated with the mouse strain, with no apparent side effects. The mouse-virus strain was subsequently used by the French government to immunize French colonials in West Africa, a hot spot for yellow fever. This so-called "scratch" vaccine was a combination of infected mouse **brain tissue** and cowpox virus and could be quickly administered by scratching the vaccine into the skin. It was used throughout Africa for nearly 25 years and led to the near total eradication of yellow fever in the major African cities.

While he was somewhat pleased with the new vaccine, Theiler considered the mouse strain inappropriate for human use. In some cases, the vaccine led to encephalitis in a few recipients and caused less severe side effects, such as **headache** or **nausea**, in many others. Theiler believed that a "killed" vaccine, which used a dead virus, wouldn't produce an immune effect, so he and his colleagues set out to find a milder live strain. He began working with the Asibi yellow-fever strain, a form of the virus so powerful that it killed monkeys instantly when injected under the skin. The Asibi strain thrived in a number of media, including chicken embryos. Theiler kept this virus alive for years in tissue cultures, passing it from embryo to embryo, and only occasionally testing the potency of the virus in a living animal. He continued making subcultures of the virus until he reached strain number 176. Then, he tested the strain on two monkeys. Both animals survived and seemed to have acquired a sufficient immunity to yellow fever. In March 1937, after testing this new vaccine on himself and others, Theiler announced that he had developed a new, safer, attenuated vaccine, which he called 17D strain. This new strain was much easier to produce, cheaper, and caused very mild side effects.

From 1940 to 1947, with the financial assistance of the Rockefeller Foundation, more than 28 million 17D-strain vaccines were produced, at a cost of approximately two cents per unit, and given away to people in tropical countries and the U.S. The vaccine was so effective that the Rockefeller Foundation ended its yellow-fever program in 1949, safe in the knowledge that the disease had been effectively eradicated worldwide and that any subsequent outbreaks could be controlled with the new vaccine. Unfortunately, almost all yellow-fever research ended around this time and few people studied how to cure the disease. For people in tropical climates who live outside of the major urban centers, yellow fever is still a problem. A major outbreak in Ethiopia in 1960–62 caused 30,000 deaths. The World Health Organization still uses Theiler's 17D vaccine and is attempting to inoculate people in remote areas.

The success of the vaccine brought Theiler recognition both in the U.S. and abroad and even from his former employer, Harvard University. Over the next ten years, he received the Chalmer's Medal of the Royal Society of Tropical Medicine and Hygiene (1939), the Lasker Award of the American Public Health Association, and the Flattery Medal of Harvard University (1945).

In 1951, Theiler received the Nobel Prize in physiology or medicine "for his discoveries concerning yellow fever and how to combat it."

After developing the yellow-fever vaccine, Theiler turned his attention to other **viruses**, including some unusual and rare diseases, such as Bwamba fever and Rift Valley fever. His other, less exotic research focused on polio and led to his discovery of a polio-like **infection** in mice known as encephalomyelitis or Theiler's disease. In 1964, he retired from the Rockefeller Foundation, having achieved the rank of associate director for medical and natural sciences and director of the Virus Laboratories. In that same year, he accepted a position as professor of epidemiology and microbiology at Yale University in New Haven, Connecticut. He retired from Yale in 1967.

Theiler married in 1938 and had one daughter. His non-scientific interests included reading (mostly history and philosophy but absolutely no fiction) and watching baseball games, especially those involving his beloved Brooklyn Dodgers. Although he immigrated to the U.S. in 1923 and remained in America for the rest of his life, he never applied for U.S. citizenship. Theiler died on August 11, 1972, at the age of 73.

See also Fever and febrile seizures; Immune system; Immunity; Immunology

THIRST

Physiological thirst is the sensation that produces a conscious drive to replace depleted body fluids. The sensation is most often described as a dryness of the mouth and throat, accompanied by a craving for liquid. Thirst is a critical component in the regulation of body fluids and electrolyte concentrations (e.g., sodium).

Body fluid balance, or hydration, is dependent on the difference between water intake and output each day. A great deal of water is lost each day via evaporation from the skin, expulsion and evaporation in respiratory air, and excretion by the **kidneys**. A balancing water intake—in whatever drink form it is delivered—is needed to prevent dehydration.

The thirst regulation center of the **brain** is located in a localized area of the **hypothalamus**. Physiological experimentation in animals establishes that electrical stimulation of this area produces drinking behavior associated with thirst. Neural cells in the thirst center are also excited by intercellular dehydration. Accordingly, any physiological state that increases osmolar concentration in the extracellular fluids—a state that produces an osmolar gradient driving water out of cells—is capable of producing thirst. Loss of some **electrolytes**, especially potassium, may also cause a loss of intercellular fluid.

The daily loss of bodily fluid is essentially a loss of **extracellular fluid**. The loss of fluid results in a rise in the concentration of electrolytes (e.g., sodium) and other osmolar molecules that, in turn, result in a rise in osmolality. The osmolar gradient than draws water out of the intercellular fluid in an attempt to restore osmotic balance.

Malnutrition, especially diets depleted in electrolytes such as potassium may also lead to a loss of intercellular volume and thirst.

The loss of intercellular fluid in cells lining the mouth that produces a dry mouth sensation does not directly produce thirst. Instead, it is probable that the same factors that lead to a dryness in the mouth are responsible for a loss of intercellular fluid in thirst center cells. However, it must be noted that individuals experiencing thirst find relief from the sensation after drinking in a much shorter time than needed for the intake of water to be absorbed from the gastro-intestinal tract and travel through the **blood** stream. Although the compensation mechanisms are not fully understood, there exists a well-documented restraint from over-drinking that allows individuals relief from thirst and the desire to intake more fluid that needed.

In contrast to a dry mouth that is associated with thirst, there are well-defined levels of sodium concentrations in the extracellular fluid that will directly induce thirst by producing a sufficient osmolar gradient to produce significant intercellular dehydration. Extracellular fluid concentrations and osmolar pressures are regulated by an interaction and coordination of thirst and antidiuretic mechanisms (mechanisms that promote fluid retention by decreasing urine output).

See also Cell membrane transport; Diarrhea; Diuretics and antidiuretic hormones; Edema; Electrolytes and electrolyte balance; Elimination of waste; Fluid transport; Interstitial fluid; Osmotic equilibria between intercellular and extracellular fluids; Renal system

THOMAS, E. DONNALL (1920-)
American physician

E. Donnall Thomas pioneered techniques for transplanting bone marrow, an operation used to treat patients with cancers of the **blood**, such as **leukemia**. For proving that such transplants could save the lives of dying patients, Thomas was awarded the Nobel Prize in physiology or medicine in 1990, a commendation he shared with Joseph E. Murray, another American physician working in the area of transplants.

Thomas spent most of his career at the Fred Hutchinson Cancer Research Center in Seattle, Washington, which he built into the world's leading center for bone marrow transplants. The Hutchinson Center has also become an important training site for doctors learning to perform such operations, and transplant centers around the world are staffed by physicians who studied with Thomas in Seattle.

Thomas was born on March 15, 1920, in the small town of Mart, Texas, to Edward E. Thomas, a doctor, and Angie Hill Donnall Thomas, a schoolteacher. After graduating from a high school class of approximately fifteen students, Thomas entered the University of Texas at Austin in 1937. He received a B.A. in 1941 and continued on for a master's degree, which was awarded in 1943. In 1942, he married another University of Texas student, Dorothy Martin, who would later help him manage his research and write medical papers.

After completing his master's degree, Thomas started medical school at the University of Texas Medical Branch in Galveston. After six months, however, he transferred to Harvard Medical School, where he received his M.D. in 1946. He became an intern and then a resident at Peter Bent Brigham Hospital in Boston and began to specialize in blood diseases. Thomas interrupted his formal medical training to serve as a physician in the United States Army from 1948 to 1950. He then returned to the Boston area and did research on leukemia treatments for a year as a postdoctoral fellow at the Massachusetts Institute of Technology. In 1953 he worked as an instructor at Harvard Medical School.

Thomas moved to New York in 1955 to take the position of physician-in-chief at the Mary Imogene Bassett Hospital in Cooperstown. The next year he became, in addition, an associate clinical professor of medicine at the College of Physicians and Surgeons at Columbia University. During the next eight years Thomas had the opportunity to develop and research his ideas about bone marrow transplants, and he applied these concepts to treating cancers of the blood.

Leukemia is a type of cancer in which certain blood cells, known generally as white blood cells, are produced in abnormally large numbers by the bone marrow. In other kinds of cancer, the diseased cells form **tumors** that can often be treated by excising or removing the mass. Leukemic blood cells, however, circulate throughout the body, making them much more difficult to eliminate. Furthermore, the white blood cells that become abnormal in leukemia are an important part of the body's **immune system**. When destroyed by means such as radiation, patients become vulnerable to infections.

In the 1950s, researchers showed that inbred laboratory mice could be irradiated, thus destroying the production of white blood cells by their bone marrow, and then saved from **infection** by a transplant of bone marrow taken from healthy mice. Inspired by these experiments, Thomas began similar studies on dogs, but he faced two important obstacles. First, the recipient animal's immune system had to be prevented from attacking and destroying the transplanted bone marrow—such immune rejection has long been a problem for bone marrow as well as organ transplant surgery. Second, if the bone marrow transplant was successful and the donated marrow began to produce white blood cells, these cells were likely to attack the recipient's other tissues, perceiving them as foreign. Both of these problems had been avoided in the earlier studies with inbred mice because the mice were genetically identical, and hence, have identical immune systems. People are not so similar genetically, with the exception of identical twins. All attempts to graft bone marrow between a donor and recipient who were not identical twins failed. In 1956, Thomas performed the first bone marrow transplant to a leukemia patient from an identical twin. Although the patient's immune system did not reject the transplant, the cancer recurred.

Many researchers gave up working on organ transplants because the problems of immune rejection seemed insurmountable, but Thomas persisted. In 1963, he moved to Seattle to become a professor at the University of Washington Medical School. There he put together a team of expert researchers and began experimenting with new drugs that could suppress the recipient's immune system and thus prevent rejection of the new **tissue**. In the meantime, new methods were being developed by other researchers to identify people whose immune systems were similar, in order to match organ donors and recipients. The new methods of tissue typing were based on molecules known as histocompatibility **antigens**. Thomas's team performed the first bone marrow transplant to a leukemia patient from a matched donor in March 1969. During the 1970s they developed and perfected a comprehensive procedure for treating leukemia patients: first the patients receive radiation, both to kill cancer cells and to weaken the immune system so that it does not reject the transplant; then their bone marrow is replaced with marrow from a compatible donor. The patients also are given drugs that continue to suppress their immune systems. Many patients had been cured of leukemia using this technique by the late 1970s. Since then, Thomas and his colleagues have improved their success rate from about 12% to about 50%. In addition to leukemia and other cancers of the blood, bone marrow transplants are used to treat certain inherited blood disorders and to aid people whose bone marrow has been destroyed by accidental exposure to radiation.

Thomas received wide recognition for his work, including the American Cancer Society's National Award for Basic Science in 1980, and the National Medal of Science of the United States in 1990. The Nobel Prize that he received in 1990, however, came as a surprise. Thomas told reporters that the award is more often given to scientists who do basic research than to those that develop clinical treatments. Thomas shared the prize with **Joseph Murray**, who performed the first kidney transplant and whose research paved the way for the transplantation of other **organs**. As reported in *Time* magazine, both men were cited by the Nobel committee for discoveries "crucial for those tens of thousands of severely ill patients who either can be cured or given a decent life when other treatment methods are without success."

See also Bone histophysiology; Hemopoiesis; B lymphocytes; Immunity; Immunology; Leukocytes; T lymphocytes; Transplantation of organs

THORACIC AORTA AND ARTERIES

The thoracic aorta is a special region of the descending aorta as it passes through the **mediastinum** of the thorax. Superiorly (upward), the thoracic aorta is continuous with the **aortic arch** and inferiorly (downward), it becomes the **abdominal aorta** as passes through the **diaphragm**.

The thoracic aorta supplies oxygenated **blood** to the **pericardium**, esophagus, **bronchi**, and **lungs** via visceral branches (i.e., vessels that supply blood to internal **organs** such as the lungs). Another set of branches, the parietal branches (i.e., vessels that supply blood to the walls of a body cavity such as the thorax or particular organ), supply oxygenated blood to the thoracic cavity.

Mediastinal branches of the thoracic aorta supply blood to the lymph nodes and surrounding **tissue** in the mediastinum. Short and small diameter phrenic branches from the lowest (most inferior) regions of the thoracic aorta supply blood to the diaphragm. A set of esophageal **arteries** arise from the front (ventral) side of the thoracic aorta.

As the thoracic aorta courses downward (inferiorly) it gives off nine sets of arteries, the posterior intercostal arteries, that supply blood to the nine lowest spaces between the ribs (intercostal spaces). Superior intercostal arteries, branches of the **subclavian arteries** supply blood to the uppermost (superior) two intercostal spaces. Below the level of the last rib, the arteries branching off the thoracic aorta are termed the subcostal arteries.

The posterior intercostal arteries themselves branch into several smaller branches that supply specialized regions of the intercostal spaces. Muscular branches run ventrally to supply blood to the intercostal and pectoral muscles. These arteries often fuse (anastomose) with branches of the axillary artery.

Regions of the thoracic nerves termed the lateral cutaneous branches receive blood from corresponding lateral cutaneous branches off the posterior intercostal arteries.

In lactating (milk-producing) females, the mammary branches of the posterior intercostals associated with breast tissue and **mammary glands** (at about the level of the second to fifth ribs) often dilate (expand in diameter) to allow an increased blood supply.

Dorsal branches of the posterior intercostal arteries lead to spinal branches that enter the vertebral canal through the intervertebral foramen to, along with spinal arteries, supply oxygenated blood to the thoracic vertebrae, **spinal cord**, and associated tissue.

The left and right subclavian arteries also supply oxygenated blood to the internal thoracic region (including the thoracic wall and internal thoracic region), parts of the upper arm (upper limb), neck, spinal cord, **meninges**, and **brain**. The internal thoracic arteries, branching from the subclavian arteries, supply blood to the diaphragm, mediastinal structures, and the anterior thoracic wall.

The bronchi receive blood from branches of the thoracic aorta, termed bronical arteries, that are often found to show considerable variations. Normally, there is one bronchial artery on the right side of the body and two bronchial arteries on the left. The right bronchial artery branches from the third posterior intercostal artery, while the left bronchial arteries split directly from the thoracic aorta.

As a region of the descending aorta, the thoracic aorta arises in the embryo from the dorsal aortae that are located on each side of the **notochord**. At about the end of the first month of development, these embryonic dorsal aortae fuse to form the descending aorta.

In some cases there are blockages (occlusions) of the thoracic aorta. Such occlusions are termed a coarctation of the aorta and force blood to flow to the lower parts of the body through collateral arteries located in the chest wall (thoracic wall). This diversion of blood through higher resistance vessels results in an initially greatly increased blood pressure in the upper part of the body that may be as great as twice the normal pressures. Blood pressures found in the lower part of the body past the diversion are lower than normal. In cases where there is a persistent, long lasting occlusion, a compensation mechanism termed long-term local blood flow regulation works to equalize the actual blood flow on both sides of the blockage. Although increased blood pressure (such as those found in the upper part of the blood) usually mean a short term (acute) corresponding increases in blood flow, over time, physiological compensating mechanisms change the vascularity (number and size of blood vessels) of tissues to return blood flow to near normal levels.

See also Angiology; Blood pressure and hormonal control mechanisms; Ductus arteriosis; Fetal circulation; Systemic circulation; Thoracic veins

THORACIC ARTICULATIONS · *see* STERNUM AND THORACIC ARTICULATIONS

THORACIC MUSCLES · *see* MUSCLES OF THE THORAX AND ABDOMEN

THORACIC SPINE · *see* VERTEBRAL COLUMN

THORACIC VEINS

Blood from the **systemic circulation** of the body is returned to the right atrium of the **heart** by the large **veins** called the superior and inferior vena cava. Deoxygenated blood used by heart **tissue** itself is returned to the right atrium by a coronary vein termed the coronary sulcus, and a number of smaller coronary veins that also return blood directly to the right atrium.

The inferior vena cava, which runs through the thorax, returns blood from the **abdomen**, legs, and hepatic circulation (the hepatic portal system of the **liver**) runs along the frontal (anterior) side of the spine and lies next to the **abdominal aorta** (a continuation of the **thoracic aorta**). After passing through the **diaphragm** that separates the thoracic from the abdominal cavity, the inferior vena cava continues onward to the heart for approximately 1 in. (2.5 cm) before it fuses with, and empties its contents into, the right atrium of the heart along with the superior vena cava.

The superior vena cava returns blood from the upper half of the body, including the head, neck, upper limbs, and the thoracic region itself. The superior vena cava is a large diameter vein that is almost 3 in. long (approximately 7 mm) running from about the level of the first rib to the right atrium of the heart. Within the right atrium of the heart, blood from the superior vena cava mixes with blood from the inferior vena cava and blood returned from the **coronary circulation**.

The superior vena cava is formed from the fusion of the brachiocephalic veins—also termed the innominate veins— are two large veins that run on each side of the neck. The left brachiocephalic vein is formed by the fusion of the left subclavian and left jugular vein. Correspondingly, the right brachiocephalic vein is formed by the fusion of the right subclavian and right jugular veins. The left and right subclavian veins return blood from the corresponding left or right arm and shoulder region (upper limb) and the jugular veins return blood from the head and neck regions. The brachiocephalic veins, also return blood from the thyroid and thorax proper (e.g., rib or intercostals veins).

Other major thoracic veins include the internal thoracic veins (also termed the mammary veins), the inferior thyroid veins, bronchial veins, vertebral veins, and intercostal veins.

The azygous vein runs down the right side of the **vertebral column**. Although the azygous vein actually begins in the lumbar region near the renal veins, it travels upward to pass through the diagram and eventually fuse with the superior vena cava. Running through the posterior **mediastinum**, the azygos vein receives venous contributions from a number of smaller posterior intercostals veins. A hemizygous vein runs alongside the left side of the vertebral column to terminate in a superior intercostal vein. Connections between the azygos and hemiazygos veins are not uncommon, and they occur at variable levels.

See also Abdominal veins; Anatomical nomenclature; Collateral circulation; Vascular system (embryonic development); Vascular system overview

THYMUS HISTOPHYSIOLOGY

The thymus is a lymphoid gland localized between the **lungs** in the anterior superior **mediastinum** (in the chest). Its cortex (i.e., external layer) is constituted by lymphatic **tissue**, with the internal portion containing lymphocytes. The thymus has also a thick reticular structure comprised of groups of granular cells enveloped by epithelial cells, known as Hassall's corpuscles. Much remains to be found about the thymus' physiological role and products. Thymine (2,4-Dihydroxy-5-methylpyrimidine), for instance, was first isolated from this organ and seems to be an important nutritional compound against macrocytic anemia. However, the better understood thymic function is related to its role in the preprocessing of **stem cells** into different lymphocytes, promoting their maturation and specialization (i.e., differentiation) as millions of functional T lymphocytes (where "T" stands for "thymus"). T lymphocytes are a family of white **blood** cells pertaining to the acquired immunity system that defends the body against infectious agents through the specific recognition and binding to foreign **antigens**.

During fetal development, millions of stem cells are formed in the bone marrow and a portion of this population migrates to the thymus to be preprocessed as T cells, whereas the other portion is preprocessed either in the fetal **liver** or in the bone marrow itself, thus forming the **B lymphocytes** or B-cells ("B" standing for "bones"). B lymphocytes also continue to be preprocessed in the bone marrow after birth, and are subsequently stored in other lymphoid **organs** throughout the body along with matured T lymphocytes, waiting to be recruited by the **immune system**. In the early stages of fetal development, stem cells migrate to the thymus in a series of sequential waves, where they divide and differentiate, originating the several kinds of T cells that populate the epithelium of the mouth, skin, **gastrointestinal tract**, etc. Each different wave of stem cells reaching the thymus of the fetus produce groups of T cells bearing different receptors on its membranes, in a specific order. The first wave receives gamma-delta 3 receptors, and these T cells are found in the skin, where they recognize and attack infected, mutated, or damaged skin-cells. The second wave of stem cells gives origin to T cells with gamma-delta 4 receptors that populate the oral epithelium, the vagina, and uterus. These two first waves of stem cells are preprocessed exclusively in the fetal thymus. The following waves give birth to T cell with gamma-delta 2 and gamma-delta 5 receptors, which will populate mainly the spleen (gamma-delta 2) and the gastrointestinal epithelium (gamma-delta 5). During the later stages of fetal development and during a few months after birth as well, the stem cells migrating to the thymus will be processed into T cells known as helper T cells, and killer T cells, carrying alpha-beta receptors. Each class of receptors, either gamma-delta or alpha-beta, contain a great variety of individual T cells, each specialized in the recognition of a different alien protein-particle, or antigen. Therefore, each T lymphocyte bears a receptor specifically constituted by two chains of different polypeptide sequences (whether gamma-delta or alpha-beta), which binds to a foreign antigen in particular.

The thymus also plays a crucial role in the selection of young T cells that are able to tell apart foreign antigen from those pertaining to the body itself, known as self-antigens. It seems that the process of adding receptors to the membrane surface of lymphocytes is rather random. Therefore, many young T lymphocytes bear receptors that react against self-antigens, and could cause tissue destruction if allowed to leave the thymus. The thymus plays a search-and-destroy action against these cells through a process known as clonal deletion. The thymus tests the young T cells by producing samples of most cell types of body tissues, whose proteins are chopped by an antigen-presenting cell and exposed to the young T cells in a protein molecule known as the major histocompatibility complex, or MHC. Once a strong reaction occurs to a self-antigen, the young T cell is induced to apoptosis (i.e., programmed cell deaths or suicide). Therefore, only T cells that show tolerance to self-antigens are allowed to survive and migrate to other tissues. However, when this system fails, T cells reacting to some type of self-antigen may survive and escape the thymus, thus causing autoimmune diseases such as lupus erythematosus, myasthenia gravis, rheumatic fever, etc. The thymus remains highly functional during fetal development and for a few months after birth; and during childhood, it plays a role in the formation of lymphoid cells. Lymphoid cells produce antibodies, i.e., globulin molecules that circulate in the blood **plasma**, bind to infectious agents, and inactivate them. However, after **puberty** the thymus progressively undergoes atrophy and loses function.

See also Antigens and antibodies; Autoimmune disorders; Bacteria and responses to bacterial infection; Immune system; Infection and resistance; Leukocytes; Viruses and responses to viral infection

THYROID HISTOPHYSIOLOGY AND HORMONES

The thyroid gland is localized below the **larynx**, occupying the anterior and lateral space around the trachea. The most abundant hormone secreted by the thyroid is thyroxine or T_4 (approximately 93%), followed by triiodothyronine or T_3 (approximately 7%). These thyroid **hormones** are formed from thyroglobulin, which is synthesized in the endoplasmatic reticulum and in the Golgi apparatus, and stored into structures of the thyroid known as follicles. Each follicle is surrounded by capillary **blood** vessel, and it is constituted by epithelial cubical cells on the outside, containing a substance secreted by the gland termed colloid. These epithelial cubical cells have the ability of trapping iodides on their membranes, which are later enzymatically combined with thyroglobulin to form the thyroidal hormones inside the follicles. A weekly dietary intake of 1mg of iodine is essential for hormone synthesis by the thyroid. Iodine, mostly under the form of iodides, is trapped by the basal membrane of the thyroid cells and pumped inside the cells where its concentrations may be up to 30 times higher than in the blood circulation. The thyroidal follicles store large amounts of T_4 and T_3 associated to molecules of thyroglobulin for two to three months, and these hormones are cleaved from the thyroglobulin molecule before they are released into the blood circulation. Both the hormone synthesis and the levels of thyroid hormones that are secreted into the circulation are controlled by the **hypothalamus** and the anterior portion of the pituitary gland.

The pituitary hormone TSH (thyroid-stimulating hormone) increases the rate of **iodine trapping** and stimulates the release of T_4 and T_3 into the circulation. On the other hand, the hypothalamus regulates the secretion of TSH through the release of TRH (thyrotropin-releasing hormone). Another important hormone synthesized by the thyroid is calcitonin, which is formed by the C cells in the **interstitial fluid** present between the follicles. Calcitonin is crucial for bone formation because it promotes the transport of **calcium** ions from the blood circulation to the skeleton. When the levels of calcium ions increase in the blood **plasma**, the thyroid is stimulated to release calcitonin. Calcitonin also decreases the loss of calcium from the bones, favoring the deposition of this mineral in the skeleton, and is essential for balanced body growth during childhood.

The thyroid hormones affect several other physiological systems. During fetal life, for example, they play an important role in the development and maturation of the **brain**; during childhood, they are essential for skeletal development and body growth. The thyroidal hormones (T_4 and T_3) increase the basal metabolic rate and promote the metabolism of **lipids** (i.e., fats), decreasing the blood levels of lipoproteins, such as LDL (low density lipoprotein), and HDL (high density lipoprotein), cholesterol, and **triglycerides**. Thyroid hormones also promote the rapid conversion into energy of body **fat** deposits. Due to the stimulatory effect upon the basal metabolism, the thyroid hormones have a systemic impact on the rate of activity of many other **organs**, such as **liver** function, respiratory rate, production of **enzymes**, gastrointestinal motility, pancreatic production of insulin, **menstruation**, arterial pressure, sex drive, muscles reaction, and the **central nervous system**.

Deregulation of thyroid hormonal synthesis and secretion may lead to two different metabolic diseases known as **hyperthyroidism and hypothyroidism**. The first condition is caused by an excess of thyroid function and abnormally high levels of T_4 and T_3 in the blood circulation. Patients with **hyperthyroidism** may present basal metabolic rates up to forty times higher than normal, which results in nervous excitability and irritability, anxiety, extreme fatigue, increased appetite, intolerance to heat, agitation, insomnia, muscle weakness, and, eventually, protrusion of the eyes (exophthalmos). Conversely, patients suffering from hypothyroidism have below normal levels of thyroid hormones leading to a slow basal metabolic rate, which may cause obesity, apathy, muscular sluggishness, excess sleeping, depression, constipation, low cardiac rate, and increased levels of cholesterol and triglycerides.

See also Child growth and development; Endocrine system and glands; Iodine pump and trapping

TISSUE

A tissue is a group of cells that are integrated and have a common **structure and function**. The term tissue comes from the Latin word meaning weave. This is reflective of the fact that tissues are often held together by an extracellular matrix that coats and weaves the cells together. In vertebrates, there are four main categories of tissue including connective tissue, **epithelial tissue**, nervous tissue, and muscle tissue.

Connective tissues function mainly to bind and support other tissues and **organs**. They are composed of a relatively small number of cells scattered throughout an extracellular matrix. Typically, this matrix is composed of some type of protein fiber embedded in a gelatinous substance. The main vertebrate connective tissues are loose connective tissue, **adipose tissue**, fibrous connective tissue, **cartilage**, **blood** and bone. Loose connective tissue is the most abundant type and it holds organs in place. It is made up of collagen and elastin fibers, fibroblast cells that secrete protein, and macrophage cells that protect and repair the tissue. Adipose tissue is a special type of loose connective tissue that stores **fat**.

The other connective tissues play important roles in the body. Fibrous connective tissue, which is dense, holds muscles, bones and **joints** together. Cartilage is found between joints. It is a rubbery tissue that is embedded with collagen fibers and composed of chondrocyte cells, which secrete chondroitin sulfate. Bone is a type of mineralized connective tissue. Bone is made up of osteoblast cells, which release **calcium** phosphate and collagen. While blood appears different from other connective tissues, it is categorized as such because it is made up of cells that are connected through an extracellular matrix. In this case the cells include red and white blood cells. The extracellular matrix is **plasma** which is a combination of water, salts and various proteins.

Epithelial tissue is composed of cells that are more tightly packed than connective tissue. It is found throughout the body lining the inner organs and covering the outside. The cells are typically connected to each other. This tight packing allows the epithelial tissues to act as a barrier to protect against injury, invading microorganisms, and regulate fluid loss. Skin and the mucous membrane are examples of epithelial tissues.

Nervous tissue is made up of nerve cells called **neurons**. They are connected to each other by structures called dendrites and axons. These structures allow the cells to transmit signals and nerve impulses throughout the body. Muscle tissue is composed of long contractible cells. Within the cells are a large number of microfilaments made up of the proteins actin and myosin. Examples of muscle tissue include **skeletal muscle**, cardiac muscle, and **smooth muscle**.

See also Histology and microanatomy

TODDLER GROWTH AND DEVELOPMENT

Toddler is a term to define a child that learns to walk. Normally the term is applied to children one to three years old.

The toddler years are ones of rapid change and represent a challenge most for parents. The most dramatic advances occur in language and **locomotion**, but progress is evident in all areas of cognitive and physical **growth and development**.

The average 18-month-old has a vocabulary of at least 20 words. Over the next few months, the child will experience a burst in vocabulary. At this time, a toddler can say nouns, names of special people, and a few action words and phrases. At this point, about 50% of what the child says should be intelligible to strangers. At two years of age, the child can combine words, forming simple sentences. Toddlers use language to convey their thoughts and needs (such as hunger and **pain**).

By the age of three years, the vocabulary increases to about 500 words, and 75% of speech is understandable to strangers. He begins to make complete sentences and experiments with speech and language, varying word usage and changing the intensity, as well as intonation, of speech. He typically now begins several daily "why" questions, characteristic of the preschool years. Toddlers sometimes get frustrated because they do not have the language skills to express themselves. Often they have difficulty separating themselves from their parents and other people who are important to them.

Progress in language development is influenced by environmental factors as well as by innate abilities. Bilingual children, for example, may mix languages initially but ultimately will "choose" in their language skills by 2 to 3 years of age. The timetable for language development is broad, and toddlers may go through different kinds of roadblocks along the way. For instance, many preschoolers think faster than they can talk, so they may have trouble saying certain words or sounds, and may even stutter. However, children often repeat words or phrases as they learn to talk or when they are excited or tired. This is normal for most toddlers and children up to 5 years of age.

Affectivity is a crucial point in the toddler's development. The transition from infancy to toddlerhood is marked by a new drive for independence. Toddlers also begin to develop impulse control. The 18-month-old may have minimal impulse control, but two-year-olds typically exhibit wide variations in impulse control, with the degree of control often varying with the struggle for autonomy. Most three-year-olds have mastered some degree of self-control, in part because they are developing the ability to delay gratification. Positive affectivity includes joy, activity, and smiling. Negative affectivity includes fear, anger, sadness, low soothability. Parental influence for affectivity is very influential at this stage.

After infancy, growth speed slows down in the toddler years. After age two, toddlers gain about 5 lb. (2.26 kg) in weight and 2.5 in. (6.4 cm) in height each year. In comparison, head circumference increases by about 1 in. (2.5 cm) from two to 12 years. Growth does not increase steadily. A toddler's weight can remain the same for some weeks. Increases in height result primarily from growth of the lower extremities and, secondarily, to a lesser extent elongation of the trunk. Body proportions change, with upper-to-lower segment ratios ranging from 1.40 at age two years to 1.15 to 1.20 at age 5 years. An average 15-month-old girl weighs about 22 lb. (9.9

kg) and stands 31 in. (78.7 cm) tall. Boys tend to be about a pound heavier at 15 months but about the same height. By age two, both will stand about 34 in. (86 cm) tall and weigh about 27 or 28 lb. (12 kg). Growth charts are very useful at this time in order to detect abnormal growth that can be a manifestation of clinical disorders. Height, weight, and head circumference are plot versus the values recorded for national averages for children of the same age and sex. By means of percentiles it is then possible to establish the relative value of the measurements. The 70th percentile for weight, for example, means that 70% of the toddlers of that age and gender, in the United States, are lighter, and 30% are heavier.

With a newly erect **posture**, a toddler stance often includes lordosis (a forward curvature of the spine) and a protuberant **abdomen**. The percentage of body **fat** steadily decreases from 22% at age one year to about 12.5% to 15% at age five years. By the end of toddlerhood, increased muscle tone and decreased body fat give the child the appearance of being more lean and muscular. Gross motor skills also develop rapidly. Complex gross motor patterns develop, while balance and coordination improve. Gross motor development milestones include the ability to stand alone well by 12 months, usually walking well by 12–14 months, the ability to kick a ball forward at about 15–18 months, and the ability to jump in place at 20–24 months. At about age 36 months, toddlers have developed their balance and can stand on one foot briefly.

Toddlers make the transition from the sensory-motor to the preoperational stage, as outlined by the Swiss psychologist Jean Piaget (1896–1980). During the sensory-motor period, knowledge of the world is limited and developing. The infant primarily learns about the world by touching, looking, and listening because its based on physical interactions and physical experiences. Intelligence is demonstrated through motor activity without the use of symbols. In preoperational stage (which has two substages), intelligence is demonstrated through the development of symbolic thinking, language use matures, and memory and imagination are developed, although thinking is done in an illogical, manner. This progression from sensory-motor to symbolic thought occurs typically between 18 and 24 months of age. The child's recognition that one object can represent another becomes more evident during playing. Older toddlers continue to develop symbolic thinking. By age three years, they draw primitive figures that represent known people and the environment. They are not able to take the viewpoint of another person, however; egocentric thinking predominates, and toddlers assume that other people think and feel as they do.

Toddlers come to clinical attention usually because of behavioral, relational, or developmental difficulties. Behavioral disturbances may include aggression, and overactivity. In addition, developmental delays and more subtle physiologic, sensory, and sensory-motor processing problems are also observable during the toddler years.

See also Child growth and development; Neonatal growth and development

TONEGAWA, SUSUMU (1939-)

Japanese molecular biologist

Japanese molecular biologist Susumu Tonegawa is a professor at the Massachusetts Institute of Technology (MIT). Tonegawa's work made important advances into understanding of the genetic mechanisms of immunological systems.

Tonegawa received his doctorate from the University of California at San Diego in 1969. In 1971, Tonegawa became a of member of the Basel Institute for Immunology in Switzerland where he conducted research until accepting the professorship at MIT in 1981.

Tonegawa's work in immunogenetics showed distinct relationships between antibodies and the genes responsible for their production and regulation. Tonegawa reported that there were alterations in patterns of chromosomal recombination that allowed the genes responsible for antibody production to move closer to one another on **chromosomes**. Tonegawa's worked gained a Nobel Prize in physiology or medicine in 1987.

The selective recombination mechanism allows organisms to enhance the production of antibodies. For example, although the human body has a limited number of chromosomes and a finite amount of **DNA** (only a portion of which is related to **immune system** function), cells are able to produce highly specific antibodies to a vast number of **antigens**. Tonegawa's work established that gene rearrangements allowed for increased variety in the production of antibodies.

Born in Nagoya, Japan, Tonegawa took his undergraduate studies in chemistry at Kyoto University in Japan, and in 1963, moved to the United States to undertake his graduate studies at the University of California. Tonegawa's Nobel Prize winning research was conducted at Basel Institute for Immunology in Basel, Switzerland.

See also Antigens and antibodies; Human genetics

TONGUE • *see* GUSTATORY STRUCTURES

TONSILS

In the human body, there are three different structures referred to as tonsils. Most commonly, tonsils refer to the palatine tonsils; a pair of ellipsoid, almond sized structures located at the back of the throat. There are, however, another set of tonsils—the lingual tonsils—that are located under the tongue. In addition, there are adenoids (pharyngeal tonsils) that are embedded in the upper rear wall of the oral cavity. The adenoids or pharyngeal tonsils are often prominent in childhood but usually diminish in size in the adult. All of the tonsillar structures are part of the lymphatic system and contain lymphoid **tissue**.

As with other lymphoid tissue, the physiological function of tonsils is to process lymphatic fluid and to aid in the resistance to bacterial infection.

Palatine tonsils are embedded in mucosal membranous tissue of the mouth and throat, lie at the back of the mouth (oral cavity). Inflammation of these tonsils is commonly referred to as tonsillitis, a condition characterized by a painful swelling of the throat often associated with fever and difficulty in **swallowing**. Although minor tonsil infections are common, more severe infections can scar the tissue or produce an abscess (termed peritonsillar quinsy). In children and in the middle part of the twentieth century, tonsillectomy (the surgical removal of the tonsils) was a common minor surgical procedure for pre-pubescent children. Although still not an uncommon procedure, improvements in drug therapies and antibiotics, along with a realization of the important role tonsil tissue can play in the function of the **immune system**, has diminished the use of tonsillectomy in modern medicine.

Inflammation of the adenoids (pharyngeal tonsils) may also obstruct the nasopharynx region and lead to inflammation and **infection** of the **Eustachian tubes** or **middle ear**. When infections are severe or chronic, the adenoids are usually surgically removed (adenoidectomy).

The term tonsil can also refer to any small, generally oval-shaped mass of tissue. Accordingly, at the base of the **cerebellum** there is a rounded neural tissue mass termed the tonsil of the cerebellum. This neural tissues should not be confused, in either structure or physiological function, with the lymphatic tonsils located in the mouth.

See also Bacteria and responses to bacterial infection; Pharynx and pharyngeal structures; Swallowing and dysphagia

TOOTH DEVELOPMENT, LOSS, REPLACEMENT, AND DECAY

Teeth are required for the **mastication** process, during which food is ground up and swallowed.

The development of teeth in humans, which begins prior to birth, is a highly orchestrated process, with over 20 genes known to be involved. At birth, the developing teeth are usually still hidden beneath the surface of the gums. But, beginning at around six months of age and extending until about 24 months of age, a succession of teeth appears. These so-called deciduous (or milk) teeth are designed for grinding (the molars located in the back of the mouth) and for cutting (incisors) and tearing (canines). A normal set of deciduous teeth comprises ten upper teeth (central incisor pair, lateral incisor pair, canine (or cuspid) pair, first molar pair, and second molar pair). This first set of teeth is successively replaced by the adult (permanent) teeth, from seven up to about 20 years of age.

In the replacement process, the anchorage of a deciduous tooth is gradually lessened until the tooth "falls out." The replacement of the lost tooth need not occur immediately. Months may go by until the space occupied by the tooth is filled with an adult tooth. A normal adult set of teeth consists of 32 teeth, arranged as 16 upper and lower pairs of incisors, canines, premolars (or bicuspids), first molar, second molar, and third molar.

The loss of an adult tooth is a permanent event. In contrast to reptiles such as alligators, where tooth loss and replacement is ongoing, humans do not generate a third set of teeth.

Teeth consist of a portion that projects above the gum line and the roots, which are below the gum line and serve to anchor the tooth in position. Each tooth is similar in construction, being composed of multiple layers and components. Reflecting the role of teeth in grinding and slicing food, the outer surface is hard. The material, called enamel, is the hardest substance in the human body. Underneath the enamel lies the dentine, a layer that is hard but not quite as hard as the overlaying enamel. Dentine extends down into the roots. Within the dentine are millions of tiny tubes that run into the next layer, called the pulp. The pulp is the central portion of the tooth, and consists of **blood** vessels and nerves. It is the layer that nourishes the tooth. Despite its appearance, teeth are living structures. A bone-like substance called cementum covers the roots. Finally, each tooth is held in position and anchored to bone by a peridontal ligament.

The health of teeth can be adversely affected by several factors. Environmental compounds can harm teeth. For example, it is now known that the compound dioxin can cause malformation in developing teeth, because of the interference with epidermal growth factor. Teeth can also e mechanically damaged, such as by a blow. Overwhelmingly, however, the health of teeth is compromised by decay (also called caries). Tooth decay is the second most common disorder in humans, next to the common cold. The basis of tooth decay is the **bacteria** that inhabit the mouth. A great number of species of bacteria live in the mouth, where they convert foods—especially sugars and starches—into acids. The bacteria combine with food debris, **saliva** and other compounds in the mouth to form a coating on teeth. The coating is referred to as plaque. Normally, plaque is removed by brushing and flossing. However, plaque that is not removed because of improper dental hygiene, or because the bacteria are able to hide in crevasses, can harden into a structure called tartar. Within tartar the acids from the metabolic activity of the bacteria can dissolve the enamel. If left untreated, the resulting cavity (literally a hole in the tooth) can reach the interior dentine and enamel, and can kill the tooth (tooth abscess). Initially, tooth decay is not painful and so can escape detection. However, once the nerves are involved, a toothache ensues, and can be very painful.

Depending on the severity of a cavity, treatment can be minimal (sealing the hole; often referred to as a filling) or more drastic (complete removal of the tooth). Preventative measures can be taken to ward off the development of dental caries. Proper tooth hygiene is important. In urban settings, drinking water often is supplemented with fluoride to retard the onset of cavities. The issue of fluoridation of drinking water continues to be debated, as some evidence exists that too much fluoride can damage tooth enamel and in fact promote tooth decay.

See also Digestion

Mirror image of top of boy's mouth, age 4, taken before 12 of his decayed teeth were pulled because of lack of fluoridation use. *Photo by Val Cheever. AP/Wide World Photos. Reproduced by permission.*

TOUCH, PHYSIOLOGY OF

The ability to sense when the body contacts something depends on the presence of so-called cutaneous **mechanoreceptors** on the surface of the skin, and on the transmittance of signals from these receptors to the **brain**. This also is referred to as the somatosensory, or body-sensing, system.

The surface of the body is covered with touch receptors. Their distribution is not uniform In humans, the fingertips and the tongue can have 100 receptors per square centimeter, while the back of the hand has less than ten receptors.

The receptors respond to the pressure of touch by initiating signals that are then routed to the brain for analysis. The **spinal cord** is important in this regard, as it functions as a conduit for sensory fibers. Then, feedback from the brain to appropriate regions of the body occurs to generate a response to the touch. The responses are varied, ranging from laughter in the case of tickling, to sexual arousal, to a rapid **muscle contraction** to pull a hand away from a painfully hot object.

Different areas of the body have different sensory pathways for touch. Light touch is sensed in the upper body by the fasciculus cuneatus and in the lower body by the fasciculus gracilis. Both these nerve fibers tend to run directly to the brain. The light touch receptors are often located close to a **hair** follicle. Even if the receptor is not touched directly, movement of the hair can be detected.

Touch stimuli are routed to the region of the brain known as the medulla, specifically to a region called the **thalamus**. This area of the brain is also called the somatosensory cortex, in recognition of its importance in the interpretation of touch stimuli.

If a touch is continuous, the stimulus of that touch will decrease. This enables the body to be aware of other touch sensations. The end of a prolonged touch will be noted and the involved receptors are ready for another response. This mechanism can go awry, so that the touch of things like clothing is constantly felt sometimes so much so that it is painful. This condition is known as mechano-allodynia (touch **pain**).

As the understanding of the **physiology** and the mechanisms of touch increases, it will possible to mimic these functions in inanimate forms. Already, rudimentary work is underway to devise robots capable of human-like touch.

See also Nerve impulses and conduction of impulses; Neurons

TRACHEA • *see* RESPIRATORY SYSTEM

TRANSCRIPTION • *see* MOLECULAR BIOLOGY

TRANSECTION • *see* ANATOMICAL NOMENCLATURE

TRANSFUSIONS

Transfusions most commonly involve the addition of whole **blood**. Blood components such as red blood cells and **plasma** are also capable of being used in transfusions. For example, the transfusion of plasma infuses the body with factors such as **platelets**, which are vital for the clotting of blood.

The intent of transfusion with whole blood or with blood components is to replace blood or its particular constituent that has been lost by injury, or to replace blood that is functionally defective with blood that contains a full functional complement of cells. Blood can become functionally defective after **chemotherapy**, because of defects in blood cell production in the bone marrow, or because of medical conditions like hemophilia (a blood clotting disorder) or **sickle cell anemia**.

Transfusion carries a risk of **immune system** reaction against a foreign antigen in the added fluid. This can occur, for example, if blood of type A is given to someone whose blood is of type B. Severe illness of even **death** can result. Those belonging to the blood group designated as O-negative are called universal blood donors. The blood from these people may be transfused to anyone regardless of their blood type.

Over 90% of the complications from blood transfusions are attributable to the presence of white blood cells, and the viral agents they may carry, in the transfused blood. Evidence is mounting that filtering blood to reduce or eliminate white blood cells prior to transfusion is a recommended course of action.

Transfusions also carry a risk of the so-called graft versus host reaction. Here, the donated blood cells attack the recipient.

Another risk from transfusions is the transmittance of an infectious agent to the patient. For example, in the 1980s, thousands of patients in the United States and Canada received blood contaminated with the viral agents of hepatitis or acquired immunodeficiency syndrome. Improved molecular-based screening methods have virtually eliminated this possibility.

The possibility of immune reaction to the transfused fluid can be reduced greatly by the use of a patient's own (also called autologous) blood or blood products. This can be an option with prior knowledge of an operation requiring transfusion. However, and unexpected or emergency need for a transfusion necessitates the use of blood or blood product from a "bank" of available supply.

See also Immunity, cell mediated; Immunity, humoral regulation; Leukocytes

TRANSLATION • *see* PROTEIN SYNTHESIS

TRANSPLANTATION OF ORGANS

Transplantation is the surgical removal of an organ from one person and the placing of that organ into another person. The transplanted organ replaces one that has stopped functioning, or is functioning so improperly as to be life threatening. Examples of solid **organs** that are transplanted are **heart**, kidney, **liver**, **pancreas**, intestine and lung. In some cases, two organs can be transplanted at once, such as with heart-lung transplants and kidney-pancreas transplants.

Transplantation of organs dates back to the years following the Second World War. Then, the immunological basis for the rejection of transplants was unraveled. By the 1960s, some organ transplantation was becoming a more routine part of surgical therapy.

Organ failure can occur because of injury, such as the trauma suffered in an automobile accident, or because of illness. Examples of illness include congestive heat failure and genetic conditions like cystic fibrosis. In the latter, where malfunctioning tissues lining the **lungs** can contribute to repeated lung infections and lung damage, the transplantation of new lungs has extended the life of cystic fibrosis patients.

Transplantation need not involve an entire organ. Tissues can also be transplanted. Examples of tissues that can be transplanted are the **cornea** of the **eye**, bone, **cartilage**, skin, heart valves and certain **veins**.

Speed is essential for successful transplantation. Once an organ is removed from its natural environment, degradation begins. For example, once removed from a donor a heart needs to be installed in the recipient in four to five hours. Other organs, such as a kidney, can be kept for up to eighteen hours before transplantation.

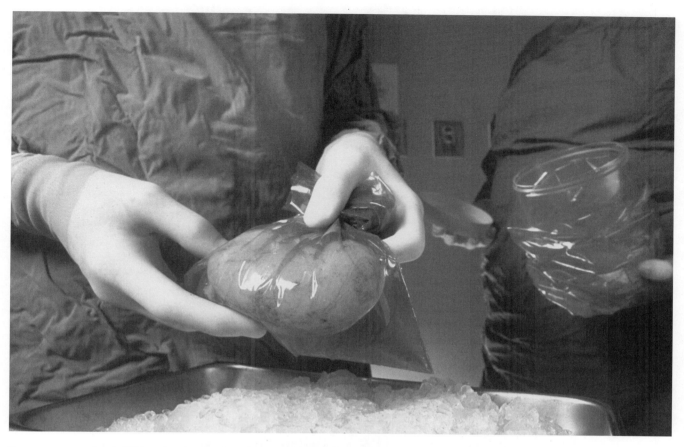

Surgeon removes donated kidney from its shipping container. © Will and Deni McIntyre. Photo Researchers, Inc. Reproduced by permission.

Some transplanted organs, the heart for example, are obtained from someone who has recently died. This is called cadaveric transplantation. But other organs, such as kidney and portions of the liver, can be taken from living donors.

Transplantation of organs is not a random process, where an organ from any donor can be placed into any recipient. Rather, the surface chemistry of the donor organ is matched as closely as possible to that of the recipient organ. This is done to avoid the rejection of the transplanted organ by the **immune system** of the recipient. Organ rejection, while devastating to the immediate health of the patient, is nevertheless part of the normal operation of the immune system. The immune system functions to protect the body from invading microorganisms or other foreign matter. In the case of organ transplants, the immune system focuses on a surface antigen called the major histocompatibility complex (HLA). If HLA **antigens** on the surface of a donor organ are distinctly different from the target organ, the T cells of the immune system of the recipient will treat the donor organ as foreign. Rejection of the organ can occur.

In an effort to circumvent the rejection of the transplanted organ, the transplantation process typically includes the administration of so-called immunosupressant drugs. These drugs act to impair the operation of the immune system.

One example of an immunosupressant is cyclosporin, which acts to inhibit the functioning of T cells. Other examples include corticosteroids (such as prednisolone), inhibitors of metabolic activity, and, as of 2001, one drug that specifically inhibits the migration of immune cells toward the transplanted organ.

The diminished efficiency of the immune system allows the transplanted organ to persist and function in the patient. Often, this immunosuppressant therapy continues for the rest of the life of the patient. Even with the use of immunosupressants, patients have an approximate 20–50% risk of organ rejection or malfunction during the first three years following transplantation. Graft failure occurs more frequently than organ rejection. Furthermore, the impaired function of the immune system leaves the patient open to **infection**, toxic side effects in the liver and kidney and even the development of some types of **cancer**. So the transplantation of organs saves a life in the short-term, but not without risks to long-term health.

In the future, organ transplantation in humans may involve xenotransplantation, the use of organs obtained from animals. To date the antigenic differences between humans and other mammals, and the risk of the transmission of infectious agents such as the human immunodeficiency virus (HIV), have limited the success of xenotransplantation.

However, research is underway to more completely understand the antigenic differences and conclusively test candidate organs for infectious agents.

See also Immunology

TRIGLYCERIDES

Triglycerides are the chemical form in which most **fat** exists in food as well as in the body. They belong to a larger group of compounds called the acylglycerols or glycerides, which are esters of saturated or unsaturated fatty acids with glycerol. Mono- and diacylglycerols usually only occur as metabolic intermediates, whereas the triacylglycerols, or triglycerides, are neutral fats.

Fat molecules are generally made up of four parts: a molecule of glycerol and three molecules of fatty acids. Each fatty acid consists of a hydrocarbon chain with a carboxyl group at one end. The glycerol molecule has three hydroxyl groups, each able to interact with the carboxyl group of a fatty acid. Removal of a water molecule at each of the three positions forms a triglyceride. The three fatty acids in a single fat molecule may be all alike or they may be different. They may contain as few as four carbon atoms or as many as 24. Because fatty acids are synthesized from fragments containing two carbon atoms, the number of carbon atoms in the chain is almost always an even number. In animal fats, 16-carbon, for example, palmitic acid and 18-carbon, for example, stearic acid fatty acids are the most common.

Some fatty acids comprising a given triglyceride have one or more double bonds between their carbon atoms. They are then said to be unsaturated because they can hold more hydrogen atoms than they do. Mono-unsaturated fats have a single double bond in their fatty acids while polyunsaturated fats, such as trilinolein, have two or more. Additionally, there are trans-fats, which are only partially hydrogenated having fewer double bonds in a *trans* (as opposed to the usual *cis*) chemical configuration, and also omega-3 fats, which have at least one double bond, three carbon atoms in from the end of the fatty acid molecule. Linolenic acid is an example and fish oils are generally a rich source of omega-3 fatty acids.

Double bonds are rigid and those in natural fats introduce a kink into the molecule. This prevents the fatty acids from packing close together and as a result, unsaturated fats have a lower melting point than saturated fats. Because most of them are liquid at room temperature, they are called oils. Corn oil, canola oil, cottonseed oil, peanut oil, and olive oil are common examples. As this list suggests, plant fats tend to be unsaturated while fats from such animals as cattle tend to be saturated.

Ingested fats provide the precursors from which we synthesize our own fat as well as cholesterol and various phospholipids. Fat provides our most concentrated form of energy. Its energy content (9 kcal/gram) is over twice as great as **carbohydrates** and proteins (4 kcal/gram).

Humans can synthesize fat from carbohydrates. However, there are two essential fatty acids that cannot be synthesized this way and must be incorporated into the diet. These are linoleic acid (an omega-6 fat, with the endmost double bond 6 carbons from the methyl end) and alphalinolenic acid (an omega-3 fat, with the endmost double bond 3 carbons from the methyl end). Many studies have examined the relationship between fat in the diet and cardiovascular disease. There is still no consensus, but the evidence seems to indicate that a diet high in fat is harmful and that mono- and poly-unsaturated fats are less harmful than saturated fats, with the exception of *trans* unsaturated fats which, according to some, are more harmful than saturated fats. It is also been suggested that ingestion of omega-3 unsaturated fats may be protective for the human body.

See also Biochemistry; Fat, body fat measurements; Protein metabolism

TUMORS AND TUMOROUS GROWTH

A tumor (also known as a neoplasm) is an abnormal **tissue** growth. Neoplasm means new formation. Tumors can be either malignant (cancerous) or nonmalignant (benign), but either type may require therapy to remove or reduce its size. In either case, the tumor's growth is unregulated by normal genetic and somatic body control mechanisms. Usually the growth is not beneficial to the organ in which it is developing.

Normally, cells are generated at a rate needed to replace those that die or are needed for an individual's **growth and development**. Moreover, cells become differentiated into specialized cell forms (muscle cells, bone cells). Genetic controls modulate the formation of any given cells. The process of some cells becoming muscle cells, some becoming nerve cells, and so on is called **cell differentiation**. Tumor formation is an abnormality in cell differentiation.

A benign tumor is a well-defined growth with smooth boundaries. This type of tumor simply grows in diameter. A benign growth compresses adjacent tissues as it grows. A malignant tumor usually has irregular boundaries and invades the surrounding tissue. This **cancer** also sheds cells that travel through the bloodstream implanting themselves elsewhere in the body and starting new tumor growth. This process is called metastasis.

It is important that the physician determine which kind of tumor is present when one is discovered. In some cases this is not a simple matter. It is difficult to determine whether the growth is benign without taking a sample of it and studying the tissue under the microscope. This sampling is called a biopsy. Biopsy tissue can be frozen quickly, sliced thinly, and observed without staining (this is called a frozen section); or it can be sliced, stained with dyes, and observed under the microscope. Cancer tissue is distinctly different from benign tissue.

A benign tumor can be lethal if it compresses the surrounding tissue against an immovable obstacle. A benign **brain** tumor compresses brain tissue against the **skull** or the bony floor of the **cranium** and results in paralysis, loss of hearing or sight, dizziness, or loss of control of the extremities. A tumor growing in the **abdomen** can compress the intestine and interfere with **digestion**. It also can prevent proper **liver** or pancreatic function. The benign tumor usually grows at a relatively slow pace and may stop growing for a time when it reaches a certain size.

A cancer may grow quite rapidly or slowly, but usually is irregular in shape. It invades the neighboring tissue instead of pressing it aside. Most importantly, a cancer often sheds cells, that is, metastasizes, so that new cancer growths can spring up in areas distant from the original cancer. The cancerous cells also can establish a cancer in tissue that is different from the original cancer. A breast cancer could spread to bone tissue or to liver.

A benign tumor can be removed surgically if it is in a location that a surgeon can reach. A tumor growing in an unreachable area of the brain can be treated using radiation. It can also be treated by inserting thin probes through the brain tissue into the tumor and circulating liquid nitrogen through the probe to freeze the tumor. This operation is called cryosurgery.

A malignancy requires steps to remove it, but consideration must be given to the possibility that the tumor has begun to metastasize. The main or primary tumor may be removed surgically, but if the tumor has been growing for some time the patient also may require treatment with powerful drugs to kill any stray cells. This treatment is called **chemotherapy**. Chemotherapy allows the antitumor drug to be circulated throughout the body to counter any small tumor growths.

See also Cell cycle and cell division; Cell differentiation; Genetic regulation of eukaryotic cells

TYMPANIC MEMBRANE

Most commonly called the eardrum, the tympanic membrane is an oval-shaped, thin, fibrous membrane that covers the ear canal separating the outer ear (external acoustic meatus) from the middle **ear** (tympanum).

The tympanic membrane is bilaminar with the mucous membrane inner portion covered by epithelial skin.

In response to sound waves funneled into the ear canal, the tympanic membrane vibrates. The degree of vibration depends upon the frequency and amplitude (energy) or the sound wave. The vibrations of the tympanic membrane are then transmitted via the malleus of the middle ear (the hammer) to the incus and stapes bones of the middle ear. The energy transmitted via the sound waves in the external ear canal is converted via the tympanic membrane and middle ear structures into wave-like fluid disturbances in the inner ear.

Excessive vibration of the tympanic membrane caused by chronic exposure to loud music, jet noise, etc., usually leads to a diminished vibratory response in the tympanic membrane and a loss or lessening of the ability to hear at certain frequencies. Although there are other causes, there is often an in ability to hear softer sounds across a range of frequencies. There is also a broader loss of hearing associated with a rupture of the tympanic membrane. In addition to a loss of hearing due directly to the rupture, there is often a persistent loss of hearing ability as a result of scar **tissue** formation on the healed or repaired tympanic membrane.

The amount of hearing loss associated with membrane perforation depends upon the exact location and size of the perforation. In most cases, bony conduction of sound wave induced vibrations in the **skull** prevents, and offers a limited compensatory mechanism, to total hearing loss. In some cases the tympanic membrane is deliberately ruptured during a surgical procedure (myringotomy) to relieve pressure on the middle ear resulting from **infection**.

See also Ear: Otic embryological development; Sense organs: Otic (hearing) structures; Valsalva maneuver

U

ULTRASOUND · *see* IMAGING

UNDERWATER PHYSIOLOGY

The underwater world is not a hospitable one for humans. Whereas on land, oxygen is freely available in the air we breathe, the oxygen dissolved in water is not available to humans. **Breathing** in water will produce **drowning**. As a result, we must hold our breath when underwater, or use artificial means of oxygen delivery like a SCUBA tank. Humans can hold their breath underwater for only a short time. A highly trained person might be capable of holding their breath for two minutes. Other mammals, such as seals and whales, are capable of holding their breath for much longer, up to about an hour. One reason for these mammals' diving prowess may be the ability of the **lungs** to collapse under the high pressures of the underwater depths and then to reinflate easily upon return to the surface. Also, the muscles of seals can hold high levels of myoglobin, an oxygen storage protein. Up to half the oxygen in the mammals body can be stored in muscles, where it can be put to use. Humans, in contrast, can store only 15% of the total body oxygen in muscles.

For humans, the environment under the water poses two challenges relative to the normal terrestrial habitat. These are pressure and temperature.

There are two effects of pressure. One effect concerns the gases that are dissolved in the **blood**. Atmospheric air is made up mainly of nitrogen (78%) and oxygen (21%). Normally, when air is inhaled the body uses the oxygen. Nitrogen is not used although some does dissolve in the blood. Underwater, pressure is greater on the body, and the pressure increases with depth. The increased pressure forces more nitrogen and oxygen into the blood. While the body uses the excess oxygen, the nitrogen still remains unused. The increased nitrogen concentration can produce a feeling of euphoria and sleepiness that is known as nitrogen narcosis.

This impairment of judgment is dangerous. Narcosis (a stupor-like state) appears suddenly and disappears as quickly once a diver ascends to a shallower depth, because the nitrogen can quickly come out of solution.

Depending on the depth of a dive, the ascent to the surface may have to be made in stages, with rests at each stage, to allow nitrogen to exit the blood gradually. If the ascent is rapid, nitrogen exits the blood so quickly that bubbles form in the bloodstream. The bubbles block small blood vessels, restricting the flow of blood. The resulting decompression sickness (also called "the bends") can cause a **heart** attack, a **stroke**, rupture blood vessels, and produce intense **pain** in **joints**. In severe cases, decompression sickness can be lethal. A more gradual and staged ascent gives the nitrogen time to eliminate itself from the bloodstream.

Another effect of nitrogen is termed residual nitrogen. This refers to nitrogen that remains in tissues after excess nitrogen has exited the blood. After reaching the surface, residual nitrogen will dissipate with time. Thus, divers need to give themselves time between dives for their body to return to the normal level of nitrogen.

The other effect of underwater pressure is felt in the ears and the sinuses. The ear canal and **skull** (including the skull bone itself) contain air spaces. As water pressure increases, the air in these spaces is compressed. This causes a feeling of pressure and even pain in the head and ears. Relief is obtained by equalizing the pressure between the inside and the outside of the head. Gently blowing the **nose** can equalize the pressure, for example. If pressure is not relieved, the eardrum can be damaged or may even burst.

Another aspect of underwater **physiology** is temperature. Often the temperature of the water is below that of the body. The net effect is the loss of heat from the body to the surrounding water. In an immense body of water, such as a lake or ocean, the heat loss will essentially be continual. The body will not be able to maintain internal temperature. **Hypothermia** can develop, which can be lethal. Humans cope with low water temperature by wearing a suit constructed of a rubber compound. A thin layer of water is held between the suit and the skin of the diver. This water is warmed by body heat, providing a zone of reasonable warmth surrounding the body. Other mammals, such as seal lions, have layers of built in insulating material.

See also Respiration control mechanisms; Space physiology; Valsalva maneuver

UPPER LIMB STRUCTURE

In general, limbs are the paired appendages of the body that are used for grasping or for locomotion. The upper limbs of the body, which are located above the waist, function in grasping. In other species, such as the gorilla, the upper limbs also participate in locomotion. In humans, the upper limbs are the two arms, including the forearm and the hand, and their components (bones, muscles, **blood** vessels, and nerves).

Each upper limb begins at the shoulder. The main skeletal support is the scapula. The eight muscles located there reflect the high degree of mobility of the shoulder. These muscles power the up-down, side-to-side and rotational movement of the shoulder.

The arm runs from the shoulder to the elbow. Skeletal support is provided by one large bone called the humerus. There are four main muscles in the upper arm. These are the biceps branchii, bicipital aponeurosis, coracobranchialis, and the branchialis. Nutrients are delivered to the muscles and waste products removed via five main routes. These are the brachial artery and brachial vein, the basilic vein, cephalic vein, and the median cuboidal vein.

Each upper limb continues with the forearm running from the elbow to the hand. There are two long bones in the forearm. These bones are known as the radius and the ulna. As they extend down the length of the forearm, they are not parallel to each other. The radius passes over one surface of the ulna and fits into a notch (the trochlear notch) located just above the hand. This union of the bones forms the wrist. The forearm is more extensively muscled than the arm. There are eight flexor muscles, which withdraw the forearm, and 13 extensor muscles, which extend the forearm. Blood flow is via the radial and ulnar **arteries**. Four nerves run through the forearm: the radial nerve, median nerve, ulnar nerve and the lateral cutaneous nerve.

The forearm connects to the hand. The hand is a complex structure, which is designed to grasp, manipulate, and even to assist in communication. Twenty-seven separate bones provide the skeletal structure. There are many **joints** present. Befitting its function, the hand is a highly articulated (capable of movement) structure. There are five separate muscles, two arteries, and five nerves present in the hand. The hand of human and primates, and particularly the articulated thumb, has propelled these species into a pathway of different evolutionary development than the other creatures on Earth.

See also Brachial system; Muscular system overview

URIC ACID

Uric acid is a nitrogen-containing end product of the metabolic processing of purines and physiological oxidation (chemical oxidation if bodily biochemical processes). Uric acid is produced principally in the gastro-intestinal tract during the **digestion** of many foods.

In birds uric acid is discharged as guano and in fish uric acid is converted into urea and then expelled in urine. In humans, a low level of uric acid is a normal component of **blood** and is excreted in the urine. An elevated level of uric acid is often associated with high blood pressure and/or elevated cholesterol levels.

Blood levels of uric acid are elevated in individuals suffering gout—a condition triggered by deposits of uric acid crystals in tissues surrounding **joints**. Gout is usually exacerbated (worsened) by diet's rich in protein—the nitrogen-rich precursor for uric acid.

Uric acid is a weak organic acid that has a low solubility in bodily fluids. Excessive levels of uric acid in the blood (uriaciduria) may lead to the formation of insoluble salts (urates) that are commonly referred to as stones if found in the urinary tract. Uric acid deposits in the kidney impair proper renal function and may eventually lead to kidney failure. Uric acid deposits may also form in tissues and joints.

The enzyme uricase catalyses the biochemical pathway that converts uric acid to allantonin. A hereditary disorder known as Lesch-Nyhan syndrome results in chronic elevation of uric acid levels.

Biomedical research has established that low levels of uric acid are also often found in individuals suffering multiple sclerosis where normal levels of uric acid may be required to inhibit or eliminate biochemical agents (e.g., peroxynitrite) that may be responsible for the production of **brain** lesions. It is interesting to note that conditions where uric acid is excessive (gout) and multiple sclerosis are seemingly exclusive of one another—individuals do not suffer both conditions. There is ongoing research indicating that uric acid may play a role in protecting neural **tissue**.

See also Acid-base balance; Elimination of waste; Protein metabolism

URINE AND URINATION · *see* ELIMINATION OF WASTE

UROGENITAL SYSTEM

The urogenital system of males and females incorporates the systems for excretion of wastes and for reproduction. These two systems originate from tissues that are next to each other in the body, and some ducts (or tubules) are used by both systems

The primary organ of the waste treatment and disposal system is the kidney. The pair of **kidneys** function in two ways. First, they remove nitrogen containing and other waste

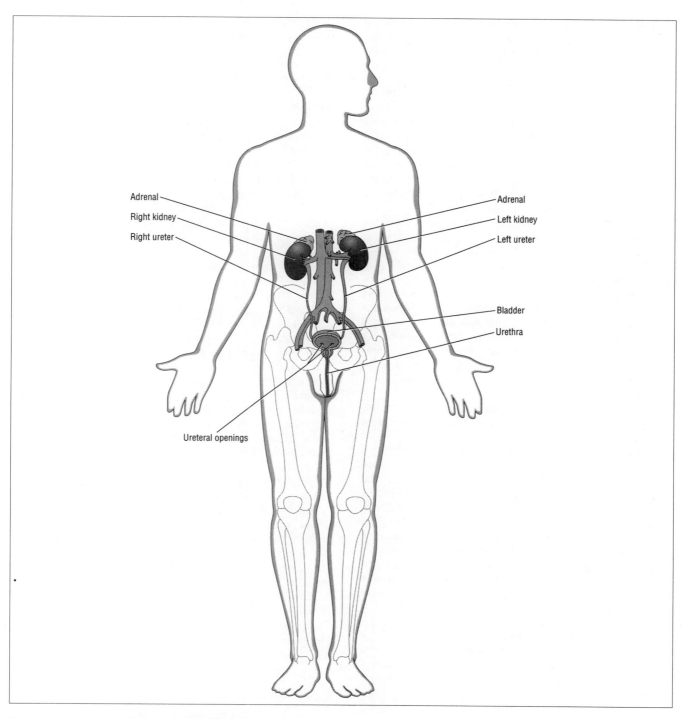

The urogenital system. *Illustration by Argosy Publishing.*

material that are the by-products of the breakdown of protein into fuel, and to eliminate these wastes as ammonia, **uric acid** or urea. Secondly, the help maintain the balance of water and salts in the body, which aids in maintaining the proper internal chemistry of the body.

Each kidney connects to the holding bag for liquid. This bag is referred to as the urinary bladder. The bladder voids liquid (urine) to the outside of the body through an exit tube called the urethra. This tube is also used for the ejaculation of **sperm** in males.

The excretion system develops at about the same time in an embryo as the reproductive system. In this parallel development, certain aspects of the two systems, such as the penis in the male, are destined to perform both excretory and reproductive functions.

In males, sperm is produced in the male **gonads** or testis. Elsewhere, the prostate glands secrete seminal fluid, which helps the sperm move more efficiently when they enter the female. Both the testes and prostate glands have tubes exiting from them. The passage of sperm takes it directly over the urinary bladder. The seminal vesicles are located immediately behind the bladder. These tubes merge, allowing the sperm and seminal fluid to mix. Finally, this common tube merges with the urethra. Thus, depending upon the physiological state of the male, the urethra can convey urine or sperm down the length of the penis to exit the body.

The penis can be flaccid or erect. An erection, which occurs for an efficient and effective delivery of sperm, is stimulated by the activity of muscles in the organ and the copious flow of **blood** to the organ. At the same time, **veins** in the penis are blocked to prevent outflow of fluid. A series of signals cause the connection between the urethra and the bladder to close. This prevents the flow of urine through the penis during ejaculation. Following ejaculation of sperm, the blood is rapidly removed from the penis and the connection betweeb the urethra and the bladder opens. At this point, urination is possible once again.

In females, the gonads are ovaries and are the site where the eggs (also called ova) are stored in oogonia clusters until final maturation into oocytes and ova before expulsion from the ovary during ovulation. Genital ducts or fallopian tubes provide a channel to ferry the egg to a site where **fertilization** can take place upon introduction of sperm during mating. The ovary and genital ducts of the female are more independent of the excretory system than are those of the male. For example, females maintain a separate duct for the passage of sex cells (ova). Indeed, in males the discipline of **urology** is concerned with the health of both the urinary and reproductive tracts.

The commonality of some aspects of excretion and reproduction have been conserved in **evolution**. The design of the human urogenital system is similar to that found in other animals.

See also Infertility (male); Metabolic waste removal; Renal system, embryological development; Renal system; Reproductive system and organs (female); Reproductive system and organs (male); Urogenital system, embryonic development

UROGENITAL SYSTEM, EMBRYONIC DEVELOPMENT

The urogenital systems in both males and females develop from early embryonic urogenital ridges compose of mesodermal cells. Although the sex of the embryo is fixed at birth by its sex **chromosomes** (XX for females and XY for males), for

a considerable period in embryonic development, male and female embryos are anatomically ambisexual and thus share a number of features in common with regard to early development. Ultimately, male and female structures differentiate into the characteristic male and female structures.

Gonads—ovaries in females and testes in males—arise from thickening areas of cells of the urogenital ridge. Initially very much alike in their path of development (the indifferent stage of development), the ultimate development of ovaries and testis is an example of the phenotypic (outward) expression of karyotype or genotype (the actual genes and chromosomes present).

Cells comprising and continuing the germ cell line are diploid cells that give rise to cells that undergo meiosis to form haploid (half the chromosome number) sex cells (i.e., male spermatids and female oocytes). **Germ cells** are differentiated early in development, possibly during the initial divisions of the zygote—but no later than in the cellular divisions that take place in or near the primitive urogenital ridge. Regardless, although the germ cells come to lie in the **gonads**, the germ cells are extra-gonadal in origin.

In males, testes begin to differentiate from the indifferent gonad stage when the embryo is just over one-half inch long (approximately 13 to 15 mm) at about 37 days post **fertilization**. Tubes appear, and a number of mesenchymal cells help to partition off the developing testis. The number of interstitial cells increases and sex cords from the indifferent stage become seminiferous tubules. Germ cells migrate into the seminiferous tubules and proliferate (duplicate). The development of the testes takes place intraperitoneally. As the **abdomen** and pelvis grow, the testes appear to descend toward an inguinal canal to the developing scrotum. At about the beginning of the eighth month of development, the testes normally descend into the scrotum. Proper descent of each testis is important to subsequent male **spermatogenesis** and fertility, because the scrotum, a site of intense metabolic and cellular activity, to remain cooler than in the main body cavity.

In females, the ovaries develop later. In the early stages when the female embryo is at the size where the male embryos begin to develop testis, the indifferent gonads do not form tubules. During the fifth month of development, a connective **tissue** partition termed the tunica albuginea begins to develop. By the end of the eighth month the partitioning and sex cord development is complete and the sex cells are divided into clusters of oogonia. The sex cords ultimately develop into primordial follicles and granulose cells that protect, support, and nurture sex cells and oogonia until the oogonia mature and continue division into oocytes.

The full development of the urogenital systems in both males and females involves a coordinated series of developmental steps that take place throughout gestation. Throughout the course of development, primitive embryonic structures give rise to sexually characteristic structures. The male and female structures are homologue because they are derived from the same early embryonic cells. For example, the testis and ovary are homologues as both are derived from the indifferent gonad.

The embryonic mesonephros develops into the efferent ductules and **appendix** of the epididymis. In females, the mesonephros develops into the paroophoron and epoophoron.

The embryonic Mullerian duct develops into the appendix of the testis in males. In females, the Mullerian duct develops into the uterus, uterine tube and the cervix that separates the vaginal canal from the uterus.

The Wolffian duct develops in the male to form the duct of the epididymis and, in females, the homologue is the duct of the epoophoron. In males, the Wolffian duct also develops into the vas deferens, seminal vesicle, and the common ejaculatory duct.

In both males and females, an anatomically similar portion of the urogenital sinus, the vesical portion, develops into the urinary bladder. In males, another corresponding region, the pelvic region of the urogenital sinus, develops into the prostatic urethra and a portion of urethra, while in females, it develops into the urethra. The pelvis region of the urogenital sinus develops into the vagina and hymen in the female, and the penile urethra in the male.

The female clitoris and male penis are derived from the same embryological tissue, the genital tubercle. In males, this tissue also develops into penile corpus spongiosum and corpora cavernosa tissue that when engorged with **blood** causes penile erection. In females, the clitoris—with smaller amounts of and less developed erectile tissue is capable of a much smaller erectile state.

The embryonic gunernaculum in males develops into the inferior scrotal ligament. In females, these cells ultimately form the ligament of the ovary.

In males, the scrotum and a portion of the penis (crus of the penis) develop from the ambisexual outer genital folds. In females, the outer genital folds ultimately form the labia majora and crus of the clitoris.

The labia minora in females forms from the inner genital folds. In males, the inner genital folds contribute to the development of the penile urethra and corpus spongiosum.

Although rare, in some cases there are abnormalities in the development of the **urogenital system** that result in varying degrees of development of both male and female **organs** (hermaphroditism). True hermaphrodite possess both ovary and testis. Pseudohermaphrodites, have either testes or ovaries but develop opposite **secondary sexual characteristics** (including genitalia). The majority of genetic hermaphrodites are genetically female (i.e., they carry two X chromosomes) and are sterile (incapable of reproducing).

See also Cell differentiation; Embryonic development, early development, formation, and differentiation; Estrogen; Genetics and developmental genetics; Genitalia (female external); Genitalia (male external); Gonads and gonadotropic hormone physiology; Homologous structures; Human development (timetables and developmental horizons); Human genetics; Infertility (female); Infertility (male); Inherited diseases; Prostate gland; Renal system, embryological development; Renal system; Reproductive system and organs (female); Reproductive system and organs (male); Sex determination; Sex hormones; Sexual reproduction; Urology

UROLOGY

Urology is the branch of medicine that deals with the urinary tract in females and with the urogenital tract in males. In both sexes, the urinary tract consists of the **kidneys**, ureters, bladder, and urethra. In males, additional structures such as the **prostate gland** are included in the **urogenital system**.

The problems with which an urologist deals tend to fall into three general categories **infection**, **cancer**, and stone formation. Cystitis is any infection of the urinary tract. The condition often appears to be centered in the urinary bladder, but is usually associated with infections of other parts of the urinary system. Cystitis is accompanied by frequent and painful urination. It is treated relatively effectively with antibiotics, although other urinary tract problems with which it is associated may require other treatments.

Enlargement of the prostate gland is now one of the most common disorders among males, especially older males. In some cases, the condition is benign and is primarily a matter of inconvenience for men who find that urination becomes more difficult and more frequent. Non-cancerous enlargement of the prostate is known as benign prostatic hyperplasia (BPH). BPH can be treated surgically by the removal of excess fatty **tissue**, although a number of urologists now recommend the use of newly-approved drugs as a way of shrinking the enlarged gland.

Cancer of the prostate has become one of the leading causes of **death** among older males in the United States and other parts of the developed world. Some physicians hypothesize that carcinogenic substances in the environment accumulate in the prostate, one of the most fat-containing **organs** in the body, and induce **tumors**. Prostate cancer can be treated with the same surgical techniques used with BPH, although the complete removal of the testicles may be recommended if the cancer has metastasized.

Kidney stones form when certain chemicals that normally dissolve in urine begin to precipitate out and form stones ranging from microscopic particles to marble-size structures. In the majority of cases, the stones are expelled from the urinary system without incident. In some cases, however, they may become lodged in various parts of the system: along the ureter, in the bladder, or in the prostate, for example. When this happens, the stone may cut into tissue and cause extreme **pain**.

Stones can be removed in a number of ways: surgically, with drugs that dissolve the stony material, or with ultrasound therapy. In the last of these treatments, high frequency sound waves are used to break apart a stone, allowing the smaller fragments to be carried away in urine.

See also Urogenital system, embryonic development

UTERUS • *see* REPRODUCTIVE SYSTEM AND ORGANS (FEMALE)

V

VAGINA • *see* GENITALIA (FEMALE EXTERNAL)

VALSALVA MANEUVER

The Valsalva maneuver is a coordinated muscular movement that increases pressure in the **Eustachian tubes** and middle ears. The maneuver produces a bilateral increase in pressure on both the left and right middle ears. The maneuver was first described by Italian anatomist Antonio Maria Valsalva (1666–1723). The Valsalva maneuver may occur as a result of natural muscle movements during defecation, coughing, or gagging, or as an induced maneuver.

The natural Valsalva maneuver occurs when exhaled air is blocked by a closed glottis. In addition to causing pressure changes in the Eustachian tubes, the Valsalva maneuver has a number of physiological side effects. The maneuver causes a decrease in the return of venous **blood** to the **heart** and decrease in blood pressure. When the airway is reopened, blood pressure increases and the heart race increases until slowed by a reflex slowing (brachycardia).

The artificial maneuver requires that the nostril passages be blocked or obstructed. In addition to mechanical devices that can pinch off the airway, a slight pressure pinch of the thumb and forefinger is usually sufficient to close the airway. Increasing pinching pressure, and a concurrent tightening of the mouth, lips and cheek muscles keep the airway closed as increasing pressure build in the **lungs** and chest. As a result of the closures and pressurization, the Eustachian tubes become pressurized.

Divers frequently use the Valsalva maneuver to counteract the effects of water pressure on the Eustachian tubes and to eliminate pressure problems associated with the middle **ear**. When subjected to pressure the tubes may collapse or fail to open unless pressurized. Eustachian tubes connect the corresponding left and right middle ears to the back of the **nose** and throat and function to allow the equalization of pressure in the middle ear air cavity with the outside (ambient) air pressure. The degree of Eustachian tube pressurization can be roughly regulated by the intensity of abdominal, thoracic, neck and mouth muscular contractions used to increase pressure in the closed airway. Divers commonly achieve pressurizations in the Eustachian tubes that are able to counteract the collapsing pressure found at a depth of about ten feet.

Although the Valsalva maneuver is one of the simplest procedures to perform to pressurize or equalize pressure in the Eustachian tubes and middle ear, there are a number of others techniques including the Frenzel maneuver, Toynbee maneuver and French Navy's *Beance Tubaire Volontaire (BTV)*. The techniques all require some closure of the airway and the manipulation of the tongue, plate or muscles associated with respiration or **swallowing** to achieve pressurization of the Eustachian tubes that can open the tubes and subsequently allow for an equalization of pressures with the middle ear.

Although a common maneuver preformed by underwater divers to pressurize or clear Eustachian tubes, the Valsalva maneuver can have adverse consequences if performed too vigorously, or if performed by individuals with certain diseases or abnormalities affecting the middle ear. The maneuver should not be attempted without proper instruction and supervision.

In individuals with some forms of heart disorders, the reflex bracycardia following the Valsalva maneuver may lead to cardiac arrest.

See also Aviation physiology; Ear (external, middle and internal); Ear: Otic embryological development; Palate (hard and soft palate); Sense organs: Balance and orientation; Swallowing and disphagia; Underwater physiology

VANE, JOHN R. (1927-)
English pharmacologist

John R. Vane's research on **prostaglandins**, hormone-like substances produced by the body, proved fundamental in the research and treatment of such illnesses as **heart** disease, strokes, ulcers and asthma. Through his studies, first at the Royal College of Surgeons, and then at the Wellcome pharmaceutical company, Vane discovered how these previously little-known secretions function. For these contributions to medicine and to physiology, Vane shared the 1982 Nobel Prize in physiology or medicine.

Vane was born March 29, 1927, in Tardebigge, Worcester, the son of Maurice Vane and the former Frances Fisher. Vane's father, the son of Russian immigrants, owned a small manufacturing company; his mother came from a family of farmers. Their Christmas gift of a chemistry set sparked Vane's interest in science when he was twelve, and his home became the site of numerous experiments. However, upon entering the University of Birmingham in 1944, he found that the work given him was not as challenging as he anticipated. After receiving his B.S. in chemistry from Birmingham in 1946, Vane decided to go to Oxford University to study **pharmacology** under Harold Burn. He obtained a B.S. in pharmacology from Oxford in 1949, and earned his doctorate in 1953. While at Oxford, Vane married. He and his wife eventually had two daughters. After leaving Oxford, Vane came to America to teach at Yale University as an instructor and assistant professor of pharmacology. He returned to England in 1955 as a senior lecturer in pharmacology at the Royal College of Surgeons, at its Institute of Basic Medical Sciences.

Vane became interested in prostaglandins in the late 1950s. Discovered in the 1930s, they were originally thought to be secreted by the **prostate gland**, which is how they got their name. Prostaglandins are natural compounds, developed from fatty acids, which control many bodily functions. Different prostaglandins regulate **blood** pressure and coagulation, allergic reactions to substances, the rate of **metabolism**, glandular secretions, and contractions in the uterus.

For many years after the discovery of prostaglandins, scientists were unaware of how they were produced and how they functioned. In the early 1960s, Vane expanded upon the procedure known as biological assay (bioassay), by which the strength of a substance is measured by comparing its effects on an organism with those of a standard preparation. Vane developed the dynamic bioassay, which allows scientists to measure more than one substance in blood or body fluids. This method enabled Vane and his colleagues at the Royal College to prove that prostaglandins are produced by many tissues and **organs** in the body. Further research led the scientists to discover that, unlike **hormones**, certain prostaglandins are effective only in the areas where they were formed.

In 1966, Vane advanced to professor of experimental pharmacology at the Institute for Basic Medical Sciences and continued his studies. An experiment he conducted in 1969 resulted in the discovery of the methods by which aspirin alleviates **pain** and reduces inflammation. Using the lung **tissue** of guinea pigs, Vane found that aspirin inhibited the production of a certain prostaglandin that causes inflammation. He published the results in a June, 1971, issue of *Nature New Biology,* a science magazine.

In 1973, Vane resigned his post at the Institute to enter the business world as director of research and development at the Wellcome Foundation, a pharmaceutical company. Following up on research by the Swedish chemist **Bengt Samuelsson** (who found that a type of prostaglandin was responsible for allowing blood to clot), Vane discovered the existence of a prostaglandin with the opposite quality, which inhibits clot formation. With the assistance of the Upjohn Chemical Corporation, Vane isolated the secretion, which he named prostacyclin. This discovery proved to be of great assistance in dissolving clots blocking the blood supply in **stroke** and heart attack victims and is useful for keeping blood from clotting during surgery. Scientists have discovered even more uses for prostaglandins, including the treatment of ulcers, alleviating pain from **menstruation** and gallstones, and stimulating contractions for childbirth.

Vane, along with Samuelsson and Swedish chemist **Sune Bergström**, received an Albert Lasker Basic Medical Research Award in 1977 for his work on prostaglandins. Five years later, the Nobel Committee gave the trio the Nobel Prize in physiology or medicine. After receiving the award, Vane predicted that future research on prostaglandins would create major breakthroughs in the areas of medicine. "In the next 20 years we should see a substantial attack on the disease process," *Time* quoted him as saying. "We will be able to find new drugs that have effects on cardiovascular disease, on asthma, on heart attack," and even health problems associated with old age, the magazine reported.

During the 1980s Vane embarked on a crusade for greater research on new drugs to fight both new diseases (such as acquired immunodeficiency syndrome, known as **AIDS**) and drug-resistant strains of old diseases, such as **malaria**. In articles for scientific and medical journals, he stressed the need for greater international cooperation in the search for a cure or vaccine for AIDS and advocated the creation of an Institute for Tropical Diseases to research new drugs to battle disease in the tropics.

Vane's professional activities also include memberships in the British Pharmacological Society and the American Academy of Arts and Sciences. A popular lecturer, he has received more than a dozen awards for his accomplishments. In addition to the many hours he devotes to his work, Vane finds time for his hobbies of photography, travel, snorkeling, and water-skiing.

See also Allergies; Blood pressure and hormonal control mechanisms; Inflammation of tissues

VASCULAR EXCHANGE

Vascular exchange refers to the movement of fluid and gas across the boundary between the **blood** and **tissue** (e.g., alveolar lung tissue) or between blood and **interstitial fluid**. The boundary of the exchange is the membrane of the network of tiny tubes called **capillaries** that transport blood

The membrane of a capillary, which in some places is only a single cell thick, is not impermeable. The **plasma** inside the capillary can move though pores to mingle with the fluid outside the capillary (the interstitial fluid). Concentration and hydrostatic forces drive the outward movement of plasma or the inward movement of water across the cell barriers. Osmotic pressure arises when there is an ion imbalance (i.e., more of a specific type of ion on one side of the membrane than on the other). Hydrostatic pressure arises because of the flow of blood through the capillary. Overall, there tends to be more movement of fluid out of a capillary than into it. Larger objects like protein and blood cells do not normally pass through the capillary membrane, and so are not subject to vascular diffusion.

Another way that substances can move across the capillary membrane is by diffusion. This is a movement that occurs passively, without the use of energy. Membranes, such as those of a capillary, are made of **lipids** interspersed with protein. Lipids are hydrophobic ("water-hating"). Thus, hydrophobic materials, such as the gases oxygen and carbon dioxide, will tend to partition into the membrane. If the concentration of a gas is low on one side of the membrane, relative to the other side, the concentration gradient (differences in concentration over a specified distance) drives gas diffusion across the membrane.

For example, the concentration of oxygen is high in the **lungs** following the intake of a breath, while the oxygen concentration in the blood entering the capillaries of the lung is low. Oxygen will diffuse from the air into the blood, to be subsequently dispersed through the body. Likewise, the concentration of carbon dioxide is higher in the blood entering the lung capillaries than it is in the air in the air sacs of the lung. So, carbon dioxide will diffuse from the blood to the lung to be exhaled.

The process of vascular exchange ensures that vital oxygen gets to the blood and that the carbon dioxide—generated as a waste product of metabolic activity—exits the body.

See also Acid-base balance; Osmotic equilibria between intercellular and extracellular fluids; Oxygen transport and exchange; Respiration control mechanisms

VASCULAR SYSTEM (EMBRYONIC DEVELOPMENT)

The early embryonic vascular system permits circulation through a series of incompletely fused and developing channels. Before the **heart** is fully formed, the flow of circulation is diffuse rather than strongly unidirectional as it is in the atrial and venous systems of later fetal development. Movement of **blood** through the early embryonic vascular system begins as soon as the primitive heart tubes form and fuse. Contractions of the primitive heart begin early in development, as early as the initial fusion of the endothelial channels that fuse to form the heart.

Primitive blood **plasma** is almost indistinguishable from **interstitial fluid** and it contains few cells. The vascular system is essentially transparent until enough blood cells form within the yolk sac. As the cells take on **hemoglobin**, they gain color and the early embryonic system becomes visible.

At the end of the first month of development, the embryonic circulation takes on specialized features outside of the primitive heart that can ultimately be traced to adult vessels. Surrounding the atrium is the sinus venosus that contains two horns (major branches) that collect and pool blood from a number of primitive **veins**. The right horn ultimately develops into the superior vena cava and the left horn into the coronary sinus.

At about this stage, blood from the umbilical cord (carrying blood from the **placenta**) also begins to flow into the embryonic **liver** to establish the hepatic portal system. Portions of the umbilical veins continue onward to become a modified vitello-umbilical system that ultimately becomes the inferior vena cava.

Initially, three systems return venous blood to the primitive heart. This venous blood pools in the sinus venosus. Vitelline veins return blood from the yolk sac (the site of blood cell development). Umbilical veins return oxygenated blood from the placenta. The right umbilical vein does not develop, but the left umbilical vein enlarges to deliver blood to the embryonic liver before continuing on to become the inferior vena cava that communicates with the sinus venosus and the right atrium of the heart. The third major venous return comes from the cardinal system. Anterior cardinals drain venous blood from the developing head region into the common cardinal veins that then empty into the sinus venous. The cardinal veins ultimately form portions of the internal jugular and superior vena cava. Subcardinal veins return venous blood from the developing renal and **urogenital system**, while supracardinals drain the developing body wall.

The supply of blood to the neural tube area is critical for proper **brain**, **spinal cord**, and cranial nerve development. As the embryonic vascular system develops, a number of temporary pathways appear to carry blood to the heart while other vessels degenerate (e.g., the brachiocephalic vein)

Blood leaves the developing heart after being pumped into a bulbous cordis that forms an early aortic sac and trunk of the arterial system, The ascending aorta and the **aortic arch** of the ascending aorta develop from the embryonic pharyngeal arterial system. A series of fusions and degenerations of embryonic **arteries** follow a head to tail (craniocaudal) scheme of vascular development. For example, the twisted and curving ascending aorta and the many branches and trunks arising from the fully formed aortic arch result from the complex development of the embryonic pharyngeal arch system.

Primitive **carotid arteries** that branch from the developing pharyngeal arterial system and a pair of dominant dorsal aortae provide early embryonic arterial blood to developing neural **tissue** in the upper spinal cord and developing brain. The complex scheme of development of the aorta and its eventual continuation as the **thoracic aorta** and the **abdominal aorta** explain the twisting course the arterial branches take around other **organs**, especially the esophagus and trachea.

Although there is a dynamic formation of vascular structure, the essential scheme of embryonic circulation remains consistent throughout development. Oxygenated blood from the placenta returns to the fetus by an umbilical vein. Before entering the right atrium of the heart, a portion of the blood passes through the embryonic liver sinusoids. The blood from the umbilical vein and developing hepatic system enters the heart via the inferior vena cave into the right atrium. A valve at the juncture of the inferior vena cava and the right atrium directs the majority of the flow of oxygenated blood through the foramen ovale (a communicating hole between the right and left atrium). Because the **lungs** are not functional, only a small amount of blood return from the **pulmonary circulation** enters the left atrium. After shunting to left atrium, the oxygenated blood flows into the left ventricle. Contractions of the heart then expel this oxygenated blood into the fetal systemic arterial system.

The fetal vascular system undergoes at least four major revisions during development. From a bilateral system of self-contractile tubes, vessels fuse to form the heart. During early embryonic stages, arteries and veins that serve transitory structures develop, and then are reabsorbed (e.g., the early posterior cardinal venous system). Ultimately, the heart develops into two separate pumps. The right pump gathers blood from the placenta and **systemic circulation** and pumps it into the pulmonary circulation; the left pump gathers blood from the pulmonary circulation and pumps it into the systemic circulation. During fetal development, a series of valves, holes, and ducts act to shunt the majority of blood from the inoperative pulmonary system to the systemic system. At birth, a series of changes radically transforms the **fetal circulation** into the adult circulatory pattern.

The streaming flow of oxygenated blood across the right atrium caused by the valve of the inferior vena cava allows the return of deoxygenated fetal blood via the superior vena cava to remain mostly directed toward the right ventricle. Because the lungs are not yet functional, another shunt, the **ductus arteriosis** (also spelled ductus arteriosus) provides a diversionary channel that allows fetal blood to cross between the pulmonary artery and aorta. In response to inflation of the lungs and pressure changes within the pulmonary system, both the foramen ovale and the ductus arteriosis normally close at birth to establish the normal adult circulatory patterns.

See also Angiology; Aortic arch; Ductus arteriosis; Heart, embryonic development and changes at birth; Placenta and placental nutrition of the embryo; Systemic circulation

VASCULAR SYSTEM OVERVIEW

The vascular system is composed of a complex network of tubular vessels allowing **blood** to circulate around the body. It consists of **arteries** (carrying oxygenated blood from **heart**), **veins** (carrying the deoxygenated blood from the tissues to the heart) and **capillaries** (the site of gas, nutrient and metabolite exchange) with veins and arteries located side by side in tissues.

Although veins and arteries have different functions, they share a number of anatomical features. The walls of the majority of vessels consist of three distinguishable layers: tunica interna (or intima), tunica media, and tunica externa (or adventitia).

The internal layer (tunica intima) is composed of the endothelial cell layer, constituting an interface with the blood. At their basal surface, the endothelial cells are connected to the underlying connective **tissue**. The endothelium in arteries and large veins is enveloped by a continuous layer of vascular **smooth muscle** cells (tunica media), while in capillaries and small veins it is surrounded by pericytes.

Pericytes contract, but not as strongly as muscle cells; their main function is to support the endothelial cells in small vessels by providing gentle constrictor effect called tone. In addition to the vascular smooth muscle cells, the larger arteries have bundles of collagen and elastic fibers interspersed between the layers of basement membrane.

The most external layer is the tunica adventitia, which forms a loose connective tissue sheath-like structure surrounding for the vessel. It is composed of elastic fibers and collagen, which help to anchor the vessels within the tissues they serve. Such layering gives vessels strength, but in very large vessels it also creates a need for additional blood supply to provide nutrients to the cells of the internal and external layer. This additional network of blood vessels is called vaso vasorum.

Muscle cells contained in the vessel walls are able to change the diameter of the vessel by contracting and relaxing. They usually surround the vessel in a circumferential pattern, but in large arteries they are also located along the walls allowing two-way contractions and stretching. Presence of the muscles is fundamental to the ability of the vessels to accommodate variations in volume and cope with alterations in the blood pressure.

Although the general plan of arteries and veins is similar, there are some differences in their general appearance. Arteries have thicker but more elastic walls than veins. Because of these differences, arteries retain their shape in tissue sections. In contrast, veins tend to collapse. The main reason for such a difference is the fact that arteries have to cope with much higher pressures and an oscillating type of blood flow resulting from the heartbeat.

The large elastic arteries stretch during systole to protect the downstream vessels from unduly high blood pressure thus providing smooth blood flow. Therefore the large vessels are replaced with smaller muscular arteries then arterioles, eventually branching to form capillaries. The structure of the walls of these vessels changes accordingly, with an increase in muscle layer (muscle arteries) and loss of tunica externa (arterioles). In comparison, the veins become larger as they lead the blood back to the heart. Therefore small venules collecting capillary blood merge into medium veins and eventually they become the superior and inferior venae cavae. The walls of veins have no tunica media in venules and a thin one in medium veins. As the venous circulation operates at significantly lower pressure than the arterial circulation, the veins are equipped with valves preventing the backflow and pooling of blood in the extremities of the body.

Blood flow in the vessels is not only dependent on the blood pressure and speed, but it also depends on the resistance offered by the vessels. Long and narrow ones (e.g. arterioles) create the highest resistance; an increase in vessel diameter results in easier flow. Increased resistance can be also caused by an increase in blood viscosity arising from interactions between the components of the blood.

Blood vessels undergo age-related (or wear-and-tear) weakening leading to appearance of varicose veins. An overload of veins with blood can occur especially in lower extremities leading to swollen and stretched veins, causing also weakening of the venous valves. Even more dangerous is an increase of blood pressure (hypertension) as it increases the shear stress on the arterial walls and can lead to damage of weaker vessels normally not exposed to high pressure.

See also Abdominal veins; Angiology; Blood pressure and hormonal control mechanisms; Brachial system; Carotid arteries; Circle of Willis; Collateral circulation; Drug treatment of cardiovascular and vascular disorders; Fetal circulation; Lymphatic system; Metabolic waste removal; Nervous system overview; Oxygen transport and exchange; Pulse and pulse pressure; Thoracic aorta and arteries; Thoracic veins; Vascular exchange; Vascular system (embryonic development)

VEGETATIVE FUNCTIONS

The vegetative functions are the operations of the body at rest, such as the workings of the **organs**. When a person is in a persistent vegetative state, the vegetative functions continue to work when other functions of the nervous system, such as consciousness and perception, have shut down.

The nervous system consists of the **central nervous system**, that is the **brain** and **spinal cord**, and the **peripheral nervous system**, the nerves that branch throughout the body. The peripheral nervous system is made up of sensory receptors and the motor division, which controls movement. The motor division includes both somatic motor **neurons**, which control voluntary movements, and autonomic motor neurons, which control involuntary movements such as the beating of the **heart**.

The **autonomic nervous system** itself has two divisions. The sympathetic neurons are involved in the "fight or flight" reactions of stress. Responding to the secretion of adrenalin by the adrenal gland, the **sympathetic nervous system** is activated as a unit. Its primary neurotransmitter is norepinephrine, and it serves to dilate the pupils and increase the heart rate, **blood** sugar, and circulation to the skeletal muscles.

The parasympathetic division of the autonomic nervous system controls the vegetative functions by releasing the neurotransmitter acetylcholine, operating on each organ separately. The effects of the parasympathetic division generally slow down the body for rest and **digestion**. The heart rate decreases, and the pupils constrict. Circulation is increased to the skin and internal organs, correspondingly decreasing the blood flow in the skeletal muscles.

Because the survival of the organism generally requires balance, most organs are innervated by both the sympathetic and parasympathetic divisions. Thus the heart rate, for example, can be increased or decreased as necessary.

See also Brain stem functions and reflexes; Homeostatic mechanisms; Nervous system overview; Parasympathetic nervous system; Peripheral nervous system; Somatic and visceral

VEINS

Veins are vessels designed to collect and return **blood**, including deoxygenated **hemoglobin**, from tissues to the **heart**. Veins and the venous vascular system can be divided in to three separate systems depending on anatomical relationships and function. Initially, veins can be divided into systemic and pulmonary systems. The veins that drain the heart, comprising the coronary venous system, may be described as an independent venous system, or be considered a subset of the systemic vascular system. The systemic veins transport venous blood—deoxygenated when compared with arterial blood—from the body to the heart. The pulmonary veins return freshly oxygenated blood from the **lungs** to the heart so that it may be pumped into the systemic arterial system.

Veins can also be described by their anatomical position. Deep veins run in **organs** or **connective tissues** that support organs, muscle, or bone. Superficial veins are those that drain the outer skin and **fascia**.

In contrast to **arteries**, veins often run a more convoluted course, with frequent branching and fusions with other veins (anastomoses) that make the tracing of the venous system less straightforward than mapping the arterial system. In addition, there are reservoirs or pools (sinus) that collect venous return from multiple sources. Many veins contain valves that assure a unidirectional (one way) flow of venous blood toward the heart.

The systemic venous system can be roughly divided into groups depending on the region they drain, and the vessel through which they return blood to the heart.

The first systemic venous group consists of veins that drain the head, neck, thorax, and upper limbs. These vein ultimately return blood to the heart through the superior vena cava.

Veins that drain the **abdomen**, pelvis, and lower limbs return blood through the inferior vena cave. Both the superior and the inferior vena cava return deoxygenated blood to the right atrium of the heart. The coronary sinus collects blood from a number of cardiac veins before returning blood to the right atrium near the point where the inferior vena cava enters the right atrium.

The pulmonary veins return blood oxygenated in the lungs to the left atrium. There are four major pulmonary veins, each lung being drained by a pair of pulmonary veins. Akin to the drainage of a land basin from streams into a larger river system, smaller venules arise from the lung alveolar capillary bed, then the venules fuse to form single veins that separately drain isolated lobes of the lung. The veins from the upper and middle lobes of the three-lobed right lung fuse to create a pair of veins, a superior and inferior pulmonary vein, that separately transport blood to the left atrium.

At a microscopic or histological level, veins have thinner walls than do arteries. They are more elastic and capable of a wider range of lower pressure volume transformations. The elasticity is a result of the fact that veins have less subendothelial connective **tissue** in their vascular walls. In addition, the tunica media and tunica adventitia are often indistinguishable layers, or are poorly developed when compared with arterial linings.

Venules drain **capillaries** and capillary beds. The venules ultimately fuse (coalesce) into veins that, as they increase in size, also increase in organization and differentiation of their vascular walls. In general, the larger the vein, the more likely it is to be invested or surrounded with **smooth muscle** tissue. The values with the venous system are formed from a cusp forming multiple folding of the tunica intima. Valves are generally absent from the largest veins and the pulmonary veins.

Veins also serve in fluid uptake and can receive lymph fluid from lymphatic vessels. The major lymphatic duct, for example, drains into the fused vein formed from the fusion of the subclavian and left internal jugular vein.

See also Abdominal veins; Angiology; Collateral circulation; Coronary circulation; Fetal circulation; Thoracic veins; Vascular exchange; Vascular system (embryonic development); Vascular system overview

VENTRAL · *see* ANATOMICAL NOMENCLATURE

VERTEBRA

The spine or backbone is made of a long series of bones called the vertebrae. Each vertebra has a function and distinctive shape that separates it from the rest of the series. There are groups of vertebrae that are easily identified by the specific shape that reflects their collective work along the spine. For instance, the neckbones are called the cervical vertebrae and have distinctive shapes based on how they support the head. The ribcage is attached to the thoracic vertebrae, and encircles the thoracic region of the body. The lumbar vertebrae are the largest and bear most of the weight of the body. The sacral vertebrae are fused and are the point of attachment for the pelvic (hip) bones. The final tiny series of bones are the coccyx. Paleobiologists believe these are the remnants of a small tail.

The general structure of a vertebra consists of a round central body. These are the regions that lie on top of another separated by a cartilaginous intervertebral disc. On the ventral surface of the vertebra there is a prominence called the neural arch. A large foramen lies between the body and the arch. Its purpose is to provide protection for the **spinal cord**. Each group of vertebrae has a different structure surrounding the neural arch and a variety of processes that meet the functional requirements of the bone.

The seven cervical bones (C1-C7) are the smallest and lightest of the spine. The first two are unique in shape and

allow for the rotation of the head. The first (C1) is called the atlas after the Greek mythological character that was condemned to carry the world on his shoulders. The atlas allows the head to rotate in a "yes" gesture. The second axis permits the "no" gesture type of rotation. All the cervical vertebrae are easily identified by their flat shape and small transverse foramen on each side of the small bone encircling the spinal cord.

The thoracic vertebrae provide support for the ribs. Because of this they are easily identified by small facets on the processes of the neural arch. These facets are small depressions in which the ends of the ribs rest. The body of the vertebrae is fairly large and the neural crest (the high point of the neural arch) is fairly tall for muscle attachment.

The lumbar vertebrae are the largest in the body. They form the bottom of the s-shaped spine and are large for weight bearing back muscle insertion. They are conspicuous by their lack of distinct features. The body is heavy and round and the prominent neural arch provides a huge surface area for muscles.

The sacral vertebrae begin to fuse at about age 16 to 18. By age 26, they have fused together to form a triangle shaped structure called the sacrum. The sacrum serves a dual purpose in that it supports the pelvic girdle and also forms the back wall of the pelvic cavity. The dorsal surface where the spinous processes of the neural arch are fused is called the medial sacral crest.

The last four to five bones are called the coccyx. They are fused by the mid-twenties into a single traingular bone. Even though it is a vestigial tail, some important ligamnets are attached to the coccyx, and it still serves an important function in the spine.

See also Skeletal system overview (morphology)

VERTEBRAL COLUMN

Composed of 33 separate vertebrae, the vertebral column, commonly called the spine, forms the rigid backbone of the body and is the major structural support and weight-bearing element in the skeletal system. The vertebral column is the axis of the spine, and in **anatomical nomenclature** is the axis dividing the body into bilaterally symmetrical halves.

The vertebral column is subdivided into five principal regions, each with their own characteristic vertebrae. Articulations between 24 of the vertebrae allow the spine to remain flexible and this flexibility provides allows a wide range of motion (e.g., flexion and extension) that would not be possible in a fused vertebral column. Nine of the vertebrae are fused and the fusions are important to spinal strength in intensely weight bearing areas.

The most superior region of the spine is the cervical region. Composed of seven cervical vertebrae, this region supports the **skull** on the first cervical vertebrae (the atlas). A unique articulation between the atlas and the second cervical **vertebra** (the axis) allows the head and neck to pivot (i.e., a limited degree of rotation about the axis).

Below or inferior to the cervical region is the thoracic region of the spine. Composed of 12 thoracic vertebrae, the thoracic region is the site of attachment for costal cartilages (ribs) and **muscles of the thorax**. The size of the thoracic vertebra increase in size from the uppermost first thoracic vertebrae to the lowest (most inferior) twelfth thoracic vertebrae.

Below the thoracic region are five larger lumbar vertebrae that compose the lumbar region. As with the thoracic region, the lumbar vertebrae increase in size as one moves inferiorly down the spine. Although large and important in weight baring the lumbar vertebrae retain independent articulations and a degree of flexibility.

Below (inferior) to the lumbar region of the spine lie the fused vertebrae of the sacrum and, most inferiorly, the coccyx. Five vertebral structures fuse to form the sacrum and four fuse to form the coccyx.

Viewed laterally the spine takes on a subtle double "S" like shape with the cervical and lumbar regions curved inward (anteriorly) while the thoracic region, sacrum, and coccyx are curved outward from the body (posteriorly).

Because the **spinal cord** runs continuously through the vertebral canal that is bounded by the vertebral foramen of each vertebra, any fracture of the vertebral column or other form of spinal injury may cause **death** or severe neurological injury or impairment (loss) of function at the level and below the level of injury. Spinal disorders may also cause a loss of neurological function. Mechanical slippage of the **cartilage** disks that help separate and cushion the individual vertebrae of the vertebral column impairs flexibility and mobility and may be painful if the slippage applies pressure to the spinal cord or causes inflammation of surrounding **tissue**.

See also Appendicular skeleton; Cerebrospinal fluid; Meninges; Posture and locomotion; Skeletal system overview (morphology)

VERTEBRAL SPINAL RAMI • *see* SPINAL NERVES

AND RAMI

VESALIUS, ANDREAS (1514-1564)
Flemish anatomist and physician

While still a young physician, Andreas Vesalius overturned the fourteen-centuries-old Galenic canon of medicine and founded modern scientific **anatomy**.

Vesalius was born in Brussels in what is today Belgium to a family established in medicine for several generations. Young Vesalius showed an early interest in anatomy. He attended the University of Louvain and then studied medicine at the University of Paris, where he became skilled at dissection under teachers who were dedicated followers of **Galen**.

After a stint as a military surgeon, Vesalius enrolled at the University of Padua, Europe's preeminent medical school, receiving his doctor of medicine degree in 1537. Immediately assuming a post as lecturer in surgery and anatomy at Padua, Vesalius proved to be an innovative teacher. Contrary to prevailing practice, he performed dissections himself during the

lectures and illustrated the lesson with large, detailed anatomical charts. The lectures were enormously popular and demand for the charts was so great that Vesalius had them printed as *Tabulae anatomicae sex* in 1538.

As Vesalius proceeded with his dissections, he increasingly noted obvious conflicts between what he saw in the human body and what Galen described. Galen's errors, Vesalius reasoned, arose because the ancient anatomist relied only on animal dissections, which often did not apply to human anatomy. Vesalius set down the principle that true, fundamental medical knowledge must come from human dissection, practiced by each individual physician.

To attract established physicians to the study of anatomy and promote the teaching of this new science, Vesalius devoted himself for five years to the production of his magnum opus, one of the most important books in medical history and the world's first textbook of anatomy: *De humani corporis fabrica*, published in 1543. Vesalius carefully supervised all aspects of the book's production. The *Fabrica* contained detailed anatomical descriptions of all parts of the human body, including directions for carrying out dissections; magnificent, meticulous illustrations, probably by students from Titian's studio; and a clear explanation of the objective, scientific method of conducting medical research.

The publication of the *Fabrica* rocked Galenism to its foundations. This shattering of the revered, supposedly infallible ideas of Galen provoked bitter controversy, which may have been why Vesalius abruptly quit anatomical research and became court physician to Emperor Charles V and, later, to Charles's son Philip II of Spain. As in anatomy, Vesalius achieved renown as a medical practitioner. In 1564 he left Spain for a trip to the Holy Land, perhaps intending then to return to teaching at Padua. On the way back from Palestine, however, his ship was wrecked, and Vesalius died on the island of Zante at the age of fifty.

See also Careers in anatomy and physiology; Comparative anatomy; History of anatomy and physiology: The Classical and Medieval periods; History of anatomy and physiology: The Renaissance and Age of Enlightenment; Medical training in anatomy and physiology

VESTIGIAL STRUCTURES

A structure or organ is vestigial if it has diminished in size or usefulness in the course of **evolution**. Vestigial structures are markers of evolutionary descent. For example, boa constrictors, which are descended from four-legged reptiles, grow tiny hind legs. Duckbill platypuses, which are descended from extinct platypus species that had teeth as adults, grow and reabsorb teeth before birth. In human beings, the vermiform **appendix** (a hollow, worm-shaped organ about the size of a pencil, attached to the beginning of the large intestine) marks descent from mammals that had a much larger sac in this position and used it to digest their high-cellulose diet (as many species, including other primates, still do).

From the late nineteenth century until the 1960s, biologists thought that the human body contained scores of useless vestigial structures, including the coccyx, ear muscles, pineal gland, thymus, vermiform appendix, wisdom teeth, and others. Most of these structures are now known to have at least minor functions, leading to controversy over whether the human body contains any vestigial structures at all. However, the discovery of a function for a structure does not necessarily mean that it is not vestigial. A vestigial structure may be completely without function, like fetal platypus teeth, or it may be changed and diminished in function. The human appendix appears to be a vestigial structure with changed and diminished function; it is attached to the digestive system just where an anatomically similar, essential digestive organ is attached in many other mammals, but performs no digestive or other essential function. (People who have had their appendixes removed suffer no known ill effects.) Some biologists assert that the appendix assists the **immune system**; if so, the appendix's tiny opening on the large intestine still appears to be a truly useless (or worse than useless) vestige of this organ's digestive origin. Hardened feces can block this opening and cause the appendix to swell and rupture, a potentially fatal disorder that afflicts about one person in every thousand annually.

Another vestigial structure with diminished usefulness is the wisdom teeth. When these succeed in coming in properly, they are useful for chewing. However, they are not essential for chewing, and often come in sideways, fail to grow at all, or remain embedded in the jawbone. They are vestiges of a more massive chewing apparatus in our primate ancestors.

Vestigial structures may also be molecular, as in the case of vestigial genes that exist in most species. For example, although humans cannot manufacture their own vitamin C, most other mammals can because they possess a gene enabling them to produce an enzyme (L-gulono-gamma-lactone oxidase) which in turn makes it possible for them to produce vitamin C. Humans possess a defective copy of this gene that does not produce the required enzyme (or any other product). This gene was presumably disabled by mutation at a time in primate evolution when its loss was not a significant disadvantage, and now remains as a vestigial genetic sequence.

Men's nipples, sometimes cited as vestigial structures, are not truly vestigial because they are not remnants of functional male nipples in ancestral species. They occur because nipple precursors are grown early in the development of the human embryo, before sexual differentiation. Later in life these structures become more fully developed in women, while remaining undeveloped in men. The navel, too, is a by-product of embryonic development rather than an evolutionary vestige.

Truly useless vestigial structures are few or absent in humans, but well-known in other species: blind cave-dwelling species of crayfish grow eye-stalks but no eyes, embryonic baleen whales grow teeth which they re-absorb before birth, and so on. However, evolution tends to eliminate useless **organs**, for every structure requires energy to grow, sustain, and transport. Individuals that expend their limited resources on useless organs are therefore less likely to leave offspring. This explains the tendency of useless anatomic structures to diminish in size during evolution. Vestigial genes, in contrast, can linger indefinitely because they can passed on to one's descendants at essentially zero cost. Vestigial genes inherited from shared ancestors have been preserved for many millions of years in species as far removed from each other today as humans, cows, and mice.

See also Evolution and evolutionary mechanisms

VIEUSSENS, RAYMOND (CA. 1635-1715)
French physician and surgeon

Raymond Vieussens advanced the understanding of the **anatomy** and **physiology** of the **brain**, **heart**, nervous system, and **circulatory system**. Many anatomical features in these systems are named after him, such as Vieussens's centrum (the white oval core of each hemisphere of the brain); Vieussens's valve (a sheet of thin white **tissue** in the brain); Vieussens's ventricle (one of the fluid-filled spaces in the brain); Vieussens's ansa (a loop in the ganglia around the subclavian artery); Vieussens's ganglion (a network of nerves between the aorta and the stomach); Vieussens's anulus, isthmus, or limbus (a ring of muscle in the right atrium of the heart); Vieussens's foramina (tiny openings in the **veins** of the right atrium of the heart); and Vieussens's veins (small veins on the surface of the heart).

Vieussens was born perhaps as late as 1641, the son of François Vieussens, a townsman of Vigan, France. After receiving his medical degree from the University of Montpellier in 1670, he became chief physician at Hôtel Dieu St.-Eloi, the main hospital in Montpellier. Although very scholarly, he never held a university appointment. His relations with the Montpellier faculty were soured by his long, bitter, public quarrel with Montpellier Professor Pierre Chirac (1650–1732) about which of the two had first discovered an acidic salt in the **blood**. Ironically, both Vieussens's and Chirac's results were incorrect.

In the late 1670s Vieussens married Elisabeth Peyret, with whom he ultimately had twelve children. Two of his sons and two of his sons-in-law became physicians. Dividing his career between Montpellier and Paris, Vieussens was favored by the French aristocracy and became rich through their patronage. He was the personal physician of the Marquis de Castries, the Archbishop of Toulouse, and the Duchess of Montpensier. He was named "Royal Physician" in 1688 and "State Councillor" in 1707, even though King Louis XIV was not his patient.

Thomas Willis and Nicolaus Steno (1638–1686) were the two main influences on Vieussens's study of anatomy, especially cerebral anatomy. Although he was a careful, observant, and generally accurate anatomist, Vieussens sometimes allowed dubious contemporary metaphysical speculation to influence his work in physiology. The dualistic iatromechanics of **René Descartes** and the mystical iatrochemistry of Franciscus de le Boe Sylvius (1614–1672) shaped his views of bodily processes, rendering his contributions to physiology less valuable than his contributions to anatomy.

Beautifully executed copperplate illustrations make Vieussens' major work, *Neurographia universalis* (General neurography, 1684), second in importance only to Willis's *Cerebri anatome* (Anatomy of the brain, 1664) among seventeenth-century books on neuroanatomy. *Neurographia universalis* contains the first precise description of the centrum ovale. This structure is still sometimes called Vieussens' centrum, but because Félix Vicq d'Azyr (1748–1794) achieved a more refined understanding of it, it is more often called the centrum semiovale or Vicq d'Azyr's centrum.

Vieussens' classic of cardiology, *Novum vasorum corporis humani systema* (New system of the vessels of the human body, 1705), includes the earliest correct descriptions of the left ventricle and some of the coronary blood vessels, as well as mitral stenosis, aortic insufficiency, aortic regurgitation, and several other heart diseases and circulatory disorders. On related topics he wrote *Epistola de sanguinis humani* (A letter about human blood, 1698); *Deux dissertations* (Two dissertations, 1698), both about blood; *Traité nouveau des liqueurs du corps humain* (New treatise on human body fluids, 1715); and *Traité nouveau de la structure et des causes du mouvement naturel du coeur* (New treatise on the structure of the heart and the causes of its natural motion, 1715).

His other books include *Tractatus duo* (Treatise on two subjects) (1688), which first discusses anatomy, then fermentation; *Dissertatio anatomica de structura et usu uteri ac placentae muliebris* (Anatomical dissertation on the structure and use of the uterus and **placenta** in women, 1712); and *Traité nouveau de la structure de l'oreille* (New treatise on the structure of the ear, 1714).

See also Aortic arch; Central nervous system (CNS); Cerebral cortex; Cerebral hemispheres; Coronary circulation; Ear (external, middle and internal ear); Nervous system overview; Placenta and placental nutrition of the embryo; Reproductive system and organs (female)

VINCI, LEONARDO DA (1452-1519)

Italian painter, anatomist, inventor, and sculptor

Although Leonardo da Vinci is best remembered for Renaissance paintings, his anatomical studies yielded over 750 drawings that have become the foundation for modern scientific illustration. By coupling new artistic techniques of perspective with experimentation and keen observational skills, Leonardo also created a systematic method of describing the natural sciences that was widely used for over 300 years.

Born in the small Tuscan village of Vinci, Italy, Leonardo was the illegitimate son of Ser Piero d'Antonio, the local notary, and a peasant girl named Caterina. Leonardo spent most of his childhood in his father's home. Although Leonardo's father and mother never married, there is evidence that Leonardo and his mother communicated throughout her life, and that he knew several of his half-brothers and half-sisters from both of his parents' later marriages. Even in childhood, Leonardo displayed exceptional talent, especially in

music (Leonardo played the lyre) and singing. Leonardo embraced his studies and excelled in mathematics, although he never mastered Latin and Greek, as was the style of the day among academics. A careful observer of nature, young Leonardo often made detailed drawings of the plants and animals surrounding him.

At the age of fifteen, Leonardo was apprenticed to the master artist Andrea del Verrocchio in Florence. Under Verrocchio's tutelage, Leonardo studied painting as well as methods of plaster and bronze casting, techniques that would later prove useful for his anatomical studies.

Leonardo's anatomical works began in earnest after 1505 and continued through the next decade, during which time he lived in Milan and often visited Florence. In Florence, Renaissance artists of the time were embracing a "new realism" when portraying the human form. Leonardo combined the new realism with scientific observation, and drew studies of the internal human form based upon information gleaned from dissecting cadavers. Leonardo's artistic ability enabled him to precisely communicate to paper the intricate details of the structure of muscles, **organs**, and the vascular and skeletal systems.

Convinced that every structure in the body had an exact function, Leonardo reasoned that his drawings must be detailed and precise; no structure was extraneous. Leonardo often illustrated his subjects from four different views, enabling the subject to be studied from different angles simultaneously. Leonardo also invented the technique of illustrating the interior of structures of the body through cross-section representation. His drawings include cross-section images of the **skull**, nerves, and vascular system.

Leonardo's earlier interest in mechanics and engineering had great influence upon his anatomical works. Muscles and **joints**, frequently illustrated, were referred to as resembling levers. The surrounding muscle tissues were often termed lines of force. Structures were compared to architecture, as Leonardo used plaster of Paris, clay, and wax to form three-dimensional molds of body cavities, as well as the **heart** and skull. Vessels of the vascular system were compared to underground streams, and **blood** flow through the heart valves was described as a vortex similar to water flowing through a narrowing channel. Such mechanistic views of the body foreshadowed **René Descartes**'s and his metaphysics of **physiology**.

Leonardo's mistakes in his drawings were largely based upon misconceptions of the day. In his famous drawing of a fetus within the uterus, Leonardo correctly represented the umbilical cord, but the **placenta** is consistent with a much larger animal. He did, however, correctly represent the uterus as single-chambered organ, rather than a multi-chambered organ as was assumed at the time. In at least one illustration, Leonardo adopted the five-lobed **liver**, an incorrect assumption left over from animal dissections conducted during medieval times.

Many historians hold that Leonardo ultimately intended to publish an **anatomy** treatise or text. His notes and observations that accompanied his drawings, however, were guarded throughout his lifetime, and often were written in code-like fashion, with mirror-image backwards text. Upon Leonardo's death at the age of 67, his studio works were willed to one of his students, and were not published for over 200 years. By that time, most of Leonardo's observations were confirmed or refuted by independent sources. It was the scientific method left by Leonardo that endured, communicated through the centuries by the artistry of Leonardo the man.

See also History of anatomy and physiology: The Renaissance and Age of Enlightenment; History of anatomy and physiology: The science of medicine

VIRCHOW, RUDOLF CARL (1821-1902)
German physician and anatomist

Rudolf Virchow was a pioneer in cellular **pathology** and one of the first to describe **leukemia**, a pathological condition involving a diminished number of white **blood** cells. Virchow described the condition as "white blood."

Early in his career, Virchow identified fibrinogen and myelin and set forth a physiological mechanism for thrombosis and pulmonary emboli. Later in his career, Virchow identified neuroglia and a number of **tumors**. Virchow also discovered the **amino acids** leucine and tyrosine. Most importantly, Virchow helped establish the viability of **cell theory** and that all cells arise from pre-existing cells.

Virchow, an only child, was born in a small rural town in Germany. His early interest in the natural sciences and broad humanistic training helped him get high marks throughout school. In 1839, his outstanding scholarly abilities earned him a military fellowship to study medicine at the Freidrich-Wilhelms Institute in Berlin, Germany. Virchow had the opportunity to study under **Johannes Müller**, gaining experience in experimental laboratory and diagnostic methods. In 1843, he received his medical degree from the University of Berlin and went on to become company surgeon at the Charité Hospital in Berlin.

As a young scientist, Virchow became a powerful speaker for the new generation of German physicians. He viewed medical progress as coming from three main sources: clinical observations, including examination of the patient; animal experimentation to test methods and drugs; and pathological **anatomy**, especially at the microscopic level. He also insisted that life was the sum of physical and chemical actions and essentially the expression of cell activity. Although these views caused some older physicians to condemn Virchow, he received his medical license in 1846.

Two years later, Virchow was sent to Prussia to treat victims of a typhus epidemic. Seeing the desperate condition of the Polish minority, he recommended sweeping educational and economic reform and political freedom. From that point on, he argued that to do any good for the sick, one must treat the sick society. Acting on his convictions, Virchow fought in the uprisings of 1848 and became a member of the Berlin Democratic Congress. Unfortunately, his strong political and social con-

science cost him his university post. Virchow finally left Berlin for the more liberal atmosphere of the University of Wurzburg.

It was at Wurburg that Virchow embarked on his highest level of scientific achievement—his development of cellular pathology. In 1855, Virchow published his journal article on cellular pathology. "*Omnis cellula e cellula,*" he wrote, meaning all cells arise from cells. Essentially, his article generalized the concept of cell theory and modernized the entire medical field. The cell became the fundamental living unit in both healthy and diseased **tissue**. He used the microscope to bring the study of disease down to a more fundamental level; disease occurred because healthy living cells were altered or disturbed. In 1859, Virchow's book *Cell Pathology* became a classic textbook that would influence generations of physicians. Although Virchow's work carries lasting significance, Virchow rejected the germ theory developed by Louis Pasteur, arguing instead that diseased tissue resulted from the breakdown of order within cells and not from the invasion of a foreign body. Scientists have since discovered that disease results from both circumstances.

See also Bacteria and responses to bacterial infection; Hemopoiesis; Immune system; Infection and resistance; Leukocytes

VIRUSES AND RESPONSES TO VIRAL INFECTION

There are a number of different viruses that challenge the human **immune system** and that may produce disease in humans. In common, a virus is a small, infectious agent that consists of a core of genetic material (either **deoxyribonucleic acid** [DNA] or **ribonucleic acid** [RNA]) surrounded by a shell of protein. Although precise mechanisms vary, viruses cause disease by infecting a host cell and commandeering the host cell's synthetic capabilities to produce more viruses. The newly made viruses then leave the host cell, sometimes killing it in the process, and proceed to infect other cells within the host. Because viruses invade cells, drug therapies have not yet been designed to kill viruses, although some have been developed to inhibit their growth. The human immune system is the main defense against a viral disease.

Bacterial viruses, called bacteriophages, infect a variety of **bacteria**, such as *Escherichia coli,* a bacteria commonly found in the human digestive tract. Animal viruses cause a variety of fatal diseases. Acquired immune deficiency syndrome (**AIDS**) is caused by the human immunodeficiency virus (HIV); hepatitis and rabies are viral diseases; and **hemorrhagic fevers**, which are characterized by severe internal bleeding, are caused by filoviruses. Other animal viruses cause some of the most common human diseases. Often, these diseases strike in childhood. Measles, mumps, and chickenpox are viral diseases. The common cold and influenza are also caused by viruses. Finally, some viruses can cause **cancer** and **tumors**. One such virus, human T cell **leukemia** virus (HTLV), was only recently discovered and its role in the development of a special kind of leukemia is still being elucidated.

Although viral structure varies considerably between the different types of viruses, all viruses share some common characteristics. All viruses contain either RNA or DNA surrounded by a protective protein shell called a capsid. Some viruses have a double strand of DNA, others a single strand of DNA. Other viruses have a double strand of RNA or a single strand of RNA. The size of the genetic material of viruses is often quite small. Compared to the 100,000 genes that exist within human DNA, viral genes number from 10 to about 200 genes.

Viruses contain such small amounts of genetic material because the only activity that they perform independently of a host cell is the synthesis of the protein capsid. In order to reproduce, a virus must infect a host cell and take over the host cell's synthetic machinery. This aspect of viruses—that the virus does not appear to be "alive" until it infects a host cell—has led to controversy in describing the nature of viruses. Are they living or non-living? When viruses are not inside a host cell, they do not appear to carry out many of the functions ascribed to living things, such as reproduction, **metabolism**, and movement. When they infect a host cell, they acquire these capabilities. Thus, viruses are both living and non-living. It was once acceptable to describe viruses as agents that exist on the boundary between living and non-living; however, a more accurate description of viruses is that they are either active or inactive, a description that leaves the question of life behind altogether.

All viruses consist of genetic material surrounded by a capsid, but within the broad range of virus types, variations exist within this basic structure. Studding the envelope of these viruses are protein "spikes." These spikes are clearly visible on some viruses, such as the influenza viruses; on other enveloped viruses, the spikes are extremely difficult to see. The spikes help the virus invade host cells. The influenza virus, for instance, has two types of spikes. One type, composed of hemagglutinin protein (HA), fuses with the host cell membrane, allowing the virus particle to enter the cell. The other type of spike, composed of the protein neuraminidase (NA), helps the newly formed virus particles to bud out from the host cell membrane.

The capsid of viruses is relatively simple in structure, owing to the few genes that the virus contains to encode the capsid. Most viral capsids consist of a few repeating protein subunits. The capsid serves two functions: it protects the viral genetic material and it helps the virus introduce itself into the host cell. Many viruses are extremely specific, targeting only certain cells within the plant or animal body. HIV, for instance, targets a specific immune cell, the T helper cell. The cold virus targets respiratory cells, leaving the other cells in the body alone. How does a virus "know" which cells to target? The viral capsid has special receptors that match receptors on their targeted host cells. When the virus encounters the correct receptors on a host cell, it "docks" with this host cell and begins the process of **infection** and replication.

Most viruses are rod- or roughly sphere-shaped. Rod-shaped viruses include tobacco mosaic virus and the filoviruses. Although they look like rods under a microscope, these viral capsids are actually composed of protein molecules arranged in a helix. Other viruses are shaped somewhat like spheres, although many viruses are not actual spheres. The capsid of the adenovirus, which infects the respiratory tract of animals, consists of 20 triangular faces. This shape is called an icosahedron. HIV is a true sphere, as is the influenza virus.

Some viruses are neither rod- or sphere-shaped. The poxviruses are rectangular, looking somewhat like bricks. Parapoxviruses are ovoid. Bacteriophages are the most unusually shaped of all viruses. A bacteriophage consists of a head region attached to a sheath. Protruding from the sheath are tail fibers that dock with the host bacterium. The bacteriophage's structure is eminently suited to the way it infects cells. Instead of the entire virus entering the bacterium, the bacteriophage injects its genetic material into the cell, leaving an empty capsid on the surface of the bacterium.

Viruses are obligate intracellular parasites, meaning that in order to replicate, they need to be inside a host cell. Viruses lack the machinery and **enzymes** necessary to reproduce; the only synthetic activity they perform on their own is to synthesize their capsids.

The infection cycle of most viruses follows a basic pattern. Bacteriophages are unusual in that they can infect a bacterium in two ways (although other viruses may replicate in these two ways as well). In the lytic cycle of replication, the bacteriophage destroys the bacterium it infects. In the lysogenic cycle, however, the bacteriophage coexists with its bacterial host, and remains inside the bacterium throughout its life, reproducing only when the bacterium itself reproduces.

An example of a bacteriophage that undergoes lytic replication inside a bacterial host is the T4 bacteriophage which infects E. coli. T4 begins the infection cycle by docking with an E. coli bacterium. The tail fibers of the bacteriophage make contact with the cell wall of the bacterium, and the bacteriophage then injects its genetic material into the bacterium. Inside the bacterium, the viral genes are transcribed. One of the first products produced from the viral genes is an enzyme that destroys the bacterium's own genetic material. Now the virus can proceed in its replication unhampered by the bacterial genes. Parts of new bacteriophages are produced and assembled. The bacterium then bursts, and the new bacteriophages are freed to infect other bacteria. This entire process takes only 20-30 minutes.

In the lysogenic cycle, the bacteriophage reproduces its genetic material but does not destroy the host's genetic material. The bacteriophage called lambda, another E. coli-infecting virus, is an example of a bacteriophage that undergoes lysogenic replication within a bacterial host. After the viral DNA has been injected into the bacterial host, it assumes a circular shape. At this point, the replication cycle can become either lytic or lysogenic. In a lysogenic cycle, the circular DNA attaches to the host cell genome at a specific place. This combination host-viral genome is called a prophage. Most of the viral genes within the prophage are repressed by a special repressor protein, so they do not encode the production of new bacteriophages. However, each time the bacterium divides, the viral genes are replicated along with the host genes. The bacterial progeny are thus lysogenically infected with viral genes.

Interestingly, bacteria that contain prophages can be destroyed when the viral DNA is suddenly triggered to undergo lytic replication. Radiation and chemicals are often the triggers that initiate lytic replication. Another interesting aspect of prophages is the role they play in human diseases. The bacteria that cause diphtheria and botulism both harbor viruses. The viral genes encode powerful toxins that have devastating effects on the human body. Without the infecting viruses, these bacteria may well be innocuous. It is the presence of viruses that makes these bacterial diseases so lethal.

Scientists have classified viruses according to the type of genetic material they contain. Broad categories of viruses include double-stranded DNA viruses, single-stranded DNA viruses, double-stranded RNA viruses, and single stranded RNA viruses. For the description of virus types that follows, however, these categories are not used. Rather, viruses are described by the type of disease they cause.

Poxviruses are the most complex kind of viruses known. They have large amounts of genetic material and fibrils anchored to the outside of the viral capsid that assist in attachment to the host cell. Poxviruses contain a double strand of DNA.

Viruses cause a variety of human diseases, including smallpox and cowpox. Because of worldwide vaccination efforts, smallpox has virtually disappeared from the world, with the last known case appearing in Somalia in 1977. The only places on Earth where smallpox virus currently exists are two labs: the Centers for Disease Control in Atlanta, Georgia, and the Research Institute for Viral Preparation in Moscow. Prior to the eradication efforts begun by the World Health Organization in 1966, smallpox was one of the most devastating of human diseases. In 1707, for instance, an outbreak of smallpox killed 18,000 of Iceland's 50,000 residents. In Boston in 1721, smallpox struck 5,889 of the city's 12,000 inhabitants, killing 15% of those infected.

Edward Jenner (1749–1823) is credited with developing the first successful vaccine against a viral disease, and that disease was smallpox. A vaccine works by eliciting an immune response. During this immune response, specific immune cells, called memory cells, are produced that remain in the body long after the foreign microbe present in a vaccine has been destroyed. When the body again encounters the same kind of microbe, the memory cells quickly destroy the microbe. Vaccines contain either a live, altered version of a virus or bacteria, or they contain only parts of a virus or bacteria, enough to elicit an immune response.

In 1797, Jenner developed his smallpox vaccine by taking pus from a cowpox lesion on the hand of a milkmaid. Cowpox was a common disease of the era, transmitted through contact with an infected cow. Unlike smallpox, however, cowpox is a much milder disease. Using the cowpox pus, he inoculated an 8-year-old boy. Jenner continued his vaccination efforts through his lifetime. Until 1976, children were vaccinated with the smallpox vaccine, called vaccinia. Reactions to the introduction of the vaccine ranged from a mild fever to severe complications, including (although very rarely) **death**.

In 1976, with the eradication of smallpox complete, vaccinia vaccinations for children were discontinued, although vaccinia continues to be used as a carrier for recombinant DNA techniques. In these techniques, foreign DNA is inserted in cells. Efforts to produce a vaccine for HIV, for instance, have used vaccinia as the vehicle that carries specific parts of HIV.

Herpesviruses are enveloped, double-stranded DNA viruses. Of the more than 50 herpes viruses that exist, only eight cause disease in humans. These include the human herpes virus types 1 and 2 that cause cold sores and genital herpes; human herpes virus 3, or varicella-zoster virus (VZV), that causes chicken pox and shingles; cytomegalovirus (CMV), a virus that in some individuals attacks the cells of the **eye** and leads to **blindness**; human herpes virus 4, or Epstein-Barr virus (EBV), which has been implicated in a cancer called Burkitt's lymphoma; and human herpes virus types 6 and 7, newly discovered viruses that infect white **blood** cells. In addition, herpes B virus is a virus that infects monkeys and can be transmitted to humans by handling infected monkeys.

Adenoviruses are viruses that attack respiratory, intestinal, and eye cells in animals. More than 40 kinds of human adenoviruses have been identified. Adenoviruses contain double-stranded DNA within a 20-faceted capsid. Adenoviruses that target respiratory cells cause bronchitis, pneumonia, and tonsillitis. Gastrointestinal illnesses caused by adenoviruses are usually characterized by **diarrhea** and are often accompanied by respiratory symptoms. Some forms of appendicitis are also caused by adenoviruses. Eye illnesses caused by adenoviruses include conjunctivitis, an infection of the eye tissues, as well as a disease called pharyngoconjunctival fever, a disease in which the virus is transmitted in poorly chlorinated swimming pools.

Human papoviruses include two groups: the papilloma viruses and the polyomaviruses. Human papilloma viruses (HPV) are the smallest double-stranded DNA viruses. They replicate within cells through both the lytic and the lysogenic replication cycles. Because of their lysogenic capabilities, HPV-containing cells can be produced through the replication of those cells that HPV initially infects. In this way, HPV infects epithelial cells, such as the cells of the skin. HPVs cause several kinds of benign (non-cancerous) warts, including plantar warts (those that form on the soles of the feet) and genital warts. However, HPVs have also been implicated in a form of cervical cancer that accounts for 7% of all female cancers.

HPV is believed to contain oncogenes, or genes that encode for growth factors that initiate the uncontrolled growth of cells. This uncontrolled proliferation of cells is called cancer. When the HPV oncogenes within an epithelial cell are activated, they cause the epithelial cell to proliferate. In the cervix (the opening of the uterus), the cell proliferation manifests first as a condition called cervical neoplasia. In this condition, the cervical cells proliferate and begin to crowd together. Eventually, cervical neoplasia can lead to full-blown cancer.

Polyomaviruses are somewhat mysterious viruses. Studies of blood have revealed that 80% of children aged 5-9 years have antibodies to these viruses, indicating that they have at some point been exposed to polyomaviruses.

However, it is not clear what disease this virus causes. Some evidence exists that a mild respiratory illness is present when the first antibodies to the virus are evident. The only disease that is certainly caused by polyomavirses is called progressive multifocal leukoencephalopathy (PML), a disease in which the virus infects specific **brain** cells called the oligodendrocytes. PML is a debilitating disease that is usually fatal, and is marked by progressive neurological degeneration. It usually occurs in people with suppressed immune systems, such as cancer patients and people with AIDS.

The hepadnaviruses cause several diseases, including hepatitis B. Hepatitis B is a chronic, debilitating disease of the **liver** and immune system. The disease is much more serious than hepatitis A for several reasons: it is chronic and long-lasting; it can cause cirrhosis and cancer of the liver; and many people who contract the disease become carriers of the virus, able to transmit the virus through body fluids such as blood, **semen**, and vaginal secretions.

The hepatitis B virus (HBV) infects liver cells and has one of the smallest viral genomes. A double-stranded DNA virus, HBV is able to integrate its genome into the host cell's genome. When this integration occurs, the viral genome is replicated each time the cell divides. Individuals who have integrated HBV into their cells become carriers of the disease. Recently, a vaccine against HBV was developed. The vaccine is especially recommended for health care workers who through exposure to patient's body fluids are at high risk for infection.

Parvoviruses are icosahedral, single-stranded DNA viruses that infect a wide variety of mammals. Each type of parvovirus has its own host. For instance, one type of parvovirus causes disease in humans; another type causes disease in cats; while still another type causes disease in dogs. The disease caused by parvovirus in humans is called erythremia infectiosum, a disease of the red blood cells that is relatively rare except for individuals who have the inherited disorder **sickle cell anemia**. Canine and feline parvovirus infections are fatal, but a vaccine against parvovirus is available for dogs and cats.

Orthomyxoviruses cause influenza ("flu"). This highly contagious viral infection can quickly assume epidemic proportions, given the right environmental conditions. An influenza outbreak is considered an epidemic when more than 10% of the population is infected. Antibodies that are made against one type of rhinovirus are often ineffective against other types of viruses. For this reason, most people are susceptible to colds from season to season.

These helical, enveloped, single-stranded RNA viruses cause pneumonia, croup, measles, and mumps in children. A vaccine against measles and mumps has greatly reduced the incidence of these diseases in the United States. In addition, a paramyxovirus called respiratory syncytial virus (RSV) causes bronchiolitis (an infection of the bronchioles) and pneumonia.

Flaviviruses (from the Latin word meaning "yellow") cause insect-carried diseases including yellow fever, an often-fatal disease characterized by high fever and internal bleeding. Flaviviruses are single-stranded RNA viruses.

The two filoviruses, Ebola virus and Marburg virus, are among the most lethal of all human viruses. Both cause severe fevers accompanied by internal bleeding, which eventually kills the victim. The fatality rate of Marburg is about 60%, while the fatality rate of Ebolavirus approaches 90%. Both are transmitted through contact with body fluids. Marburg and Ebola also infect primates.

Rhabdoviruses are bullet-shaped, single-stranded RNA viruses. They are responsible for rabies, a fatal disease that affects dogs, rodents, and humans.

Retroviruses are unique viruses. They are double-stranded RNA viruses that contain an enzyme called reverse transcriptase. Within the host cell, the virus uses reverse transcriptase to make a DNA copy from its RNA genome. In all other organisms, RNA is synthesized from DNA. Cells infected with retroviruses are the only living things that reverse this process.

The first retroviruses discovered were viruses that infect chickens. The Rous sarcoma virus, discovered in the 1950s by **Peyton Rous**, was also the first virus that was linked to cancer. However, it was not until 1980 that the first human retrovirus was discovered. Called human T cell leukemia virus (HTLV), this virus causes a form of leukemia called adult T cell leukemia. In 1983–4, another human retrovirus, human immunodeficiency virus, the virus responsible for AIDS, was discovered independently by two researchers. Both HIV and HTLV are transmitted in body fluids.

See also Bacteria and bacterial infection; Functional characteristics of living things.; History of anatomy and physiology: The science of medicine; Interferon actions; Protein synthesis

VISCERAL • *see* SOMATIC AND VISCERAL

VISION: HISTOPHYSIOLOGY OF THE EYE

Histophysiology refers to the collective functioning of cells and tissues. In the case of the **eye**, these cells and tissues function in vision.

The human eye is sensitive to light having wavelengths between 397 and 723 nanometers (a nanometer is 10^{-9} meters). The light entering the eye is focused by a lens to produce an image. The image is projected onto the back surface of the eye, a region called the **retina**. Here, specialized cells known as rods convert the light signal into an electrical signal. Chemically, this is accomplished by the bleaching out by the light of a rod protein called rhodopsin. Rods are important in lower-light vision, where color and detail are not as prominent. Color vision and the addition of visual detail is the concern of other specialized cells in the retina, which are known as cones. A trio of photopigments in cones cells called iodopsin are responsible for the selective detection of light whose wavelength corresponds to blue, red and green.

The light to electrical energy conversion in the rod and cone cells is transmitted to a bundle of nerves in one region of the retina (the optic nerve). The signals representing the visual image are conveyed by the optic nerve to the **brain** for interpretation.

The eye is designed to function in varying intensities of light. A lens operates very much like a camera lens, to let in a greater amount of light in dark conditions or less light in brighter conditions. Unlike a camera lens, the human eye is capable of adaptation to a change from relatively bright to less intense light, and visa versa.

Another difference between the human eye and a camera concerns the lens. While the focal length of a camera lens can be changed, the shape of the lens is permanent. However, the lens of the eye can be changed in shape, through the ciliary muscles that hold the lens in place. Contraction of the muscles shortens the lens and causes it to bulge, giving it a more convex shape. This permits the focusing of objects that are closer to the eye. When focusing on an object that is further away, the ciliary muscles relax. This relaxation allows the lens to adopt a thinner, more concave shape, which increases the focal length.

The information passing through the left eye is focused on the rightward region of the retina while the right eye focuses information on the leftward side of the retina. To provide a complete picture, the cells and tissues of the eye are designed so that the visual field of one eye overlaps the visual field of the other eye. Furthermore, as visual information passes to the brain, the information from the left retina flows along both of the two optical nerves leading to the brain, while information from the right retina flows along only one of the optic nerves. This different routing provides one of the physiological bases of depth perception.

Physiological abnormalities in the eye can cause visual impairment. Abnormal focusing, due to the shape of the lens, which is termed astigmatism, cane corrected by external lenses. Another condition called presbyopia decreases the ability of the eye to adapt, because the lens becomes less elastic. The loss of water in the **cornea** or the lens can produce opaque regions termed cataracts.

See also Blindness and visual impairments; Eye and ocular fluids; Retina and retinal imaging

VITAMINS

Vitamins are organic molecules that are needed in small amounts in the diet. They are frequently molecules that bind in the active site of an enzyme and thereby alter its structure in a way that permits it to react more readily. Vitamins serve nearly the same role in all forms of life and many are essential in the **metabolism** of all living organisms. They are synthesized by plants and micro-organisms and the absolute requirement for vitamins in the diet of higher animals is the result of the loss of this biosynthetic capability during **evolution**. The biosynthetic abilities and thus the dietary requirement of different species vary. For example, ascorbic acid (vitamin C) is a vitamin only for primates and a few other animals, such as the guinea pig, but most other animals can synthesize it, so for them it is not a vitamin. Certain

vitamins can be synthesized from provitamins obtained from the diet. Some of the vitamin requirements of humans and higher animals are supplied by the intestinal flora, for example most of the vitamin K required by humans is provided in this way.

Several diseases resulting from vitamin deficiencies were prevalent until the last century and sailors on long sea voyages, where fresh vegetables were not readily available, were often victims. In the Orient, the disease beriberi was rampant and millions died of its associated polyneuritis. The condition could be relieved by feeding the patients rice polishings. The founder of the vitamin concept was Lumin (1853–1937). During subsequent decades, the importance of accessory food factors for normal **growth and development** was gradually recognized. The Polish biochemist, Casimir Funk formulated the vitamin theory in 1912 and proposed that several common diseases such as beriberi, pellagra, rickets and scurvy resulted from lack in the diet of essential nutrients. It was Funk who suggested the name "vitamin" for these accessory factors, from the Latin *vita* + amine, the "amine" reflecting the fact that the first of these factors to be studied, vitamin B_1, contained nitrogen.

The metabolic role of vitamins is largely catalytic. Most vitamins serve as coenzymes and prosthetic groups of **enzymes**. For most of these, the nature of the biocatalytic function has been elucidated. Vitamin D, however, acts as a regulator of bone metabolism and is thus has an activity similar to **hormones**. As a component of the visual pigments, vitamin A acts as a prosthetic group, however, it is not known whether it is associated with catalytic proteins in its other functions. Nicotinamide and riboflavin are constituents of the hydrogen-transferring enzymes, such as those in the respiratory electron transport chain. Biotin, folic acid, pantothenic acid, pyridoxine, cobalamin and thiamine are coenzymes, or precursors of coenzymes, of group transfer reactions. The low daily requirements for vitamins reflect their catalytic and/or regulatory roles. Thus vitamins are nutritionally quite different from **fat**, carbohydrate or protein, which are required in the diet in considerable quantities as substrates of **tissue** synthesis and energy metabolism.

Vitamins can be grouped according to whether they are soluble in water or polar solvents. The water-soluble vitamins are ascorbic acid, the vitamin B series (thiamain, B_1, riboflavin, B_2, pyridoxine, B_6, cobalamin, B_{12},), folic acid, niacin and pantothenic acid. Ascorbate, the ionised form of ascorbic acid, is essential in the prevention of scurvy and acts as a reducing agent (an antioxidant). It serves, for example, in the hydroxylation of proline residues in collagen. The vitamin B series are components of coenzymes. For example, riboflavin (vitamin B_2) is a precurser of FAD, and pantothenate is a component of coenzmye A. Vitamin B_1 (thiamine) was found to cure beriberi.

Much is known about the molecular actions of the fat-soluble vitamins, which are designated by the letters A, D, E and K. Vitamin K, which is required for normal **blood** clotting, participates in the carboxylation of γ-carboxyglutamate, which makes it a much stronger chelator of Ca^{2+}. Vitamin A (retinol) is the precurser of retinal, the light sensitive group in

ESSENTIAL VITAMINS

Vitamin	What It Does For The Body
Vitamin A (Beta Carotene)	Promotes growth and repair of body tissues; reduces susceptibility to infections; aids in bone and teeth formation; maintains smooth skin
Vitamin B-1 (Thiamin)	Promotes growth and muscle tone; aids in the proper functioning of the muscles, heart, and nervous system; assists in digestion of carbohydrates
Vitamin B-2 (Riboflavin)	Maintains good vision and healthy skin, hair, and nails; assists in formation of antibodies and red blood cells; aids in carbohydrate, fat, and protein metabolism
Vitamin B-3 (Niacinamide)	Reduces cholesterol levels in the blood; maintains healthy skin, tongue, and digestive system; improves blood circulation; increases energy
Vitamin B-5	Fortifies white blood cells; helps the body's resistance to stress; builds cells
Vitamin B-6 (Pyridoxine)	Aids in the synthesis and breakdown of amino acids and the metabolism of fats and carbohydrates; supports the central nervous system; maintains healthy skin
Vitamin B-12 (Cobalamin)	Promotes growth in children; prevents anemia by regenerating red blood cells; aids in the metabolism of carbohydrates, fats, and proteins; maintains healthy nervous system
Biotin	Aids in the metabolism of proteins and fats; promotes healthy skin
Choline	Helps the liver eliminate toxins
Folic Acid (Folate, Folacin)	Promotes the growth and reproduction of body cells; aids in the formation of red blood cells and bone marrow
Vitamin C (Ascorbic Acid)	One of the major antioxidants; essential for healthy teeth, gums, and bones; helps to heal wounds, fractures, and scar tissue; builds resistance to infections; assists in the prevention and treatment of the common cold; prevents scurvy
Vitamin D	Improves the absorption of calcium and phosphorous (essential in the formation of healthy bones and teeth) maintains nervous system
Vitamin E	A major antioxidant; supplies oxygen to blood; provides nourishment to cells; prevents blood clots; slows cellular aging
Vitamin K (Menadione)	Prevents internal bleeding; reduces heavy menstrual flow

Table representing the essential vitamins and their role in physiological processes. *Illustration by Stanley Publishing.*

rhodopsin and other visual pigments. A deficiency of this vitamin leads to night **blindness**. Furthermore it is required for growth by young animals. Retinoic acid, which contains a terminal carboxylate in place of the alcohol terminus of retinal, activates the transcription of specific genes that mediate growth and development. The metabolism of **calcium** and phosphorus is regulated by a hormone derived from vitamin D. A deficiency of vitamin D impairs bone formation in growing animals and causes the disease rickets. Infertility in rats is a consequence of vitamin E (α-tocopherol) deficiency and this vitamin also protects unsaturated membrane **lipids** from oxidation.

Most vitamins were purified between 1920 and 1950. The last one was vitamin B_{12}, in 1948, whose chemical structure was elucidated by A. R. Todd in 1955. Chemical syntheses are known for all vitamins.

See also Biochemistry; Nutrition and nutrient transport to cells

Vᴏᴄᴀʟ ᴄᴏʀᴅs • *see* Lᴀʀʏɴx ᴀɴᴅ ᴠᴏᴄᴀʟ ᴄᴏʀᴅs

Vᴏʟᴛᴀ, Aʟᴇssᴀɴᴅʀᴏ Gɪᴜsᴇᴘᴘᴇ (1745-1827)

Italian physicist

Alessandro Giuseppe Volta became one of the most widely recognized scientists of his time day for his work with electrical currents. Volta extended his physics work to include the electrical interactions of living organisms, and the electrical phenomena associated with various physiological processes.

Volta was born on February 18, 1745, in Como, Lombardy, Italy. Most of his eight brothers and sisters entered the church, but Alessandro became engrossed in the study of with electricity. Influenced by a history on the subject written by **Joseph Priestley**, fourteen-year-old Volta announced his intention of becoming a physicist, and became educated on the subject.

In 1774, Volta was appointed professor of physics at the high school in Como. There he created one of his most significant inventions; the electrophorous. This device had the ability to store significant electrical charges, and replaced the Leyden jar which, up until that time, had been used for storing smaller charges.

The discovery that led Volta to the invention of the electrophorous actually began with the work of French physicist Charles-Augustine Coulomb. Coulomb had discovered that electrical charges were located on the surface of a charged body, and not in its interior. Volta's electrophorous used two metal discs; one was rubbed to produce a negative electrical charge, the second disk was brought close enough to the first to establish a positive charge on the one side, leaving a negative charge on the other. Volta used Coulomb's discovery to draw off the negative charge from one side of his charged disc, leaving just a positive charge on the opposite side. The electrophorous was the predecessor of the modern condenser, which stores electricity in circuits.

In 1776, Volta became involved with an entirely different subject. By studying the components of marsh gas, he was able to discover methane gas. He also exploded hydrogen gas to remove oxygen from air and was able to make the first accurate determination of the proportion of oxygen in the air. Later, around 1796, Volta discovered that the vapor pressure of a given liquid had nothing to do with the pressure of the surrounding atmosphere; it was solely dependent on temperature.

Meanwhile, the electrophorous had established Volta's reputation and, in 1779, he was appointed a professor at the University of Pavia. There he developed an electrometer that allowed him to measure electric currents. Then, in 1791, he was drawn into a controversy that erupted when his compatriot **Luigi Galvani** announced the existence of "animal electricity" that caused muscles in frog's legs to twitch when touched with metal probes of different composition. Volta did not believe such a thing possible and proceeded to experiment himself.

Volta experimented on numerous animals and ultimately discovered there was no such thing as "animal electricity," and that it was the action of the two different types of metal probes that was the source of the current. This fact was made painfully certain when Volta placed the two metal probes on his tongue and, along with an experiencing an electrical shock, discovered his own tongue was a more sensitive detector of electricity than his electrometer.

The controversy raged on until 1800. In that year, Volta built a device that produced a large flow of electricity. He filled bowls with a saline solution and "connected" them with strips of different metals. One end of the strip was copper; the other end was tin or zinc. By bending his strips from one bowl into another, Volta was able to create a constant flow of electrical current; the world's first electric battery had been invented. Volta had proven that the metal was the source of the electricity, and animal electricity did not exist.

In the interest of making his battery smaller, Volta used round discs of copper, zinc, and cardboard that had been soaked in a saline solution. He stacked his discs one on top of the other. Attaching a wire to the top and bottom of his pile allowed the electric current to flow. The invention of this Voltaic pile marked the apex of Volta's career.

The Voltaic pile came to the attention of English chemist William Nicholson (1753-1815), who proceeded to build his own in the same year. Nicholson placed the ends of his wires in water and discovered the flowing current "electrolyzed" the water, breaking it up into hydrogen and oxygen. Henry Cavendish had shown those two elements could form water; Nicholson reversed the procedure.

The invention of the Voltaic pile, the earliest form of an electric battery, was the high point of Volta's life. Volta died on March 5, 1827, at the age of 82. In his honor, the unit of force that moves electric current was named the volt.

See also Electrocardiogram (ECG); Electroencephalograph (EEG); Electrolytes and electrolyte balance; Heart: rhythm control and impulse conduction; History of anatomy and physiology: The Renaissance and Age of Enlightenment; Nerve impulses and conduction of impulses

W

WAKSMAN, SELMAN ABRAHAM (1888-1973)

Russian-born American microbiologist

Selman Waksman discovered of life-saving antibacterial compounds and his investigations spawned further studies for other disease-curing drugs. Waksman isolated streptomycin, the first chemical agent effective against tuberculosis. Prior to Waksman's discovery, tuberculosis was often a lifelong debilitating disease, and was fatal in some forms. Streptomycin effected a powerful and wide-ranging cure, and for this discovery, Waksman received the 1952 Nobel Prize in physiology or medicine. In pioneering the field of antibiotic research, Waksman had an inestimable impact on human health and well-being.

The only son of a Jewish furniture textile weaver, Selman Abraham Waksman was born in the tiny Russian village of Novaya Priluka on July 22, 1888. Life was hard in late-nineteenth-century Russia. Waksman's only sister died from diphtheria when he was nine. There were particular tribulations for members of a persecuted ethnic minority. As a teen during the Russian revolution, Waksman helped organize an armed Jewish youth defense group to counteract oppression. He also set up a school for underprivileged children and formed a group to care for the sick. These activities prefaced his later role as a standard-bearer for social responsibility.

Several factors led to Waksman's immigration to the United States. He had received his diploma from the *Gymnasium* in Odessa and was poised to attend university, but he doubtless recognized the very limited options he held as a Jew in Russia. At the same time, in 1910, his mother died, and cousins who had immigrated to New Jersey urged him to follow their lead. Waksman did so, and his move to a farm there, where he learned the basics of scientific farming from his cousin, likely had a pivotal influence on Waksman's later choice of field of study.

In 1911 Waksman enrolled in nearby Rutgers College (later University) of Agriculture, following the advice of fellow Russian immigrant Jacob Lipman, who led the college's bacteriology department. He worked with Lipman, developing a fascination with the **bacteria** of soil, and graduated with a B.Sc. in 1915. The next year he earned his M.S. degree. Around this time, he also became a naturalized United States citizen and changed the spelling of his first name from Zolman to Selman. Waksman married Bertha Deborah Mitnik, a childhood sweetheart and the sister of one of his childhood friends, in 1916. Deborah Mitnik had come to the United States in 1913, and in 1919 she bore their only child, Byron Halsted Waksman, who eventually went on to a distinguished career at Yale University as a **pathology** professor.

Waksman's intellect and industry enabled him to earn his Ph.D. in less than two years at the University of California, Berkeley. His 1918 dissertation focused on proteolytic **enzymes** (special proteins that break down proteins) in fungi. Throughout his schooling, Waksman supported himself through various scholarships and jobs. Among the latter were ranch work, caretaker and night watchman, and tutor of English and science.

Waksman's former advisor invited him to join Rutgers as a lecturer in soil bacteriology in 1918. He was to stay at Rutgers for his entire professional career. When Waksman took up the post, however, he found his pay too low to support his family. Thus, in his early years at Rutgers he also worked at the nearby Takamine Laboratory, where he produced enzymes and ran toxicity tests.

In the 1920s Waksman's work gained recognition in scientific circles. Others sought out his keen mind, and his prolific output earned him a well-deserved reputation. He wrote two major books during this decade. *Enzymes: Properties, Distribution, Methods, and Applications,* coauthored with Wilburt C. Davison, was published in 1926, and in 1927 his thousand-page *Principles of Soil Microbiology* appeared. This latter volume became a classic among soil bacteriologists. His laboratory produced more than just books. One of Waksman's students during this period was René Dubos, who would later discover the antibiotic gramicidin, the first chemotherapeutic agent effective against gram-positive bacteria (bacteria that hold dye in a stain test named for Danish bacteriologist Hans Gram). Waksman became an associate professor at Rutgers in the mid–1920s and advanced to the rank of full professor in 1930.

During the 1930s Waksman systematically investigated the complex web of microbial life in soil, humus, and peat. He was recognized as a leader in the field of soil microbiology, and his work stimulated an ever-growing group of graduate students and postdoctoral assistants. He continued to publish widely, and he established many professional relationships with industrial firms that utilized products of microbes. These companies that produced enzymes, pharmaceuticals, **vitamins**, and other products were later to prove valuable in Waksman's researches, mass-producing and distributing the products he developed. Among his other accomplishments during this period was the founding of the division of Marine Bacteriology at Woods Hole Oceanographic Institution in 1931. For the next decade he spent summers there and eventually became a trustee, a post he filled until his death.

In 1939, Waksman was appointed chair of the U.S. War Committee on Bacteriology. He derived practical applications from his earlier studies on soil microorganisms, developing antifungal agents to protect soldiers and their equipment. He also worked with the Navy on the problem of bacteria that attacked ship hulls. Early that same year Dubos announced his finding of two antibacterial substances, tyrocidine, and gramicidin, derived from a soil bacterium (*Bacillus brevis*). The latter compound, effective against gram-positive bacteria, proved too toxic for human use but did find widespread employment against various bacterial infections in veterinary medicine. The discovery of gramicidin also evidently inspired Waksman to dedicate himself to focus on the medicinal uses of antibacterial soil microbes. It was in this period that he began rigorously investigating the antibiotic properties of a wide range of soil fungi.

Waksman set up a team of about fifty graduate students and assistants to undertake a systematic study of thousands of different soil fungi and other microorganisms. The rediscovery at this time of the power of penicillin against gram-positive bacteria likely provided further incentive to Waksman to find an antibiotic effective against gram-negative bacteria, which include the kind that causes tuberculosis.

In 1940, Waksman became head of Rutgers' department of microbiology. In that year too, with the help of Boyd Woodruff, he isolated the antibiotic actinomycin. Named for the actinomycetes (rod- or filament-shaped bacteria) from which it was isolated, this compound also proved too toxic for human use, but its discovery led to the subsequent finding of variant forms (actinomycin A, B, C, and D), several of which were found to have potent anti-cancer effects. Over the next decade Waksman isolated ten distinct antibiotics. It is Waksman who first applied the term antibiotic, which literally means against life, to such drugs.

Among these discoveries, Waksman's finding of streptomycin had the largest and most immediate impact. Not only

did streptomycin appear nontoxic to humans, however, it was highly effective against gram-negative bacteria. (Prior to this time, the antibiotics available for human use had been active only against the gram-positive strains.) The importance of streptomycin was soon realized. Clinical trials showed it to be effective against a wide range of diseases, most notably tuberculosis.

At the time of streptomycin's discovery, tuberculosis was the most resistant and irreversible of all the major infectious diseases. It could only be treated with a regime of rest and nutritious diet. The tuberculosis bacillus consigned its victims to a lifetime of invalidism and, when it invaded **organs** other than the **lungs**, often killed. Sanatoriums around the country were filled with persons suffering the ravages of tuberculosis, and little could be done for them.

Streptomycin changed all of that. From the time of its first clinical trials in 1944, it proved to be remarkably effective against tuberculosis, literally snatching sufferers back from the jaws of death. By 1950, streptomycin was used against seventy different germs that were not treatable with penicillin. Among the diseases treated by streptomycin were bacterial meningitis (an inflammation of membranes enveloping the **brain** and **spinal cord**), endocarditis (an inflammation of the lining of the **heart** and its valves), pulmonary and urinary tract infections, leprosy, typhoid fever, bacillary dysentery, cholera, and bubonic plague.

Waksman arranged to have streptomycin produced by a number of pharmaceutical companies, since demand for it soon skyrocketed beyond the capacity of any single company. Manufacture of the drug became a $50-million-per-year industry. Thanks to Waksman and streptomycin, Rutgers received millions of dollars of income from the royalties. Waksman donated much of his own share to the establishment of an Institute of Microbiology there. He summarized his early researches on the drug in *Streptomycin: Nature and Practical Applications* (1949). Streptomycin ultimately proved to have some human toxicity and was supplanted by other antibiotics, but its discovery changed the course of modern medicine. Not only did it directly save countless lives, but its development stimulated scientists around the globe to search the microbial world for other antibiotics and medicines.

In 1949, Waksman isolated neomycin, which proved effective against bacteria that had become resistant to streptomycin. Neomycin also found a broad niche as a topical antibiotic. Other antibiotics soon came forth from his Institute of Microbiology. These included streptocin, framicidin, erlichin, candidin, and others. Waksman himself discovered eighteen antibiotics during the course of his career.

Waksman served as director of the Institute for Microbiology until his retirement in 1958. Even after that time, he continued to supervise research there. He also lectured widely and continued to write at the frenetic pace established early in his career. He eventually published more than twenty-five books, among them the autobiography *My Life with the Microbes,* and hundreds of articles. He was author of popular pamphlets on the use of thermophilic (heat-loving) microorganisms in composting and on the enzymes involved in jelly-making. He wrote biographies of several noted microbiologists, including his own mentor, Jacob Lipman. These works are in addition to his numerous publications in the research literature.

On August 16, 1973, Waksman died suddenly in Hyannis, Massachusetts, of a cerebral hemorrhage. He was buried near the institute to which he had contributed so much over the years. Waksman's honors over his professional career were many and varied. In addition to the 1952 Nobel Prize, Waksman received the French Legion of Honor, a Lasker award for basic medical science, elected a fellow of the American Association for the Advancement of Science, and received numerous commendations from academies and scholarly societies around the world.

See also Antigens and antibodies; Bacteria and responses to bacterial infection; Enzymes and coenzymes; Infection and resistance

WALD, GEORGE (1906-1997)

American biochemist

George Wald received a Nobel Prize for his discovery of the way in which hidden biochemical processes in the retinal pigments of the **eye** turn light energy into sight. Among Wald's important experiments were the effects of vitamin A on sight and the roles played by rod and cone cells in black and white and color vision. Outside the laboratory, his lectures at Harvard to packed audiences of students generated great intellectual excitement. Wald gained additional fame as a political activist during the turbulent 1960s. Wald's personal belief in the unity of nature and the kinship among all living things was evidenced by the substantial roles he played in the scientific world as well as the political and cultural arena of the 1960s.

Wald's father, Isaac Wald, a tailor and later a foreman in a clothing factory, emigrated from Austrian Poland, while his mother, Ernestine Rosenmann Wald, emigrated from Bavaria. Most of Wald's youth was spent in Brooklyn, New York, where his parents moved after his birth on the lower east side of Manhattan on November 18, 1906. He attended high school at Brooklyn Tech, where he intended to study to become an electrical engineer. College changed his mind, however, as he explained for the *New York Times Magazine* in 1969, "I learned I could talk, and I thought I'd become a lawyer. But the law was man-made; I soon discovered I wanted something more real."

Wald's bachelor of science degree in zoology, which he received from New York University in 1927, was his ticket into the reality of biological research. He began his research career at Columbia University, where he was awarded a master's degree in 1928, working under Selig Hecht, one of the founders of the field of biophysics and an authority on the **physiology** of vision. Hecht exerted an enormous influence on Wald, as both an educator and a humanist. The elder scientist's belief in the social obligation of science, coupled with the conviction that science should be explained so the general public could understand it, made a great impression on the young

Wald. Following Hecht's sudden death in 1947 at the age of 55, Wald wrote a memorial as a tribute to his colleague.

In 1932, Wald earned his doctorate at Columbia, after which he was awarded a National Research Council Fellowship in Biology. The two-year fellowship helped to support his research career, which first took him to the laboratory of **Otto Warburg** in Berlin. It was there, in 1932, that he discovered that vitamin A is one of the major constituents of retinal pigments, the light sensitive chemicals that set off the cascade of biological events that turns light into sight.

Warburg sent the young Wald to Switzerland, where he studied **vitamins** with chemist Paul Karrer at the University of Zurich. From there Wald went to Otto Meyerhof's laboratory of cell **metabolism** at the Kaiser Wilhelm Institute in Heidelberg, Germany, finishing his fellowship in the department of physiology at the University of Chicago in 1934. His fellowship completed, Wald went to Harvard University, first as a tutor in **biochemistry** and subsequently as an instructor, faculty instructor, and associate professor, finally becoming a full professor in 1948. In 1968, he became Higgins Professor of Biology, a post he retained until he became an emeritus professor in 1977.

Wald did most of his work in eye physiology at Harvard, where he discovered in the late 1930s that the light-sensitive chemical in the rods—those cells in the **retina** responsible for night vision—is a single pigment called rhodopsin (visual purple), a substance derived from opsin, a protein, and retinene, a chemically modified form of vitamin A. In the ensuing years, Wald discovered that the vitamin A in rhodopsin is "bent" relative to its natural state, and light causes it to "straighten out," dislodging it from opsin. This simple reaction initiates all the subsequent activity that eventually generates the sense of vision.

Wald's research moved from rods to cones, the retinal cells responsible for color vision, discovering with his co-worker Paul K. Brown, that the pigments sensitive to red and yellow-green are two different forms of vitamin A that co-exist in the same cone, while the blue-sensitive pigments are located in separate cones. They also showed that color **blindness** is caused by the absence of one of these pigments.

For much of his early professional life, Wald concentrated his energy on work, both research and teaching. By the late 1950s, Wald began to be showered with honors, and during his career, he received numerous honorary degrees and awards. After Wald was awarded (with Haldan K. Hartline of the United States and **Ragnar Granit** of Sweden) the Nobel Prize in physiology or medicine in 1967 for his work with vision, John E. Dowling wrote in *Science* that Wald and his team formed "the nucleus of a laboratory that has been extraordinarily fruitful as the world's foremost center of visual-pigment biochemistry."

As Wald's reputation flourished, his fame as an inspiring professor grew as well. He lectured to packed classrooms, inspiring an intense curiosity in his students. The energetic professor was portrayed in a 1966 *Time* article that summarized the enthusiasm he brought to teaching his natural science course: "With crystal clarity and obvious joy at a neat explanation, Wald carries his students from protons in the fall to living organisms in the spring, [and] ends most lectures with some philosophical peroration on the wonder of it all." That same year, the *New York Post* said of his lectures, "His beginnings are slow, sometimes witty.... The talk gathers momentum and suddenly an idea "pings" into the atmosphere—fresh, crisp, thought-provoking."

Six days after he received the Nobel Prize, Wald wielded the status of his new prestige in support of a widely popular resolution before the city council of Cambridge, Massachusetts—placing a referendum on the Vietnam War on the city's ballot of November 7, 1967. Echoing the sentiments of his mentor Hecht, he asserted that scientists should be involved in public issues.

The Cambridge appearance introduced him to the sometimes-stormy arena of public politics, a forum from which he never retired. The escalating war in Vietnam aroused Wald to speak out against America's military policy. In 1965, during the escalation of that war, Wald's impromptu denunciation of the Vietnam war stunned an audience at New York University, where he was receiving an honorary degree. Shortly afterward, he threw his support and prestige behind the presidential campaign of Eugene McCarthy. His offer to speak publicly on behalf of McCarthy was ignored, however, and he became a disillusioned supporter, remaining on the fringe of political activism.

Then on March 4, 1969, he gave an address at the Massachusetts Institute of Technology (MIT) that, "upended his life and pitched him abruptly into the political world," according to the *New York Times Magazine*. Wald gave "The Speech," as the talk came to be known in his family, before an audience of radical students at MIT The students had helped to organize a scientists' day-long "strike" to protest the influence of the military on their work, a topic of much heated debate at the time.

Although much of the MIT audience was already bored and restless by the time Wald began, even many of those students who were about to leave the room stopped to listen as the Nobel laureate began to deliver his oration, entitled, "A Generation in Search of a Future." "I think this whole generation of students is beset with a profound sense of uneasiness, and I don't think they have quite defined its source," Wald asserted as quoted in the *New York Times Magazine*. "I think I understand the reasons for their uneasiness even better than they do. What is more, I share their uneasiness."

Wald's discourse evoked applause from the audience as he offered his opinion that student unease arose from a variety of troublesome matters. He pointed to the Vietnam War, the military establishment, and finally, the threat of nuclear warfare. "We must get rid of those atomic weapons," he declared. "We cannot live with them." Speaking to the students as fellow scientists, he sympathized with the their unease at the influence of the military establishment on the work of scientists, intoning, "Our business is with life, not death...."

The speech was reprinted and distributed around the country by the media. Through these reprints, Wald told readers that some of their elected leaders were "insane," and he referred to the American "war crimes" enacted in Vietnam. In the furor that followed, Wald was castigated by critics, many

of whom were fellow academics, and celebrated by sympathizers. A letter writer from Piney Flats, Tennessee was quoted in the *New York Times Magazine* as saying, "So good to know there are still some intellects around who can talk downright horse sense." Wald summed up his role as scientist-political activist in that same article by saying, "I'm a scientist, and my concerns are eternal. But even eternal things are acted out in the present." He described his role as gadfly as putting certain controversial positions into words in order to make it, "easier for others to inch toward it."

His role as a Vietnam war gadfly expanded into activism in other arenas of foreign affairs. He served for a time as president of international tribunals on El Salvador, the Philippines, Afghanistan, Zaire, and Guatemala. In 1984, he joined four other Nobel Prize laureates who went with the "peace ship" sent by the Norwegian government to Nicaragua during that country's turmoil.

In addition to his interests in science and politics, Wald's passions included collecting Rembrandt etchings and primitive art, especially pre-Columbian pottery. This complex mixture of science, art, and political philosophy was reflected in his musings about religion and nature in the *New York Times Magazine*. "There's nothing supernatural in my mind. Nature is my religion, and it's enough for me. I stack it up against any man's. For its awesomeness, and for the sense of the sanctity of man that it provides."

In addition to the Nobel Prize, Wald received numerous awards and honors, including the Albert Lasker Award of the American Public Health Association in 1953, the Proctor Award in 1955 from the Association for Research in Ophthalmology, the Rumford Premium of the American Academy of Arts and Sciences in 1959, the 1969 Max Berg Award, and the **Joseph Priestley** Award the following year. In addition, he was elected to the National Academy of Science in 1950 and the American Philosophical Society in 1958. He is also a member of the Optical Society of America, which awarded him the Ives Medal in 1966. In the mid–1960s Wald spent a year as a Guggenheim fellow at England's Cambridge University, where he was elected an Overseas fellow of Churchill College for 1963–64. Wald also held honorary degrees from the University of Berne, Yale University, Wesleyan University, New York University, and McGill University.

Wald died on April 12, 1997, at his home in Cambridge, Massachusetts, at the age of 90.

See also Blindness and visual impairments; Central nervous system (CNS); Enzymes and coenzymes; Eye and ocular fluids; Eye: Ocular embryological development; Nerve impulses and conduction of impulses; Sense organs: ocular (visual) structures; Vision: histophysiology of the eye

WALDEYER-HARTZ, HEINRICH WILHELM GOTTFRIED VON (1836-1921)

German anatomist

Heinrich Wilhelm Gottfried von Waldeyer-Hartz was a professor of **anatomy** and **histology** who coined the word "chromosome" in 1886. Waldeyer, noting the ability of thread-like structures in the nucleus to be stained by the dye fuchsin, named them "chromo" meaning color, and "soma", meaning body. Waldeyer also coined the term "neuron." In 1884, he described an area in the **pharynx** near the **tonsils** that has come to be known as Waldeyer's tonsillar ring. His contributions were critical to the science of **neurology** and to the understanding and treatment of **cancer**.

Waldeyer was born in Germany in Hehlen, a small village near Braunschweig. In 1856, at the age of twenty, he attended Göttingen University where he studied mathematics and then medicine, specializing in anatomy. Waldeyer then taught at the university in Königsberg, Germany, for two years and then at the University of Breslau, Germany, for eight years. In 1872, he became professor of anatomy at the University of Strasbourg (at that time in Germany), where he remained for eleven years. In 1883, Waldeyer joined the faculty at the University of Berlin, Germany, as professor of anatomy and later served as the director of the department of anatomy for more than thirty years. Additionally, Waldeyer served as rector at the University of Berlin from 1889-1899; he retired in October 1916.

Waldeyer's first main contribution to modern science was in 1867, when he postulated that abnormal **cell division** led to cancer. Waldeyer's theory countered the contemporary authority **Rudolf Virchow**, the German pathologist and founder of cellular **pathology**. Virchow had written in his three-volume work, *Die Krankhaften Geschwulste* (1863–67), that cancer originated in changes in the connective **tissue**. Waldeyer's explanation that cancer begins on a cellular level wasn't accepted until 1872, when Virchow's theory on the origin of cancer was finally recognized as erroneous and the legitimacy of Waldeyer's work was confirmed. Waldeyer also explained that the relocation of a single cancer cell could lead to the formation of secondary **tumors** elsewhere in the body. He wrote that early detection and treatment of cancer offered the best cure. His recommendations were especially significant when radiation and **chemotherapy** became standard cancer treatments.

In 1884, Waldeyer identified a ring of lymphoid tissue formed by the pharyngeal, palatine and lingual tonsils. This tissue, now referred to as Waldeyer's throat ring, forms a protective ring at the opening of the pharynx. The function of Waldeyer's ring is to provide immunity from certain **antigens** and to protect against oropharyngeal (throat and mouth) cancer. Waldeyer's ring is frequently involved in a form of cancer, termed non-Hodgkin's lymphoma, which originates in the lymphatic system.

In 1888, Waldeyer suggested a name for the threads within the nucleus of a cell. German embryologist Oskar Hertwig had discovered the sphere of microscopic threads in the 1860's, and approximately twenty years later Waldeyer named the threadlike structures **chromosomes**.

Waldeyer's contributions to neurology followed upon the work of **Camillo Golgi**, an Italian pathologist. Golgi's work on the nervous system, which earned him the Nobel Prize in 1906, had laid the groundwork for further study of the nervous system. After Golgi's discovery of a particular type of nerve cell, now referred to as Golgi cells, Waldeyer was the first to hypothesize that the nervous system was comprised of individual cells. In 1891, he named these cells **neurons** and thus paved the way for the neuron theory, which is the **cell theory** of the nervous system.

WARBURG, OTTO HEINRICH (1883-1970)
German biochemist

One of the world's foremost biochemists, Otto Warburg's achievements include discovering the mechanism of cell oxidation and identifying the iron-enzyme complex, which catalyzes this process. He also made great strides in developing new experimental techniques, such as a method for studying the respiration of intact cells using a device he invented. In 1931, Warburg received a Nobel Prize in medicine and physiology in 1931.

Otto Heinrich Warburg was born on October 8, 1883, in Freiburg, Germany, to Emil Gabriel Warburg and Elizabeth Gaertner. Warburg was one of four children and the only boy. His father was a physicist of note and held the prestigious Chair in Physics at University of Berlin. The Warburg household often hosted prominent guests from the German scientific community, such as physicists Albert Einstein, Max Planck, Emil Fischer—the leading organic chemist of the late-nineteenth century, and Walther Nernst—the period's leading physical chemist.

Warburg studied chemistry at the University of Freiburg beginning in 1901. After two years, he left for the University of Berlin to study under Emil Fischer, and in 1906 received a doctorate in chemistry. His interest turned to medicine, particularly to **cancer**, so he continued his studies at the University of Heidelberg where he earned an M.D. degree in 1911. He remained at Heidelberg, conducting research for several more years and also making several research trips to the Naples Zoological Station.

Warburg's career goal was to make great scientific discoveries, particularly in the field of cancer research, according to the biography written by Hans Adolf Krebs, one of Warburg's students and winner of the 1953 Nobel Prize in medicine and physiology. Although he did not take up problems specifically related to cancer until the 1920s, his early projects provided a foundation for future cancer studies. For example, his first major research project, published in 1908, examined oxygen consumption during growth. In a study using sea urchin eggs, Warburg showed that after **fertilization**, oxygen consumption in the specimens increased 600%. This finding helped clarify earlier work that had been inconclusive on associating growth with increased consumption of oxygen and energy. A number of years later, Warburg did some similar tests of oxygen consumption by cancer cells.

Warburg was elected in 1913 to the Kaiser Wilhelm Gesellschaft, a prestigious scientific institute whose members had the freedom to pursue whatever studies they wished. He had just begun his work at the institute when World War I started. He volunteered for the army and joined the Prussian

Horse Guards, a cavalry unit that fought on the Russian front. Warburg survived the war and returned to the Kaiser Wilhelm Institute for Biology in Berlin in 1918. Now 35 years old, he would devote the rest of his life to biological research, concentrating on studies of energy transfer in cells (cancerous or otherwise) and photosynthesis.

One of Warburg's significant contributions to biology was the development of a manometer for monitoring cell respiration. He adapted a device originally designed to measure gases dissolved in **blood** so it would make measurements of the rate of oxygen production in living cells. In related work, Warburg devised a technique for preparing thin slices of intact, living **tissue** and keeping the samples alive in a nutrient medium. As the tissue slices consumed oxygen for respiration, Warburg's manometer monitored the changes.

During Warburg's youth, he had become familiar with Einstein's work on photochemical reactions as well as the experimental work done by his own father, Emil Warburg, to verify parts of Einstein's theory. With this background, Warburg was especially interested in the method by which plants converted light energy to chemical energy. Warburg used his manometric techniques for the studies of photosynthesis he conducted on algae. His measurements showed that photosynthetic plants used light energy at a highly efficient 65%. Some of Warburg's other theories about photosynthesis were not upheld by later research, but he was nevertheless considered a pioneer for the many experimental methods he developed in this field. In the late 1920s, Warburg began to develop techniques that used light to measure reaction rates and detect the presence of chemical compounds in cells. His "spectrophotometric" techniques formed the basis for some of the first commercial spectrophotometers built in the 1940s.

His work on cell respiration was another example of his interest in how living things generated and used energy. Prior to World War I, Warburg discovered that small amounts of cyanide could inhibit cell oxidation. Since cyanide forms stable complexes with heavy metals such as iron, he inferred from his experiment that one or more catalysts important to oxidation must contain a heavy metal. He conducted other experiments with carbon monoxide, showing that this compound inhibits respiration in a fashion similar to cyanide. Next he found that light of specific frequencies could counteract the inhibitory effects of carbon monoxide, at the same time demonstrating that the "oxygen transferring enzyme," as Warburg called it, was different from other **enzymes** containing iron. He went on to discover the mechanism by which iron was involved in the cell's use of oxygen. It was Warburg's work in characterizing the cellular catalysts and their role in respiration that earned him a Nobel Prize in 1931.

Nobel Foundation records indicate that Warburg was considered for Nobel Prizes on two additional occasions: in 1927 for his work on **metabolism** of cancer cells, then in 1944 for his identification of the role of flavins and nicotinamide in biological oxidation. Warburg did not receive the 1944 award, however, because a decree from Hitler forbade German citizens from accepting Nobel Prizes. Two of Warburg's students

also won Nobel Prizes in medicine and physiology: **Hans Krebs** (1953) and Axel Theorell (1955).

In 1931 Warburg established the Kaiser Wilhelm Institute for Cell Physiology with funding from the Rockefeller Foundation in the United States. During the 1930s, Warburg spent much of his time studying dehydrogenases, enzymes that remove hydrogen from substrates. He also identified some of the cofactors, such as nicotinamide derived from vitamin B3 (niacin), that play a role in a number of cell biochemical reactions.

Warburg conducted research at the Kaiser Wilhelm Institute for Cell Physiology until 1943 when the Second World War interrupted his investigations. Air attacks targeted at Berlin forced him to move his laboratory about 30 mi. (48 km) away to an estate in the countryside. For the next two years, he and his staff continued their work outside the city and out of the reach of the war. Then in 1945, Russian soldiers advancing to Berlin occupied the estate and confiscated Warburg's equipment. Although the Russian commander admitted that the soldiers acted in error, Warburg never recovered his equipment. Without a laboratory, he spent the next several years writing, publishing two books that provided an overview of much of his research. He also traveled to the United States during 1948 and 1949 to visit fellow scientists.

Although Warburg was of Jewish ancestry, he was able to remain in Germany and pursue his studies unhampered by the Nazis. One explanation is that Warburg's mother was not Jewish and high German officials "reviewed" Warburg's ancestry, declaring him only one-quarter Jewish. As such, he was forbidden from holding a university post, but allowed to continue his research. There is speculation that the Nazis believed Warburg might find a cure for cancer and so did not disturb his laboratory. Scientists in other countries were unhappy that Warburg was willing to remain in Nazi Germany. His biographer Hans Krebs noted, however, that Warburg was not afraid to criticize the Nazis. At one point during the war when Warburg was planning to travel to Zurich for a scientific meeting, the Nazis told him to cancel the trip and to not say why. "With some measure of courage," wrote Krebs, "he sent a telegram [to a conference participant from England]: 'Instructed to cancel participation without giving reasons.'" Although the message was not made public officially, the text was leaked and spread through the scientific community. Krebs believed Warburg did not leave Germany because he did not want to have to rebuild the research team he had assembled. The scientist feared that starting over would destroy his research potential, Krebs speculated.

In 1950 Warburg moved into a remodeled building in Berlin which had been occupied by U.S. armed forces following World War II. This new site was given the name of Warburg's previous scientific home—the Kaiser Wilhelm Institute for Cell Physiology—and three years later renamed the Max Planck Institute for Cell Physiology. Warburg continued to conduct research and write there, publishing 178 scientific papers from 1950 until his death in 1970.

For all of his interest in cancer, Warburg's studies did not reveal any deep insights into the disease. When he wrote about the "primary" causes of cancer later in his life, Warburg's proposals failed to address the mechanisms by which cancer cells undergo unchecked growth. Instead, he focused on metabolism, suggesting that in cancer cells "fermentation" replaces normal oxygen respiration. Warburg's cancer studies led him to fear that exposure to food additives increased one's chances of contracting the disease. In 1966 he delivered a lecture in which he stated that cancer prevention and treatment should focus on the administration of respiratory enzymes and cofactors, such as iron and the B **vitamins**. The recommendation elicited much controversy in Germany and elsewhere in the Western world.

Warburg's devotion to science led him to forego marriage, since he thought it was incompatible with his work. According to Karlfried Gawehn, Warburg's colleague from 1950 to 1964, "For him [Warburg] there were no reasonable grounds, apart from death, for not working." Warburg's productivity and stature as a researcher earned him an exemption from the Institute's mandatory retirement rules, allowing him to continue working until very near to the end of his life. He died at his Berlin home on August 1, 1970.

See also Antioxidents; Biochemistry; Cell structure; Enzymes and coenzymes

WATSON, JAMES D. (1928-   )
American molecular biologist

James D. Watson won the 1962 Nobel Prize in physiology or medicine, along with **Francis Crick** and Maurice Wilkins, for discovering the structure of **DNA**, or deoxyribonucleic acid—the carrier of genetic information at the molecular level. Watson and Crick had worked as a team since meeting in the early 1950s, and their research ranks as a fundamental advance in **molecular biology**. More than thirty years later, Watson became the director of the Human Genome Project, an enterprise devoted to a difficult goal: the description of every human gene, the total of which numbers over thirty thousand. This project would not be possible without Watson's groundbreaking work on DNA.

James Dewey Watson was born in Chicago, Illinois, to James Dewey and Jean (Mitchell) Watson. He was educated in the Chicago public schools, and during his adolescence became one of the original Quiz Kids on the radio show of the same name. Shortly after this experience in 1943, Watson entered the University of Chicago at the age of fifteen.

Watson graduated in 1946, but stayed on at Chicago for a bachelor's degree in zoology, which he attained in 1947. During his undergraduate years, Watson studied neither **genetics** nor biochemistry—his primary interest was in the field of ornithology. In 1946, Watson spent a summer working on advanced ornithology at the University of Michigan's summer research station at Douglas Lake. During his undergraduate career at Chicago, Watson had been instructed by the well-

known population geneticist Sewall Wright, but he did not become interested in the field of genetics until he read Erwin Schrödinger's influential book *What is Life?* It was then, Horace Judson reports in *The Eighth Day of Creation: Makers of the Revolution in Biology,* that Watson became interested in finding out the secret of the gene.

Watson enrolled at Indiana University to perform graduate work in 1947. Indiana had several remarkable geneticists who could have been important to Watson's intellectual development, but he was drawn to the university by the presence of the Nobel laureate Hermann Joseph Muller, who had demonstrated twenty years earlier that x rays cause mutation. Nonetheless, Watson chose to work under the direction of the Italian biologist Salvador Edward Luria, and it was under Luria that he began his doctoral research in 1948.

Watson's thesis was on the effect of x rays on the rate of phage lysis (a phage, or bacteriophage, is a bacterial virus). The biologist Max Delbrück and Luria—as well as a number of others who formed what was to be known as "the phage group"—had demonstrated that phages could exist in a number of mutant forms. A year earlier, Luria and Delbruck had published one of the landmark papers in phage genetics, in which they established that one of the characteristics of phages is that they can exist in different genetic states so that the lysis (or bursting) of bacterial host cells can take place at different rates. Watson's Ph.D. degree was received in 1950, shortly after his twenty-second birthday.

Watson was next awarded a National Research Council fellowship grant to investigate the molecular structure of proteins in Copenhagen, Denmark. While Watson was studying enzyme structure in Europe, where techniques crucial to the study of macromolecules were being developed, he was also attending conferences and meeting colleagues.

From 1951 to 1953, Watson held a research fellowship under the support of the National Foundation for Infantile Paralysis at the Cavendish Laboratory in Cambridge, England. Those two years are described in detail in Watson's 1965 book, *The Double Helix: A Personal Account of the Discovery of the Structure of DNA.* An autobiographical work, *The Double Helix* describes the events—both personal and professional—that led to the discovery of DNA. Watson was to work at the Cavendish under the direction of Max Perutz, who was engaged in the x-ray crystallography of proteins. However, he soon found himself engaged in discussions with Crick on the structure of DNA. Crick was twelve years older than Watson and, at the time, a graduate student studying protein structure.

Intermittently over the next two years, Watson and Crick theorized about DNA and worked on their model of DNA structure, eventually arriving at the correct structure by recognizing the importance of x-ray diffraction photographs produced by Rosalind Franklin at King's College, London. Both were certain that the answer lay in model-building, and Watson was particularly impressed by Nobel laureate Linus Pauling's use of model-building in determining the alpha-helix structure of protein. Using data published by Austrian-born American biochemist Erwin Chargaff on the symmetry between the four constituent nucleotides (or bases) of DNA

In their Cambridge laboratory, James Watson (right) and Francis Crick use a large physical model of DNA to calculate and confirm specific base pairing and the double helical structure for their DNA model. *Archive Photos, Inc. Reproduced by permission.*

molecules, they concluded that the building blocks had to be arranged in pairs. After a great deal of experimentation with their models, they found that the double helix structure corresponded to the empirical data produced by Wilkins, Franklin, and their colleagues. Watson and Crick published their theoretical paper in the journal *Nature* in 1953 (with Watson's name appearing first due to a coin toss), and their conclusions were supported by the experimental evidence simultaneously published by Wilkins, Franklin, and Raymond Goss. Wilkins shared the Nobel Prize with Watson and Crick in 1962.

After the completion of his research fellowship at Cambridge, Watson spent the summer of 1953 at Cold Spring Harbor, New York, where Delbruck had gathered an active group of investigators working in the new area of molecular biology. Watson then became a research fellow in biology at the California Institute of Technology, working with Delbruck and his colleagues on problems in phage genetics. In 1955, he joined the biology department at Harvard and remained on the faculty until 1976. While at Harvard, Watson wrote *The Molecular Biology of the Gene* (1965), the first widely used university textbook on molecular biology. This text has gone through seven editions, and now exists in two large volumes as a comprehensive treatise of the field. In 1968, Watson

became director of Cold Spring Harbor, carrying out his duties there while maintaining his position at Harvard. He gave up his faculty appointment at the university in 1976, however, and assumed full-time leadership of Cold Spring Harbor. With John Tooze and David Kurtz, Watson wrote *The Molecular Biology of the Cell,* originally published in 1983.

In 1989, Watson was appointed the director of the Human Genome Project of the National Institutes of Health, but after less than two years he resigned in protest over policy differences in the operation of this massive project. He continues to speak out on various issues concerning scientific research and is a strong presence concerning federal policies in supporting research. In addition to sharing the Nobel Prize, Watson has received numerous honorary degrees from institutions, including one from the University of Chicago, which was awarded in 1961, when Watson was still in his early thirties. He was also awarded the Presidential Medal of Freedom in 1977 by President Jimmy Carter.

Most of Watson's professional life has been spent as a professor, research administrator, and public policy spokesman for research. More than any other location in Watson's professional life, Cold Spring Harbor (where he is still director) has been the most congenial in developing his abilities as a scientific catalyst for others. Watson's work there has primarily been to facilitate and encourage the research of other scientists.

See also Human genetics

WHIPPLE, GEORGE HOYT (1878-1976)
American pathologist

George Hoyt Whipple advanced medical research into the creation and breakdown of oxygen-carrying **hemoglobin** in the **blood**; this research resulted in not only a treatment for pernicious anemia, but also in a share of the 1934 Nobel Prize. An industrious physician-scientist, Whipple authored more than 200 publications on anemia, pigment **metabolism**, **liver** injury and repair, and other related subjects.

Whipple was born on August 28, 1878, in Ashland, New Hampshire, the son of Frances Anna Hoyt Whipple and Ashley Cooper Whipple, a general practitioner held in high esteem by his patients and colleagues. Whipple's father died of typhoid fever just two years after the birth of his son, and Whipple and his sister Ashley were brought up by their mother and grandmothers. His was an outdoor life in rural New Hampshire, and he took a love of hunting, fishing, and camping with him into adulthood. Whipple knew he would be a physician from the time he was in elementary school. At the age of fourteen Whipple entered Phillips Academy in Andover, Massachusetts, enrolling at Yale College (now Yale University) as a premedical student four years later. At Yale, he was a star baseball player and was on the gymnastics and rowing teams, as well as an outstanding student. Though versed in the humanities in these years of public and private schools, he had always been attracted by science and mathematics. After graduating with high standing in 1900, Whipple spent a year teaching and coaching at Holbrook Military Academy in New York to earn money for medical studies, and in 1901 he entered Johns Hopkins University's School of Medicine.

During his years as a student at Johns Hopkins, Whipple earned his way with a paying instructorship. Initially Whipple had considered going into pediatrics, but upon receiving his M.D. in 1905 instead joined the Johns Hopkins staff as an assistant in **pathology**, working under the renowned pathologist William Henry Welch. It was as a 29-year-old assistant performing an autopsy on a missionary doctor that Whipple made his first notable medical contribution, describing a rare condition in the intestinal tissues, which has since come to be called Whipple's disease. A year spent at a hospital in the Panama Canal Zone led to further notable advances in **malaria** and tuberculosis research.

When he returned to Johns Hopkins in 1908, Whipple turned his attention to studies in liver damage and the way in which liver cells repair themselves. Studies with dogs led Whipple to realize the importance of **bile**, a substance manufactured in the liver by the breakdown of hemoglobin, a complex pigment in red corpuscles. In normal concentrations, bile helps to break down fats during **digestion**, but can produce

jaundice when present in excessive amounts. Beginning his assistant professorship at Johns Hopkins in 1911, Whipple came to focus on the interrelationship of bile, hemoglobin, and the liver. In 1913, along with a talented medical student, Charles W. Hooper, Whipple was able to show that bile pigments could be produced outside of the liver, solely from the breakdown of hemoglobin in the blood. Using this experiment as a starting point, Whipple set a new course for his studies. Since bile pigments are formed from hemoglobin, Whipple reasoned that he should tackle the question of hemoglobin itself, beginning with how it is manufactured. It was a fateful decision.

In 1914 Whipple accepted a position as director of the Hooper Foundation for Medical Research at the University of California in San Francisco. In that same year, he also married his long-time sweetheart, Katharine Ball Waring, and the couple moved to California. Though burdened with administrative duties, Whipple continued his research into hemoglobin production. His assistant, Hooper, came with him to California and together with a new assistant, Frieda Robscheit-Robbins, they began experiments that would lead to a major breakthrough. By systematically bleeding laboratory dogs, Whipple and his team were able to induce a controlled anemic condition. They then tested various foods and their effects upon hemoglobin regeneration, finding that a diet of liver produced a pronounced increase in hemoglobin regeneration. While such short term effects were encouraging, they were still far from conclusive.

Though in 1920 Whipple was named dean of the University of California Medical School, he remained in California for just a year before accepting (somewhat reluctantly) a similar position at a new medical complex at the University of Rochester in New York—a facility heavily endowed by Kodak founder George Eastman and the Rockefeller Foundation. Courted enthusiastically by Eastman and university president Rush Rhees, Whipple moved home and laboratory to New York, bringing Robscheit-Robbins and the group of anemic dogs with him.

The next decade proved busy for Whipple: he directed the building and staffing of the University of Rochester School of Medicine and Dentistry, all the while directing further hemoglobin research. Perfecting their technique of bleeding the dogs, Whipple and Robscheit-Robbins induced long-term anemia and were able to prove conclusively that a liver diet was successful in counteracting its effects by increasing the production of hemoglobin. His results were published in 1925, and the pharmaceutical firm of Eli Lilly, with Whipple's cooperation, began producing a commercially available liver extract within a year. Whipple refused to patent his findings, and directed all royalties from the sales of the extract to fund additional research. Whipple's experiments paved the way for further studies by two Boston researchers, George Richards Minot and William P. Murphy, who used liver therapy to successfully treat pernicious anemia in 1926.

Whipple's work soon won international repute and in 1934 he received word that he, along with Minot and Murphy, was going to receive the Nobel Prize in physiology or medicine for their separate work in liver therapy. Whipple did not let fame slow his work. He continued his hemoglobin experiments, turning now to the study of iron in the body and utilizing the new technology of radioisotope elements to follow the distribution of iron in the body. He also made important contributions to the study of an anemic disorder peculiar to people of Mediterranean extraction, a disorder for which Whipple suggested the name *thalassemia*. Other studies involved the use of **plasma** or **tissue** proteins to rebuild hemoglobin in cases of anemia. A spin-off of this latter research was the development of intravenous feeding.

Despite the administrative and research duties that pressed upon him, Whipple did not forget his students, and took real pleasure in teaching. When in later years he was offered the position of Director of the Rockefeller Institute, he politely but adamantly declined, preferring his classes and his research. Whipple finally relinquished his chair as dean in 1953 at the age of 75, after a long and distinguished career that had seen the once-small university grow to more than 12,000 graduates in medicine and other related fields. He remained on the faculty of the University of Rochester teaching pathology until 1955. In 1963, he established a medical and dental library for the university valued at $750,000. In addition to the Nobel Prize, Whipple was also a trustee of the Rockefeller Foundation from 1927-43, a Kober Medal winner in 1939, and a recipient of the Kovalenko Medal of the National Academy of Sciences in 1962.

Whipple's life was long and productive. He was an active outdoorsman well into his ninth decade. With his wife Katharine, he had two children one who followed in the Whipple tradition of medicine. Whipple died in Rochester on February 1, 1976, in the hospital he had helped to build.

See also Hemopoiesis; Liver; Necrosis; Oxygen transport and exchange; Spleen histophysiology; Transfusions; Vascular exchange

WIECHAUS, ERIC FRANCIS (1947-   )
American molecular biologist

Eric Wieschaus, by studying the fruit fly *Drosophila melanogaster*, made important discoveries concerning genetic mechanisms of control of early embryonic development. For this research, Wieschaus, along with colleagues **Christiane Nüsslein-Volhard** and Edward B. Lewis received the 1995 Nobel Prize in physiology or medicine.

Wieschaus was born in South Bend, Indiana, but grew up in Alabama. He received his bachelor's degree in biology from the University of Notre Dame in 1969 and his doctorate from Yale in 1974. His doctoral dissertation involved using genetic methods to label the progeny (offspring) of single cells in fly embryos. He showed that even at the earliest cellular stages, cells were already determined to form specific regions of the body called segments.

Wieschaus began his Nobel-winning work in the latter part of the 1970s. He spent three years with Christiane Nüsslein-Volhard in the European Molecular Biology Lab at the University of Heidelberg, Germany, tackling the question of why individual cells in a fertilized egg develop into various specific tissues. They elected to study *Drosophila*, or fruit flies, because of their extremely fast embryonic development. New generations of fruit flies can be bred in a week. In addition, fruit flies have only one set of genes controlling development compared to the four sets humans possess. This means that testing each fruit fly gene individually takes a quarter of the time it would involve to test human genes.

To begin their experiment, Nüsslein-Volhard and Wieschaus damaged male fruit fly **deoxyribonucleic acid (DNA)** by applying ultraviolet light to the genes or by feeding the flies sugar water laced with chemicals. Then the team "knocked out" one gene from the fly, breeding generations of fruit flies without that particular piece of code. In this way, Nüsslein-Volhard and Wieschaus were able to isolate all the genes crucial to the early stages of embryonic development. When the flies were bred, the females produced dead embryos. These lifeless embryos resulted from only 150 different mutations of the 40,000 mutations applied. These 150 genes proved to be essential to the proper development of the fly embryo because, when damaged, the genes caused extraordinary deformities that killed the embryo. By viewing the fly embryos with a two-person microscope, Wieschaus and Nüsslein-Volhard were able to simultaneously view and classify a large quantity of malformations caused by gene mutations. Next, they identified 15 different genes, that, when mutated, eliminate specific body segments in the fly embryos. Wieschaus also established that systematic categorizing of genes that control the various stages of development could be accomplished.

Their first research results reported that the number of genes controlling early development was not only limited, but could also be classified into specific functional groups. They also identified genes that cause severe congenital defects in flies. After additional experimentation, the principles involved with the fruit fly genes were found to apply to higher animals and humans. This led to the realization that many similar genes control **human development**, and this finding could have a tremendous impact on the medical world. The applications of their research extend to *in vitro* **fertilization**, identifying congenital **birth defects**, and increased knowledge of substances that can endanger early stages of **pregnancy**.

It wasn't until 1995, however, that Wieschaus won the Nobel Prize in physiology or medicine, along with Edward B. Lewis and Christiane Nüsslein-Volhard, for his work on identifying key genes that make a fertilized fruit fly egg develop into a segmented embryo. His research could help improve knowledge of how genes control embryonic development in higher organisms, including identifying genes that cause human birth defects.

See also Embryonic development: early development, formation, and differentiation; Human genetics

WIESEL, TORSTEN N. (1924-   )

Swedish American neurophysiologist

Torsten Wiesel, in collaboration with David H. Hubel, provided fundamental insight into **physiology** of vision. Wiesel's work on charting the visual or striate cortex, the posterior section of the **cerebral cortex**, provided new insights into the complexity of the visual process that also proved to have direct clinical applications. Wiesel's discovery of critical periods in childhood development for learning to see led to earlier clinical intervention in visual problems in children. In 1981 Wiesel, along with Hubel and another **brain** researcher, Roger W. Sperry, shared the Nobel Prize in physiology or medicine.

Torsten Nils Wiesel was born on June 3, 1924, in Uppsala, Sweden, the son of Anna-Lisa Bentzer Wiesel and Fritz S. Wiesel, the chief psychiatrist at the Beckomberg Mental Hospital in Stockholm. Wiesel lived at his father's hospital as a youth, attending a private school where he was more interested in sports than academics. However, this attitude changed in 1941 when Wiesel entered medical school at the Karolinska Institute in Stockholm and studied neurophysiology under Carl Gustaf Bernhard. He also studied psychiatry during this time, and in 1954 he received his medical degree, becoming an instructor at the institute as well as an assistant in the Department of Child Psychiatry at Karolinska Hospital. Wiesel then came to the United States in 1955 to do postdoctoral work at the Wilmer Institute of Johns Hopkins School of Medicine in Baltimore, Maryland.

At Johns Hopkins, Wiesel worked under Stephen Kuffler, a researcher in visual physiology who had studied the nerve activity in the **retina** of the cat as well as in animals of other classes. Kuffler's exhaustive work had proved that the vision of mammals is distinctly different from that of non-mammals. Research with frogs had shown that their vision occurred in the optical nerve: that they had **neurons**, or nerve cells, sensitive not only to light and dark, but also to shapes, movements, and the boundaries between light and dark. Cats have no such specification in their ganglia, and lack the ability to give the detailed boundary information found in frogs. Yet, mammalian vision is stereoscopic, whereas non-mammalian appears to be in most cases binocular but lacking three dimensions. Wiesel became interested in the direction in which such investigations must logically lead: namely that the critical level of visual perception must take place in the brain of mammals. In 1958 **David Hubel**, a graduate of McGill University, returned to the institute from military service, and together Wiesel and Hubel set off on research that would result in a new theory of visual perception.

The striate or visual cortex is located at the back of the brain, an area of about 5.9 in.^{2} (15 cm^2) in some of the monkeys Wiesel and Hubel would study. It had long been known, from accident victims, that this region of the cortex was involved with vision, and it is here that Wiesel and Hubel began their studies. They painstakingly measured electrical discharge of cells in the visual cortex with the aid of a microelectrode, a microscopic needle with an electrode built in to measure electrical impulses. Initially using anesthetized cats whose sight was trained on various patterns of light and dark, lines and circles, and probing the animal's visual cortex with their microelectrode at various angles, they discovered which cells in the cortex responded to which pattern or level of light. They also conducted experiments in which they injected the eyes of experimental animals with radioactively labeled amino acid. These **amino acids** would be taken up by the cell bodies of the retina and transported to cells in the visual cortex, giving a map of the pathway of vision. In some cases the laboratory animals were sacrificed and their visual cortexes dissected in order to see, by the use of autoradiographs or x-ray-like photos, where the labeled amino acids actually ended up. Such experiments, begun in 1959, used both cats and macaque monkeys. That same year Kuffler was appointed a professor at the Harvard University Medical School, and Wiesel and Hubel joined him there. Wiesel was appointed assistant professor of physiology, and became a full professor in 1964.

The Wiesel-Hubel team soon began publishing the results of their experimental method, and it was clear that they had uncovered new complexities to the visual process. Mapping the path of vision with radioactive amino acid, they showed that vision passed in coded signals from neuron to neuron through the optic nerve and split at the optic chiasm so that a representation of each half of the visual scene is projected deep in the brain on a nest of cells called the lateral geniculate nucleus, a way station to the cells in the cortex. From here, the path of vision continues to the back of the brain to various parts of the visual cortex, depending on the specialization of each cell. However, the pathway does not end there; indeed, the visual cortex was shown to be an early step in the processing of visual information. Information is sent back to parts of the brain from the cortex as well as back to the geniculate nuclei.

Within the visual cortex itself, Wiesel and Hubel made two important discoveries. First, they showed that there is a hierarchy of types of cells in the cortex, ranking from simple to complex to hypercomplex, depending on the information each is able to process. They termed the process of putting the millions of building blocks of visual information back together into a picture convergence. Various cells have preferences for the bits of visual information they process: size, shape, light, and sharpness of boundary differentiation, as well as which **eye** is sending the information. Such a complexity of visual processing destroys the old notion of sight being simply a film played in the mind. Instead, the accretion of bits of visual information into visual representation appears more similar to language processing than to an analogy of a film. Cells in the visual cortex "read" neuron messages. Their second major discovery was a further organization of the cortical cells into roughly vertical divisions of two types: orientation columns and ocular dominance columns. The orientation columns transform what is essentially circular information from the retina and geniculate nerve cells into linear information, while the ocular combine the neural information from both eyes to provide three-dimensional vision. Within these columns are simple, complex, and hypercomplex cells working toward a progressive convergence of visualization. Until the time of Wiesel's and Hubel's work, it was assumed that all cells of the cerebral cortex were more or less uniform. Wiesel and Hubel showed that the visual cortex is constituted of a cell pattern, which appears to be designed specifically for vision. As a result of their discovery, current theory now posits that the rest of the cerebral cortex may follow this form-follows-function rule.

Wiesel and Hubel researched another experimental model in which they used kittens to study the effect of various visual impairments on development. They discovered that if one eye were deprived of visual stimuli during the third to fifth postnatal weeks, the central functioning of that eye would always be suppressed from cortical processing. Kittens, and by extension mammals in general, though born with a complete visual cortex, must still "learn" to see. Even if an early impairment is later corrected, the repaired eye will remain functionally impaired as far as the visual cortex is concerned. The realization that there is a critical stage for visual development revolutionized the field of pediatric ophthalmology, calling for the earliest possible intervention in cases of strabismus, or crossed eyes, and congenital cataracts.

By 1973, Wiesel succeeded Kuffler as chair of the Department of Neurobiology at Harvard, and was named the Robert Winthrop Professor of Neurobiology in 1974. Wiesel became a naturalized U.S. citizen in 1990. Wiesel has been the recipient of many awards over the years, including the Lewis S. Rosentiel Award in 1972, the Jonas S. Friedenwald Memorial Award in 1975, the Karl Spencer Lashley Prize in 1977, the Louisa Gross Horowitz Prize in 1978, and the George Ledlie Prize in 1980. Wiesel shared a 1981 Nobel Prize, with Hubel and Sperry (who worked at California Institute of Technology). The Karolinska Institute in Stockholm, which administers the prize and where Wiesel began his professional career, praised Hubel and Wiesel for their discoveries concerning information processing in the visual system. Wiesel and Hubel continued their close working relationship until Wiesel left Harvard in 1984 to head the neurobiology lab at Rockefeller University where he continued his researches on vision. In 1992 he was named president of Rockefeller University.

See also Blindness and visual impairments; Central nervous system (CNS); Eye and ocular fluids; Eye: Ocular embryological development; Nerve impulses and conduction of impulses; Sense organs: Ocular (visual) structures; Vision: Histophysiology of the eye

WILLIS, THOMAS (1621-1675)

English physician, anatomist, and chemist

Thomas Willis advanced the anatomical and physiological knowledge of the nervous and circulatory systems, and the interaction between these two systems. He coined the term "neurology" and was the first to understand the physiological function of the cerebral arterial circle (circulus arteriosus cerebri), now commonly known as the "circle of Willis."

Born a farmer's son in Great Bedwyn, Wiltshire, England, Willis moved with his family before 1631 to the village of North Hinksey, then in Berkshire, since 1974 in Oxfordshire, and only one-and-a-half miles from Oxford. He matriculated at Christ Church College, Oxford University, in 1637, receiving his B.A. in 1639, his M.A. in 1642, and his B.Med. in December 1646. The English Civil War interrupted his studies and threatened his career. As a Royalist in Oxford, a Royalist stronghold, Willis had to be careful not to run afoul of the ascendant Parliamentary forces, especially after they captured Oxford in June 1646. Nevertheless, Willis harbored secret Anglican religious services during the dangerous time when Puritans banned the Anglican rite, and married Mary Fell, daughter and sister of Anglican clergymen.

Willis's fortunes improved immediately upon the restoration of the Stuart monarchy in 1660. Willis and several scientific colleagues co-founded the Royal Society. Oxford gave him his D.Med. and appointed him Sedleian Professor of Natural Philosophy, a post which he held until his death, even though he moved to London in 1666. Among his students at Oxford were the cardiologist **Richard Lower** and the microscopist **Robert Hooke**. From 1667 until he died in London, he enjoyed the most lucrative medical practice in England.

In connection with his studies of both the circulatory and nervous systems, Willis paid special attention to the circle of **arteries** at the base of the **brain**. Anatomists before Willis had noticed the circle, but his name attaches to it because he was the first to understand its purpose to protect the brain from local anemia, or ischemia, when one or two of the three arteries feeding the circle are blocked. The **circle of Willis** works by a process called anastomosis, whereby several arteries communicate freely with one another and share their contents. Thus it ensures adequate and even **blood** supply to the brain by distributing blood from the internal **carotid arteries** and the basilar artery throughout the entire brain via the pairs of anterior, middle, and posterior cerebral arteries. He presented these findings in his masterpiece *Cerebri anatome* [Cerebral Anatomy] (1664). The main illustrator of this book was Christopher Wren (1632–1723), later the architect of St. Paul's Cathedral in London. Lower and Thomas Millington (1628–1704) also assisted Willis with this project.

Willis adhered to iatrochemistry, sometimes called chemiatry, a medical philosophy derived from the teachings of Paracelsus (1493–1541) and Jean Baptiste van Helmont (1577–1644), which held that the secrets of medicine were contained within those of chemistry or alchemy. The iatrochemists were the intellectual rivals of the iatromechanists or iatromathematicists, such as **René Descartes** and Giovanni Alfonso Borelli (1608–1679), who claimed that **physiology** could be understood purely in terms of the laws of physics and mathematics. Iatrochemistry and iatromechanics each made significant progress in physiology, but iatrochemistry, being more flexible and more empirical, probably made the greater. Willis and Descartes were the only important believers in brain function localization before the late eighteenth century. In *Cerebri anatome*, Willis claimed that the cerebrum is the organ of thought and that the **cerebellum** controls the vital functions.

As a leader of the "Oxford physiologists," an informal group dedicated to furthering the investigations of **William Harvey**, Willis's research resulted in several eponyms. Besides his circle, they include Willis's centrum nervosum, the ganglia celiaca; Willis's cords, fibrous bodies traversing the superior sagittal sinus; Willis's nerve; the ophthalmic branch of the fifth cranial nerve; Willis's paracusis, improved hearing in a noisy environment; Willis's **pancreas**, a portion of the head of the pancreas; Willis's pouch, a fold in the **peritoneum** near the **liver**; and Willis's gland, the corpus luteum.

Willis published many scientific papers and several longer works, including *Diatribae duae medico-philosophicae* [Two Medical-Philosophical Discussions] (1659), one on fermentation, the other on fevers; *Pathologiae cerebri et nervosi generis specimen* [A Model of the Pathology of the Brain and Nervous Tissues] (1667); and *De anima brutorum* [On the Spirit of Animals] (1672), a work of **comparative anatomy**.

See also History of anatomy and physiology: The Renaissance and Age of Enlightenment; History of anatomy and physiology: The science of medicine; Nervous system overview

Y

YAWN REFLEX

The yawn reflex (pandiculation) is not a classical neural reflex arc, but rather a coordinated neural and muscular arousal reflex mediated by the **brain** stem. A yawn is a sequence of events that begins with a deep inspiration of air and ends with a forced and deep expiration of air concurrent with a general contraction of several muscle groups, especially those associated with mouth, throat, and face. Yawns last four to eight seconds and the yawn reflex is well established by the second trimester of fetal development. Male yawns tend to be more vigorous than female yawns. The exact mechanisms stimulating the yawn reflex remain scientifically contentious.

Anatomically, the yawn is a coordinated movement of thoracic muscles, **diaphragm**, **larynx**, and palate. The chest wall expands, the diaphragm lowers, the palate rises and there is a tendency to move the tongue downward and to the rear of the mouth. Groups of facial muscles contract, there is an abduction of the **vocal cords** during expiration. Yawning may also be associated with a general stretching of muscles in the neck, arms, and legs. During yawning, the eyes usually close and tearing (lacrimation) may occur.

At one time, physiologists argued that the yawn reflex was a brain stem reaction to high levels of carbon dioxide (CO_2) in the **blood** stream. The deep inspiration and expiration was argued to allow the intake of extra oxygen during the deep inspiration and the explosive venting of built up carbon dioxide during the deep expiration. That yawning is often associated with boredom was explained by the fact that the slower metabolic rate in sedentary or bored individuals resulted in slower and shallower respiration that allowed the build-up of carbon dioxide. Recent research involving controlled inspiration of mixtures of air with varying amounts of oxygen failed to clearly indicate or support the inspiration-ventilation hypothesis. Moreover, researchers have also discovered that there is not necessarily a direct link between the level of alertness and yawning. In some studies, the number of yawns was not altered by exercise. Accordingly, some physiologists now argue that different physiological mechanisms control the respiration and yawning.

Yawning may be mediated or controlled by the **hypothalamus**. Increased levels of **neurotransmitters**, neuropeptide proteins, and certain **hormones** in the hypothalamus are associated with yawning.

The yawn reflex does have other important physiological and anatomical effects. During a yawn, there is often an increase in blood pressure and **heart** rate. A yawn reflex also serves as a mechanism to initiate the contraction of several muscles and muscle groups

Yawning occurs in many species, and there are social implications to the yawn reflex. Accordingly, there are sociological and evolutionary biology based explanations of some aspects of the yawn reflex, especially the "contagious" aspects where a yawn by one person in a group is followed by a chain of yawning in other members of the group. One hypothesis is that contagious yawning served to coordinate group behavior and acted as a signal initiated by external stimuli (or lack thereof). Some evolutionary biologists argue that the pronounced jaw movements associated with yawning are a remnant of primitive teeth display behavior.

Excessive yawning can also be a symptom of various neurological problems including encephalitis and tumor growth.

See also Brain stem function and reflexes; Respiration control mechanisms; Respiratory system

Z

ZINKERNAGEL, ROLF M. (1944-)
Swiss microbiologist

Rolf M. Zinkernagel, a Swiss microbiologist who won the Nobel Prize in physiology or medicine for research concerning structures on the surface of the cell that alert the **immune system** to the presence of foreign invaders, was born in Basel, Switzerland. Zinkernagel attended medical school in 1962 at the University of Basel and graduated in 1968. From 1969 to 1970, he worked as a postdoctoral fellow on electron microscopy at the Institute of Anatomy, University of Basel, and from 1971 to 1973 at the Institute of Biochemistry, University of Lausanne, Switzerland.

In 1973, Zinkernagel joined the Department of Microbiology at the Australian National University in Canberra to study immunity of infectious disease. Zinkernagel worked with **Peter Doherty** who studied inflammatory processes of the **brain**. Utilizing mice with lymphocytic choriomeningitis virus (LCMV), Zinkernagel and Doherty researched the immune responses that led to the discovery of major histocompatibility complex (MHC) restriction. They determined that T lymphocytes (white **blood** cells) must recognize the foreign microorganism, in this case the virus, as well as the self molecules, in order to effectively kill virus-infected cells. This discovery came about when one strain of mice with LCMV developed killer T lymphocytes that were able to protect the mice from the virus. However, when the T lymphocytes were placed with virus-infected cells from another strain of mice in vitro, the T lymphocytes did not kill the cells infected with the virus. The self molecules that are necessary in order for T lymphocytes to recognize the foreign microorganism became known as major histocompatibility **antigens**. Zinkernagel and Doherty proposed a structural model to explain how the T lymphocytes recognize both foreign microorganisms and major histocompatibility antigens. A peptide from a foreign microorganism becomes bound with a major histocompatibility antigen that forms a complex recognized by T cell receptors, recognition molecules of T lymphocytes.

In 1996, Zinkernagel was awarded the Nobel Prize in physiology or medicine in conjunction with Doherty for their discovery on the specificity of the cell mediated immune defense. Their discovery has provided further understanding of how the immune system can determine the difference between foreign microorganisms and major histocompatibility antigens, and is relevant to certain diseases such as **cancer**, multiple sclerosis, diabetes, and rheumatic conditions. Understanding how the immune system works provides new avenues for the development of vaccines.

Since 1992, Zinkernagel has served as the head of the Institute of Experimental Immunology in Zurich, Switzerland.

See also Antigens and antibodies

SOURCES CONSULTED

Books

Ackerknecht, Erwin H. "Contributions of Gall and the Phrenologists to Knowledge of Brain Function," in *The History and Philosophy of Knowledge of the Brain and its Functions*, F. N. L. Poynter, ed. Oxford: Blackwell, 1958, pp. 149–153.

Ackerknecht, Erwin H., and Henri Victor Vallois. *Franz Joseph Gall, Inventor of Phrenology and his Collection*. Madison: University of Wisconsin, 1956.

Adelmann, H. B. *The Embryological Treatises of Hieronymus Fabricius of Aquapendente*. Ithaca, NY: Cornell University Press, 1942.

Alberts, B., et al. *Essential Cell Biology*. New York: Garland Publishing, Inc., 1998.

Alberts, B., et al. *Molecular Biology of the Cell,* 2nd ed. New York: Garland Publishing, Inc., 1989.

Allen, Garland E., William E. Castle, Charles C. Gillispie, eds. *Dictionary of Scientific Biography,* Vol. 3, New York: Scribner, 1971.

Allen, Garland Edward. *Thomas Hunt Morgan: The Man and His Science*. Princeton, NJ: Princeton University Press, 1978.

American Council of Learned Societies, Marshall DeBruhl, ed. *Dictionary of Scientific Biography*. New York: Scribner, 1980.

Audesirk, T., and G. Audesirk. *Biology: Life on Earth,* 4th ed. New Jersey: Prentice Hall Publishing, Inc., 1996.

Baltzer, F. *Theodor Boveri: Life and Work of a Great Biologist*. Berkeley, CA: University of California Press, 1967.

Bear, M., et al. *Neuroscience: Exploring the Brain*. Baltimore: Williams & Wilkins, 1996.

Berman, E.R. *Biochemistry of the Eye: Perspectives in Vision Research*. New York: Plenum Press, 1991.

Beurton, Peter, Raphael Falk, Hans-Jörg Rheinberger., eds. *The Concept of the Gene in Development and Evolution*. Cambridge: Cambridge University Press, 2000.

Bock, G.R., Cardew, G. *Transport and Trafficking in the Malaria-infected Erythrocyte*. John Wiley, Chichester, New York, 1999, pp. 20–36 and pp. 55–73.

Bodmer, W. F., L. L. Cavalli-Sforza. *Genetics, Evolution, and Man*. San Francisco: W.D. Freeman, 1976.

Bonner, J. T. *First Signals: The Evolution of Multicellular Development*. Princeton, NJ: Princeton University Press, 2000.

Bonner, J. T. *The Ideas of Biology*. New York: Harper & Row, 1962.

Bowler, Peter J. *The Mendelian Revolution: The Emergence of Hereditarian Concepts in Modern Science and Society*. Baltimore: Johns Hopkins University Press, 1989.

Boylan, M. *Method and Practice in Aristotle's Biology*. Lanham, MD: University Press of America, 1983.

Brazier, Mary A. B., *A History of Neurophysiology in the 17th and 18th Centuries*. New York: Raven, 1984.

Brooker, R. *Genetics Analysis and Principals*. Menlo Park: Benjamin Cummings, 1999.

Brown, Theodore M. "Lower, Richard." *Dictionary of Scientific Biography* 8, 1973, pp. 523–527.

Brown, Theodore M., "Descartes, Dualism, and Psychosomatic Medicine," in *The Anatomy of Madness*, ed. by W.F. Bynum, Roy Porter, and Michael Shepherd, v. 1, *People and Ideas* London: Tavistock, 1985, pp. 40–62.

Burghard, Charles Fredrick. *"Mechanism and Mind in the Philosophy of René Descartes.* Nashville: Vanderbilt University, 1988.

Campbell, N., J. Reece, and L. Mitchell. *Biology,* 5th ed. Menlo Park: Benjamin Cummings, Inc. 2000.

Carlson, Bruce M. *Human Embryology & Developmental Biology,* 2nd ed. St. Louis: Mosby, 1999.

Carlson, E. A. *The Gene: A Critical History.* Philadelphia: Saunders, 1966.

Carter, Richard B. *Descartes' Medical Philosophy: The Organic Solution to the Mind-Body Problem.* Baltimore: Johns Hopkins University Press, 1983.

Cartmill, Matt, et al. *Human Structure.* Cambridge, MA: Harvard University Press, 1987.

Castiglioni, Arturo. *A History of Medicine.* New York: Knopf, 1941.

Cecil, R.L., et al. *Cecil Textbook of Medicine,* 21st ed. Philadelphia: W.B. Saunders Co., 2000.

Clarke, C. A.*Human Genetics and Medicine.* 3rd ed. Baltimore, MD: E. Arnold, 1987.

Clarke, Edwin, and C.D. O'Malley. *The Human Brain and Spinal Cord.*Berkeley: University of California Press, 1968.

Cooper, Geoffrey M. *The Cell: A Molecular Approach.* Washington, DC: ASM Press, 1997.

Cooter, Roger J. "Phrenology: The Provocation of Progress." *History of Science*14, 1976, pp. 211–234.

Coren, S. *The Left-hander Syndrome: the Causes and Consequences of Left-handedness.* NewYork: Free Press, 1992.

Creese, I., and Claire M. Fraser, eds. *Dopamine Receptors.* New York: A.R. Liss, 1987.

Crouch, James E. *Functional Human Anatomy,* 4th ed. Philadelphia: Lea & Febiger, 1985.

Daintith, John and D. Gjertsen, eds. *A Dictionary of Scientists.* New York: Oxford University Press, 1999.

Darnell, J., H. Lodish, and D. Baltimore. *Molecular Cell Biology.* New York: Scientific American Books, Inc., 1986.

Darwin, C.R. *The Origin of the Species.* London: John Murray, 1859.

Dawkins, R. *The Selfish Gene.* Oxford: Oxford University Press. 1989.

DeGrood, David H. *Haeckel's Theory of the Unity of Nature: A Monograph in the History of Philosophy* Boston: Christopher, 1965.

Des Chene, Dennis. *Spirits and Clocks: Machine and Organism in Descartes.* Ithaca: Cornell University Press, 2001.

Desowitz, R. S. *The Malaria Capers. More Tales of Parasites and People, Research and Reality.* W.W. Norton & Company, New York, 1991.

Duffin, Jacalyn Mary. *To See with a Better Eye: A Life of Laennec.* Princeton: Princeton University Press, 1998.

Elseth, G. D., and K. D. Baumgardner. *Principles of Modern Genetics.* Minnesota: West Publishing Co., 1995.

Emde, Robert N., and John K. Hewitt, eds. *Infancy to Early Childhood: Genetic and Environmental Influences on Developmental Change.* New York: Oxford University Press, 2001.

Emery, A. E. H. *Neuromuscular Disorders: Clinical and Molecular Genetics.* Chicester: John Wiley & Sons, 1998.

Engs, R. C., ed. *Controversies in the Addiction's Field.* Dubuque: Kendal-Hunt, 1990.

Erhart, E. A. *Elementos de Anatomia Humana,* 8th ed. São Paulo: Atheneu Editora São Paulo Ltda., 1992.

Feldman, M, Scharschmidt B.F., Sleisenger, M., Zorab,R., *Sleisenger and Fordtran's Gastrointestinal and Liver Disease,* 6th ed., New York: W.B. Saunders Co., 1998.

Finlayson, C. P. "Monro, Alexander (Secundus)." *Dictionary of ScientificBiography*1974, pp. 482–484.

Frank, Robert G. "Willis, Thomas." *Dictionary of Scientific Biography*1976, pp. 404–409.

Frank, Robert G. *Harvey and the Oxford Physiologists: Scientific Ideas and Social Interaction.* Berkeley: University of California Press, 1980.

Friedman, J., F. Dill, M. Hayden, B. McGillivray. *Genetics.* Maryland: Williams & Wilkins, 1996.

Fruton, J. S. *Molecules and Life. Historical Essays on the Interplay of Chemistry and Biology.* New York: Wiley-Interscience, 1972.

Futuyama, D. J. *Evolutionary Biology.* Sunderland, MA: Sinauer Associates, Inc., 1979.

Ganong W. F. *Review of Medical Physiology,* 16th ed. Prentice-Hall International, Inc., 1993.

Ganong, W. F. *Fisiologia Médica. 2nd ed.* São Paulo: Atheneu Editora, 1975.

Gariepy, Thomas Peter. *Mechanism Without Metaphysics: Henricus Regius and the Establishment ofCartesian Medicine.* New Haven: Yale University, 1990.

Garrett, Laurie. *The Coming Plague.* New York: Random House, 1995.

Garrison, Fielding H. *An Introduction to the History of Medicine.* Philadelphia: Saunders, 1922.

Garrison, Fielding H., *History of Neurology*. Springfield, IL: Charles C. Thomas, 1969.

Gaskings, E. *Investigations into Generation 1651–1828*. Baltimore, MD: Johns Hopkins University Press, 1967.

Gasman, Daniel. *The Scientific Origins of National Socialism: Social Darwinism in Ernst Haeckel and the German Monist League*. London: Macdonald, 1971.

Gasman, Daniel. *Haeckel's Monism and the Birth of Fascist Ideology*. New York: Peter Lang, 1998.

Gauvain, Mary, and Michael Cole. *Readings on the Development of Children*. New York: Scientific American Books, 1993.

Gennaro, A., A. Hart-Nora, J. Nora, R. Stander, L. Weiss, Eds. *Gould Medical Dictionary*, 4th ed., McGraw-Hill, New York, 1979.

Gilbert, S. F., ed. *A Conceptual History of Modern Embryology*. Baltimore, MD: Johns Hopkins Press, 1991.

Giordano, Davide. *Giambattista Morgagni*. Torino: Unione Tipografico, 1998.

Goetz C. G., et al. *Textbook of Clinical Neurology* Philadelphia:W.B. Saunders Company, 1999.

Goldman, *Cecil Textbook of Medicine*, 21st ed. New York: B. Saunders Co., 2000.

Goodman Gilman, A. "Opioid Analgesia and Antagonist", in *The Pharmacological Basis of Therapeutics*, 8th ed. New York: Pergamon Press, 1991.

Gould, Stephen Jay. *Bully for Brontosaurus: Reflections in Natural History*. New York: W.W. Norton & Company, 1992.

Gould, Stephen Jay. *Ever Since Darwin: Reflections in Natural History*. New York: W.W. Norton & Co., 1977.

Gray, H. *Anatomy; Descriptive and Surgical*. New York, Bounty Books, 1997.

Gray, Henry. Carmine D.Clemente (editor). *Anatomy of the Human Body*, 30th ed. Philadelphia: Lea & Febiger, 1984.

Griffiths, A. et al. *Introduction to Genetic Analysis*, 7th ed. New York, W.H. Freeman and Co., 2000.

Grmek, M.D. "Vieussens, Raymond." *Dictionary of Scientific Biography*1976, pp. 25–26.

Guyton & Hall. *Textbook of Medical Physiology*, 10th ed. New York: W.B. Saunders Company, 2000.

Haigh, Elizabeth. *Xavier Bichat and the Medical Theory of the Eighteenth Century*. London: Wellcome Institute for the History of Medicine, 1984.

Haines, Duane E., ed. *Fundamental Neuroscience*. Singapore: Churchill Livingstone, 1997.

Halliwell, B., Aruoma, O. I. ed. *DNA and Free Radicals*. New York: Ellis Horwood, 1993.

Hamburger, V. *The Heritage of Experimental Embryology: Hans Spemann and the Organizer*. Oxford: Oxford University Press, 1988.

Haraway, D. J. *Crystals, Fabrics and Fields: Metaphors of Organicism in Twentieth-Century Developmental Biology*. New Haven: Yale University Press, 1976.

Harrington, Anne. *Medicine, Mind, and the Double Brain*. Princeton:Princeton University Press, 1987.

Herskowitz, I. H. *Genetics,* 2nd ed. Boston: Little, Brown and Company, 1965.

Hintzsche, Erich. "Rudolf Albert von Koelliker." *Dictionary of Scientific Biography*1973, pp. 437–440.

Hollinshead, W. Henry, and Cornelius Rosse. *Textbook of Anatomy,* 4th ed. Philadelphia: Harper & Row, Publishers, 1985.

Horder, T. J., J. A. Witkowski, and C. C. Wylie, eds. *A History of Embryology*. New York: Cambridge University Press, 1986.

Hornof, Zdenek. "Skoda, Josef." *Dictionary of Scientific Biography.* 1975, pp. 450–451.

Hughes, Arthur. *A History of Cytology*. London: Abelard-Schuman, 1959.

Hughes, John Trevor. *Thomas Willis, 1621-1675: His Life and Works*. London: Royal Society of Medicine Services, 1991.

Hughes, S. *The Virus: A History of the Concept*. New York: Science History Publications, 1977.

Ingram, V. M. *Hemoglobins in Genetics & Evolution*. New York: Columbia University Press, 1963.

Isler, Hansruedi. *Thomas Willis, 1621–1675: Doctor and Scientist*. New York: Hafner, 1968.

Jacob, François *The Logic of Life: A History of Heredity*. New York: Pantheon, 1973.

Jenkins, John B. *Human Genetics,* 2nd ed. New York: Harper & Row, 1990.

Johnson, George, and Peter Raven. *Biology: Principles & Explorations*. Austin: Holt, Rinehart, and Winston, Inc., 1996.

Johnson, L. R. *Essential Medical Physiology,* 2nd ed. Philadelphia: Lippincott-Raven Publishers, 1998.

Jones, R. E. *Human Reproductive Biology*. San Diego: Academic Press, Inc., 1991.

Jorde, L. B., et al. *Medical Genetics,* 2nd ed. St. Louis: Mosby, Inc., 2000.

Jorde, L. B., J. C. Carey, M. J. Bamshad, and R. L. White. *Medical Genetics,* 2nd ed. St. Louis: Mosby-Year Book, Inc., 2000.

Kandel, Eric R., et al. *Essentials of Neural Science and Behavior.* Norwalk, CT: Appleton & Lange, 1995.

Kay, L. E. *Who Wrote the Book of Life? A History of the Genetic Code.* Palo Alto: Stanford University Press.

Keenan, Katherine. "Lilian Vaughan Morgan (1870–1952)." *Women in the Biological Sciences: A Biobibliographic Sourcebook* Ed. Grinstein, Louise A., Carol A. Biermann, and Rose K. Rose. Westport, CT: Greenwood Press, 1997.

Kendrew, J., et al.*The Encyclopedia of Molecular Biology.* Oxford: Blackwell Science Ltd., 1994.

Kervran, Roger. *Laennec: His Life and Times.* New York: Pergamon, 1960.

King, Barry G. and Mary Jane Showers. *Human Anatomy and Physiology.* Philadelphia: W.B. Saunders, 1969.

Klaassen, Curtis D. *Casarett and Doull's Toxicology,* 6th ed. McGraw-Hill, Inc. 2001.

Knobel, E., Neill, J.D., *The Physiology of Reproduction.* New York: Raven Press, 1994.

Kopal, Zdenek. "Bichat, Marie-François-Xavier." *Dictionary of Scientific Biography*, v. 2 1970: 122–123.

Kruta, Vladislav. "Flourens, Marie-Jean-Pierre." *Dictionary of Scientific Biography* 1972, pp. 44–45.

Lanteri-Laura, Georges. *Historie de la phrénologie: l'homme et son cerveau selon F.J. Gall.* 2nd ed. Paris: PUF, 1993.

Lee, Victor, and Prajna Das Gupta., eds. *Children's Cognitive and Language Development.* Cambridge, MA: Blackwell Publishers, 1995.

Lens, A., et al., *Ocular Anatomy and Physiology.* Thorofare: SLACK Inc., 1999.

Lesch, John E. *Science and Medicine in France: The Emergence of ExperimentalPhysiology, 1790–1855.* Cambridge, Mass.: Harvard University Press, 1984.

Lesch, John E. *The Origins of Experimental Physiology and Pharmacology in France: 1790–1820: Bichat and Magendie.* Ph.D. thesis, Princeton University, 1977.

Lewin, B. *Genes VII.* New York, Oxford University Press Inc., 2000.

Lewis, Ricki. *Human Genetics: Concepts and Applications,* 2nd ed. IA: William C. Brown, Publishers, 1997.

Lindeboom, Gerrit Arie. *Descartes and Medicine.* Amsterdam: Rodopi, 1978.

Louro, I. D. et al. *Genética Molecular do Câncer,* 1st ed. São Paulo; MSG Produção Editorial; 2000.

Louro, Iuri D., Juan C. Llerena, Jr., Mario S.Vieira de Melo., Patricia Ashton-Prolla, Gilberto Schwartsmann, Nivea Conforti-Froes, eds. *Genética Molecular do Cancer.* São Paulo: MSG Produçao Editorial Ltda., 2000.

Luft, Eric von der, "The Birth of Spirit for Hegel out of the Travesty of Medicine," in *Hegel's Philosophy of Spirit*, ed. by Peter G. Stillman.Albany: SUNY, 1987, pp.25 –42.

Mader, S.S. *Inquiry into Life,* 7th ed., Dubuque: Wm. C. Brown Publisher, 1994.

Madigan M.T., Martinko J.M. and Parker J."*Brock Biology of Microorganisms*, 8th ed. Prentice-Hall International, London, U.K., 1997.

Magner, L. *A History of the Life Sciences.* New York: Marcel Dekker, Inc., 1994.

Mange, E. and A. Mange. *Basic Human Genetics,* 2nd ed. Massachusetts: Sinauer Associates, Inc., 1999.

Martin R. B., D. B. Burr, and N. A. Sharkey. *Skeletal Tissue Mechanics.* New York: Springer-Verlag, 1998.

Martini, F. H., et al. *Fundamentals of Anatomy and Physiology,* 3rd ed. New Jersey: Prentice Hall, Inc., 1995.

Martini, F.H. *Fundamentals of Anatomy and Physiology,* 5th ed. New Jersey: Prentice Hall International, 2001.

Masters, R.D., and McGuire, M. T., ed. *The Neurotransmitter Revolution.* Carbondale: Southern Illinois University Press, 1994.

Mathematics and Physics of Emerging Biomedical Imaging. Washington, DC: National Academy Press, 1996.

Mayr, E. *The Growth of Biological Thought.* Cambridge, MA: Harvard University Press, 1982.

Mayr, E., and P. D. Ashlock. *Principles of Systematic Zoology,* 2nd ed. New York: McGraw-Hill, Inc., 1991.

McClatchey, K. *Clinical Laboratory Medicine.* Baltimore: Williams & Wilkins, 1994.

McClintic, J.R. *Physiology of the Human Body.* New York: John Wiley & Sons, 1985.

Mertz, L. *Recent Advances and Issues in Biology.* Phoenix, Arizona: Oryx Press, 2000.

Mettler, F.A. and Guiberteau M.J. *Essentials of Nuclear Medicine Imaging.* 3rd ed., WB Saunders Company, 1991.

Meyer, A.W. *An Analysis of William Harvey's Generation of Animals.* Stanford, CA: Stanford University Press.

Meyer, Alfred. *Historical Aspects of Cerebral Anatomy.* London: Oxford University Press, 1971.

Miller, Mark Steven and M.T. Cronin, eds. *Genetic Polymorphisms and Susceptibility to Disease.* London: Taylor and Francis, 2000.

Moore K. L. and Persaud T. V. N. *The Developing Human, Clinically Oriented Embryology,* 5th ed. Philadelphia: W.B. Saunders Company, Inc., 1994.

Morgan, Thomas Hunt. *The Theory of The Gene.* New Haven: Yale University Press, 1926.

Motulsy, G. A. and W. Lenz. *Birth Defects.* Amsterdam: Excerpta Medica, 1974.

Mueller, R. F. and I. D. Young. *Emery's Elements of Medical Genetics,* 11th ed. Edinburgh: Churchill Livingstone, 2001.

Needham, J. *The Rise of Embryology.* New York: Cambridge University Press.

Nei, M. *Molecular Evolutionary Genetics.* New York: Columbia University Press, 1987.

Nelkin, D., and M. S. Lindee. *The DNA Mystique.* New York: Freeman, 1995.

Norman, Jeremy M., eds. "Anatomy and Physiology," in *Morton's Medical Bibliography* Brookfield, Vt.: Gower, 1991, pp. 68–249.

Nussbaum, R. L., et al. *Thompson and Thompson Genetics in Medicine,* 6th ed. Philadelphia: W.B. Saunders Co., 2001.

Olby, R. *The Path to the Double Helix.* Seattle, WA: University of Washington Press, 1974.

Olby, Robert C. *Origins of Mendelism,* 2nd ed. Chicago: University of Chicago Press, 1985.

Oppenheimer, J. M. *Essays in the History of Embryology and Biology.* Cambridge, MA: MIT Press, 1967.

Orr, R. *Vertebrate Biology.* Saunders College Publishing, Philadelphia, 1982.

Ottaway, J. H., D. K. Apps. *Biochemistry,* 4th ed. Edinburgh: Baillier Tindall, 1986.

Parent, A. *Carpenter's Human Neuroanatomy.* London: Williams & Wilkins, 1996.

Patten, B. M., et al. *Human Embryology,* 3rd ed. NewYork: McGraw-Hill, Inc., 1968.

Pence, G.E. *Who's Afraid of Human Cloning?* New York: Rowman & Littlefield, 1998.

Persaud, T. V. N. *Early History of Human Anatomy.* Springfield, IL: Charles C. Thomas, 1984.

Piaget, J. *The Child's Conception of the World.* New York: Littlefield Adams, 1990.

Piaget, Jean. *The Psychology of the Child.* New York: Basic Books, 1972.

Pinto-Correia, C. *The Ovary of Eve: Egg and Sperm and Preformation.* Chicago, IL: University of Chicago Press, 1997.

Portugal, F. H. and J. S. Cohen. *A Century of DNA.* Cambridge, MA: The MIT Press.

Preston, Richard. *The Hot Zone.* New York: Random House, 1995.

Primrose, S.P. *Principles of Genome Analysis.* Oxford: Blackwell, 1995.

Rand, H. P. *Peptides as Mediators in Pharmacology.* New York: Churchill Livingstone, 1995.

Reiser, Stanley Joel. "The Stethoscope and the Detection of Pathology by Sound," in *Medicine and the Reign of Technology* Cambridge: Cambridge University Press, 1978, pp. 23–44.

Rhoades, R. and R. Pflanzer. *Human Physiology,* 3rd ed. New York: Saunders College Publishing, 1996.

Rieger, R, A. Michaelis, and M. M. Green. *Glossary of Genetics and Cytogenetics,* 4th ed. Berlin: Springer Verlag, 1976.

Riese, Walther. *A History of Neurology.* New York: MD Publications, 1959.

Rifkin, J. *The Biotech Century.* Putnam Publishing Group. 1998.

Ritter, B., et al. *Biology,* B.C. ed. Scarborough: Nelson Canada, 1996.

Robeck, Mildred C. *Infants and Children: Their Development and Learning.* New York: McGraw-Hill.

Roe, S. A. *Matter, Life, and Generation. Eighteenth Century Embryology and the Haller-Wolff Debate.* New York: Cambridge University Press.

Romer, A., T. Parsons, *The Vertebrate Body; the Shorter Version.* W.B. Saunders Company, Philadelphia, 1978.

Rosenzweig, M., et al. *Biological Psychology.* Sunderland: Sinauer Associates, Inc., 1996.

Rothschuh, K.E. "Pflüger, Eduard Friedrich Wilhelm." *Dictionary of Scientific Biography* 10, 1974, pp. 578–581.

Rothwell, Norman V. *Understanding Genetics,* 4th ed. New York: Oford University Press, 1988.

Rothwell, Norman V. *Human Genetics.* New Jersey: Prentice-Hall, 1977.

Russell, P. J. *Genetics,* 3rd ed. New York: Harper Collins, 1992.

Sager, R and F. J. Ryan. *Cell Heredity.* New York: John Wiley & Sons, 1961.

Saladin, B. *Anatomy and Physiology; The Unity of Form and Function.* Boston: WCB/McGraw Hill, 1998.

Sanders, M. J. *Mosby's Paramedic Textbook,* 2nd ed. St. Louis: Mosby Inc., 2001.

Sauer, Gordon C. *Manual of Skin Diseases,* 5th ed. Philadelphia: J.B. Lippincott Company, 1985.

Sayre, A. *Rosalind Franklin & DNA.* New York: Norton, 1975.

Schaler, J.A. *Addiction is a Choice.* Chicago: Open Court Publishers, 2000.

Schottenfeld, D., and J.F. Fraumeni Jr., eds. *Cancer Epidemiology and Prevention* New York: Oxford University Press, 1996.

Schuster, John Andrew. *Descartes and the Scientific Revolution, 1618–1634: An Interpretation.* Princeton, NJ: Princeton University, 1977.

Scott, Joseph Frederick. *The Scientific Work of René Descartes(1596–1650).*. London: Taylor & Francis, 1952.

Scriver, Charles R. et al. *The Metabolic and Molecular Bases of Inherited Disease,* 8th ed. New York: McGraw-Hill Professional Book Group, 2001.

Seashore, M. and R. Wappner. *Genetics in Primary Care & Clinical Medicine.* Stamford: Appleton and Lange, 1996.

Sebba, Gregor. *Bibliographia Cartesiana: A Critical Guide to the Descartes Literature, 1800–1960.* Hague: Martinus Nijhoff, 1964.

Seely, R., Stephens, T., and P. Tate. *Essentials of Anatomy and Physiology.* Boston: WCB/McGraw Hill, 1999.

Shephard R. J. *Aging, Physical Activity, and Health.* Human Kinetics, 1997.

Shine, Ian, and Sylvia Wrobel. *Thomas Hunt Morgan: Pioneer of Genetics.* Lexington: University Press of Kentucky, 1976.

Simpson, G. G. *Principles of Animal Taxonomy.* New York: Columbia University Press, 1961.

Singer, Charles. *A Short History of Anatomy and Physiology from the Greeks to Harvey.* New York: Dover, 1957.

Singer, M. and P. Berg. *Genes and Genomes.* Mill Valley, CA: University Science Books, 1991.

Smith, A.D., et al. *Oxford Dictionary of Biochemistry and Molecular Biology.* New York: Oxford University Press, Inc., 1997.

Snustad, D. Peter, Michael J. Simmons, and John B. Jenkins. *Principles of Genetics.* New York: John Wiley, 1997.

Solomon, Eldra Pearl, Linda R. Berg, and Diana W. Martin. *Biology,* 5th ed. New York: Saunders College Publishing, 1999.

Sorenson, J.A. and M.E. Phelps. *Physics in Nuclear Medicine,* 2nd ed. Grune & Stratton Inc, 1987.

Spector, D.L., R.D. Goldman, and L.A. Leinwand. *Cells: A Laboratory Manual.* Plainview, NY: Cold Spring Harbor Laboratory Press, 1998.

Spemann, H. *Embryonic Development and Induction.* New York: Hafner Publishing Company, 1967.

Spicker, Stuart F. *The Philosophy of the Body: Rejections of Cartesian Dualism.* Chicago: Quadrangle, 1970.

Stadtman, E. R. and P. B. Chock, eds. *Current Topics in Cellular Regulation,* Vol. 36, London: Academic Press Ltd., 2000.

Stanbury, J.B. et al. *The Metabolic Basis of Inherited Diseases.* New York: McGraw-Hill. 1988.

Steinberg, L. *Adolescence,* 5th ed. New York: McGraw-Hill, 1999.

Stent, G. S., ed. Watson, J. D. *The Double Helix. A Personal Account of the Discovery of the Structure of DNA.* New York: Norton, 1980.

Stern, M. B., and H. I. Hurtig. *The Comprehensive Management of Parkinson's Disease.* New York: PMA Publishing Corp., 1988.

Steudel, Johannes. "Auenbrugger, Joseph Leopold." *Dictionary of Scientific Biography*, v. 1 1970, 332–333.

Strachan, T., and A. Read. *Human Molecular Genetics,* 1st ed. Oxford: Bios Scientific Publishers Ltd., 1996.

Stryer, L. *Biochemistry,* 3rd ed., New York, W.H. Freeman and Co., 1988.

Stryer, L. *Biochemistry,* 4th ed. New York: W.H. Freeman and Co., 1995.

Sturtevant, A. H. *A History of Genetics.* New York: Harper & Row, 1965.

Su, X., Wellems, T.E. *Genome Discovery and Malaria Research: Current Status and Promise. Malaria: Parasite Biology, Pathogenesis, and Protection.* Washington D.C.: American Society of Microbiology Press, 1998.

Tasman A.,et al. *Psychiatry.* Philadelphia: W.B. Saunders Company, 1997.

Thompson, M., et al. *Genetics in Medicine.* Philadelphia: Saunders, 1991.

Tomei, L. David, F.O. Cape, eds. *Apoptosis: The Molecular Basis of Cell Death.* Cold Spring Harbor, N.Y.: Cold Spring Harbor Laboratory Press, 1991.

Tortora G., and S. Grabowski. *Principles of Anatomy and Physiology,* 7th ed. New York: Harper Collins College Publishers, 1993.

Tortora, G. J., and S. R., Grabowski. *Principles of Anatomy and Physiology,* 9th ed. New York: John Wiley and Sons Inc., 2000.

Van De Graaff, K. M. *Human Anatomy,* 6th ed. Boston: McGraw Hill, 2002.

Van De Graaff, K. M., I. S. Fox. *Concepts of Human Anatomy and Physiology.* WCB Publishers, 1992.

Vassiliki, Betty Smocovitis. *Unifying Biology: The Evolutionary Synthesis and Evolutionary Biology.* Princeton, NJ: Princeton University Press, 1996.

Verma, R. S., and A. Babu. *Human Chromosomes Principles and Techniques,* 2nd ed. McGraw-Hill Inc., Health Professions Division. 1995.

Voet, D., and J. Voet, *Biochemistry,* 2nd ed. New York: John Wiley and Sons, Inc., 1995..

Vogelstein, B & Kinzler, K., eds.. *The Genetic Basis of Human Cancer.* New York: McGraw-Hill, 1998.

Waterson, A. and L. Wilkinson. *An Introduction of the History of Virology.* Cambridge: Cambridge University Press, 1978.

Watson, J. D., et al. *Recombinant DNA,* 2nd ed. New York: Scientific American Books, Inc., 1992.

Watson, Richard A. *The Downfall of Cartesianism, 1673–1712: A Study of Epistemological Issues in Late 17th-Century Cartesianism.* Hague: Martinus Nijhoff, 1966.

West, J.B. *Respiratory Physiology - the Essentials,* 5th ed. Baltimore, MD: Williams & Wilkins, 1995.

White, N. J., Breman, J. G. *Harrison's Principles of Internal Medicine,* 14th ed. Mc Graw Hill 1998, pp. 1180–1188.

Willier, B. H. and J. M. Oppenheimer, eds. *Foundations of Experimental Embryology.* New York: Hafner Press, 1974.

Wilson J. D. et al. *Williams Textbook of Endocrinology,* 9th ed. Philadelphia: W.B. Saunders Company, 1998.

Wilson, E. O. *The Diversity of Life.* Cambridge, MA: The Belknap Press of Harvard University Press, 1992.

Wright-St. Clair, Rex Earl. *Doctors Monro: A Medical Saga.* London: Wellcome Historical Medical Library, 1964.

Young, Robert M. "Gall, Franz Joseph." *Dictionary of Scientific Biography* 5, 1972, pp. 250–256.

Zigler, Edward, and Matia Finn-Stevenson. *Children in a Changing World: Development and Social Issues,* 2d ed. Pacific Grove, CA: Brooks/Cole Publishing Company, 1993.

Journals

Acierno, L.J., and T. Worrell. "Profiles in Cardiology: James Bryan Herrick." *Clinical Cardiology* 23 (2000): 230–232.

Ackermann U. "Essentials of Human Physiology." *Mosby Year Book* (1992).

American Judicial Society. "Genes and Justice: The Growing Impact of the New Genetics on the Courts," *Judicature* 83(November-December 1999).

Artlett, Carol M. et al. "Identification of Fetal DNA and Cells in Skin Lesions From Women With Systemic Sclerosis." *The New England Journal of Medicine* 338: 1186–1191.

Associated Press. "New Cancer Treatment Starves Tumors." *The Detroit News* (28 November 1997).

Axel, R. "The Molecular Logic of Smell." *Scientific American* 273 (October 1995): 154–159.

Battaglia F. C., Regnault T. R. "Placental Transport and Metabolism of Amino Acids." *Placenta* 22, 2 (February-March 2001):145–61.

Baxevanis, A. D. "The Molecular Biology Database Collection: an Updated Compilation of Biological Database Resources." *Nucleic Acids Research* 29 (January 2001): 1–10.

Berg, P., et al. "Asilomar Conference on Recombinant DNA Molecules." *Science* 188 (6 June 1975): 991–994.

Boguski, M. S. "The Turning Point in Genome Research." *Trends in Biochemical Sciences* 20 (August 1995): 295–296.

Bouwens, L., De Bleser, P., Vanderkerken, K., Geerts, B., and E. Wisse, "Liver Cell Heterogeneity." *Enzyme* 46 (1992):155–168.

Brinster, R. "The Effect of Cells Transferred Into Mouse Blastocyst on Subsequent Development." *Experimental Medicine* (1974): 1049–1056.

Brownlee, B. "Inside the Teen Brain." *U.S. News & World Report* (August 9, 1999).

Burton GJ, Hempstock J, Jauniaux E. *Nutrition of the Human Fetus During the First Trimester-a Review."* Placenta. 22 (April 2001): Supplement A:S70 –7.

Cantor, Geoffrey N. "A Critique of Shapin's Interpretation of the Edinburgh Phrenology Debate."*Annals of Science* 32 (1975): 245–256.

Cantor, Geoffrey N. "The Edinburgh Phrenology Debate: 1803–1828." *Annals of Science* 32 (1975): 195–218.

Caplan, A. L. "If Gene Therapy is the Cure, What is the Disease?" *Gene Mapping* (1992): 128–141.

Chagnon, Y.C., et al. "The Human Obesity Gene Map: the 1999 Update." *Obesity Research* 8 (2000): 89–117.

Chakravarti, A. "To a Future of Genetic Medicine" *Nature* 409 (2001): 822–823.

Collins F. S., and V. A. McKusick. "Implications of the Human Genome Project for Medical Science." *JAMA* 285 (7 February 2001): 540–544.

Colson E. R., Dworkin P. H. "Toddler development." *Pediatr Rev.* 1997 Aug:18(8): 255 –9. Review.

Condic, M. L. "Adult Neuronal Regeneration Induced by Transgenic Integrin Expression." *Journal of Neuroscience* 21, 13 (July 1, 2001): 4782–4788.

Conforti-Froes, N., et al. "Predisposing Genes and Increased Chromosome Aberrations in Lung Cancer Cigarette Smokers." *Mutation Research* 379 (1997): 53–59.

Crick, F. H. C. "The Origin of the Genetic Code" *Journal of Molecular Biology* 38 (1968): 367–379.

Dalton, P., N. Doolittle, H. Nagata, and P. A. S. Breslin. "The Merging of the Senses: Integration of Subthreshold Taste and Smell." *Nature Neuroscience* 3 (2000): 431–432.

Darnell, J. E., Jr. "The Processing of RNA." *Scientific American* 249 (1983): 90–100.

Deitsch, K.W., Wellems, T.E. "Membrane Modifications in erythrocytes Parasitized by *Plasmodium falciparum.*" *Mol. Biochem. Parasitol.* 76 (1996) 1–10.

Devroey, P., B. Mannaerts, J. Smitz, C. Bennink, and A.V. Steirteghem. "First Established Pregnancy and Birth After Ovarian Stimulation with Recombinant Follicle Stimulating Hormone." *Human Reproduction* 8 (1993): 863–865.

Donnelly, S., C.R. McCarthy, and R. Singeleton, Jr. "The Brave New World of Animal Biotechnology." *Special Supplement, Hastings Center Report* (1994).

Douglas, J. T., and D. T. Curie. "Targeted Gene Therapy." *Tumor Targeting* (1995): 67–84.

Easton, D. F., D. Ford, D. T. Bishop, et cols. "Breast and Ovarian Cancer Incidence in BRCA1 Mutation Carriers." *American Journal of Human Genetics* 56 (1995): 265–271.

Editor. "Understanding the Causes of Schizophrenia." *The New England Journal of Medicine* 8 (25 February 1999).

Edmondson, D. G. and Roth, S. Y. "Interactions of Transcriptional Regulators with Histones." *Methods* 15 (1998): 355 – 384.

Farina A, and Bianchi DW. "Fetal Cells in Maternal Blood as a Second Non-invasive Step for Fetal Down Syndrome Screening." *Prenatal Diagnosis* 18 (September 1998): 983–984.

Farina A., et al. "A Latent Class Analysis Applied to Patterns of Fetal Sonographic Abnormalities: Definition of Phenotypes Associated With Aneuploidy." *Prenatal Diagnostics* 19 (September 1999): 840–845.

Farrell P. G. et al. "Interferon Action: Two Distinct Pathways for Inhibition of Protein Synthesis by Double-stranded RNA." *Proc. Nat. Acad. Sci.* 75 (1978): 5896.

Folkman, Judah. "Fighting Cancer by Attacking its Blood Supply." *Scientific American* (September 1996).

Frauenfelder, H., and H. C. Berg. "Physics and Biology." *Physics Today* 2 (February 1994): 20–21.

Freiher, G. "The Race to Develop a Painless Blood Glucose Monitor." *Medical Devices and Diagnostic Industry* (March 1997): 58–64.

Fung, J. J. "Transplanting Animal Organs into Humans is Feasible." *USA Today* (November 1999).

Glickstein, Mitchell. "The Discovery of the Visual Cortex." *Scientific American* 259, 3 (September 1988): 118–127.

Gobbi H., Barbosa A. J. A., Teixeira V. P. A., and Almeida, H. O. "Immunocytochemical Identification of Neuro-endocrine Markers in Human Cardiac Paraganglion-like Structures." *Histochemistry* 95 (1991): 337–340.

Golden, Frederic. "Mental Illness: Probing the Chemistry of the Brain." *Time* 157 (January 2001).

Greenwood, M. R. C., and P. R. Johnson. "Genetic differences in adipose tissue metabolism and regulation." *Annals of the New York Academy of Science* 676 (1994): 253– 269.

Grundfast, K. M., J. L. Atwood, and D. Chuong. "Genetics and Molecular Biology of Deafness." *Otolaryngologic Clinics of North America* 32 (December 1999):1067–1088.

Halim, N. S. "Research: New Approaches to Drug Addiction Therapy." *The Scientist* 13, 19 (September 1999): 12.

Hamon, P., M. Robert, P Schamasch, M. Pugeat. "Sex Testing at the Olympics." *Nature* 358 (1992): 447.

Hayflick, Leonard. "How and Why We Age." *Experimental Gerontology* 33 (1998): 639–653.

Hayward, C., Killen, J., Wilson, D., and Hammer, L. "Psychiatric Risk Associated with Early Puberty in Adolescent Girls." *Journal of the American Academy of Child & Adolescent Psychiatry* 36, 2 (1997): 255–262.

Herrick, J. B. "Peculiar Elongated and Sickle-shaped Red Blood Corpuscles in a Case of Severe Anemia." *Archives of Internal Medicine* 6 (1910): 517–521.

Hersher, Leonard, "On the Absence of Revolutions in Biology." *Perspectives in Biology and Medicine* 1, 3 (Spring 1988): 318–323.

Hoskins, K. P., J. E. Stopfer, K. A. Calzone, et cols. "Assessment and Counseling For Women With a Family History of Breast Cancer—A Guide For Clinicians." *JAMA* 273 (1995): 577–585.

Humayun, M. S., de Juan, E Jr., Dagnelie, G. "Visual Perception Elicited by Electrical Stimulation of Retina in Blind Humans." *Archives of Ophthalmology* 114 (1996): 40–46.

Illsley, N. P. "Glucose Transporters in the Human Placenta." *Placenta* 1, (January 2000):14–22.

Ingram, S. L., Vaughn, C. W., Bagley, E. E., Connor, M., and Christie, M.J. "Enhanced Opiod Efficacy in Opiod Dependence is due to an Additional Signal Transduction Pathway." *Journal of Neuroscience* 18 (1998): 10269–10276.

Ingram, V. M. "Gene Mutations in Human Hemoglobin: the Chemical Difference Between Normal and Sickle Hemoglobin." *Nature* 180 (1957): 326–328.

International Genome Sequencing Consortium. "Initial Sequencing and Analysis of the Human Genome." *Nature* 409 (2001): 860–921.

Jeffords, J. M. and Tom Daschle. "Political Issues in the Genome Era," *Science* 291 (16 February 2001): 1249–50.

Jimenez P. A., and M. A. Rampy. "Keratinocyte Growth Factor-2 Accelerates Wound Healing in Incisional Wounds." *Journal Surg. Res.* 81, 2 (Febrary 1999): 238–242.

Johnson, P., and D. A. Hopkinson. "Detection of ABO Blood Group Polymorphism by Denaturing Gradient Gel Electrophoresis." *Human Molecular Genetics* 1 (1992): 341 – 344.

Kalfas, I. H. "Principles of bone healing." *Neurosurgery Focus* 10, 4 (2001): 1–4.

Keisell, D. P., J. Dunlop, H. P. Stevens, et al. "Connexin-26 mutations in hereditary non-syndromic sensorineural deafness." *Nature* (1997): 80–83.

Kerr, J. F. K., A. Wyllie, and A. H. Currie. "Apoptosis, a Basic Biological Phenomenon With Wider Implications in Tissue Kinetics." *British Journal of Cancer* 26 (1972): 239–45.

Kincardon, J., and C. Boutzale. "The Physiology of Pigmented Nevi." *Pediatrics* 104, 4 (October 1999): 1043–1049.

Knight, J. A. "The biochemistry of ageing." *Adv. Clin. Chem.* 35 (2000):1–62

Kobayashi, M., T. Takatori, K. Iwadate, and M. Nakajima. "Reconsideration of the Sequence of Rigor Mortis Through Postmortem Changes in Adenosine Nucleotides and Lactic Acid in Different Rat Muscles." `1Forensic Science International/1 82 (1996): 243–253.

Kyogoku, Y., R. C. Lord, and A. Rich. "Hydrogen Bonding Specificity of Nucleic Acid Purines and Pyrimidines in Solution." *Science* 154 (28 October 1966): 518–520.

Lacayo, Richard. "For Whom the Bell Curves." *Time* (October 4, 1994).

Lawrence Christopher. "Alexander Monro *Primus* and the Edinburgh Manner ofAnatomy." *Bulletin of the History of Medicine* 62, 2 (Summer1988): 193–214.

Lesky, Erna. "Structure and Function in Gall." *Bulletin of the History of Medicine* 44 (1970): 297–314.

Maassen, J. A., J. J. Jansen, T. Kadowaki, J. M. van den Ouweland, L. M. 't Hart and H. H. Lemkes. "The Molecular Basis and Clinical Characteristics of Maternally Inherited Diabetes and Deafness (MIDD), a Recently Recognized Diabetic Subtype." *Experimental and Clinical Endocrinology and Diabetes* 104, 3 (1996): 205–11.

Marrack, F. and Kappler, J. W. "How the Immune System Recognizes the Body" *Scientific American* (September 1993): 81–87

Maxam, A., and W. Gilbert. "A New Method of Sequencing DNA." *Proceedings of the National Academy of Sciences of the United States of America* 74 (1998): 560–564.

McCartney, G., and P. Hepper. "Development of Lateralized Behaviour in the Human Fetus From 12 to 27 Weeks' Gestation." *Developmental Medicine and Child Neurology* 41, 2 (February 1999): 83–86.

McClearn, G. E. "Biogerontologic theories." *Exp. Gerontol* 32 (1997): 3–10

Merten, Thomas R. "Introducing Students to Population Genetics and the Hardy-Weinberg Principle." *The American Biology Teacher* 54 (1992): 103–107.

Mihic, S.J., and Harris, R.A. "Molecular Mechanisms of Aaesthetic Actions on Ligand-gated Ion Channels." *Neurotransmissions* 13, 1 (February 1997): 1–12.

Miller, T., et al. "Exercise and Its Role in the Prevention and Rehabilitation of Cardiovascular Disease." *Annals of Behavioural Medicine* (March 1997): 220–229.

Mitchell CB: *Ethics & Medicine*17, 3 (2001).

Pennisi, E.: "After Dolly, a Pharming Frenzy." *Science* 279 (1998) 646–648.

Pennisis, E. "Architecture of Hearing." *Science* 278 (1997): 1223.

Perera, F. P., and I. B. Weinstein. "Molecular Epidemiology: Recent Advances and Future Directions." *Carcinogenesis* 21 (2000): 517–524.

Pertl, B., and D. W. Bianchi. "First Trimester Prenatal Diagnosis: Fetal Cells in the Maternal Circulation." *Seminar Perinatology* 23 (October 1999): 393–402.

Pugsley, M. K., and R. Tabrizchi "The Vascular System. An Overview of Structure and Function." *J.Pharmacol. Toxicol.Meth.* 44 (2000): 333–340.

Rabinowicz, T., J. MacDonald-Comber Petot, P. Gartside, D. Sheyn, G. de Courten-Meyers. "Structure of the Cerebral Cortex in Both Men and Women." *Journal of Neuropathy and Experimental Neurology* 61 (January 2002): 46.

Shapin, Steven. "Phrenological Knowledge and the Social Structure of Early Nineteenth-CenturyEdinburgh." *Annals of Science* 32 (1975): 219–243.

Sill, Geoffrey M. "Neurology and the Novel: Alexander Monro *Primus* and*Secundus*, *Robinson Crusoe*, and the Problem ofSensibility." *Literature & Medicine* 16, 2 (Fall 1997): 250–265.

Smaglik, P. "Gene Therapy–The Next Generation." *The Scientist* 12 (1998): 4.

Smith, D. V., and S. J. St. John. *"Neural Coding of Gustatory Information."Current Opinion in Neurobiology"* 9 (April 1999): 427-435.

The British Society of Cell Biology, *Symposium Notes and Newsletter*, (Summer 2000).

Thomson, J., et al. "Embryonic Stem Cell Lines Derived From Human Blastocysts." Venter, J. C., et al. "The Sequence of the Human Genome." *Science* 291 (2001): 1304–1351.

Vogan, K. J., and C. J. Tabin. "A new spin on handed symmetry." *Nature* 397 (1999): 295–298.

Walsh, K., and L. J. Alexander, "Update on Chronic Viral Hepatitis." *Postgrad. Med. J.* 77 (2001): 498–505

Walsh, T., and Devlin, M. *"Eating Disorders: Progress and Problems."* Science 280 (1998): 1387–1390.

Watson, J. D., and F. H. C. Crick. "Genetical Implications of the Structure of Deoxyribonucleic Acid." *Nature* 171 (1953): 964–969.

Watson, J. D., and F. H. C. Crick. "Molecular structure of nucleic acids." *Nature* 171 (1953): 737–738.

Weissman, I. L. and Cooper, M. D. "How the Immune System Develops." *Scientific American* (September 1993): 67–69.

White, L. E., G. Lucas, A. Richards, and D. Purves. "Cerebral asymmetry and handedness." *Nature* 368 (1994): 197–198.

Wilmut, I., A. E. Schnieke, J. McWhir, A. J. Kind, and K. H. S. Campbell. "Viable Offspring Derived From Fetal and Adult Mammalian Cells." *Nature* 385 (1997): 810–813.

Wilmut, Ian, and Keith Kasnot. "Cloning for Medicine." *Scientific American* 279 (1 December 1998): 58–63.

Yaspo, M. L. et al. "The DNA Sequence of Human Chromosome 21." *Nature* 6784 (May 2000): 311 –319.

Yuspa, S. H. "Overview of Carcinogenesis: Past, Present and Future." *Carcinogenesis* 21 (2000): 341–344.

Web Sites

(Editor's note: As the World Wide Web is constantly expanding, the URLs listed below may be altered or nonexistent as of February 14, 2002.)

"A History of the University of California San Francisco. " Biographies: Choh Hao Li (1913–1987)." <http://www.library.ucsf.edu/ucsfhistory/biographies.li.html>

A Science Odyssey: People and Discoveries. "Role of Endorphins Discovered" 1975. <http://www.pbs.org/wgbh/aso/databank/entries/dh75en.htm>

Agder University College. "Maximal Oxygen Consumption—The VO$_2$ MAX." 1996. (06 January 2002) <http://home.hia.no/~stephens/vo2max.html>

Alliance for Lung Cancer (October 5, 2001) <http://www.alcase.org>/sourcetxt>

Altrius Biomedical network. "Structure and function of skin." (November 5, 2001). <http://www.skin-information.com>

Altruis Biomedical Network. "Gynecology." (2000). <http://www.e-gynecologic.com>

Altruis Biomedical Network. "Testicles." (2000). <http://www.e-testicles.com>

Altruis Biomedical Network. "The Human Breast." (2000). <www.e-breasts.net>

Altruis Biomedical Network. "Sweat Glands." (2002). <http://www.sweating.net>

American Academy of Family Physicians. "Plantar Fasciitis: A Common cause of Heel Pain." 2000 (December 7, 2001). <http://familydoctor.org/handouts/140.html>

American Academy of Orthopaedic Surgeons. "Dupuytren's contracture." 2000. (November 6, 2001). <http://orthoinfo.aaos.org/fact/thr_report.cfm?Thread_ID=140&topcategory=Hand.HTML>

American Association for the Advancement of Science. "Stem Cell Research and Applications: Monitoring the Frontiers of Biomedical Research." 1999. <http://www.aaas.org/spp/dspp/sfrl/projects/stem/report.pdf>

American Heart Association, " Stroke Risk Factors" (2001). <http://www.americanheart.org/Heart and Stroke A Z Guide/strokeri.html>

American Heart Association. "Children and Heart Disease." 2001 (December 4, 2001). <http://www.amercianheart.org/presenter.html>

American Heart Association. "Pericardium and Pericarditis." 2001 (January 13, 2002). <http://www.americanheart.org/presenter.jhtml>

American Lung Association "How our lungs work." (January 6, 2002). <http://www.lungusa.org/learn/>

American Museum of Natural History. "Infection!" 2000. (December 6, 2001) <http://www.amnh.org/nationalcenter/infection/03_inf/03_inf.html>

American Psychological Association. "APA Task Force Examines the Knowns and Unknowns of Intelligence." (1996). <http://www.apa.org/releases/intell.html>.>

American Psychological Association. "Concern follows psychological testing." *APA Monitor* (December 1999). <http://www.apa.org/monitor/dec99/ss4.html>

American Sleep Foundation, "ABCs of ZZZs," (December 2, 2001). <http://www.sleepfoundation.org/publications/ZZZs.html>

American Sleep Foundation, "Helping Yourself to a Good Night's Sleep," 1999 (02 December 2001). <http://www.sleepfoundation.org/publications/goodnights.html>

Apoptosis Interest Group. "About Apoptosis." (15 June 2001). <http//www.nih.gov/sigs/aig/Aboutapo.html>

ASPCA National Animal Poison Control Center. "List of the most dangerous types of house and garden plants." 1998 (November 5, 2001). <http://www.vetprof.com/clientinfo/toxicplants.html>

Athabasca University. "Tutorial 29: Sense of Taste." 2001 (10 January 2002). <http://psych.athabascau.ca/html/Psych402/Biotutorials/29/intro.shtml>

Australian Academy of Science " Video Histories of Australian Scientists - Teachers notes: Jacques Miller" <http://www.science.org.au/scientists/notesjm.html>

Australian National University. "The Physiology of Hearing." 2001 (05 January 2002). <http://www.anu.au/ITA/ACAT/drw/PpofM/hearing/hearing1.html>

BaylorCollege of Medicine. "Review of Anatomy: The Neck." 1996 (10 January 2002). <http://www.bcm.tmc.edu/oto/studs/anat/neck>

BBC News. "Q & A: Anthrax Infection." 2001 (06 December 2001). <http://news.bbc.co.uk/hi/english/health/newsid_1580000/1580930.stm>

BBC News. "The Nature of Addiction." 2000 (04 December 2001). <http://news.bbc.co.uk/hi/english/health/newsid_784000/784033.stm>

Brescia University. "The Wobble Hypothesis." (29 April 2001). <http://www.med.unibs.it/∽marchesi/protsyn.html>

British Council. "Alec Jeffreys—genetic evidence." (1 March 2001).<http://www.britishcouncil.org/science/science/personalities/text/ukperson/jeffreys.htm>

Brown University. "VSLI Models of Neural Systems." (02 May 2001). <http://landow.stg.brown.edu/cpace/sciencedgneuro/present/systems.html>>

Caceci, T. Veterinary Histology. <http://education.vetmed.vt.edu/Curriculum/VM8054/Labs/Lab17/EXAMPLES/Excheek.htm>/sourcetxt>

Canadian Space Agency. "Human Physiology." 2000 (03 December 2001). <http://www.space.gc.ca/csa_sectors/human_presence/missions/sts78/s78sc3e.htm>

Centers for Disease Control. "Viral Hemorrhagic Fevers: Fact Sheets." 2001 (05 December 2001). <http://www.cdc.gov/ncidod/dvrd/spb/mnpages/dispages/vhf.htm>.

Centers for Disease Control. "Antibiotic Resistance." 2001 (06 December 2001). <http://www.cdc.gov/ncidod/dbmd/antibioticresistance>

Christian Brothers University. Bio 211: Vertebrate Embyology. Spring, 2002. <http://www.cbu.edu/~aross/embhome.htm>

CNN.com. "Pre-1990 transfusions may have infected thousands with hepatitis C." 1998 (14 January 2002). <http://www.cnn.comHEALTH/9807/08/hepatitis.notify>.

Colorado State University, Anatomy and Physiology, 2002. <http://arbl.cvmbs.colostate.edu/hbooks/pathphys/misc_topics/peritoneum.html>

Colorado State University. " Histology of the Adrenal Medulla" 2002. <http://arbl.cvmbs.colostate.edu/hbooks/pathphys/endocrine/adrenal/histo_medulla.html>

Colorado State University. "Biliary Excretion of Waste Products: Elimination of Bilirubin." 2001. <http://arbl.cvmbs.colostate.edu/hbooks/pathphys/digestion/liver/bilirubin.html>

Colorado State University. "Gonadotrophins: Luteinizing and Follicle Stimulating Hormones." 1998 (08 January 2002). <http://arbl.cvmbs.colostate.edu/hbooks/pathphys/endocrine/hypopit/lhfsh.html>

Colorado State University. "Oxytocin." 1998 (05 November 2001). <http://arbl.cvmbs.colostate.edu/hbooks/pathphys/endocrine/hypopit/oxytocin.html>

Colorado State University. "Pathophysiology of the Digestive System". 1995 (5 August 2000). <http://arbl.cvmbs.colostate.edu/hbooks/pathphys/digestion/index.html>

Colorado State University. "Secretion of bile and the role of bile acids in secretion." 1999 (21 January 2001). <http://arbl.cvmbs/colostate.edu/hbooks/pathphys/digestion/liver/bile.html>

Colorado State University. "Pathophysiology of the Digestive System". 1995 (5 August 2000). <http://arbl.cvmbs.colostate.edu/hbooks/pathphys/digestion/index.html>

Colorado State University. "Prehension, Mastication, Swallowing." 2001 (10 December 2001). <http://arbl.cvmbs.colostate.edu/hbooks/pathphys/digestion/pregastric/mastication.html>

Columbia University. "Disorders of blood coagulation." 2001 (30 November 2001). <http://cpmcnet.columbia.edu/texts/guide/hmg23_0009.html>

Contact a Family. "Heart Defects." 1997 (04 December 2001). <http://www.cafamily.org.uk/Direct/h27.html>

Cornell University. "Breast Feeding and Breast Cancer Risk." 1999. <http://www.cfe.cornell.edu/bcerf/FactSheet/Diet/fs29.brfeed.cfm>

Craig, Sandy. "Appendicitis, Acute." eMedicine Journal, December 13 2001, Volume 2, Number 12 (29 January 2002). <http://www.emedicine.comEMERG/topic41.htm>

Crime Library. "Serology: It's in the Blood." 2001(13 January 2002). <http://www.crimelibrary.com> forensics/serology/3.htm>

Darnell, J., H. Lodish, and D. Baltimore. *Molecular Cell Biology*, New York: Scientific American (2001). <http://www.americanheart.org/Heart and Stroke A Z Guide/riskfact.html>

Deafness Research Foundation. "Causes of Hearing Loss." 2001 (07 January 2002). <http//www.drf.org/cms/index.cfm?displayArticle=15>

Division of Physiology, GKT School of Biomedical Sciences, King's College London, Sherrington Building, St Thomas' Campus. "Physiology." 2002 <http://www.umds.ac.uk/physiology/rbm/hypoxfx.htm>

Dräger Aerospace. "Human Breathing Physiology." 2001. <http://www.drager.com>com/DAE/o2_in_ae/human_breath.jsp >

Drug Base."Drowning and Near-Drowning." 2001 (08 January 2002). <http://www.drugbase.co.za/data/med_info/drowning.htm>

e-Kidneys.net. "The Kidneys." 2000 (06 December 2001). <http://www.e-kidneys.net>

eMedicine Journal. "Submersion Injury, Near Drowning." 2001 (08 January 2002).. <http://www.emedicine.com>emerg/topic744.htm>

eMedicine Journal. 2001 (02 December 2001). "Face Embryology." <http://www.emedicine.com>ent/topic30.htm>

Emory University. "Pharynx and Larynx." 1999 (12 January 2002). <http://www.emory.edu/ANATOMY/AnatomyManual/pharynx.html>

e-Ophthalmology net. "Physiology of vision." 2000 (07 January 2002). <http://www.e-ophthalmology.net/vision.html>

Food and Drug Administration. "The Mechanics of Breathing." 1999 (01 December 2001). <http://www.fda.gov/fdac/features/1999/emphside.html>

Food and Drug Interactions, FDA, 1998 (Dec 31, 2001). <http://vm.cfsan.fda.gov/~lrd/fdinter.html>

Franklin Institute. "Poison Protection." 2000 (02 December 2001). <http://sln.fi.edu/biosci/systems/excretion.html>

George Washington University, Department of Biological Sciences. Washington, DC "The Human Brain Stem & Cranial Nerves" 2002. <http://gwis2.circ.gwu.edu/~atkins/Neuroweb/brainstem.html>

George Washington University. "Sense of Touch." 1998 (14 January 2002). <http://gwis2.circ.gwu.edu/~atkins?neuroweb/touch..html>

Halcyon.com>. "Cycling performance tips: Forms of dietary fat." 2001 (05 November 2001). <http://www.halcyon.com>gasman/fattype.htm>

Harvard University. "Deep Underwater, the Breath of Life." 1997 (07 January 2002). <http//www.oeb.harvard.edu/courses/bio21/Underwater.html>

Harvard University. "Peering at a machine that pries apart DNA." (02 April 2001). <http//www.hms.harvard.edu/news/releases/1099helicase.html>

Headache.net. "Headaches: An Overview." 2001 (07 December 2001). <http://headache.net/focus_article.asp>

Hollinger, T. et al. Gold Standard Multimedia. Microcopy Anatomy. 1994–2000. <http://imc.gsm.com>integrated/maonline/maonline/ma/chapters/oral/283934.htm>

Hyman-Newman Institute for Neurology and Neurosurgery Beth Israel Medical Center, Singer Division New York, NY, "Pediatric Neurosurgery" 2002. <http://nyneurosurgery.org/neuroanatomy/neuroanatomy.htm>

Hyperthyroidism, The Merck Manual (December 05, 2001). <http://www.merck.com>pubs/mmanual/section2/chapter8/8d.htm>

Imaginis Corporation. "General Information on Breast Cancer." 1997 (22 Januaruy 2001). <http://www.imaginis.com>breasthealth/breast_anatomy.asp>

Imaginis. "Women's Health. Heart Disease: Congenital Heart Defects." 2001 (04 December 2001). <http://www.imaginis.com>heart-disease/congenital.asp>

Index Medico. "Adipose Tissue Metabolism: An Overview. 2001 (07 December 2001). <http://indexmedico.com>english/obesity/obadtis.htm>

Indiana State University. "Fatty Acid Synthesis." 2001(06 December 2001). <http://web.indstate.edu/theme/mwking/lipid-synthesis.html>

Indiana State University. "Blood Coagulation." 2001(30 November 2001). <http://web.indstate.edu/theme/mwking/blood-coagulation.html>

Indiana University. "Genetic factors weigh-in as clues to obesity." (02 April 2001). <http//www.medicine.indiana.edu/mini_med/1999/mms6.htm>

Karolinska Institutet "Eye diseases" (28 January 2002). <http://www.mic.ki.se/Diseases/c11.html>

Leeds University. "Introductory Anatomy: Joints." 1999 (14 January 2002). <http://www.leeds.ac.uk/chb/lectures/anatomy.html>

Loyola University Medical Center. "The Process of Ossification." 1999. <http://www.meddean.luc.edu/lumen/meded/cellbio/98lab5b.htm>

Loyola University Medical Education Network. "Epidermis." 1996 (08 December 2001). <http://www.meddean.luc.edu/lumen/MedEd/medicine/dermatology/skinsn/epider.htm>

Loyola University. "The biliary system." 2000 (30 November 2001). <http://www.luhs.org/health/topics/liver.htm>

Macalester College. "The Olfactory System: Anatomy and Physiology." 1998 (January 13 2002). <http://www.macalester.edu/~psych/whathap/UBNRP/Smell/nasal.html>

Massachusetts Institute of Technology. "Lipids" 2001(06 December 2001). <http://esg-www.mit.edu:8001/esgbio/lm/lipids/lipids.html>

Mayo Clinic. "Stem Cells: Medicine's New Frontier." 2001 (10 August 2001). <http://www.mayoclinic.com>invoke.cfm?id=CA00013>

McGill University. "Development of Female External Genitalia." 1999 (04 December 2001). <http://sprojects.mmi.mcgill.ca/embryology. Reproductives/Normal/Female_external.html>

McGill University. "Development of Male External Genitalia." 1999. <http://sprojects.mmi.mcgill.ca/embryology/ug/Reproductives/Normal/Male_external.html>

Medical Information, Drugbase.cxom. <http://www.drugbase.co.za/data/med_info/heatstr.htm>

Medical Pages. "Surgical Anatomy of the Liver." 1999-2001 (28 January 2002). <http://www.liver.co.uk/anatomy.htm>

MedicineNet. "What is a fracture?" 2000 (01 December 2001). <http://www.medicinenet.com/Script/Main/Art.asp>

MedMedia. "Hyaline cartilage." 1996 (03 December 2001). <http://www.medmedia.com/o2/59.htm>

Medscape Health. "Allodynia: When Even Touch is Painful." 2001 (14 January 2002). <http://health.medscape.com/cx/viewarticle/403212>

Memorial University of Newfoundland. 1999 (06 December 2001). <http://www.mun.ca/biochem/courses/1430/lipmethints.html>

Merck Manual of Diagnosis and Therapy, Section 6, 1995–2001 (December 19, 2001). <http://www.merck.com>pubs/mmanual/section6/chapter80/80a.htm>

National Aeronautics and Space Administration. "Mixed Up in Space." 2001 (06 January 2002). <http://science.nasa.gov/headlines/y2001/ast07aug_1.htm>

National Bioethics Advisory Commission. "Executive Summary of Cloning Human Beings." (2001). <http://bioethics.gov/pubs/cloning1/executive.htm>

National Center for Chronic Disease Prevention and Health Promotion, "Physical Activity and Good Nutrition: Essential Elements for Good Health" (2001). <http://www.cdc.gov/nccdphp/dnpa/dnpaaag.html>

National Human Genome Research Institute (2001). <http://www.nhgri.nih.gov/>

National Human Genome Research Institute. "Ethical, Legal and Social Implications of Human Genetic Research." (October 2000). <http://www.nhgri.nih.gov/ELSI/>

National Human Genome Research Institute. "Twenty Questions About DNA Sequencing (and the Answers)." (28 April 2001). <http://www.nhgri.nih.gov/NEWS/Finish_sequencing_early/twenty_questions_about_DNA.html>

National Human Genome Research Institute. "Promoting Safe and Effective Testing in the United States." 2000 (11 January 2002). <http://www.nhgri.nih.gov/ELSI/TFGT_final.html>

National Institute of Neurological Disorders and Stroke. "NINDS Anoxia/Hypoxia Information Page." 2002. <http://www.ninds.nih.gov/health_and_medical/disorders/anoxia_doc.htm>/sourcetxt>

National Institute of Neurological Disorders and Stroke: "Neural Repair and Maintenance" <http://www.ninds.nih.gov/news_and_events/pressrelease>

National Institute on Drug Abuse. "Section II: The Reward Pathway and Addiction." 2001(04 December 2001). <http://165.112.78.61/Teaching2/teching3.html>

National Institutes of Health. "Stem Cells: A Primer". 2000. <http://www.nih.gov/news/stemcell/primer.htm>

National Institutes of Health. "Serology." 2002(13 January 2002). <http://www.nlm.nih.gov/medlineplus/ency/article/003511.htm>

National Institutes of Health. "Stem Cells: Scientific Progress and Future Research Directions." 2001. <http://www.nih.gov/news/stemcell/scireport.htm>

National Institutes of Health. 1997. "Genetic 'Short Circuit' Leads to Cleft Palate." (02 December 2001). <http://www.nih.gov/news/pr/dec97/nidr-22.htm.>

National Science Foundation. "Engineering Sight: Advances in Artificial Retina Development." 1997(04 January 2002). <http://www.nsf.gov/od/lpa/news/publicat/frontier/7-97/7retina.htm>

National Space Biomedical Research Institute. "Space Physiology." 2001(06 January 2002). <http//www.nsbri.org/HumanPhysSpace/focus6/spacephy.html>

Nephrology Channel. "Electrolytes Imbalance." 2001 (07 December 2001). <http://www.nephrologychannel.com>electrolytes/>

New York Times. "The Uvula." 2000 (13 January 2002). <http://www.nytimes.com>learning/students/scienceqa/archive001226.html>

Newcastle University, England, " Medulla, Pons & Cerebellum" 2002. <http://www.newcastle.edu.au/fmhs/disciplines/anatomy/subjects/anat333/aug30.htm>

Nobel e-Museum. "Ferid Murad - Autobiography." 1998 (18 October 2001). <http//www.nobel.se/medicine/laureates/1998/murad-autobio.html>

Nobel e-Museum. "Biography of Niels Ryberg Finsen." <http://www.nobel.se/medicine/laureates/1903/finsen-bio.html>

Nobel e-Museum. "Louis J. Ignarro - Autobiography" 1998. <http://www.nobel.se/medicine/laureates/1998/ignarro-autobio.html>

Nobel e-Museum. "The 2000 Nobel Prize in Physiology or Medicine." <http://www.nobel.se/medicine/laureates/2000/press.html>

Northern Arizona University. "Arthroogy." 1999 (09 January 2002). <http://jan.ucc.nau.edu/~kkt/EXS334/Arthrology.html>

Nottingham Trent University. "Non-specific Mechanisms of Resistance to Infection." 2000 (06 December 2001).

<http://www2ntu.ac.uk/life/sh/modules/hlf349/Lectures/349-3.htm>

Novartis Transplant. "Milestones in Transplantation." 2000 (06 January 2002). <http//www.novartis-transplant.com> medpro/symposia/milestones_in_TX.html>

Novatech. "Ischemia versus Hypoxia." 2002 <http://novatech.on.ca/medical/ischem.html>

Oak Ridge National Laboratory. "Genetics Privacy and Legislation." (2001). <http://www.ornl.gov/hgmis/elsi/legislat.html>

Oak Ridge National Laboratory. "Potential Benefits of Human Genome Project Research." (21 January 2001). <http//www.ornl.gov/hgmis/project/benefits.html>

Ohio State University. "The Conduction System." 2002. <http://oak.cats.ohiou.edu/~jr888793/conduction.html>

Oncology Channel. "Anatomy of the Breast." 1998 (3 December 2001). <http://www.oncologychannel.com> breastcancer/breastanatomy.shtml>

Oregon State University. "Manifestations of toxic effects." 1993 (05 November 2001). <http://ace.orst.edu/info/extonet/tibs/manifest.htm>

Overview of Drug Interactions, US Pharmacist, Vol 25:5, (Dec 31, 2001). <http://www.uspharmacist.com>NewLook/DisplayArticle.cfm?item_num=522>

Penn Valley Community College. "Homeostasis." 1999 (21 November 2001). <http://www.kcmetro.cc.mo.us/pennvalley/biology/lewis/homeo.htm>

Physicians and Scientists for Responsible Application of Science and Technology. "How Are Genes Engineered?" (04 March 2001).<http//www.psrast.org/whisge.htm>

Princeton University. "Outdoor action guide to high altitude: acclimatization and illness." 1999 (04 November 2001). <http://www.princeton.edu/~oa/safety/altitude.html>

Princeton University. "Physical maps and positional cloning." (31 March 2001). <http//www.princeton.edu/~lsilver/book/MG10.html>

Principia Cybernetica Web. "Homeostasis." 1997 (02 December 2001). <http://pespmc1.vub.ac.be/HOMEOSTA.html>

Pulmonary Channel, "COPD Causes." 2002. <http://www.pulmonologychannel.com>copd/causes.shtml>

Purdue University. "Evolutionary Homologs." 1996 (20 November 2001). <http://sdb.bio.purdue.edu/fly/neural.achaete2.htm>

Quickcare.org. "Nausea and vomiting." 1997 (08 January 2002). <http://www.quickcare.org/gast/nausea.html>

Quickcare. "Poison Ivy." 1997 (05 November 2001). <http//www.quickcare.org/skin/poison.html>

Retina Source.com. "New Infrared Camera Aids in Retinal Imaging." 2001 (04 January 2002). <http://www.theretinasource.com/news/articles/InfraredCamera_0801.htm>

Rice University. " Language and Brain: Neurocognitive Linguistics" 2002. <http://www.owlnet.rice.edu/~ling306/cglidden/myelen.html>

Rochester Institute of Technology. "Sodium-potassium pump" 1996 (3 January 2002). <http://www.rit.edu/~has7647/nacl.html>

Sangstat. "Acute Rejection." 2001(06 January 2002). <http://www.sangstat.com>transplantation/trans_back-rejection.html>

Sargant, D. "Proprioception: how does it work?" Australian Journal of Podiatric Medicine, vol 34, no. (2000): 86–92. percnt;lt;http://www.apodc.com>.au/AJPM/Contents/Full%20text/Vol34/Vol34%203%2086–92.pdf%gt;

Sawyer H., C. Clay, K. Bodensteiner, C. Moeller, ARBL Research: Ovarian Function, College of Veterinary Medicine and Biological Function, Colorado State University, 2002. <http://www.cvmbs.colostate.edu/physio/arbl_ovarian.html>

Science News Online. "Dioxin can harm tooth development." 1999. <http://www.sciencenews.org/sn_arc9922099/fob7.htm>

Science Week. "Drug Addiction and the Glutamate Neurotransmitter." 1998 (04 December 2001). <http://scienceweek.com>swfr026.htm>

Scientific American. "What is homeostasis?" 2000 (21 November 2001). <http://www.sciam.com>askexpert/biology/biology38/>

Scientific American: Explore! "Explaining General Anesthesia." 2000 (30 November 2001). <http://www.sciam.com>explorations/2000/102300gas/ >

Scientific American: News In Brief: "Neurons from Bone Marrow" (December 1, 2000). <http://www.sciam.com>1998/0698issue/0698infocus.html>

Scientific American: News In Brief: "Wiring the Brain" (March 12, 2001). <http://www.sciam.com>1998/0698issue/0698infocus.html>

Slovak Academy of Sciences. "Regulation and control of body temperature." 1995 (14 January 2002). <http://nic.savba.sk/logos/books/scientific/node45..html>

Society for Neuroscience. "Deafness Genes." 1999 (07 January 2002). <http://www.sfn.org/briefings/deafness.html>

Sports Science. "Adipose Tissue." 1998 (07 December 2001). <http://www.sportsci.org/encyc/adipose/adipose.html>

Springfield Technical Community College. "Bone Injury and Recovery." 2001(01 December 2001). <http://faculty.stcc.mass.edu/tamarkin/AP/APIpages/bone.htm>

Springfield Technical Community College. "Exchange across Capillaries." 2001 (14 January 2002). <http://faculty.stcc.mass.edu/tamarkin/AP/AP2pages/vessels/exchange.htm>

Springfield Technical Community College. "Synovial Joint Movements." 2001 (14 January 2002). <http://faculty.stcc.mass.edu/tamarkin/AP/APIpages/joints/synovial.htm>

Stanford University, The Visible Female, 2002. <http://esap.stanford.edu/ourwork/PROJECTS/LUCY/lucywebsite/infofrperit.html>

Swanson, Jason R., Loyola University Medical Education Network, "Dermis." 1996 (3 December 2001). <http://www.meddean.luc.edu/lumen/MedEd/medicine/dermatology/skinlsn/dermis.htm>

Texas A & M University. "Amino Acid Composition and Protein Sequencing." (28 April 2001). <http://www.ntri.tamuk.edu/graduate/sequence.html>

TMJ Association Ltd. "Changing the face of TMJ," 2002. <http://www.tmj.org/>

Tufts University. "Upper Limb Structure List." 2000 (09 January 2002). <http://iris3.med.tufts.edu/dentgross/labguide/Upper_Limb_Structure_List.html>.

Tulane University. "Acid-Base physiology." 2001 (04 November 2001). <http://www.tme.tulane.edu/departments/anesthesiology/acid/physiology.ssi>

UniSci. "New Synthetic Copies Structure, Function of Cartilage." 2001 (07 December 2001). <http://unisci.com>stories/20014/1008015.htm>

United Medical and Dental Schools . "Taste & Smell." 1998 (14 January 2002). <http://www.umds.ac.uk/physiology/jim/tasteolf.htm>

United Medical and Dental Schools. "Smell (Olfaction)." 1998.(13 January 2002). <http://www.umds.ac.uk/physiology/jim/tasteolf.htm>

University of Bristol. "Endochondral ossification." 2000 (10 December 2001). <http://www.bris.ac.uk/Depts/Anatomy/calnet/bones/page5.htm>

University of British Columbia. "Evolution Makes Sense of Homologs." 2000 (20 November 2001). <http://www.zoology.ubc.ca/~bio336/Bio336/Lectures/Lecture5/Overheads.html>

University of British Columbia. Cartilaginous Joints. <http://www.science.ubc.ca/~biomania/tutorial/bonejt/jt01ab.htm>

University of Calgary. "Musculo-Skeletal System: Osteoporosis." 1999 (07 December 2001). <http://rehab.educ.ucalgary.ca/courses/cor...sculo-Skeletal/osteoporosis.html>

University of California at Berkeley. "The Parathyroid and Thyroid Cells." 1999 (07 December 2001). <http://mcb.berkeley.edu/courses/mcb135e/parathyroid.html> .

University of Connecticutt. "Homeostasis in multi-celled organisms." 2001 (21 November 2001). <http://www.sp.uconn.edu/~bi102vc/102f01/taigen/homeostasis.html>

University of Dallas. "Urogenital System." 1999 (9 February 2002). <http://www.udalls.edu/bioloy/Brown/Anatomy/14Urogenital.html>

University of Florida. "Epithelial Tissue." 2001 (02 December 2001). <http://www.medinfo.ufl.edu/pa/chuck/summer/handouts/epi.htm>

University of Iowa. "Virtual Hospital Chapter 5 The Cerebral Hemispheres" 2002. <http://www.vh.org/Providers/Textbooks/BrainAnatomy/Ch5Text/Section03.html>

University of Maryland. "Dental caries." 2001 (07 January 2002). <http://umm.drkoop.com/conditions/ency/article/001055.htm>

University of Massachusetts Amherst. "Exercise Induced Muscle Damage & Repair." <http://www-unix.oit.umass.edu/~excs597k/sullivan/>

University of Michigan. "Cartilage." 2000 (07 December 2001). <http://www-personal.umich.edu/≈ shelden/cartilage.html>

University of Missouri. "What is a Membrane Lipid?" 2001 (06 December 2001). <http://www.biochem.missouri.edu/~lesa/LIPIDS/what_membrane.html>

University of North Carolina, School of Medicine. "Embryo Images. Normal and abnormal mammalian development." (28th January 2002). <http//www.med.unc.edu/embryo_images/unit-eye/eye_htms/>

University of Pennsylvania. "The Retina: gross anatomy." 1995 (04 January 2002). <http://retina.anatomy.upenn.edu/lance/eye/retina_gross.html>

University of Tasmania. "Overview of lower limb anatomy." 1999 (09 January 2002). <http://www.healthsci.utas.edu.au/DCL/Lecture1.html>

University of Texas Southwestern Medical Center. 1996 (04 December 2001)."Discovering genetics of heart defects points to new directions in cardiac care." <http://www.swmed.edu/home_pages/news/eoheart.htm.>

University of Toledo. "MBC 3320 Posterior pituitary hormones." (05 November 2001). <http://www.neurosci.pharm.utoledo.edu/MBC3320/vasopressin.htm>

University of Vermont. "Exchange between intracellular and extracellular compartments." 2000 (13 January 2002). <http://physioweb.med.uvm.edu/bodyfluids/icf-ecf.htm>

University of Virginia. "Mothering and Oxytocin or Hormonal Cocktails for Two." (05 November 2001). <http://www.people.virginia.edu/~rjh9u/oxytocin.html> .

University of Washington. "The Larynx." 2000 (08 January 2002). <http://depts.washington.edu/otoweb/larynx.html>

University of Washington. "The Autonomic Nervous System." 2001(04 January 2002). <http://faculty.washington.edu/chudler/auto.html>

University of Western Australia. Department of Anatomy and Human Biology. "Blue histology-vascular system." (2000). <http://www.lab.anhb.uwa.edu.au/mb140/CorePages/Vascular/Vascular.htm>

University of Wisconsin, "Pontine Nucleui and Middle Cerebellar Peduncle" 2002. <http://www.anatomy.wisc.edu/BS/text/p16/s/xs.htm>

University of Wisconsin. "McArdle Laboratory for Cancer Research." (02 April 2001). <http//www.mcardle.oncoogy.wisc.edu/donations.html>

University of Wisconsin. "Mutagenesis *in vitro*." (2000). <http//www.bact.wisc.edu/microtextbook/BactGenetics/mutainvitro.html>

University of Wisconsin-Madison. "Embryonic Stem Cells." 2001. <http://www.news.wisc.edu/packages/stemcells/index.html/?get=facts#>

University of New South Wales Embryology. Development of the organs of Vision, Audition, Equilibrium and Olfaction, September, 2001. <http://anatomy.med.unsw.edu.au/cbl/embryo/Notes/senses.htm>

US Transplant. "About Transplants: Transplant Primer." 2001(06 January 2002). <http://www.ustransplant.org/primer.html>

Vanderbilt University. "Blood: Leukocytes." 1995 (08 January 2002). <http://www.me.vanderbilt.edu/histo/blood/leukocytes.html>

Vanderbilt University. "Epithelial Tissue." 2001 (02 December 2001). <http://www.mc.vanderbilt.edu/histo/BasicTissue.html>

Victoria University of Manchester. "Anatomy." 1999 .(12 January 2002). <http://www.teaching-biomed.man.ac.uk/student_projects/1999/selvonn/newpage1.htm>

Virtual Hospital. "Functional Anatomy of Basal Ganglia." 1999 (03 December 2001). <http://www.vh.org/Providrs/Textbooks/BasalGanglia/01Definitions.html>.

Virtual Hospital. "Infectious Diseases of the Central Nervous System. 1996(10 February 2002). <http://www.vh.org/Providers/TeachingFiles/CNSInfDisR2/IDCNSHomePg.html>

Vitech America. "Potassium" 2002 (5 January 2002). <http://www.a2zvita.com>Potassium.htm>

Washington University in Saint Louis. "Basal Ganglia and Cerebellum." 2000 (03 December 2001). <http://thalamus.wustl.edu/course/cerebell.html>

WEHI Home Page. " Professor Jacques Miller is awarded the Faulding Florey Medal 2000." <http://www.wehi.edu.au/information/press/12oct00.html>

Wellcome Trust Tour of the Human Genome (2001). <http://www.wellcome.ac.uk/en/genome/>

Wheeless, C.R. Wheeless' Textbook of Orthopaedics. 1996. <http://www.medmedia.com>med.htm.>

Whitehead Center for Genome Research at WIMR(2001). <http://www.wi.mit.edu/news/genome/gc.html>

Women's Information Network Against Breast Cancer. "Breast Anatomy and Physiology." 2001 (2 January 2001). <http://www.winabc.org/yourbody.html>

World Health Organization. "Dengue." 2001 (05 December 2001). <http://www.who.int/ctd/dengue/index.html>

Xavier University. "Cartilage Tissue." 1999 (09 January 2002). <http://xavier.xula.edu/cdoumen/HistoFolder/Cartilage.html>

c. 50,000 B.C.

Homo sapiens sapiens emerges as a conscious observer of nature.

c. 10,000 B.C.

Neolithic Revolution: transition from a hunting and gathering mode of food production to farming and animal husbandry, that is, the domestication of plants and animals.

c. 3500 B.C.

Sumerians describe methods of managing the date harvest.

c. 700 B.C.

The use of anatomical models is established in India.

c. 600 B.C.

Thales, the founder of the Ionian school of Greek philosophy, identifies water as the fundamental element of nature. Other Ionian philosophers construct different theories about the nature of the universe and living beings.

c. 500 B.C.

Alcmaeon, Pythagorean philosopher and naturalist, pursue anatomical research, conclude that humans are fundamentally different from animals, and establish the foundations of comparative anatomy. Alcmaeon differentiates arteries from veins.

c. 450 B.C.

Empedocles, Greek philosopher, asserts that the universe and all living things are composed of four fundamental elements: earth, air, fire, and water.

c. 400 B.C.

Democritus, Greek philosopher, argues that atoms are the building blocks of the universe and all living things. Democritus is an early advocate of the preformation theory of generation (embryology). Democritus is one of the first to formally identify the human brain as the organ of thought.

c. 400 B.C.

Hippocrates, Greek physician, founds a school of medicine on the Aegean island of Cos. According to Hippocratic medical tradition, the four humors that make up the human body correspond to the four elements that make up the universe. Hippocrates suggests using the developing chick egg as a model for embryology. Hippocrates notes that offspring inherit traits from both parents.

c. 350 B.C.

Aristotle, the Greek philosopher, attempts to classify animals and describes various theories of generation, including sexual, asexual, and spontaneous generation. Aristotle argues that the male parent contributes "form" to the offspring and the female parent contributes "matter." Aristotle discusses preformation and epigenesis as possible theories of embryological development, but argues that development occurs by epigenesis.

c. 300 B.C.

Theophrastus, Aristotle's disciple and the founder of botany, attempts to establish a classification system for plants based upon differences between plant and animal morphology.

c. 300 B.C.

Herophilus (c. 325–c. 255), Alexandrian physician, studies the circulatory system; his investigation of the nervous system prompts him to identify the brain as the organ of thought.

c. 275 B.C.

Herophilus's younger colleague, Eristratus (c. 310–c. 250), asserts that veins and arteries are connected.

c. 50 B.C.

Lucretius proposes a materialistic, atomistic theory of nature in his poem *On the Nature of Things*. He favors the preformation theory of embryological development.

c. 70 Roman author and naturalist Pliny the Elder (23–79) writes his influential *Natural History*, a vast compilation combining observations of nature, scientific facts, and mythology. Naturalists will use his work as a reference book for centuries.

c. 200 Galen, the preeminent medical authority of late Antiquity and the Middle Ages, creates a philosophy of medicine, anatomy, and physiology that remains virtually unchallenged until the sixteenth and seventeenth centuries. Galen argues that embryological development is epigenetic, although he disagrees with Aristotle about which organs are formed first and which are most important.

529 Byzantine Emperor Justinian closes the Academy (founded by Plato) in Athens and forbids pagan scientists and philosophers to teach. This causes an exodus of scientists to Persia.

c. 850 Arab scholar Yaqub ibn–Ishaq al–Kindi (c. 800–870) advances an anatomical and physiological explanation of vision.

c. 980 Abu Al–Qasim Al–Zahravi (Abucasis) creates a system and method of human dissection along with the first formal specific surgical techniques

c. 1020 Abu Ali Al–Hussain Ibn Abdallah Ibn Sina (Avicenna) describes the structure of the thorax, heart and eye. Avicenna offers specific descriptions of the heart valves.

c. 1025 Abu Ali Hasan Ibn Al–Haitham (Alhazen) advances techniques and theory in the study of light and optics.

c. 1150 Hildegard of Bingen (1098–1179), Germanic author publishes *The Book of Simple Medicine*, a treatise on the medicinal qualities of plants and minerals.

c. 1240 Ibn Al–Nafis Damishqui offers a description of the major functions and organization of the circulatory system.

c. 1250 Arab scientist al–Qurashi ibn an–Hafis (d. 1288) first describes the pulmonary circulatory system.

c. 1267 Roger Bacon (1214–1292), English philosopher and scientist, asserts that natural phenomena should be studied empirically.

c. 1275 William of Saliceto creates the first established record of a human dissection.

1451 Nicholas of Cusa's research and experimentalism with optics allow the development of a concave lenses.

1490 Leonardo da Vinci (1452–1519), Italian artist and scientist, describes patterns of capillary action.

1540 Servetus offers the first description of the pulmonary circulation of blood

1505 Leonardo da Vinci adds to a series of anatomical studies by creating the first wax cast of oxen brain ventricles.

c. 1525 Paracelsus (1493–1541), Swiss physician and alchemist, uses mineral substances as medicines. Denying Galen's authority, Paracelsus teaches that life is a chemical process.

1543 Andreas Vesalius publishes his epoch–making treatise *The Fabric of the Human Body*. Although Vesalius generally accepts Galenic physiological doctrines and ideas about embryology, Vesalius is later regarded by many as the founder of modern anatomy because he corrected many of Galen's misconceptions regarding the human body.

1553 Publication of *On the Restoration of Christianity* by the Spanish theologian and naturalist Michael Servetus (1511–1553). In this work, Servetus describes pulmonary circulation. Persecuted by the Catholic Inquisition, Servetus is ultimately arrested and burned at the stake.

1561 Gabriello Fallopio (Fallopius) publishes *Observationes anatomicae,* which greatly increases available scholarly knowledge of the sexual organs. The Fallopian tubes are so named to honor him.

1568 Italian physicisn Constnzo Varolio (1543–1575) makes available to other scholars his independent studies of the structure of the human brain.

1568 Zacharias and Hans Janssen development of the first compound microscope opens new opportunities for the study of structural detail.

1600 Girolamo Fabrizzi (Fabricus ab Aquapendente) publishes *De formato foetu (On the formation of the fetus)*. The illustrations in the embryological works of Fabricus stir academic debate.

1603 Girolamo Fabrizzi (Fabricus ab Aquapendente) (1533–c. 1619) demonstrates that veins contain valves.

1604 German astronomer and mathematician Johannes Kepler (1571–1630) writes a treatise on optics.

1610 Jean Beguin (1550–1620) publishes the first textbook on chemistry.

1614   Italian physician Santorio Santorio (1561–1636) publishes studies on metabolism.

1621   Girolamo Fabrizzi (Fabricus ab Aquapendente) publishes *De formatione ovi et pulli* (On the formation of the egg and the chick).

1628   William Harvey (1578–1657), English physician, publishes his *Anatomical Treatise on the Movement of the Heart and Blood.* This scientific classic traces the course of blood through the heart, arteries, and veins, presents the first accurate description of blood circulation.

1651   William Harvey publishes treatise on embryology titled *On the Generation of Animals,* in which Harvey asserts that all living things come from eggs. Harvey argues that oviparous and viviparous generation are analogous. Regardless, Harvey maintains support for the Aristotelian doctrine that generation occurs by epigenesis.

1658   Dutch naturalist Jan Swammerdam publishes records of observations of red blood cells.

1660   Marcello Malpighi makes publishes works decribing vascular capillary beds and individual capillaries.

1663   Nicholas Steno asserts that the heart is a muscle or muscular organ.

1664   The idea of reflex action, formulated by René Descartes (1596–1650), French philosopher and mathematician, is made public. The assertion is included in a French edition of his posthumously published work on animal physiology. In his analysis Descartes applied his mechanistic philosophy to the analysis of animal behavior and first used the concept of reflex to denote any involuntary response the body makes when exposed to a stimulus.

1665   Robert Hooke publishes *Micrographia,* an account of observations made with the new instrument known as the microscope. Hooke presents his drawings of the tiny box–like structures found in cork and calls these tiny structures "cells." Although the cells he observes are not living, the name is retained. He also describes the streaming juices of live plant cells.

1667   Nicolaus Steno discovers the "female testicles" of the shark and introduces the term "ovary." He argues that the female testicles contain ova (eggs).

1668   Regnier de Graaf publishes his treatise on the human sex organs, and describes the structures that are now known as the Graafian follicles. He also confirmes Harvey's analogy between oviparous and viviparous reproduction.

1669   Jan Swammerdam begins his pioneering work on the metamorphosis of insects and the anatomy of the mayfly. Swammerdam suggests that new individuals

are embedded, or preformed, in their predecessors. Ultimatey, Nicolas de Malebranche reformulates Swammerdam's preformationist ideas into a more sophisticated philosophical doctrine that involves a series of embryos preexisting within each other like a nest of boxes.

1677   Antoni van Leeuwenhoek discovers "little animals" (spermatozoa) in semen. His observations are published in *The Philosophical Transactions of the Royal Society* in 1679.

1680   Posthumous publication of *On Motion in Animals* by Giovanni Alfonso Borelli (1608–1679), Italian mathematician and physicist. Borelli studied the human body from the standpoint of Descartes' mechanistic philosophy, describing physiology as a branch of physics and offering a mechanical analysis of the skeletomuscular system.

1681   Nehemiah Grew introduces the term "comparative anatomy."

1683   Antoni van Leeuwenhoek discoveres different types of infusoria (minute organisms found in decomposing matter and stagnant water). He also describes protozoa and bacteria.

1726   Stephen Hales 1677–1761), English botanist and chemist, offers a measurement of blood pressure.

1727   Stephen Hales studies plant nutrition and measures water absorbed by plant roots and released by leaves; argues that something in the air(carbon dioxide) is converted into food and that light is a necessary element of this process.

1735   Carl Linnaeus publishes his *Systema Naturae, or The Three Kingdoms of Nature Systematically Proposed in Classes, Orders, Genera, and Species,* a methodical and hierarchical classification of all living beings. He develops the binomial nomenclature for the classification of plants and animals. In this system, each type of living being is classified in terms of genus (denoting the group to which it belongs) and species (its particular, individual name). His classification of plants is primarily based on the characteristics of their reproductive organs.

1740   Abraham Trembley asserts that the fresh water hydra, or "polyp," appears to be an animal rather than a plant. When the hydra is cut into pieces, each part could regenerate a complete new organism. These experiments raised many philosophical questions about the "organizing principle" in animals and the nature of development.

1745   Charles Bonnet publishes *Insectology,* in which he describes his experiments on parthenogenesis in aphids (the production of offspring by female aphids in the absence of males).

1746   Pierre–Louis Moreau de Maupertuis publishes *Venus Physique.* Maupertuis criticizes preformationist theories because offspring inherit characteristics of both parents. He proposes an adaptationist account of organic design. His theories suggests the existence of a mechanism for transmitting adaptations.

1748   Jean-Antoine Nollet describes osmosis.

1754   Pierre–Louis Moreau de Maupertuis suggests that species change over time, rather than remaining fixed.

1757   Albrecht von Haller 1757–1766), publishes the first volume of his eight–volume *Elements of Physiology of the Human Body*, subsequently to become a landmark in the history of modern physiology.

1759   Kaspar Friedrich Wolff publishes *Theory of Generation,* that argues that generation occurs by epigenesis, that is, the gradual addition of parts. This book marks the beginning of modern embryology.

1762   Marcus Anton von Plenciz, Sr. suggests that all infectious diseases are caused by living organisms and that there is a particular organism for each disease.

1765   Lazzaro Spallanzani publishes his *Microscopical Observations.* Spallanzani's experiments refutes the theory of the spontaneous generation of infusoria.

1765   Abraham Trembley observes and publishes drawings of cell division in protozoans and algae.

1771   Luigi Galvani (1737–1798), Italian anatomist, discovers the electric nature of nervous impulses.

1772   Joseph Priestley (1733–1804), English theologian and chemist, discovers that plants give off oxygen.

1774   Antoine–Laurent Lavoisier (1743–1794), French chemist, discovers that oxygen is consumed during respiration.

1779   Jan Ingenhousz (1739–1799), Dutch physician and plant physiologist publishes his *Experiments upon Vegetables* and shows that light is necessary for the production of oxygen, and that carbon dioxide is taken in by plants in the daytime and given off at night.

1779   Domenico Cotugno (1736–1822), Italian anatomist, asserts that cerebrospinal fluid, and not "animal spirit," as previously argued, fills the brain's cavities and ventricles.

1780   Antoine–Laurent Lavoisier (1743–1794), French chemist, and Pierre–Simon Laplace (1749–1827), French astronomer and mathematician, collaborate to demonstrate that respiration is a form of combustion. Breathing, like combustion, liberates heat, carbon dioxide and water.

1780   George Adams (1750–1795), English engineer, devises the first microtome. This mechanical instrument cuts thin slice for examination under a microscope, thus replacing the imprecise procedure of cutting by s hand–held razor.

1780   Lazzaro Spallanzani carries out experiments on fertilization in frogs and attempts to determine the role of semen in the development of amphibian eggs.

1796   Erasmus Darwin, grandfather of Charles Darwin and Francis Galton, publishes his *Zoonomia.* In this work, Darwin argues that evolutionary changes are brought about by the mechanism primarily associated with Jean–Baptiste Lamarck, that is, the direct influence of the environment on the organism.

1797   Georges–Léopold–Chrétien–Frédéric Dagobert Cuvier establishes modern comparative zoology with the publication of his first book, *Basic Outline for a Natural History of Animals.* Cuvier studies the ways in which an animal's function and habits determine its form. He argues that form always followed function and that the reverse relationship did not occur.

1800   Marie–François–Xavier Bichat publishes his first major work, *Treatise on Tissues,* which establishes histology as a new scientific discipline. Bichat distinguishes 21 kinds of tissue and relates particular diseases to particular tissues.

1802   Jean–Baptiste–Pierre–Antoine de Monet de Lamarck and Gottfried Reinhold Treviranus propose the term "biology" to denote a new general science of living beings that would supercede studies in natural history.

1802   John Dalton introduces modern atomic theory into the science of chemistry.

1809   Jean–Baptiste–Pierre–Antoine de Monet de Lamarck introduces the term "invertebrate" in his *Zoological Philosophy,* which contains the first influential scientific theory of evolution. He attempts to classify organisms by function rather than by structure and is the first to use genealogical trees to show relationships among species.

1810   Franz Joseph Gall (1758–1828), German physician, first lays the basis for modern neurology with his dissections of the brain and his correct suggestions about nerve organization.

1811   Julien–Jean–César Legallois (1770–1814), French physiologist, locates the first physiological center in the brain.

1812   Charles Bell (1774–1842), Scottish surgeon, differentiates between sensory and motor roots of spinal nerves in his *New Idea of Anatomy of the Brain.* He asserts that each nerve carries either a motor or a sensory stimulus, and not both simultaneously.

1812 Georges–Léopold–Chrétien–Frédéric Dagobert Cuvier founds vertebrate paleontology with his *Investigations of the Fossil Bones of Quadrupeds.*

1812 Gustav Kirchoff identifies catalysis and mechanisms of catalytic reactions.

1817 Georges–Léopold–Chrétien–Frédéric Dagobert Cuvier publishes his major work, *The Animal Kingdom,* which expands and improves Linnaeus's classification system by grouping related classes into broader groups called phyla. He is also the first to extend this system of classification to fossils.

1818 William Charles Wells suggests the theory of natural selection in an essay dealing with human color variations. Wells notes that dark skinned people are more resistant to tropical diseases than lighter skinned people. Wells also calls attention to selection carried out by animal breeders. Jerome Lawrence, James Cowles Prichard, and others make similar suggestions. However, these individuals do not develop their ideas into a coherent and convincing theory of evolution.

1820 First United States *Pharmacopoeia* is published.

1821 Jean–Louis Prévost and Jean–Baptiste–André Dumas jointly publish a paper that demonstrates that spermatozoa originate in tissues of the male sex glands. Three years later they publish the first detailed account of the segmentation of a frog's egg.

1822 François Magendie (1783–1855), French physiologist, publishes his paper "Functions of the Roots of the Spinal Nerves," that lays the foundation for the Bell–Magendie Law.

1824 René–Joachim–Henri Dutrochet suggests that tissues are composed of living cells.

1824 Marie–Jean–Pierre–Flourens (1794–1867), French physiologist, first proves that the brain's respiratory center lies low in the brainstem. A pioneer of the idea of nervous coordination, he is also the first the recognize the role of the inner ear in maintaining body equilibrium and coordination.

1825 Jan Evangelista Purkinje describes the "germinal vesicle," or nucleus, in a hen's egg.

1826 James Cowles Prichard presents his views on evolution in the second edition of his book *Researches into the Physical History of Man* (first edition 1813). These ideas about evolution are suppressed in later editions.

1827 Karl Ernst von Baer publishes *On the Mammalian Egg,* documenting his 1826 discovery of the mammalian ovum.

1828 In his book *On the Developmental History of Animals* (2 volumes, 1828–1837), Karl Ernst von Baer demonstrates that embryological development follows essentially the same pattern in a wide variety of mammals. Early mammalian embryos are very similar, but they diverge at later stages of gestation. von Baer's work establishes the modern field of comparative embryology.

1828 Friedrich Wöhler synthesizes urea. This is generally regarded as the first organic chemical produced in the laboratory and an important step in disproving the idea that only living organisms can produce organic compounds. Work by Wöhler and others establishes the foundations of organic chemistry and biochemistry.

1828 Robert Brown observes a small body within the cells of plant tissue and calls it the "nucleus." He also discovers what becomes known as Brownian movement.

1828 Luigi Rolando (1773–1831), Italian anatomist, achieves the first synthetic electrical stimulation of the brain.

1830 Brown identifies cell nuclei (in plants).

1830 Hermann von Hemholtz designs experiment to measure the speed of nerve impulses.

1831 Charles Robert Darwin begins his historic voyage on the H.M.S. *Beagle* (1831–1836). His observations during the voyage lead to his theory of evolution by means of natural selection.

1832 Anselme Payen (1795–1871), French physiologist, first isolates diastase, which he separates from barley. This is a substance that had the property of hastening the conversion of starch into sugar, and is an example of the organic catalysts within living tissue that eventually come to be called enzymes.

1833 Johannes Peter Müller (1801–1858), German physiologist, first proposes his law of specific nerve energies. According to this law, every sensory nerve gives rise to one form of sensation, even it is excited by stimuli outside a normal range.

1836 Félix Dujardin describes the "living jell" of the cytoplasm. He calls it "sarcode."

1836 Theodor Schwann carries out experiments that refutes the theory of the spontaneous generation of infusoria. He also demonstrates that alcoholic fermentation depends on the action of living yeast cells. The same conclusion is reached independently by Charles Caignard de la Tour.

1837 René–Joachim Dutrochet (1776–1847), French physiologist, publishes his research on plant physiology, which include pioneering work on osmosis. The first scientist to systematically investigate the process of osmosis, which he names, Dutrochet recognizes the importance of this phenomenon, as well as the fact that chlorophyll is necessary for photosynthesis.

1837   Karl Theodor von Siebold and Michael Sars describe the division of invertebrate eggs.

1837   Robert Remak describes the relationship between nerve cells and nerve fibers. Remak describes and names the neurolemma (the myelin sheath around many nerve fibers). He later describes and names the three germinal layers in the developing embryo: ectoderm (outer skin), mesoderm (middle skin), and endoderm (inner skin).

1838   Matthias Jakob Schleiden notes that the nucleus, which had been described by Robert Brown, is a characteristic of all plant cells. Schleiden describes plants as a community of cells and cell products. He helps establish cell theory and stimulates Theodor Schwann's recognition that animals are also composed of cells and cell products.

1838   Jan Evangelista Purkinje (1787–1869), Czech physiologist, first describes a large group of distinct cells in the cerebellum. These cells become known as "Purkinje cells."

1839   Theodore Schwann extends the theory of cells to include animals and helpes establish the basic unity of the two great kingdoms of life. In *Microscopical Researches into the Accordance in the Structure and Growth of Animals and Plants,* Schwann asserts that all living things are made up of cells, each of which contains certain essential components. He also coins the term "metabolism" to describe the overall chemical changes that take place in living tissues.

1839   Jan Evangelista Purkinje uses the term "protoplasm" to describe the substance within living cells.

1840   Rudolf Albert von Kölliker establishes that spermatozoa and eggs are derived from tissue cells. He attempts to extend the cell theory to embryology and histology.

1840   Karl Bogislaus Reichert introduces the cell theory into embryology. He demonstrates that the segments observed in fertilized eggs develop into individual cells, and that organs develop from cells.

1840   Justus von Liebig (1803–1873), German chemist, shows that plants synthesize organic compounds from carbon dioxide in the air, but take their nitrogenous compounds from the soil. He also states that ammonia (nitrogen) is needed for plant growth.

1840   Friedrich Gustav Jacob Henle publishes the first histology textbook, *General Anatomy.* This work includes the first modern discussion of the germ theory of communicable diseases.

1842   Charles Robert Darwin writes out an abstract of his theory of evolution, but does not plan to have this theory published until after his death.

1842   Theodor Ludwig Wilhelm Bischoff publishes the first textbook of comparative embryology, *Developmental History of Mammals and Man.*

1843   Martin Berry observes the union of sperm and egg of a rabbit.

1844   Robert Chambers anonymously publishes *Vestiges of the Natural History of Creation,* which advocates the theory of evolution. This controversial book becomes a best seller and introduces the general reading public to the theory of evolution.

1848   Karl Theodor Ernst von Siebold publishes his *Textbook of Invertebrate Comparative Anatomy,* one of the first major textbooks on invertebrate anatomy.

1849   Rudolf Wagner and Karl Georg Friedrich Rudolf Leuckart report that spermatozoa are a definite and essential part of the semen, and that the liquid merely keeps them in suspension. They also reject the old hypothesis that spermatozoa are parasites, and argue that spermatozoa are essential for fertilization.

1851   Hugo von Mohl publishes his *Basic Outline of the Anatomy and Physiology of the Plant Cell,* in which he proposes that new cells are created by cell division.

1854   George Newport performs the first experiments on animal embryos. He suggests that the point of sperm entry determines the planes of the segmentation of the egg.

1854   Rudolf Ludwig Carl Virchow (1821–1902), German pathologist, first names the neuroglia, or supportive "glue cells" in the brain.

1854   Gregor Mendel begins his study of 34 different strains of peas. Eventually, Mendel selects 22 kinds for further experiments. From 1856 to 1863, Mendel will grow and test over 28,000 plants and analyze seven specific pairs of traits.

1855   Alfred Russell Wallace writes an essay entitled *On the Law Which has Regulated the Introduction of New Species* and sends it to Charles Darwin. Wallace's essay and one by Darwin are published in the 1858 *Proceedings of the Linnaean Society.*

1855   Barolomeo Panizza (1785–1867), Italian anatomist, first proves that parts of the cerebral cortex are essential for vision.

1856   Nathanael Pringsheim observes the sperm of a freshwater algae plant enter the egg.

1856   Hermann von Hemholtz offers an explanation of the physiological basis of vision.

1857   Louis Pasteur demonstrates that lactic acid fermentation is caused by a living organism. Between 1857 and 1880, he performs a series of experiments that refute the doctrine of spontaneous generation. He also introduces vaccines for fowl cholera, anthrax, and rabies, based on attenuated strains of viruses and bacteria.

1858   Rudolf Ludwig Carl Virchow publishes his landmark paper "Cellular Pathology" and establishes the field of cellular pathology. Virchow asserts that all cells arise from preexisting cells (*Omnis cellula e cellula*). He argues that the cell is the ultimate locus of all disease.

1858   Charles Darwin and Alfred Russell Wallace agree to a joint presentation of their theory of evolution by natural selection.

1859   Charles Robert Darwin publishes his landmark book *On the Origin of Species by Means of Natural Selection.*

1860   Ernst Heinrich Haeckel describes the essential elements of modern zoological classification.

1860   Max Johann Sigismund Schultze describes the nature of protoplasm and shows that it is fundamentally the same for all life forms.

1861   Carl Gegenbaur confirms Theodor Schwann's suggestion that all vertebrate eggs are single cells.

1861   Pierre–Paul Broca (1824–1880), French surgeon and anthropologist, first identifies a location in the brain's left hemisphere that (in most people) is associated with speech. It is later called "Broca's area."

1861   Rudolf Albert von Kölliker publishes *Developmental History of Man and Higher Animals,* the first treatise on comparative embryology.

1862   Hermann von Hemholtz describes the physiological mechanism of auditory senses (hearing and sound transmission).

1863   Thomas Henry Huxley extended Darwin's theory of evolution to include humans in his book *Evidence As to Man's Place in Nature*. He became the champion and defender of Darwinism in England.

1865   Gregor Mendel presented his work on hybridization of peas to the Natural History Society of Brno, Czechoslovakia. The paper is publishes in the 1866 issue of the Society's *Proceedings*. In a series of papers on "Experiments on Plant Hybridization" published between 1866 and 1869, Mendel presents statistical evidence that hereditary factors are inherited from both parents. His experiments provide evidence of dominance, the laws of segregation, and independent assortment. Mendel's work is generally ignored until 1900.

1865   Franz Schweiger–Seidel proves that spermatozoa consist of a nucleus and cytoplasm.

1865   Jules–Bernard Luys (1826–1897), French physician, describes a nucleus in the hypothalamus, forming a part of the descending pathway from the corpus striatum. It becomes known as the "nucleus of Luys."

1866   Ernst Heinrich Haeckel publishes his book *A General Morphology of Organisms*. Haeckel summarizes his ideas about evolution and embryology in his famous dictum "ontogeny recapitulates phylogeny." Haeckel suggests that the nucleus of a cell transmits hereditary information. He introduces the use of the term "ecology" to describe the study of living organisms and their interactions with other organisms and with their environment.

1866   Johann Gregor Mendel (1822–1884), Austrian botanist and monk, discovers the laws of heredity and writes the first of a series of papers on heredity (1866–1869), which formulate the laws of hybridization. His work is disregarded until 1900, when de Vries rediscovers it. Unbeknownst to both Darwin and Mendel, Mendelian laws provide the scientific framework for the concepts of gradual evolution and continuous variation.

1868   Charles Darwin publishes *The Variation of Animals and Plants under Domestication* (2 volumes).

1868   Francis Galton publishes his book *Hereditary Genius*. Galton argues that the study of human pedigrees proves that intelligence is a hereditary trait.

1868   Thomas Henry Huxley introduces the term "protoplasm" to the general public in a lecture entitled "The Physical Basis of Life."

1869   Johann Friedrich Miescher discovers nuclein, a new chemical isolated from the nuclei of pus cells. Two years later, Miescher isolates nuclein from salmon sperm. This material is now known as nucleic acid.

1869   Paul Langerhans (1847–1888), German physician, discovers irregular islands of cells in the pancreas which produce insulin. Ultimately, the islands of cells become known as the "Isles of Langerhans."

1870   Lambert Adolphe Jacques Quetelet showes the importance of statistical analysis for biologists and provides the foundations of biometry.

1870   Gustav Theodor Fritsch (1838–1927), German anatomist and anthropologist, and Eduard Hitzig (1838–1907), German physiologist and neurologist, discover that electric shocks to one cerebral hemisphere of a dog's brain produces movement on the other side of the animal's body. This is the first clear demonstration of the existence of cerebral hemispheric lateralization.

1871 Charles Robert Darwin publishes *The Descent of Man, and Selection in Relation to Sex.* This work introduces the concept of sexual selection and expands his theory of evolution to include humans.

1871 Ferdinand Julius Cohn coins the term, bacterium.

1872 Ferdinand Julius Cohn publishes the first of four papers entitled "Research on Bacteria," which establishes the foundation of bacteriology as a distinct field. He systematically divides bacteria into genera and species.

1872 Franz Anton Schneider observes and describes the behavior of nuclear filaments (chromosomes) during cell division in his study of the platyhelminth Mesostoma. His account is the first accurate description of the process of mitosis in animal cells.

1873 Walther Flemming discovers chromosomes, observes mitosis, and suggests the modern interpretation of nuclear division.

1873 Franz Anton Schneider describes cell division in detail. His drawings included both the nucleus and chromosomal strands.

1873 Camilo Golgi discoveres that tissue samples could be stained with an inorganic dye (silver salts). Golgi uses this method to analyze the nervous system and characterizes the cells known as Golgi Type I and Golgi Type II cells and the "Golgi Apparatus." Golgi subsequently wins a Nobel Prize in 1906 for his studies of the nervous system.

1874 Carl Wernicke (1848–1905), German neurologist, discovers the area of the brain associated with word comprehension, eventually to be named "Wernicke's area."

1874 Francis Galton demonstrates the usefulness of twin studies for elucidating the relative influence of nature (heredity) and nurture (environment).

1874 Wilhelm August Oscar Hertwig concludes that fertilization in both animals and plants consists of the physical union of the two nuclei contributed by the male and female parents. Hertwig carries out pioneering studies of reproduction of the sea urchin.

1874 Eduard Strasburger, German embryologist, accurately describes the processes of mitotic cell division in plants.

1875 Eduard Adolf Strasburger publishes *Cell–Formation and Cell–Division,* in which he describes nuclear division in plants. Strasburger accurately describes the process of mitosis and argues that new nuclei can only rise from the division of preexisting nuclei. His treatise helps establish cytology as a distinct branch of histology.

1875 Theodor Wilhelm Engelmann (1843–1909), German physiologist, proves experimentally that the heartbeat is myogenic, which mean means that it originates in the heart muscle itself, and not from an external impulse.

1876 Edouard G. Balbiani observes the formation of chromosomes.

1876 Wilhelm August Oscar Hertwig observes the fertilization of a sea urchin egg and establishes that both parents contribute genetic material to the offspring. He also proves that fertilization is due to a fusion of the sperm nucleus with the nucleus of the egg.

1876 Robert Koch describes new techniques for fixing, staining, and photographing bacteria.

1877 Wilhelm Friedrich Kühne proposes the term enzyme (meaning "in yeast"). Kühne establishes the critical distinction between enzymes, or "ferments," and the microorganisms that produce them.

1878 Charles–Emanuel Sedillot introduces the term "microbe." The term becomes widely used as a term for a pathogenic bacterium.

1879 Hermann Fol observes the penetration of the egg of a sea urchin by a sperm. He demonstrates that only one spermatozoon is needed for fertilization and suggests that the nucleus of the sperm is transferred into the egg.

1879 Walther Flemming describes and names chromatin, mitosis, and the chromosome threads. Flemming's drawings of the longitudinal splitting of chromosomes provide the first accurate counts of chromosome numbers.

1880 The basic outlines of cell division and the distribution of chromosomes to the daughter cells are established by Walther Flemming, Eduard Strasburger, Edouard van Beneden, and others.

1880 David Ferrier (1843–1928), Scottish scientist, maps the region of the brain called the motor cortex and discovers the sensory strip.

1881 Eduard Strasburger coines the terms cytoplasm and nucleoplasm.

1881 Walther Flemming discoveres the lampbrush chromosomes.

1881 Wilhelm Roux, the founder of experimental embryology, publishes *The Struggle of the Parts in the Organism: A Contribution to the Completion of a Mechanical Theory of Teleology.* Roux argues that his experimental approach to embryonic development, based on mechanistic principles, provides evidence that development proceeds by means of self–differentiation.

1882 Pierre Émile Duclaux suggests that enzymes should be named by adding the suffix "-ase" to the name of their substrate.

1882 Edouard van Beneden outlines the principles of genetic continuity of chromosomes and reports the occurrence of chromosome reduction during the formation of the germ cells.

1882 Wilhelm Roux offers a possible explanation for the function of mitosis.

1882 Walther Flemming publishes *Cell Substance, Nucleus, and Cell Division,* in which he describes his observations of the longitudinal division of chromosomes in animal cells. Flemming observes chromosome threads in the dividing cells of salamander larvae.

1882 Robert Koch (1843–1910), German bacteriologist, discovers the tubercle bacillus and enunciates "Koch's postulates," which define the classic method of preserving, documenting, and studying bacteria.

1883 Walther Flemming, Eduard Strasburger and Edouard Van Beneden demonstrate that chromosome doubling occurs by a process of longitudinal splitting. Strasburger describes and names the prophase, metaphase, and anaphase stages of mitosis.

1883 Wilhelm Roux suggests that chromosomes in the nucleus carry the hereditary factors.

1883 August F. Weismann begins work on his germplasm theory of inheritance. Between 1884 and 1888, Weismann formulates the germplasm theory that argues that the germplasm is separate and distinct from the somatoplasm. Weismann argues that the germplasm is continuous from generation to generation and that only changes in the germplasm are transmitted to further generations. Weismann proposes a theory of chromosome behavior during cell division and fertilization and predicts the occurrence of a reduction division (meiosis) in all sexual organisms.

1884 Louis Pasteur and coworkers publishes a paper entitled "A New Communication on Rabies." Pasteur proves that the causal agent of rabies could be attenuated and the weakened virus could be used as a vaccine to prevent the disease. This work serves as the basis of future work on virus attenuation, vaccine development, and the concept that variation is an inherent characteristic of viruses.

1884 Elie Metchnikoff discovers the antibacterial activity of white blood cells, which he calls "phagocytes," and formulates the theory of phagocytosis.

1884 Oscar Hertwig, Eduard Strasburger, Albrecht von Kölliker, and August Weismann independently report that the cell nucleus serves as the basis for inheritance.

1884 Karl Rabl suggests the concept of the individuality of the chromosomes. He argues that each chromosome originates from a preexisting chromosome in the mother cell that is like it in form and size.

1884 Walther Flemming observes sister chromatids passing to opposite poles of the cell during mitosis.

1885 Louis Pasteur (1822–1895), French chemist, inoculates a boy, Joseph Meister, against rabies. Meister had been bitten by An infected dog, and the treatment saves his life. This is the first time Pasteur uses an attenuated germ on a human being.

1886 Edouard van Beneden proves that chromosomes persist between cell divisions. He discovers that the cell nucleus of each species has a fixed and specific number of chromosomes. He demonstrates chromosome reduction in gamete maturation, thereby confirming August Weismann's predictions. That is, he showed that the division of chromosomes during one of the cell divisions that produces the sex cells is not preceded by the doubling of chromosomes

1887 Theodor Boveri observes the reduction division during meiosis in *Ascaris* and confirmes August Weismann's predictions of chromosome reduction during the formation of the sex cells.

1888 Francis Galton publishes *Natural Inheritance,* considered a landmark in the establishment of biometry and statistical studies of variation. Galton also proposes the Law of Ancestral Inheritance, a statistical description of the relative contributions to heredity made by previous generations.

1888 Heinrich Wilhelm Gottfried Waldeyer coins the term "chromosome." Waldeyer introduces the use of hematoxylin as a histological stain.

1888 Theodor Heinrich Boveri discoveres and names the centrosome (the mitotic spindle that appears during cell division).

1888 Woods Hole Marine Biological Station, which later became the headquarters of the Woods Hole Oceanographic Institution and the Marine Biological Laboratory, is established in Massachusetts.

1889 Theodor Boveri and Jean–Louis–Léon Guignard establish the numerical equality of the paternal and maternal chromosomes at fertilization.

1889 Richard Altmann develops a method of preparing nuclein that is apparently free of protein. He calls his protein–free nucleins "nucleic acids."

1891 Charles–Edouard Brown–Sequard suggests the concept of internal secretions (hormones).

1891 Hermann Henking distinguishes between the sex chromosomes and the autosomes.

1892 August Weismann publishes his landmark treatise *The Germ Plasm: A Theory of Heredity,* which emphasizes the role of meiosis in the distribution of chromosomes during the formation of gametes.

1892 George M. Sternberg publishes his *Practical Results of Bacteriological Researches.* Sternberg's observations that a specific antibody is produced after infection with vaccinia virus and that immune serum is able to neutralize the virus becomes the basis of virus serology. The neutralization test provides a technique for diagnosing viral infections, measuring the immune response, distinguishing antigenic similarities and differences among viruses, and conducting retrospective epidemiological surveys.

1893 Hans Adolf Eduard Driesch discoveres that he could separate sea urchin embryos into individual cells and that the separated cells continued to develop. Driesch concludes that all the cells of the early embryo are capable of developing into whole organisms. Therefore, embryonic cells are equipotent, but differentiated and developed in response to their position within the embryo.

1893 William Bateson publishes *Materials for the Study of Variation,* which emphasizes the importance of discontinuous variations (the kinds of variation studied by Mendel).

1896 Edmund Beecher Wilson, American zoologist, publishes the first edition of his highly influential treatise *The Cell in Development and Heredity.* Wilson calls attention to the relationship between chromosomes and sex determination.

1897 John Jacob Abel (1857–1938), American physiologist and chemist, isolates epinephrine (adrenalin). This is the first hormone to be isolated.

1898 The First International Congress of Genetics is held in London.

1898 William Bateson publishes a paper on the importance of hybridization and crossbreeding experiments.

1898 Carl Benda discover and names mitochondria, the subcellular entities previously seen by Richard Altmann.

1898 Martin Wilhelm Beijerinck publishes his landmark paper "Concerning a Contagium Vivum Fluidum as Cause of the Spot Disease of Tobacco Leaves." Beijerinck argues that the etiological agent, which could pass through a porcelain filter that removes known bacteria, might be a new type of invisible organism that reproduces within the cells of diseased plants. He realizes that a very small amount of the virus could infect many leaves and that the diseased leaves could infect others.

1898 Friedrich Loeffler and Paul Frosch publish their *Report on Foot–and–Mouth Disease.* They proved that this animal disease is caused by a filterable virus and suggests that similar agents might cause other diseases.

1899 Jacques Loeb proves that it is possible to induce parthenogenesis in unfertilized sea urchin eggs by means of specific environmental changes.

1900 Carl Correns, Hugo de Vries, and Erich von Tschermak independently rediscover Mendel's laws of inheritance. Their publications mark the beginning of modern genetics. Using several plant species, de Vries and Correns perform breeding experiments that paralleled Mendel's earlier studies and independently arrive at similar interpretations of their results. Therefore, upon reading Mendel's publication, they immediately recognized its significance. William Bateson describes the importance of Mendel's contribution in an address to the Royal Society of London.

1900 Hugo Marie de Vries describes the concept of genetic mutations in his book *Mutation Theory.* He uses the term mutation to describe sudden, spontaneous, and drastic alterations in the hereditary material.

1900 Karl Landsteiner discovers the blood–agglutination phenomenon and the four major blood types in humans.

1900 Thomas H. Montgomery studies spermatogenesis in various species of Hemiptera. He concludes that maternal chromosomes only pair with corresponding paternal chromosomes during meiosis.

1901 Clarence E. McClung argues that particular chromosomes determine the sex of the individual carrying them. Although his work is done with insects, he suggests that this might be true for human beings and other animals.

1901 Theodor Boveri discovers that in order for sea urchin embryos to develop normally, they must have a full set of chromosomes. He concludes that this meant that the individual chromosomes must carry different hereditary determinants.

1901 Jokichi Takamine (1854–1922), Japanese–American chemist, and T. B. Aldrich first isolate epinephrine from the adrenal gland. Later known by the trade name Adrenalin, it is eventually identified as a neurotransmitter. This is also the first time a pure hormone has been isolated.

1901 Archibald Edward Garrod reports that a human disease, alkaptonuria, seems to be inherited as a Mendelian recessive disease.

1901 William Bateson coins the terms genetics, F1 and F2 generations, allelomorph (later shortened to allele), homozygote, heterozygote, and epistasis.

1902 Wilhelm Ludwig Johannsen introduces and defines the terms phenotype, genotype, and selection in terms of the new science of genetics.

1902 Carl Neuberg introduces the term biochemistry.

1902 Santiago Ramón y Cajal (1852–1911), Spanish histologist, first discovers the nature of the connection between nerves, showing that the nervous system consists of a maze of individual cells. He demonstrates that neurons do not touch but that the signal somehow crosses a gap (now called a synapse.)

1902 Walter Sutton presents evidence that chromosomes have individuality, that chromosomes occur in pairs (with one member of each pair contributed by each parent), and that the paired chromosomes separate from each other during meiosis. Sutton concludes that the concept of the individuality of the chromosomes provides the link between cytology and Mendelian heredity.

1902 Ernest H. Starling (1866–1927) and William H. Bayliss (1860–1924), both English physiologists, discover and isolate the first hormone ("secretin," found in the duodenum).

1903 Walter S. Sutton publishes a paper in which he presents the chromosome theory of inheritance. The theory, which states that the hereditary factors are located in the chromosomes, is independently proposed by Theodor Boveri and is generally referred to as the Sutton–Boveri hypothesis.

1903 Archibald Edward Garrod provides evidence that errors in genes caused several hereditary disorders in human beings. His book *The Inborn Errors of Metabolism* (1909) is the first treatise in biochemical genetics.

1903 Einthoven invents the electrocardiograph (EKG).

1903 Ruska develops a primitive electorn microscope.

1903 Tiselius offers electrophoresis techniques that become the basis for the separation of biological molecules by charge, mass, and size.

1905 Nettie Maria Stevens, American geneticist, discoveres the connection between chromosomes and sex determination. She determines that there are two basic types of sex chromosomes, which are now called X and Y. Stevens proves that females are XX and males are XY. Stevens and Edmund B. Wilson independently describe the relationship between the so–called accessory or X chromosomes and sex determination in insects.

1905 William Bateson and Reginald C. Punnett reportes the discovery of two new genetic principles: linkage and gene interaction.

1907 William Bateson urges biologists to adopt the term "genetics" to indicate the importance of the new science of heredity.

1907 Godfrey Harold Hardy proposes that Mendelian mechanisms acting alone would have no effect on allele frequencies. This suggestion forms the mathematical basis for population genetics and what became known as the Hardy–Weinberg Law.

1907 Ivan Petrovich Pavlov (1849–1910), investigates the conditioned reflex (1904–1907). A great stimulus for behaviorist psychology, his work establishes physiologically–oriented psychology.

1908 Thomas H. Morgan publishes a paper expressing doubts about Mendelian explanations for inherited traits.

1909 Archibald E. Garrod publishes *Inborn Errors of Metabolism*, in addition to Garrod's papers, one of the earliest discussions of biochemical genetics.

1908 Godfrey Harold Hardy and Wilhelm Weinberg independently publish similar papers describing a mathematical system that accounts for the stability of gene frequencies in succeeding generations of population. Their resulting Hardy–Weinberg law links the Mendelian hypothesis with population studies.

1908 Margaret A. Lewis successfully cultured mammalian cells *in vitro*.

1909 Wilhelm Ludwig Johannsen argues the necessity of distinguishing between the appearance of an organism and its genetic constitution. He invents the terms "gene" (carrier of heredity), "genotype" (an organism's genetic constitution), and "phenotype" (the appearance of the actual organism).

1909 Thomas Hunt Morgan selects the fruit fly *Drosophila* as a model system for the study of genetics. Morgan and his coworkers confirm the chromosome theory of heredity and realizes the significance of the fact that certain genes tend to be transmitted together. Morgan postulates the mechanism of "crossing over." His associate, Alfred Henry Sturtevant demonstrates the relationship between crossing over and the rearrangement of genes in 1913.

1909 Phoebus Aaron Theodore Levene (1869–1940), Russian–American chemist, discovers the chemical difference between DNA (deoxyribonucleic acid) and RNA (ribonucleic acid).

1909   Jean de Mayer, French physiologist, first suggests the name "insulin" for the hormone of the islet cells.

1909   Korbinian Bordmann (1868–1918), German neurologist, publishes a "map" of the cerebral cortex, assigning numbers to particular regions.

1910   Thomas Hunt Morgan discovers the *Drosophila* mutant called white eye. Research on this mutant led to the discovery of sex–linked traits. Morgan suggests that the genes for white eyes, yellow body, and miniature wings in *Drosophila* are linked together on the X chromosome.

1910   Harvey Cushing (1869–1939), American surgeon, and his team present the first experimental evidence of the link between the anterior pituitary and the reproductive organs.

1911   (Francis) Peyton Rous publishes the landmark paper "Transmission of a Malignant New Growth by Means of a Cell–Free Filtrate." His work provides the first rigorous proof of the experimental transmission of a solid tumor and suggests that a filterable virus is the causal agent.

1912   Alfred H. Sturtevant working with Thomas H. Morgan at Columbia, provides experimental evidence of the linkage of genes and began constructing the first genetic map for *Drosophila* chromosomes.

1912   Casimir Funk (1884–1967), Polish–American biochemist, coins the term "vitamine." Since the dietary substances he discovers are in the amine group he calls all of them "life–amines" (using the Latin word *vita* for "life").

1913   Elmer Verner McCollum (1879–1967), American biochemist, produces the first chromosome map, showing five sex–linked genes.

1913   Calvin Blackman Bridges discovers evidence of nondisjunction of the sex chromosomes in Drosophila. This evidence helps support Thomas H. Morgan's new chromosome theory of heredity.

1913   Alfred H. Sturtevant publishes the landmark paper that contains the first chromosome map. "The Linear Arrangement of Six Sex–Linked Factors in *Drosophila,* as Shown by Their Mode of Association" is published in the *Journal of Experimental Zoology.* Within two years Thomas H. Morgan's group describe four groups of linked factors; these groups corresponded to the four pairs of *Drosophila* chromosomes. The chromosome theory replaces "beanbag" genetics with the image of genes as beads on a string.

1914   Thomas Hunt Morgan, Alfred Henry Sturtevant, Calvin Blackman Bridges, and Hermann Joseph Muller publish the classic treatise of modern genetics, *The Mechanism of Mendelian Heredity.*

1914   Edward Calvin Kendall (1886–1972), American biochemist, extracts thyroxin from the thyroid gland (in crystalline form).

1914   Frederick William Twort (1877–1950), English bacteriologist, and Felix H. D'Herelle (1873–1949), Canadian–Russian physician, independently discover bacteriophages, viruses which destroy bacteria.

1915   Katherine K. Sanford isolates a single mammalian cell *in vitro* and allowes it to propagate to form identical descendants. Her clone of mouse fibroblasts is called L929, because it took 929 attempts before a successful propagation was achieved. Sanford's work is important step in establishing pure cell lines for biomedical research.

1917   D'Arcy Wentworth Thompson publishes *On Growth and Form,* which suggests that the evolution of one species into another occurs as a series of transformations involving the entire organism, rather than a succession of minor changes in parts of the body.

1918   Thomas Hunt Morgan and coworkers publish *The Physical Basis of Heredity,* a survey of the remarkable development of the new science of genetics.

1920   Frederick Grant Banting (1891–1941), Canadian physician, Charles Best (1899–1878), Scottish–American physiologist, and James B. Collip (1892–1965), Canadian biochemist, discover insulin. They develop a method a method of extracting insulin from the human pancreas. The insulin is then injected into the blood of diabetics to lower their blood sugar.

1921   Otto Loewi (1873–1961), German–American physiologist, discovers that acetylcholine functions as a neurotransmitter. It is the first such brain chemical to be so identified.

1922   Frederick Banting and Charles Best make the first clinical adaptation of insulin for the treatment of diabetes.

1922   Herbert McLean Evans (1882–1971), American physician, and colleagues discover Vitamin E.

1922   Elmer Verner McCollum (1879–1967), American biochemist, discovers Vitamin D.

1923   A. E. Boycott and C. Diver describe a classic example of "delayed Mendelian inheritance." The direction of the coiling of the shell in the snail *Limnea peregra* is under genetic control, but the gene acts on the egg prior to fertilization. Thus, the direction of coiling is determined by the egg cytoplasm, which is controlled by the mother's genotype.

1925   Johannes Hans Berger (1873–1941), German neurologist, records the first human electroencephalogram (EEG).

1926   James B Sumner publishes a report on the isolation of the enzyme urease and his proof that the enzyme is a protein. This idea is controversial until 1930 when John Howard Northrop confirms Sumner's ideas by crystallizing pepsin. Sumner, Northrop, and Wendell Meredith Stanley ultimately share the Nobel Prize for chemistry in 1946.

1927   Hermann Joseph Muller induces artificial mutations in fruit flies by exposing them to x rays. His work proves that mutations result from some type of physical–chemical change. Muller goes on to write extensively about the danger of excessive x rays and the burden of deleterious mutations in human populations.

1928   Alexander Fleming (1881–1955), Scottish bacteriologist, discovers penicillin. He observes that the mold *Penicillium notatum* inhibits the growth of some bacteria. This is the first anti–bacterial, and it opens a new era of "wonder drugs" to combat infection and disease.

1928   Wilder Graves Penfield (1891–1976), Canadian neurosurgeon, first uses microelectrodes to map areas in the human cerebral cortex.

1929   Willard Myron Allen, American physician, and George Washington Corner, American Anatomist, discover progesterone. They demonstrate that it is necessary for the maintenance of pregnancy.

1930   Ronald A. Fisher publishes *Genetical Theory of Natural Selection,* a formal analysis of the mathematics of selection.

1930   Curt Stern, and, independently, Harriet B. Creighton and Barbara McClintock, demonstrates cytological evidence of crossing over.

1931   Phoebus A. Levene publishes a book that summarizes his work on the chemical nature of the nucleic acids. His analyses of nucleic acids seemed to support the hypothesis known as the tetranucleotide interpretation, which suggest that the four bases are present in equal amounts in DNAs from all sources. Perplexingly, this indicated that DNA is a highly repetitious polymer that is incapable of generating the diversity that would be an essential characteristic of the genetic material.

1931   Joseph Needham publishes his landmark work *Chemical Embryology,* which emphasizes the relationship between biochemistry and embryology.

1931   Alice Miles Woodruff and Ernest W. Goodpasture demonstrate the advantages of using the membranes of the chick embryo to study the mechanism of viral infections.

1932   Thomas H. Morgan receives the Nobel Prize in Medicine or Physiology for his development of the theory of the gene. He is the first geneticist to receive a Nobel Prize.

1932   Hans Adolf Krebs (1900–1981), German–British biochemist, first describes and names the citric acid cycle.

1934   J.B.S. Haldane presents the first calculations of the spontaneous mutation frequency of a human gene.

1935   Wendall Meredith Stanley (1904–1971), American biochemist, discovers that viruses are partly protein–based. By purifying and crystallizing viruses, he enables scientists to identify the precise molecular structure and propagation modes of several viruses.

1936   Theodosius Dobzhansky publishes *Genetics and the Origin of Species,* a text eventually considered a classic in evolutionary genetics.

1938   Krebs identifies and defines the TCA cycle.

1937   Richard Benedict Goldschmidt postulates that the gene is a chemical entity rather than a discrete physical structure.

1937   James W. Papez (1883–1958), American anatomist, suggests the name "limbic system" for the old mammalian part of the human brain that produces human emotions.

1939   Moses Kunitz reports the purification and crystallization of ribonuclease from beef pancreas.

1940   Kenneth Mather coines the term "polygenes" and describes polygenic traits in various organisms.

1941   George W. Beadle and Edward L. Tatum publish their classic study on the biochemical genetics of *Neurospora*, "Genetic Control of Biochemical Reactions in *Neurospora*." Beadle and Tatum irradiat red bread mold, *Neurospora,* and prove that genes produce their effects by regulating particular enzymes. This work led to the one gene–one enzyme theory.

1941   Lipmann describes and identifies the biochemical and physiological role of high energy phospahates (e.g., ATP).

1944   Oswald Theodore Avery (1877–1947), Canadian biologist, Macklin McCarthy, Canadian physician, and Colin Munro Macleod (1909–1972), Canadian physician and physician and microbiologist, discover the "blueprint" function of DNA (that DNA carries genetic information).

1945 New techniques and instruments, such as partition chromatography on paper strips and the photoelectric ultraviolet spectrophotometer, stimulates the development of biochemistry after World War II. New methodologies made it possible to isolate, purify, and identify many important biochemical substances, including the purines, pyrimidines, nucleosides, and nucleotides derived from nucleic acids.

1945 Joshua Lederberg and Edward L. Tatum demonstrate genetic recombination in bacteria.

1946 Hermann J. Muller is awarded the Nobel Prize in Medicine or Physiology for his contributions to radiation genetics.

1946 James B. Sumner, John H. Northrop, and Wendell M. Stanley are awarded the Nobel Prize in chemistry for their independent work on the purification and crystallization of enzymes and viral proteins.

1946 Joshua Lederberg and Edward L. Tatum demonstrate that genetic recombination occurs in bacteria as the result of sexual matings. Lederberg and Tatum announced their discovery at the 1946 Cold Spring Harbor Symposium on Microbial Genetics, an event eventually considered a landmark event in the development of molecular biology.

1946 Bloch and Purcell develop Nuclear magnetic resonance (NMR) as viable tool for observation and analysis.

1948 James V. Neel reports evidence that the sickle–cell disease caused by a Mendelian autosomal recessive trait.

1949 John F. Ender, Thomas H. Weller, and Frederick C. Robbins publish "Cultivation of Polio Viruses in Cultures of Human Embryonic Tissues." The report by Enders and coworkers is a landmark in establishing techniques for the cultivation of poliovirus in cultures on non–neural tissue and for further virus research. The technique leads to the polio vaccine and other advances in virology.

1949 The role of mitochondria is finally revealed. These slender filaments within the cell, that participate in protein synthesis and lipid metabolism, are the cell's source of energy.

1949 Walter R. Hess (1881–1973), Swiss physiologist, receives the Nobel Prize for his experiments involving probes of deep–brain functions. Using microelectrodes to stimulate or destroy specific areas of the brain in experimental animals, he discovers the role played by particular brain areas in determining and coordinating the functions of internal organs.

1950 Erwin Chargaff demonstrates that the Tetranucleotide Theory is incorrect and that DNA is more complex than the model that developed by

Phoebus A. Levene. Chargaff proves that the nucleic acids are not monotonous polymers. Chargaff also discovers interesting regularities in the base composition of DNA; these findings are later known as "Chargaff's rules." Chargaff discovers a one–to–one ratio of adenine to thymine and guanine to cytosine in DNA samples from a variety of organisms.

1950 Douglas Bevis, British physician, demonstrates that amniocentesis could be used to test fetuses for Rh–factor incompatibility.

1951 Rosalind Franklin obtains sharp x–ray diffraction photographs of DNA.

1951 Alan Hodgkin, Andrew Huxley, and Bernard Katz offer modern analysis of the mechanisms of nerve impulse transmission.

1952 Alfred Hershey and Martha Chase publish their landmark paper "Independent Functions of Viral Protein and Nucleic Acid in Growth of Bacteriophage." The famous "blender experiment" suggests that DNA is the genetic material. When bacteria are infected by a virus, at least 80% of the viral DNA enter the cell and at least 80% of the viral protein remain outside.

1952 Rosalind Franklin completes a series of x–ray crystallography studies of two forms of DNA. Her colleague, Maurice Wilkins, gives information about her work to James Watson.

1952 Alan L. Hodgkin and Andrew F. Huxley, both English physiologists, first work out the mechanism of nerve–impulse transmission, showing that a "sodium pump" system works to carry impulses.

1953 James D. Watson and Francis H. C. Crick publish two landmark papers in the journal *Nature*: "Molecular structure of nucleic acids: a structure for deoxyribose nucleic acid" and "Genetical implications of the structure of deoxyribonucleic acid." Watson and Crick proposes a double helical model for DNA and call attention to the genetic implications of their model. Their model is based, in part, on the x–ray crystallographic work of Rosalind Franklin and the biochemical work of Erwin Chargaff. Their model explains how the genetic material is transmitted.

1954 Frederick Sanger determines the entire sequence of the amino acids in insulin.

1956 Arthur Kornberg demonstrates the existence of DNA polymerase in *E. coli.*

1956 Vernon M. Ingram reports that normal and sickle cell hemoglobin differ by a single amino acid substitution.

1956   Joe Hin Tijo and Albert Levan prove that the number of chromosomes in a human cell is 46, and not 48, as argued since the early 1920s.

1956   Mary F. Lyon proposes that one of the X chromosomes of normal females is inactivated. This concept became known as the Lyon hypothesis and helped explain some confusing aspects of sex–linked diseases. Females are usually "carriers" of genetic diseases on the X chromosome because the normal gene on the other chromosome protects them, but some X–linked disorders are partially expressed in female carriers. Based on studies of mouse coat color genes, Lyon proposes that one X chromosome is randomly inactivated in the cells of female embryos.

1957   Francis Crick proposes that during protein formation each amino acid is carried to the template by an adapter molecule containing nucleotides, and that the adapter is the part that actually fits on the RNA template. Later research demonstrates the existence of transfer RNA.

1958   Frederick Sanger is awarded the Nobel Prize in chemistry for his work on the structure of proteins, especially for determining the primary sequence of insulin.

1958   Matthew Meselson and Frank W. Stahl publish their landmark paper "The replication of DNA in *Escherichia coli,*" which demonstrated that the replication of DNA follow the semiconservative model.

1959   Severo Ochoa and Arthur Kornberg are awarded the Nobel Prize in Medicine or Physiology for their discovery of the mechanisms in the biological synthesis of ribonucleic acid and deoxyribonucleic acid.

1959   Jerome Lejeune, Marthe Gautier, and Raymond A. Turpin report that Down syndrome is a chromosomal aberration involving trisomy of a small telocentric chromosome. Patients with Down's syndrome have 47 chromosomes instead of the normal 46, because they have three copies of chromosome 21.

1961   François Jacob and Jacques Monod published "Genetic regulatory mechanisms in the synthesis of proteins," a paper which describes the role of messenger RNA and proposes the operon theory as the mechanism of genetic control of protein synthesis.

1961   Marshall Warren Nirenberg synthesizes a polypeptide by using an artificial messenger RNA, a synthetic RNA containing only the base uracil, in a cell–free protein–synthesizing system. The resulting polypeptide contains only the amino acid phenylalanine, indicating that UUU is the codon for phenylalanine. This important step in deciphering the genetic code is described in the landmark paper by Nirenberg and J. Heinrich Matthaei, "The Dependence of Cell–Free Synthesis in *E. coli* upon

Naturally Occurring or Synthetic Polyribonucleotides." This work establishes the messenger concept and a system that could be used to work out the relationship between the sequence of nucleotides in the genetic material and amino acids in the gene product.

1962   James D. Watson, Francis Crick, and Maurice Wilkins are awarded the Nobel Prize in physiology or medicine for their work in elucidating the structure of DNA.

1963   John Carew Eccles (1903–1997), Australian neurophysiologist, shares a Nobel Prize for his work on the mechanisms of nerve–impulse transmission. He also suggests that the mind is separate from the brain. The mind, he affirms, acts upon the brain by effecting subtle changes in the chemical signals that flow among brain cells.

1964   Barbara Bain publishes a classic account of her work on the mixed leukocyte culture (MLC) system that is critical in determining donor–recipient matches for organ or bone marrow transplantation. Bain showed that the MLC phenomenon is caused by complex genetic differences between individuals.

1965   François Jacob, André Lwoff, and Jacques Monod are awarded the Nobel Prize in Medicine or Physiology for their discoveries concerning genetic control of enzymes and virus synthesis.

1966   Marshall Nirenberg and Har Gobind Khorana lead teams that decipher the genetic code. All of the 64 possible triplet combinations of the four bases (the codons) and their associated amino acids are determined and described.

1967   Charles T. Caskey, Richard E. Marshall, and Marshall Warren Nirenberg suggest that there is a universal genetic code shared by all life forms.

1967   Charles Yanofsky demonstrates that the sequence of codons in a gene determines the sequence of amino acids in a protein.

1968   Robert W. Holley, Har Gobind Khorana, and Marshall W. Nirenberg are awarded the Nobel Prize in Medicine or Physiology for their interpretation of the genetic code and its function in protein synthesis.

1969   Jonathan R. Beckwith, American molecular biologist, and colleagues isolate a single gene.

1970   Har Gobind Khorana and colleagues announce the first complete synthesis of a gene

1971   Christian B. Anfinsen, Stanford Moore, and William H. Stein are awarded the Nobel Prize in chemistry. Anfinsen is cited for his work on ribonuclease, especially concerning the connection between the amino acid sequence and the biologically active conformation, and Moore and Stein are cited for their contribution to the understanding of the connection between chemical structure and catalytic activity of the active center of the ribonuclease molecule.

1972 Paul Berg and Herbert Boyer produce the first recombinant DNA molecules.

1973 Joseph Sambrook and coworkers refine DNA electrophoresis by using agarose gel and staining with ethidium bromide.

1973 Herbert Wayne Boyer and Stanley H. Cohen create recombinant genes by cutting DNA molecules with restriction enzymes. These experiments mark the beginning of genetic engineering.

1973 First report is made claiming a circadian variation in blood melatonin levels (pineal hormone) in humans. These variations affect mood and may cause the type of depression association with seasonal affective disorder (SAD).

1975 David Baltimore, Renato Dulbecco, and Howard Temin share the Nobel Prize in Medicine or Physiology for their discoveries concerning the interaction between tumor viruses and the genetic material of the cell and the discovery of reverse transcriptase.

1975 Scientists at an international meeting in Asilomar, California, called for the adoption of guidelines regulating recombinant DNA experimentation.

1975 John R. Hughes, Scottish physiologist and others discover enkephalin. This first known opioid peptide, popularly called "brain morphine," occurs naturally in the brain, indicating that the brain's chemical block the transmission of pain signals.

1976 Michael J. Bishop, Harold Elliot Varmus, and coworkers established definitive evidence of the oncogene hypothesis. They discovered that normal genes could malfunction and cause cells to become cancerous.

1978 Scientists clone the gene for human insulin.

1980 Paul Berg, Walter Gilbert, and Frederick Sanger share a Nobel Prize in Chemistry. Berg is honored for his fundamental studies of the biochemistry of nucleic acids, with particular regard to recombinant–DNA. Gilbert and Sanger are honored for their contributions to the sequencing of nucleic acids. This is Sanger's second Nobel Prize.

1982 The United States Food and Drug Administration approves the first genetically engineered drug, a form of human insulin produced by bacteria.

1985 Alec Jeffreys develops "genetic fingerprinting," a method of using DNA polymorphisms (unique sequences of DNA) to identify individuals. The method, which has been used in paternity, immigration, and murder cases, is generally referred to as "DNA fingerprinting."

1986 Robert A. Weinberg and coworkers isolate a gene that inhibits growth and appears to suppress retinoblastoma (a cancer of the retina).

1986 The United States FDA approves the first genetically engineered vaccine for humans, for hepatitis B.

1986 The United States Department of Energy officially initiates the Human Genome Initiative.

1986 First gene known to inhibit growth is produced by an American team led by molecular biologist Robert A. Weinberg. The gene is able to suppress the cancer retinoblastoma.

1987 The United States Congress charters a Department of Energy advisory committee, Health and Environmental Research Advisory Committee (HERAC), recommends a 15–year, multidisciplinary, scientific, and technological undertaking to map and sequence the human genome. DOE designates multidisciplinary human genome centers. National Institute of General Medical Sciences at the National Institutes of Health (NIH NIGMS) began funding genome projects.

1987 David C. Page and colleagues discover the gene responsible for maleness in mammals. It is a single gene on the Y chromosome that causes the development of testes instead of ovaries.

1988 The Human Genome Project officially adopts the goal of determining the entire sequence of DNA comprising the human chromosomes.

1988 The Human Genome Organization (HUGO) is established by scientists in order to coordinate international efforts to sequence the human genome.

1989 James D. Watson is appointed head of the National Center for Human Genome Research. The agency is created to oversee the $3 billion budgeted for the American plan to map and sequence the entire human DNA by 2005.

1989 Cells from one embryo are used to produce seven cloned calves.

1990 Michael R. Blaese and French W. Anderson conduct the first gene replacement therapy experiment on a four–year–old girl with adenosine deaminase (ADA) deficiency, an immune–system disorder. T cells from the patient are isolated and exposed to retroviruses containing an RNA copy of a normal ADA gene. The treated cells are returned to her body where they help restore some degree of function to her immune system.

1991   Mary–Claire King concludes, based on her studies of the chromosomes of women in cancer–prone families, that a gene on chromosome 17 causes the inherited form of breast cancer and also increases the risk of ovarian cancer.

1991   The gender of a mouse is changed at the embryo stage.

1992   The United States Army began collecting blood and tissue samples from all new recruits as part of a "genetic dog tag" program aimed at better identification of soldiers killed in combat.

1992   American and British scientists develop a technique for testing embryos *in vitro* for genetic abnormalities such as cystic fibrosis and hemophilia.

1993   George Washington University researchers clone human embryos and nurtured them in a Petri dish for several days. The project provokes protests from ethicists, politicians and critics of genetic engineering.

1993   An international research team, led by Daniel Cohen, of the Center for the Study of Human Polymorphisms in Paris, produces a rough map of all 23 pairs of human chromosomes.

1993   Scientists identify p53, a tumor suppressor gene, as the crucial factor preventing uncontrolled cell growth. In addition, scientists find that p53 performs a variety of functions ensuring cell health.

1994   Biologists discover that both vertebrates and invertebrates share certain developmental genes.

1994   Researchers identify a metastasis–suppressor gene and determine that Tamoxifen, an anti–cancer drug, blocked the blood supply that supported the growth of malignant tumors.

1995   Researchers at Duke University Medical Center reported that they transplanted hearts from genetically altered pigs into baboons. All three transgenic pig hearts survived at least a few hours, suggesting that xenotransplants (cross–species organ transplantation) might be possible.

1996   Chris Paszty and co–workers successfully employ genetic engineering techniques to create mice with sickle–cell anemia, a serious human blood disorder.

1996   American geneticist Gerard Schellenberg and colleagues discover the gene that causes Werner's syndrome, a condition which leads to premature aging.

1996   Researchers C. Cheng and L. Olson demonstrate that the spinal cord can be regenerated in adult rats. Experimenting on rats with a severed spinal cord, Cheng and Olson use peripheral nerves to connect white matter and gray matter.

1996   Researchers find that abuse and violence can alter a child's brain chemistry, placing him or her at risk for various problems, including drug abuse, cognitive disabilities, and mental illness, later in life.

1996   Scientists discover a link between autoptosis (cellular suicide, a natural process whereby the body eliminates useless cells) gone awry and several neurodegenerative conditions, including Alzheimer's disease.

1997   Ian Wilmut of the Roslin Institute in Edinburgh, Scotland, announces the birth of a lamb called Dolly, the first mammal cloned from an adult cell (a cell in a pregnant ewe's mammary gland).

1997   Donald Wolf and co–workers announce that they had cloned rhesus monkeys from early stage embryos, using nuclear transfer methods.

1997   While performing a cloning experiment, Christof Niehrs, a researcher at the German Center for Cancer Research, identifies a protein responsible for the creation of the head in a frog embryo.

1997   Researchers identify a gene that plays a crucial role in establishing normal left–right configuration during organ development.

1997   Researchers report progress in using the study of genetic mutations in humans and mice to decipher the molecular signals that lead undeveloped neurons from inside the brain to their final position in the cerebral cortex.

1997   The National Center for Human Genome Research (NCHGR) at the National Institutes of Health becomes the National Human Genome Research Institute (NHGRI).

1997   Mickey Selzer, neurologist at the University of Pennsylvania, and co–workers, finds that in lampreys, which have a remarkable ability to regenerate a severed spinal cord, neurofilament messenger RNA effects the regeneration process by literally pushing the growing axons and moving them forward.

1998   Two research teams succeed in growing embryonic stem cells.

1998   Scientists find that an adult human's brain can replace cells. This discovery heralds potential breakthroughs in neurology

1998   Immunologist Ellen Heber–Katz, researcher at the Wistar Institute in Philadelphia, reports than a strain of laboratory mice can regenerate tissue in their ears, closing holes which scientists had created for identification purposes. This discovery reopens the discussion on possible regeneration in humans.

1998 Scientists in Korea claim to have cloned human cells.

1998 Craig Venter forms a company, later named Celera, and predicts that the company will decode the entire human genome within three years. Celera plans to use a "whole genome shotgun" method, which would assemble the genome without using maps. Venter says that his company would not follow the Bermuda principles concerning data release.

1999 The public genome project responds to Craig Venter's challenge with plans to produce a draft genome sequence by 2000. Most of the sequencing is done in five centers, known as the "G5": the Whitehead Institute for Biomedical Research in Cambridge, MA; the Sanger Centre near Cambridge, UK; Baylor College of medicine in Houston, TX; Washington University in St. Louis, MO; the DOE's Joint Genome Institute (JGI) in Walnut Creek, CA.

1999 Scientists announce the complete sequencing of the DNA making up human chromosome 22. The first complete human chromosome sequence is published in December 1999.

2000 The first volume of *Annual Review of Genomics and Human Genetics* is published. Genomics is defined as the new science dealing with the identification and characterization of genes and their arrangement in chromosomes and human genetics as the science devoted to understanding the origin and expression of human individual uniqueness.

2001 In February 2001, the complete draft sequence of the human genome is published. The public sequence data is published in the British journal *Nature* and the Celera sequence is published in the American journal *Science*.

2001 In July 2001, scientists from the Whitehead Institute announce test results that show patterns of errors in cloned animals that might explain why such animals die early and exhibit a number of developmental problems. The report stimulates new debate on ethical issues related to cloning. In the journal *New Scientist* Ian Wilmut, the scientist who headed the research team that cloned the sheep "Dolly" argues that the findings argue for "a universal moratorium against copying people."

2001 In August 2001, United States President George Bush announces the United States will allow and support limited forms of stem cell growth and research.

2001 In November 2001, the company Advanced Cell Technology announces that its researchers created cloned human embryos that grew to the six cell stage.

2002 The company Advanced Cell Technology announces that it has developed a chip that can automatically create hundreds of cloned embryos.

2002 In February 2002, researchers at the Whitehead Institute for Biomedical Research publish results that indicate fully differentiated adult cells can form clones. The study appears in the British journal *Nature*.

GENERAL INDEX

•

Antibiotic resistance. *See* Drug resistance
Antibiotics, 102, 103, 317
    for burns, 77
    Doisy, Edward Adelbert on, 146, 147
    Fleming, Alexander on, 208–210
    for heart defects, 260
    for infertility, 318, 319
    penicillin, 101–102
    Waksman, Selman Abraham on, 592
Antibodies, **25–27,** 309–313, 317
    allergy, 15
    in autoimmune diseases, 33–34
    B lymphocytes, 39
    Benacerraf, Baruj on, 49
    in blood, 62
    Bordet, Jules on, 67
    in breast milk, 373–374
    Burnet, Frank Macfarlane on, 77
    Dausset, Jean on, 134
    Doherty, Peter C. on, 146
    for drug addiction, 150
    Edelman, Gerald M. on, 158–159
    Ehrlich, Paul on, 162
    genetics role in, 562
    HIV, 13–14
    infertility and, 319
    in inflammation, 320
    interferons and, 324–325
    Jerne, Niels Kaj on, 329–330
    Köhler, Georges on, 340, 341
    Landsteiner, Karl on, 350–351
    lymphatic system and, 366
    Milstein, César on, 389–390
    plasma, 214, 453–454
    Porter, Rodney on, 456–457
    prion, 468
    serology, 512
    synthesis of, 159
    See also Immunoglobulins; Monoclonal antibodies
Antibody-antigen binding, 25, 309–312
    B lymphocytes, 39
    Jerne, Niels Kaj on, 329
Antibody-antigen reactions. *See* Immune reactions
Antibody-mediated immunity. *See* Humoral immunity
Antibody tests, for autoimmune diseases, 34
Anticancer agents. *See* Antineoplastic agents
Anticholesteremic agents, 151, 241
Anticoagulants, 151
Anticodons, 279
Antidepressants, 86, 468
Antidiuretic hormone (ADH), **145,** 176, 178, 284, 420
    blood pressure regulation, 64
    vs. oxytocin, 435
    pituitary gland and, 452
    renal system and, 336–337, 479
Antidotes, **454–455**
    See also Antitoxins
Antigen-antibody binding, 25, 309–312
    B lymphocytes, 39
    Jerne, Niels Kaj on, 329
Antigen-antibody reactions. *See* Immune reactions
Antigen-presenting cells (APC), 310, 311, 324–325
Antigens, **25–27,** 309–313

    in autoimmune diseases, 33–34
    B lymphocytes and, 39
    Benacerraf, Baruj on, 49
    in blood, 62, 111, 214–215
    Burnet, Frank Macfarlane on, 76–77
    cancer and, 104
    Dausset, Jean on, 134, 135
    diabetes mellitus and, 141
    Doherty, Peter C. on, 146
    Edelman, Gerald M. on, 158–159
    genetics and, 296
    HIV, 13–14
    Köhler, Georges on, 341
    Landsteiner, Karl on, 350, 351
    Milstein, César on, 390
    in placenta, 452
    self, 559
    serology, 512
Antihemorrhagic agents. *See* Hemostatic agents
Antihistamines, 15, 57
Antimalarial drugs, 447–448
Antimitotic signals, 91–92, 231
Antineoplastic agents, 103–104, 111
Antioxidants, 2, **27–28**
Antiphospholipid antibody, 33
Antiplatelets, 151
Antipsychotic drugs, 468
Antipyretic drugs, 202
Antisense drugs, 104
Antisepsis, 339, 340
Antiseptics, 208
Antiserum, 312
    Bordet, Jules on, 67
    MacLeod, Colin Munro on, 369
    Richet, Charles Robert on, 494–495
Antitoxins
    Behring, Emil von on, 46–47
    in blood, 62
    Ehrlich, Paul on, 162, 164
    immune system and, 309
    See also Antidotes
Antitubercular agents, 148
Antiviral agents, 103, 170
Antral follicles, 383
Antrum, 538
Anus, 226
Anvil. *See* Incus
Anxiety disorders, drugs for, 468
Aorta, 28, 30–31, 109, 111
    abdominal, **1–2**
    blood pressure, 63
    coronary circulation, 121, 259
    ductus arteriosis and, 152
    fetal development, 201, 261
    in systemic circulation, 545
    See also Thoracic aorta
Aortic arch, **28,** 30–31
    carotid arteries and, 86–87
    fetal circulation, 201
    ligamentum arteriosum and, 152
Aortic arch syndromes, 28
    See also Atherosclerosis
Aortic hiatus, 31

Katz, Bernard on, 334
motor neuron, 396
neurotransmitters, 420
olfactory, 510
Ramón y Cajal, Santiago on, 476
in reflexes, 122, 477
regrowth, 417–418
in spinal cord, 532–533
Ayurvedic medicine, 272
Azathioprine
Elion, Gertrude Belle on, 169
Hitchings, George Herbert on, 277
Murray, Joseph E. on, 402
Azidothymidine (AZT). *See* Zidovudine (AZT)
Azoospermia, 320
AZT. *See* Zidovudine (AZT)
Azygous vein, 558

B

B-estradiol. *See* Estradiol
B fibers. *See* Nerve fibers
B lymphocytes, **39,** 309, 310–313, 355, 366
Doherty, Peter C. on, 146
hemopoiesis and, 264
Jerne, Niels Kaj on, 330
Babinski reflex, 478
Baby teeth. *See* Milk teeth
Bacilli
Behring, Emil von on, 46–47
Ehrlich, Paul on, 162
Koch, Robert on, 338–339
Bacon, Roger, 273
Bacteremia, 41
Bacteria, **39–41,** *40*
chromosomes in, 107
diarrhea from, 142
fever and, 202
Fleming, Alexander on, 208–209
Florey, Howard Walter on, 211, 212
Hopkins, Frederick Gowland on, 282
immune system and, 308–309
intestinal, 226, 326
Koch, Robert on, 337–339
Leeuwenhoek, Anton van on, 353, 354
lymphatic system and, 366
MacLeod, Colin Munro on, 369–370
resistance, 317
tooth decay from, 563
Bacteria-host relationships, 39–41
Bacterial biochemistry. *See* Biochemistry
Bacterial fermentation, 207–208, 326
Bacterial infections, **39–41,** *40*
Bordet, Jules on, 67
diarrhea from, 142
Domagk, Gerhard on, 147
epidermis and, 180
fever from, 202
Fibiger, Johannes on, 203
Fleming, Alexander on, 208–210
Florey, Howard Walter on, 211, 212
Hitchings, George Herbert on, 276, 277
immune system and, 308–309

nails, 409
resistance, 317
septic shock from, 516
Bacterial meningitis, 98, 257, 381
Bacterial toxins, 455
Bacterial transformation, 369–370
Bacterial viruses. *See* Bacteriophages
Bacteriology
Behring, Emil von on, 46–47
Bordet, Jules on, 67–68
Eijkman, Christiaan on, 163–164
Fleming, Alexander on, 208–210
Koch, Robert on, 337–339
Landsteiner, Karl on, 350–351
Nicolle, Charles Jules Henri on, 421
Ross, Ronald on, 496–497
Bacteriolysis, 67
Bacteriophages, 76, 128, 584, 585
Bactrim. *See* Trimethoprim
Baer, Karl Ernst von, **41–42,** 275
Baillie, Matthew, 274
Baker's yeast. *See* Saccharomyces cerevisiae
Balance, *507*–508
in altitude sickness, 35
cerebellum and, 99
ear and, 156, 190
The Balance Within: The Science Connecting Health and Emotions
(Sternberg), 537
Ball and socket joints, 32, 545
Balloon catheterization, 260
Band amniotic syndrome. *See* Amniotic band syndrome
Banting, Frederick Grant, **42–44,** 276, 288
Bárány caloric test, 44
Bárány, Robert, *44*–45
Barbiturates, 24, 150
Barometers, 281–282
Barondontalgia, 35
Baroreceptors, 486
Barosinusitis, 35
Barotitis media, 35
Barry, Martin, 275
Bartolino's glands, 482
Basal bodies, of cilia, 108
Basal body temperature. *See* Body temperature
Basal cell carcinoma, 80–81, 323
Basal ganglia, **45,** 101
Carlsson, Arvid on, 86
in limbic system, 359
in motor function, 396
Basal lamina, 182
Basal nuclei. *See* Basal ganglia
Basal plate, 452
Base pairs. *See* Nucleotide base pairs
Basement membrane, 182, 364
See also Hypodermis
Basilar artery, 108
Basilar membrane, 48, 155, 258, 259, 512
Basophilic leukemia. *See* Leukemia
Basophils, 355, 451
Bateson, William, 232, 293
Batteries, electric, 590
Bayliss, William Maddock, 131, 276
Bcl genes, 417–418

Cytoskeleton, 60
Cytosol, 476
Cytotoxic T cells. *See* Killer T cells
Cytotrophoblast, 175, 452, 462

D

Dale, Henry Hallett, **131–132,** 546
Dalton, John Call, **133,** 275
Dam, Henrick, **133–134,** 146–147
Daraprim. *See* Pyrimethamine
DARPP-32, 246
Darwin, Charles
 Huxley, Thomas Henry and, 299–300
 Kölliker, Albert von and, 342
Darwinism, 299–300
Daughter cells, 91–94
 chromosomes, 107
 genetic regulation, 229, 230–231
Dausset, Jean Baptiste, 49, **134–135**
d'Azyr, Félix Vicq, 274
DDT. *See* Dichlorodiphenyltrichloroethane (DDT)
De humani corporis fabrica (Vesalius), 581
Deafness, 190, 259
Deamination, 58
Death, **135–136**
 drowning, **149**
 osteology, 432
 rigor mortis, 495
 from shock, 516
Decidual reaction, 313
Deciduous teeth, 562
Decompensated shock, 516
Decompression sickness, **49–50**
 altitude-induced, 34–35
 Eustachian tubes and, 191
 underwater-induced, 569
 See also Altitude sickness
Decongestants, 15
Deep. *See* Anatomical nomenclature
Deep fascia. *See* Fascia
Deep tendon reflex, 478
Deep veins, 579
Defecation, 327
Defibrillation, 123
Deficiency diseases, 163–164
Degenerative diseases, **405–406**
Dehydration
 antidiuretic hormone and, 145
 from diarrhea, 142
 integument, 323
 in sports, 534
 thirst and, 556
 tissue, 272
Dehydroepiandrosterone, 9
Dehydroepiandrosterone sulfate, 9
Dehydrogenase, 180
Delayed onset muscle soreness (DOMS), 403
Delbrück, Max, 597, 598
Delivery (childbirth), 441–442
Delta cells, 439
Delta waves, 167
Deltoids, 69

Dementia, 382
Demineralization (bone), 65
Democritus, 272
Dendrites, 245, 413, 416, 419
 in central nervous system, 97
 cerebral cortex, 99
 olfactory, 510
 Ramón y Cajal, Santiago on, 476
 in spinal cord, 532
Dengue hemorrhagic fever, 264–265
Dengue virus, 264–265
Dense connective tissue. *See* Connective tissue
Dental caries, **562–*563***
Dental enamel, 563
Dental hygiene, 563
Dental plaque, 563
Dentate gyrus, 417
Dentine, 563
Deoxy nucleotide-triphosphates (dNTP), 296
Deoxyribonucleic acid (DNA), **136–138,** *137,* 276
 aging and, 12
 AIDS, 13, 14
 amino acids, 19
 cancer, 80, 104
 cell cycle, 90–93
 cell differentiation, 94
 cell structure, 95
 chromosomes, 105
 Crick, Francis on, 127–129
 Elion, Gertrude Belle on, 170
 enzymes, 136–137, 138
 ethical issues, 187, 189
 evolution and, 192
 fertilization and, 201
 fingerprinting, 138, 214
 forensic pathology, 214
 genetics, 227–229, 232–233, 294, 296
 Gilbert, Walter on, 234–236
 Holley, Robert on, 278, 279
 in inherited diseases, 320–321
 Kornberg, Arthur on, 342–343
 MacLeod, Colin Munro on, 369–370
 Milstein, César on, 389, 390
 in mitochondria, 391
 Nirenberg, Marshall Warren on, 422–423
 protein synthesis, 466–467
 radiation effect on, 475, 476
 vs. RNA, 490
 study of, 392–393
 visual impairment and, 58–59
 Watson, James D. on, 597–598
Deoxyribose, 105, 136
Dephosphorylation, 246
Depolarization. *See* Neural excitation
Depressants. *See* Central nervous system depressants
Depression
 Carlsson, Arvid on, 86
 depressants and stimulants, 139
 drugs for, 468
Dermal papillae, 323
Dermatomes, 324, 527
Dermatosis, 140

Digestive system. *See* Gastrointestinal tract

Digital rays, 358

Digitalis, 479

Digitalis (drug). *See* Digitalis glycosides

Digitalis glycosides, 151

Digits, development of, 358, 359

Digoxin, for heart defects, 260

Dihydrotestosterone, 234, 513, 553

Dihydroxyacetone phosphate, 239

Dihydroxycolecalciferol. *See* Vitamin D

Dihydroxyphenylalanine (DOPA), 379

Dilatation
    blood vessel, 217
    bronchial tree, 75
    in parturition, 441

Dioptrics, 250

Diphosphopyridine nucleotide (DPN), 342

Diphtheria
    Behring, Emil von on, 46–47
    Ehrlich, Paul on, 162, 164
    Fibiger, Johannes on, 203

Diploid cells
    cell cycle, 92–93
    chromosomes, 105, 107
    oogenesis, 427
    sex determination, 512

Direct auscultation. *See* Auscultation

Directional selection. *See* Natural selection

Disaccharides, 81

Discontinuous capillaries. *See* Capillaries; Sinusoids

Disease
    Hippocrates on, 271
    immune system and, 309
    vs. infection, 39
    screening, 315
    study of, **442–443**
    See also specific diseases and conditions

Disease etiology, 442–443

Disease prevention
    AIDS, 14
    antioxidants for, 27
    Behring, Emil von on, 47

Disease progression. *See* Pathogenesis

Disease transmission, 316–317
    AIDS, 13, 14
    Ross, Ronald on, 497
    See also Sexually transmitted diseases

Disruptive selection. *See* Natural selection

Dissection, human, 581

Disseminated intravascular coagulation (DIC), 454

Distal. *See* Anatomical nomenclature

Diuresis, 145, 193

Diuretics, **145**
    for cardiovascular disease, 151
    for edema, 160

Diverticulum, 224

Diving, 49–50, 569–570, 575

DMBA. *See* 7,12-dimethylbens(a)anthracene (DMBA)

DMD. *See* Duchenne muscular dystrophy (DMD)

DNA. *See* Deoxyribonucleic acid

DNA adducts, 384

DNA fingerprinting. *See* DNA identification

DNA fragmentation, 235

DNA identification, 138, 214

DNA ligase, 136, 138

DNA polymerase, 136, 138, 296, 342–343

DNA probes, 54

DNA regulation, 392

DNA repair, 138, 393
    cell cycle, 91, 92, 231
    evolution and, 192
    Hartwell, Leland H. on, 256
    radiation and, 476
    waste removal and, 384

DNA replication, 90–93, 136–138, 280, 296, 393

DNA sequence, 392, 393
    Gilbert, Walter on, 235, 236
    Holley, Robert on, 279
    Khorana, Har Gobind on, 336
    Nirenberg, Marshall Warren on, 423
    protein synthesis, 466

DNA synthesis, 136–138, 296
    cell cycle, 91, 231
    Khorana, Har Gobind on, 336
    Kornberg, Arthur on, 342–343

DNA viruses, 584–587
    chemotherapy for, 103
    Elion, Gertrude Belle on, 170
    Kornberg, Arthur on, 343

dNTP. *See* Deoxy nucleotide-triphosphates

Doctors, women, 12, 499–501

Doherty, Peter C, **145–146,** 607

Doisy, Edward Adelbert, 133, **146–147**

Domagk, Gerhard, 102–103, **147–148**

DOMS. *See* Delayed onset muscle soreness (DOMS)

Donders, Frans Cornelis, **148–149,** 275

Donnan equilibrium. *See* Equilibrium

DOPA. *See* Dihydroxyphenylalanine (DOPA)

DOPA quinone, 379

Dopamine, 8, 9, 10, 416, 420
    basal ganglia and, 45
    Carlsson, Arvid on, 86
    deficiency in, 86
    depressants and stimulants, 139
    drugs and, 149, 150, 468
    Greengard, Paul on, 246
    in synapses, 544

Doppler techniques, 308
    See also Ultrasonic doppler techniques

Dorsal. *See* Anatomical nomenclature

Dorsal interossei. *See* Interosseus muscles

Dorsalis pedis artery, 472

Dorsoventral axis, 358

Dorsoventral flattening. *See* Paddle stage (Limbs)

Double helix. *See* Deoxyribonucleic acid (DNA)

The Double Helix (Watson), 597

Doublets. *See* Microtubules

Down syndrome, 20, 57, 107, 321, 428

DPN. *See* Diphosphopyridine nucleotide (DPN)

Drowning, **149**

Drug absorption, 151

Drug abuse. *See* Substance abuse

Drug addiction, **149–150**
    See also Substance abuse

Drug-drug interactions, 150–151

Drug-food interactions, 150, 151

Drug hypersensitivity, 447–448

Drug interactions, **150–151, 448–449**

Drug metabolism, 151, 447–448

Drug resistance, 210, 317

Drug therapy

for cardiovascular disease, **151–152**

Elion, Gertrude Belle on, 169–170

patient response, 447–448

study of, **448–449**

See also Chemotherapy

Drugs

Black, James Whyte on, 57–58

effects on nervous system, **150**

fever from, 202

Magendie, François on, 370

osteoporosis and, 432

plasma clearance, 454

respiration, 486

See also Pharmaceutical drugs; specific drugs

Dualism (Philosophy), 141

Dubos, René, 591–592

Duchenne muscular dystrophy (DMD), 188, 321, 405

Ducrey's bacillus. *See Hemophilus ducreyi*

Ductal carcinoma. *See* Breast cancer

Ductules, 52

Ductulus efferens, 108

Ductus arteriosis, 152, **152**, 201, 261

Ductus venosus, 201

Duggar, Benjamin Minge, 103

Dumdum fever. *See* Kala azar

Duodenum, 226, 326

bile and, 52

embryonic development, 224

flatus and, 207

Herophilus on, 266

Dura matter, 97–98, 100, 381

Dural septa, 381

Dural sinus, 381

Dutrochet, René-Joachim-Henri, 275

Duvé, Christian de, **152–154**

Dwarfism, 56, 177, 248, 297

Li, Choh Hao on, 357, 358

pituitary hormones and, 452

Dye reactions. *See* Stains (Microscopy)

Dynamic bioassay, 576

Dynein, in cilia, 108

Dynorphins, 179, 437

Dysfunctional uterine bleeding, 381

Dysphagia, **542**

Dysplasia, 94, 230

Dystrophin, 405

Dystrophinopathies. *See* Muscular dystrophies

E

E. coli. See Escherichia coli

Ear, **155**, 258, 510–512

Békésy, Georg von on, 47–48

bones, 155

ectoderm, 158

elastic cartilage in, 166

embryonic development, **156**, 508

Eustachio, Bartolomeo on, 191

external, **155**, 156

Fallopius, Gabriel on, 198

Helmholtz, Hermann on, 262, 263

water pressure and, 569

See also Inner ear; Middle ear

Ear auricle, 155, 156, 258, 508, 510

Ear canal, 155, 258, 508, 510

Ear infection. *See* Otitis media

Ear pressure, 155, 190, 423, 511–512

Eardrum. *See* Tympanic membrane

Eating disorders, 8

Ebola fever, 264–265

Ebola virus, 264–265

EC. *See* Enzyme catalogue (EC)

Eccles, John Carew, **156–157,** 275, 298, 515

Eccrine glands, 542

ECG. *See* Electrocardiogram (ECG)

Echocardiogram, for PDA, 152

Echolocation, 528–529

Ectoderm, **158**, 173, 247, 461, 462

Baer, Karl Ernst von on, 41

embryonic, 175–176, 202, 224–225

endoderm and, 178–179

glucose and, 239

integumentary system, 324

limbs, 358

nervous system, 127, 414

sense organs, 156, 193, 508

Ectopic pregnancy, 313, 318

Edelman, Gerald M, **158–160,** 456–457

Edema, *160*

diuretics and, 145

endocrine system and, 178

extracellular fluid, 193

hormones and, 513

plasma and, 453

See also Cerebral edema; Pulmonary edema

EDRF. *See* Nitric oxide

Edwards, Robert G, **160–161,** 536

EEG. *See* Electroencephalogram (EEG)

EF-TU. *See* Elongation factor

Effacement. *See* Cervical effacement

Effectors, peripheral nervous system and, 445

Efferent neurons. *See* Motor neurons

EGF. *See* Epidermal growth factor (EGF)

Egg cells. *See* Ovum

Ehrlich, Paul, **161–163,** 276

Behring, Emil von and, 47

Dale, Henry Hallett and, 131, 132

Domagk, Gerhard and, 147

on immune system, 309, 311

Ehrreich, Stuart, 218

Eijkman, Christiaan, *163–164*

Einthoven, Willem, **164–166,** *165,* 275

Eisenmenger syndrome, 152

Ejaculation, 234, 320, 435, 465, 506–507, 571, 572

See also Semen

EKG. *See* Electrocardiogram (ECG)

Elastic cartilage, 90, 116, **166**

Elasticity

of abdominal aorta, 2

aging and, 11

of blood vessels, 30, 31, 63, 109–110

Furchgott, Robert F. on, 217
liver, 361
Endothelial tubes, 261
Endothelium-derived relaxing factor (EDRF). *See* Nitric oxide
Endotoxins, bacterial, 41
Endurance training, 534
Energy
    from adipose tissue, 6
    conservation of, 263
    from fat, 566
    from nutrition, 425–426
Engagement (labor). *See* Fetal engagement
Enhancement gene therapy. *See* Gene therapy
Enkephalins, 179, 437
Enteric nervous system, 445–446
Enterocytes, 326
Entoderm. *See* Endoderm
Environmental factors
    birth defects and, 198
    cancer and, 79, 81
    diabetes mellitus and, 141–142
    embryology and, 174
    evolution and, 192
    genetics and, 296
    growth and development, 104–105
    inherited diseases and, 321
Environmental Protection Agency (EPA), 475
Environmental stimuli. *See* Stimuli
Enzyme activation, 385
    mitochondria and, 392
    Murad, Ferid on, 400
Enzyme catalogue (EC), 179
Enzyme degradation
    necrosis from, 411
Enzyme latency, 153, 154
Enzyme-linked immunosorbent assay, 13
Enzyme repression, 385
Enzymes, 19, **179–180,** 217
    aging, 12
    antioxidants and, 27
    Axelrod, Julius on, 37
    in bacteria, 41
    Bergström, Sune Karl on, 50
    biochemistry and, 53
    in bone reabsorption, 67
    in cell cycle, 92–94, 231
    cell differentiation, 94
    Chain, Ernst Boris on, 101
    chromosomes and, 105
    Claude, Alfred on, 112
    Cori, Carl Ferdinand on, 116
    Cori, Gerty T. on, 117–118
    DNA, 136–137, 138
    Duvé, Christian de on, 153, 154
    Edelman, Gerald M. on, 159
    Elion, Gertrude Belle on, 170
    in embryonic development, 174, 202
    in endergonic reactions, 5
    epinephrine and, 181
    in fertilization, 200
    Fischer, Edmond H. on, 207
    Fleming, Alexander on, 209
    Florey, Howard Walter on, 211

    in forensic pathology, 214, 215
    genetics, 232
    Holley, Robert on, 279
    Hopkins, Frederick Gowland on, 282, 283
    Huggins, Charles on, 289, 290
    Hunt, Tim on, 297
    interferons and, 325
    Khorana, Har Gobind on, 335, 336
    Kornberg, Arthur on, 342–343
    in protein synthesis, 466
    salivary, 251
    in Tay-Sachs disease, 56–57
    in waste removal, 384
    See also specific enzymes
Enzymes: Properties, Distribution, Methods, and Applications (Waksman), 591
Eosinophic leukocytes. *See* Leukocytes
Eosinophilia-myalgia syndrome (EMS), 537
Eosinophilic leukemia. *See* Leukemia
Eosinophils, 355
EPA. *See* Environmental Protection Agency (EPA)
Ependyma, 108, 416
Epiblast, 175, 202, 461–462
Epicardium, 260
    See also Pericardium
Epidermal growth factor (EGF)
    in cell cycle, 91, 230–231
    Cohen, Stanley on, 113–114
Epidermis, 140, **180,** 323
    burns, 77
    cell differentiation, 94
    ectoderm, 158, 324
    hair, 253
    melanocytes in, 378
Epididymis, 234
Epigastric iliac artery. *See* Iliac arteries
Epigastric region, 1
Epigenesis, 171
Epiglottis
    in cough reflex, 122
    elastic cartilage in, 166
Epilepsy
    antioxidants and, 28
    EEG for, 167, 168
Epinephrine, 8, 9–10, 178, **180–181,** 283–285
    Black, James Whyte on, 57
    cardiac cycle and, 262
    Dale, Henry Hallett on, 131–132
    depressants and stimulants, 139
    drugs and, 150
    fight or flight reflex, 204–205
    Loewi, Otto on, 362, 363
    Murad, Ferid on, 400
    from nicotine, 422
    from paraganglia, 440
    Sutherland, Earl W. on, 541–542
    sympathetic nervous system and, 544
Epiphysis, 90, 519
Epitendineum, 552
Epithalamus, **181,** 414
Epithelial cells, 181–182
    in amniotic fluid, 202
    in bile ducts, 52

F

gonads and, 243
  pituitary gland and, 452
  puberty, 469, 470
Follicular phase, 382–383, 470
Fontanels, 127, 522–523
Food-drug interactions, 150, 151
Food poisoning, 142, 410
Foot, 29, 363, 521
Foot plates, 358
Foramen, 201, 261, 394, 522
Forebrain. *See* Prosencephalon
Foregut. *See* Primitive gut
Forelimb, 358, 521
Forensic pathology, **214,** 442
  Hopkins, Frederick Gowland on, 282
  pubic symphysis, 471
Foreskin, 234
Forssmann, Werner, 122, 123, **215–216**
Fossa, 29
Fossa ovalis, 201, 260
Foster, Michael, 515
Fovea, 195
Fracture repair, **65–66**
Fractures (Injuries), **65–66,** 447
Fragile X syndrome, 321
Fragment antigen-binding (FAB) pieces, 457
Fragment crystallizable, 457
Frame shifts. *See* Gene mutations
Frameshift mutation, 229
Framingham study, 83
Franek, Frantisek, 159
Franklin, Rosalind, 597, 598
  Crick, Francis and, 128, 129
  DNA, 138
Fraternal twins, 376–377
Fred Hutchinson Cancer Research Center, 556
Fredrickson, Donald S., 240
Free energy, 5–6
Free nerve endings, 337
Free paraganglia. *See* Paraganglia
Free Radical: Albert Szent-Györgyi and the Battle over Vitamin C (Moss), 546
Free radicals, 27–28
  melanocytes and, 379
  Porter, George on, 456
  in tissue damage, 403, 475
FRH. *See* Follicle releasing hormone (FRH)
Friction injury. *See* Burns
Frontal bone, 127
Frontal leucotomy, 393–394
Frontal lobe, 99, 100
  cranial nerves and, 126
  Moniz, Egas on, 393
Frontal plane. *See* Coronal plane
Fructose, 81, 238, 239
Fruit flies, 357, 600
FSH. *See* Follicle stimulating hormone (FSH)
Fuchs' dystrophy, 120
Full-thickness burns. *See* Burns
Function and structure. *See* Structure and function
Functional magnetic resonance imaging (fMRI). *See* Magnetic resonance imaging
Fundus, 482, 538

Fungal nail infections, 409
Funk, Casimir, 276, 588
Funke, Otto, 276
Furchgott, Robert F., **217–218,** 305

G

G1 phase. *See* Interphase
G2 phase. *See* Interphase
G forces. *See* Gravitational forces
G proteins, 236, 237, 495, 496, 524, 551
GABA. *See* Gamma-aminobutyric acid (GABA)
GABA receptors, 150, 468
Gait, 458
Galactose, 81, 238, 426
Galen, *219*–220, 273, 274
  on digestion, 143
  Harvey, William and, 257, 364
  Luzzi, Mondino de and, 365, 366
  on myology, 408
  Vesalius, Andreas on, 581
Gall, Franz Joseph, 212, **220,** 275
Gallbladder, 225, 226, 430
  bile in, 52
  embryonic development, 224
  in pregnancy, 461
  referred pain, 477
  Rous, Peyton on, 497, 498
Gallo, Robert, 12–13
Gallstones, 52
Gally, Joseph, 159
Galvani, Luigi, *221,* 274
Galvanometer, 164–165, 270, 275
Gamete intra-fallopian transfer (GIFT), 200, 314, 319
Gametes, 95, 178, 234, 482
  cell cycle, 91, 92–93
  chromosomes, 105–107
  embryonic development, 174
  evolution, 192
  genetic regulation, 229
  gonads and, 243
  Mendel, Gregor on, 380
  oogenesis, 427–428
  secondary sexual characteristics, 506
  in sexual reproduction, 514
Gametocytes, 372
Gametogenesis, 234, 243, 428
  See also Oogenesis; Spermatogenesis
Gamma-aminobutyric acid (GABA), 19, 420, 468
  depressants and stimulants, 139, 150
  in muscles, 406
  See also GABA receptors
Gamma-endorphins. *See* Endorphins
Gamma globulins, 25, 62, 309
  See also Immunoglobulins
Gamma interferon, 324–325
Gamma macroglobulin, 62
Gamma rays, 475
Gamow, George, 229
Ganglia, 417, 440
  in autonomic nervous system, 34, 150, 440
  basal, **45**
  in cranial nerves, 127

insulin in, 238, 239
  in pancreas, 439
Isoflurane, 24
Isoleucine, 18, 19
Isomerase, 179
Isomers, 239
    See also Structural isomers
Isoprenaline, 57
Isoprene, 61
Itching, **327–328**
ITP. *See* Idiopathic thrombocytopenic purpura (ITP)
IUBMB. *See* International Union of Biochemistry and Molecular Biology (IUBMB)
IUD. *See* Intrauterine device (IUD)
IVF. *See* in vitro fertilization (IVF)

J

Janssen, Hans, 274
Janssen, Zacharias, 274
Jaundice, 52, 362
    bilirubin and, 53
    rheumatoid arthritis and, 265–266
Jaw. *See* Mandible
Jejunum, 224, 226, 326
Jenner, Edward, 586
    on immune system, 308, 311
    on vaccination, 27
Jerne, Niels Kaj, **329–330**
Jerne plaque assay, 329
Johnson, William Arthur, 346
Joint kinesthetic receptors, 337
Joint malformations, *32*
Joint movement, 32, 330
Joints, *330*–331
    in altitude sickness, 35
    cartilaginous, **90,** 116, 300
    embryonic development, 358–359
    fascia in, 198
    fibrous, **204**
    study of, 32
    See also specific joints and types of joints
Junctional complexes, 182
Juvenile diabetes, 249
Juvenile onset diabetes. *See* Insulin dependent diabetes mellitus
Juxtaglomerular apparatus, 9
Juxtaposition. *See* Gene juxtaposition

K

Kala azar, 353
Kandel, Eric R., 85, 86, **333**
Kaps. *See* Karyopherins
Karyopherins, 60
Karyotype, 107
Katz, Bernard, 36, 37, **333–334**
Kendall, Edward Calvin, 265, 266, 276, **334–335**
Keratin, 180, 253, 323, 324
    in epithelial cells, 182
    mucocutaneous junctions and, 396–397
    in nails, 409
Keratinocyte cells, 180, 323, 379
Keratoconus, 58, 120

Keratocytes, 379
Keratoplasty. *See* Corneal transplantation
Kernicterus, 53
Ketamine, 24
Ketham, Johannes de, 273
Keto acids, 4
Ketones, 4
Khachadurian, Avedis K., 240
Khorana, Har Gobind, **335–336**
Kidney cancer, 80
Kidney damage, 62
Kidney diseases, 441
    extracellular fluid and, 193
    waste elimination and, 169
Kidney failure
    blood pressure and, 64
    Murray, Joseph E. on, 402
Kidney filtration. *See* Glomerular filtration
Kidney induction, 480
Kidney stones, 253, 573
Kidney transplantation, 169, 170, 401–402
Kidneys, **336–337,** 430, 479–480, 570
    blood pressure regulation, 63–64
    circulatory system and, 109
    disease of, 169, 193
    diuresis and, 145
    electrolyte balance, 168
    Eustachio, Bartolomeo on, 191
    filtration, **237–238**
    hormone action on, 284, 441
    Malpighi, Marcello on, 373
    in pregnancy, 461
    sympathetic stimulation of, 544
    waste elimination, 168–169, 384
Killer T cells, 309, 310, 311, 366, 550
    Doherty, Peter C. on, 146
    interferons and, 324
    Jerne, Niels Kaj on, 330
    thymus and, 559
Kinase A, 246
Kinesthetics, **337**
Kinetochore, 231
King, Charles Glen, 547
Kircher, Athanasius, 274
Kitasato, Shibasaburo, 47
Klinefelter syndrome, 315
Knee, 90, 363
Knee injuries, 90
Knee-jerk reflex, 476
Kneecap. *See* Patella
Knoll, Max, 275
Knudson, Alfred, 322
Koagulations Vitamine. See Vitamin K
Koch, Robert, 275, **337–339**
    Behring, Emil von and, 46–47
    Ehrlich, Paul and, 161, 162
    Eijkman, Christiaan and, 163–164
    Sherrington, Charles Scott and, 514
Kocher, Emil Theodor, **339–340**
Koch's postulates, 338
Köhler, Georges, **340–341**
    on antibodies, 27
    Milstein, César and, 390

in digestion, 226
fever and, 202
in fracture repair, 66
gene therapy and, 226–227
hemopoiesis, 264
immune system and, 309, 310, 311
lymphatic system and, 366
Metchnikoff, Élie on, 386–387
in mucus, 397
Leukopenia, 355
Leukotrienes, 503
Levatores costarum, 404
Levene, Poebus Aaron Theodore, 276
Levi, Giuseppe, 355–356
Levi-Lorain dwarfism. *See* Dwarfism
Levi-Montalcini, Rita, 113, **355–356**
Levine, Philip, 350, 351
Levodopa, 86
Lewis, Edward B., **356–357**
Leydig cells, 213, 243, 484, 513
LH. *See* Luteinizing hormone (LH)
Li, Choh Hao, 296, **357–358**
Li-Fraumeni syndrome, 322
Libido, 469
Liddell, E.G.T., 515
Liebig, Justus von, 275
LIF. *See* Leukemia inhibitory factor (LIF)
Life (biology), characteristics of, **216–217**
Life support, CPR, **123–126,** *124*
Ligaments, 429
epithelium and, 182
eye, 195
gastrointesinal, 224
joints and, 204, 330
Ligamentum arteriosum, 152, 201
Ligamentum venosum, 201
Ligase, 179, 343
See also DNA ligase
Light chain, 403
Light therapy. *See* Phototherapy
Lima, Almeida, 393
Limb buds, **358–359,** 414
Limb morphogenesis. *See* Morphogenesis
Limbic lobe. *See* Limbic system
Limbic system, **359**
depressants and stimulants, 138
epithalamus and, 181
hippocampus, **271**
hypothalamus and, 301
intellectual functions and, 70, 71
Limbs, embryonic development of, **358–359**
See also Lower limbs; Upper limbs
Lindbergh, Charles A., 89
Lingual tonsils, 449
Linkage analysis, 141–142
Linkage disequilibrium, 192
Linolenic acid, 359, 566
Lipase
in digestion, 143, 226
necrosis from, 411
Lipid metabolism, **359–360,** 453
Lipids, **359–360**
in amniotic fluid, 202

digestion, 225, 226, 425, 426
endocrine system, 176
fluid transport, 213, 238
study of, 392
visual impairment and, 58–59
Lipman, Jacob, 591
Lipmann, Fritz, **360–361**
Lipolysis, 544
Lipopolysaccharides, 41, 359
Lipoproteins, 359
fluid transport, 213
Goldstein, Joseph L. on, 240–241
See also High-density lipoproteins; Low-density lipoproteins
Liquifactive necrosis. *See* Colliquative necrosis
Lister, Joseph, 340
Lithium, 139
Liver, 225, 226, *361*–362, 430
bile in, 52
bilirubin in, 52–53
Dalton, John Call on, 133
embryonic development, 224
endoderm, 179
epinephrine in, 181
estrogen in, 185
fetal circulation and, 201
glucose in, 238, 239
Goldstein, Joseph L. on, 240, 241
heart disease and, 85, 151
Herophilus on, 266
protein metabolism, 465, 466
Rous, Peyton on, 497, 498
sympathetic stimulation of, 544
waste elimination, 169
Liver cancer, 362
Liver cells, 372
Liver cirrhosis, 114, 362
Liver damage, 362, 599
blood and, 62
coma from, 114
Liver diseases, 362
melanin and, 379
osteoporosis and, 432
Liver enzymes
Cori, Carl Ferdinand on, 116
Cori, Gerty T. on, 118
Duvé, Christian de on, 153
glucose and, 238
Liver extract, 599
Forssmann, Werner on, 215
Minot, George Richards on, 390, 391
Murphy, William P. on, 401
Liver sinusoids, 361
Liver transplantation, 76, 362
Lobar brochi. *See* Bronchi
Lobular carcinoma. *See* Breast cancer
Lobules, 243, 361
Local anesthesia, 24
Localization (brain). *See* Cerebral localization
Lock and key model, 180
Locomotion, **457–458,** 545
Loewi, Otto, 131, 132, 275, **362–363**
Lohmann, Karl
on ATP, 5

Mucocutaneous junctions, **396–397**

Mucoproteins. *See* Glycoproteins

Mucosa
    in bronchi, 75
    in digestive tract, 226, 325–327

Mucous membranes
    infection, 39–40, 316–317
    junctions, 396–397
    in nose, 423

Mucus, **397**
    cervical, 318, 319
    in digestion, 225, 326
    Fleming, Alexander on, 209
    olfactory, 510

Mucus-associated lymphoid tissue (MALT), 310

Muller, Hermann Joseph, **397–398,** 525

Müller, Johannes, 275, **398–399**

Müller, Paul Hermann, **399–400**

Müllerian duct, 399, 480, 481, 573

Müller's cells, 399

Müller's law, 399

Multifactorial inheritance. *See* Complex inheritance

Multifidi, 410

Multiple birth, 319

Multiple myeloma
    antibodies and, 27
    chemotherapy for, 104

Multiple sclerosis, 33

Multipolar neurons, 419
    See also Motor neurons

Murad, Ferid, **400–401**
    Furchgott, Robert F. and, 217
    Ignarro, Louis J. and, 305

Murphy, William P., **401**

Murray, Joseph E., **401–402**

Muscle agonists, 478

Muscle biopsy, 405

Muscle cells
    Eccles, John C. on, 157
    epithelial cells and, 182
    glucose and, 238
    Katz, Bernard on, 334
    Meyerhof, Otto on, 387, 388
    Purkinjě system, 473

Muscle contractions, 396, **403,** 408, 517
    ATP in, 5, 360, 361
    brain stem and, 71
    calcium metabolism in, 79
    in cardiac cycle, 82
    in diaphragm, 142
    digestive tract, 326–327
    endocrine system and, 178
    Fabrici, Girolamo on, 197
    fascia in, 198
    with fever, 202
    Fischer, Edmond H. on, 206, 207
    Galvani, Luigi on, 221
    in gonads, 243
    Haller, Albrecht von on, 254
    Hill, Archibald on, 269, 270
    Hopkins, Frederick Gowland on, 282–283
    Huxley, Andrew Fielding on, 298
    innervation, 406

    kinesthetics and, 337
    Krebs, Edwin G. on, 344
    lactic acid in, 349
    mechanorecptors and, 375
    Meyerhof, Otto on, 387
    reciprocal innervation, 476–477
    in reflexes, 122, 477–478
    of smooth muscle, 524
    Szent-Györgyi, Albert on, 547
    in vascular system, 578
    See also Smooth muscle contraction; Uterine contractions

Muscle cramps, 54–55

Muscle damage, **403–404**

Muscle disease, 408

Muscle fibers, 396, 406
    contraction, 403
    dystrophy, 405
    innervation, 406
    kinesthetics and, 337
    Purkinjě system, **473**
    regeneration, 404

Muscle precursor cells. *See* Satellite cells

Muscle regeneration, **403–404**

Muscle relaxants, 139

Muscle relaxation
    capillaries and, 347–348
    Furchgott, Robert F. on, 217, 218
    neurotransmitters and, 420
    in reflexes, 476–477, 478

Muscle repair, **403–404**

Muscle soreness, 403

Muscle spindles, 375
    Granit, Ragnar Arthur on, 244–245
    kinesthetics and, 337

Muscles, 269, 272
    atrophy of, 528
    Haller, Albrecht von on, 254
    Helmholtz, Hermann on, 263
    Hill, Archibald on, 269
    interosseus, **325**
    metabolism, 282–283, 495
    neck, **410**
    in sports physiology, 534
    study of, **408**
    thorax and abdomen, **404–405**
    upper arm, 570
    See also specific types of muscles

Muscular atrophy, 87

Muscular dystrophies, 188, *405*–406
    See also Duchenne muscular dystrophy

Muscular innervation, **406, 476–477**

Muscular system, 22, 23, **406–407,** 429
    cardiac muscle, **85**
    embryonic development, **518–519**
    fascia in, 198–199
    infant, 315–316

Muscular tissues. *See* Muscles

Muscularis, 325–326

Musculo-membranous diaphragm. *See* Diaphragm

Musculoskeletal system, 457–458

Mutation. *See* Gene mutations

Mutation theory, 395

My Life with the Microbes (Waksman), 592

Myasthenia gravis, 33
Mycobacterium tuberculosis, 500–501
Myelencephalon, **407–408,** 414
　　See also Medulla oblongata
Myelin sheath, 97, 245, 246, 416, 417
Myelinated axon, 245, 417
Myelinated preganglionic neurons, 34, 440, 441
Myeloblasts, 264
Myelogenous cells, 354
Myelogenous leukemia. *See* Leukemia
Myeloid precursors. *See* Precursors
Myeloma, 62, 312
　　Edelman, Gerald M. on, 159
　　Milstein, César on, 390
Myeloma cells
　　antibodies and, 27, 312
　　Köhler, Georges on, 341
Myenteric plexus, 326, 418
Myers, Ronald, 531
Myoblasts, 404
Myocardial infarction. *See* Heart attack
Myocarditis, 166
Myocardium, 23, 85, 260
　　fetal, 201
　　Purkinjě system, 473
Myocytes. *See* Muscle cells
Myofibrils, 403, 406, 517
Myofilaments, 403
Myoglobin, 85
Myology, **408**
Myopia, 58
　　infant, 193
　　surgery for, 121
Myosin, 403, 517, 518, 524
Myotomes, 518, 527
Myxedema, 286, 340
Myxomatous degeneration, 260

N

N-acetyl-transferase (NAT), 384
NAD. *See* Nicotinamide adenine dinucleotide (NAD)
Nail bed, 409
Nails, 323, 324, *409*–410, 429
Nalaxone, 150
Narcosis, nitrogen, 570
Narcotics, 24
Nares. *See* Nostrils
Nasal cavities, 423, 439, 523
Nasal conchae, 423
Nasal pharynx, 449
Nasal septum, 423
Nasal sinuses, 423
Nasolacrimal ducts, 423
Nasopharyngeal tonsils. *See* Adenoids
Nasopharynx region, 423
　　Eustachian tubes and, 190, 511
　　notochord and, 424
NAT. *See* N-acetyl-transferase
National Bioethics Advisory Commission, 187
National Foundation for Cancer Research, 547–548
National Institutes of Health (NIH)
　　on diet, 240

　　on embryological research, 187
　　on gene therapy, 226–227
　　on heart disease, 85
Natural selection, 192, 233, 395
The Nature of Life (Szent-Györgyi), 548
Nausea, **410,** 458, 460
Near-sightedness. *See* Myopia
Neck fascia. *See* Cervical fascia
Neck muscles, **410**
Neckbone. *See* Cervical vertebrae
Necrosis, **410–411**
　　from embolism, 171
　　from hypoxia, 303
Needle aspiration. *See* Needle biopsy
Needle biopsy, 320
Negative feedback, 280, 281, 385
　　hormones, 283, 285
　　in menstrual cycle, 383
Neher, Erwin, **411–412,** 502
Neiseria gonorrhoeae, 40
Nematode infections. *See* Roundworm infections
Neocerebellum, 99
Neocortex, 99
　　with Alzheimer's disease, 16
　　limbic system and, 359
Neomycin, 592
Neonatal growth and development, **412–413**
　　See also Infant growth and development
Neonatal hypoglycemia, 322
Neonatal oxidative stress, 28
Neonate. *See* Newborns
Neonatology, 412
Neoplasms. *See* Tumors
Neospinothalamic tract, 437
Neovascular vessels, 58
Nephrogenic differentiation, 480
Nephrons, 237, 336, 480
Nernst equation, 431
Nerve cells. *See* Neurons
Nerve compression, 87–88
Nerve fibers, 245–246
　　damage, 417
　　Gasser, Herbert Spencer on, 223
　　Hartline, Haldan Keffer on, 255
　　Hodgkin, Alan Lloyd on, 277, 278
　　Huxley, Andrew Fielding on, 298
　　paraganglia, 440
　　sense organs, 507
Nerve growth factor (NGF), 417
　　Cohen, Stanley on, 113
　　depressants and stimulants, 139
　　Levi-Montalcini, Rita on, 355, 356
Nerve impulse, 4–5, 245, 246, 379, **413–414,** 416, 419
　　cardiac muscle, 85, 261–262
　　Gasser, Herbert Spencer on, 222–223
　　Granit, Ragnar Arthur on, 244
　　Hartline, Haldan Keffer on, 255
　　in hearing, 259
　　Helmholtz, Hermann on, 263
　　Hodgkin, Alan Lloyd on, 277–278
　　hormones and, 285
　　Hubel, David H. on, 289
　　Huxley, Andrew Fielding on, 298

Paraxial mesoderm. *See* Mesoderm
Parenchyma, 101
Parietal arteries, 31
Parietal bones, 127
Parietal lobe, 99, 100
Parietal pericardium. *See* Pericardium
Parietal pleura. *See* Pleura
Parieto-occiptotemporal cortex, 70
Parkinson's disease
    antioxidants and, 28
    basal ganglia and, 45
    Carlsson, Arvid on, 86
    dopamine, 468
    gene therapy for, 226
    stem cells for, 185–186, 188, 291, 535
    thalamus and, 553–554
Paronychia, 409
Parotid gland, 126
Parrafin, 272
Pars intermedia. *See* Intermediate pituitary gland
Parthenogenesis, 187
Partial pressures, 222, 263, 434
Partial-thickness burns. *See* Burns
Parturition, **441–442**
    fetal membranes in, 202
    homeostatic mechanisms in, 281, 284
    pubic symphysis in, 471
Parvoviruses, 587
Passive diffusion. *See* Gradient diffusion
Pasteur, Louis, 27, 276, 308, 311, 528, 584
Patch clamp technique, 411–412, 502
Patella, 363, 521
Patent ductus arteriosis (PDA), 152
Patent eustachian tubes. *See* Eustachian tubes
Patents (genes). *See* Gene patents
Pathogenesis, 39–40, 442–443
Pathogens
    bacteria, 39–41
    immune system and, 310–312
    Koch, Robert on, 338–339
Pathologic fracture. *See* Fractures (Injuries)
Pathological anatomy, 51, 276, 394–395
Pathology, **442–443**
    Avicenna on, 36
    of cancer, 80
    cellular, 584
    Florey, Howard Walter on, 211
    forensic, **214**
    Laënnec, René-Théophile-Hyacinthe on, 350
    Landsteiner, Karl on, 350, 351
    medical training, 377–378
    Metchnikoff, Élie on, 386–387
    Morgagni, Giovanni Battista on, 394–395
    muscle, 408
    tissue, 272
    See also specific conditions and disorders
Patient isolation, for burns, 77
Patulous Eustachian tubes. *See* Eustachian tubes
Pauling, Linus, 128, 138
Pavlov, Ivan Petrovich, 275, **443–445**
Pavlovian psychology, 444
PDA. *See* Patent ductus arteriosis (PDA)
PDGF. *See* Platelet-derived growth factor (PDGF)

PE. *See* Pulmonary embolism
Pectoral girdle, 28, 519
Pectoral nerves, 69
Pedal pulse. *See* Pulse
Pedigree analysis, 296
Peduncles, 100
Pelvic adhesions, 318, 319
Pelvic bone, 29
Pelvic diaphragm, 142
Pelvic examination, 318
Pelvic girdle, 29, 363, 521
Pelvic inflammatory disease (PID), 317, 318
Pelvic ultrasound, 318
Pelvis, 1, 22, 521
    bones of, 29
    iliac arteries, 306
    in parturition, 442
    pubic symphysis, 471
Penetrating keratoplasty. *See* Corneal transplantation
Penicillin, 103
    Chain, Ernst Boris on, 101–102
    Fleming, Alexander on, 208
    Florey, Howard Walter on, 210–212
Penicillium notatum, 103
    Chain, Ernst Boris on, 101, 102
    Fleming, Alexander on, 209
Penile erection, 234
Penis, 234, 430, 572
    Fallopius, Gabriel on, 198
    puberty, 469–470, 506
Penninger, Joseph, 260
Pentose phosphate cycle, 385
Pepsin
    in digestion, 143, 225, 425
    Schwann, Theodor on, 505
Peptidase, 425
Peptide bond, 18
Peptide hormones, 283, 420
    Holley, Robert on, 280
    Li, Choh Hao on, 357–358
Peptide site, 466
Peptide-tRNA complex, 466
Peptidyl transferase, 466
Percussion technique, 32–33
Perforating canals. *See* Volkmann's canals
Perfusion, 89, 516
Pericardial cavity, 259–260, 445
Pericardial fluid, 259–260, 445
Pericarditis, 445
Pericardium, 109, 259–260, **445**
Perichondrium, 90, 300, 431
Pericytes, 404, 578
Perilymph, 155, 512
Perimenopause, 381–382
Perineum, 233, 482
Periodontal ligament, 204, 563
Periosteal layer, 381
Periosteum, 552
Peripheral nervous system (PNS), 97, 416, **445–446**
    ATP and, 6
    damage, 417–418
    embryonic development, 414

•

General Index

Sutures, cranial. *See* Cranial sutures
Suturing, blood vessels and, 88, 89
Swallowing, **542**
    cranial nerves, 126
    Eustachian tubes and, 190, 191
    medulla oblongata and, 378
    mucus in, 397
    palate in, 439
    pharynx in, 449
Swammerdam, Jan, 171, 274
Sweat glands, 180, 323, 429, **542**
    embryonic development, 324
    in fight or flight reflex, 205
    in forensic pathology, 214
Sweating. *See* Perspiration
Swelling. *See* Edema
Sylvian fissure, 543
Sylvius, Franciscus dele Bo, 143, 275, **543**
Sympathetic chain ganglia, 262, 414, 418
Sympathetic nervous system, 34, 416, 445–446, **543–544**
    adrenal medulla and, 9, 10
    Axelrod, Julius on, 36–37
    cardiac cycle and, 261–262
    depressants and stimulants, 139
    effect of drugs on, 150
    erythema and, 64
    Euler, Ulf von on, 190
    fight or flight reflex, 204
    hormones and, 284
    Loewi, Otto on, 363
    vs. parasympathetic nervous system, 440
    reflexes and, 71
Sympathetic postganglionic neurons, 262
Sympathetic stimulation, 262
Sympathoadrenal system, 10, 204–205
Symphyses, 90, 330
    See also Pubic symphysis
Symphysis pubis. *See* Pubic symphysis
Synapses, 413, 416, 419, **544**
    Carlsson, Arvid on, 85–86
    in central nervous system, 97, 242
    depressants and stimulants, 138–139
    development of, 104
    drugs and, 149, 150
    Greengard, Paul on, 246
    Hodgkin, Alan Lloyd on, 277–278
    Kandel, Eric R. on, 333
    motor function, 396, 406
    neurotransmitters, 419–420
    Ramón y Cajal, Santiago on, 476
    in reflexes, 122, 477
    sense organs, 507–508
    Sherrington, Charles Scott on, 515
Synaptic cleft, 150
Synaptic knob, 413
Synaptic vesicles, 150, 413
Synarthrotic joints, 32
Synchondroses, 90, 330
Syncope, 108
    See also Carotid sinus syncope
Syncytiotrophoblast, 313, 452, 462

Syncytium
    cardiac muscle, 85
    embryo implantation, 313
Syndesmosis joints, 204
Syngameon, 191
Synovia. *See* Synovial fluid
Synovial bursa. *See* Synovial capsule
Synovial capsule, 331
Synovial fluid, 331, 545
Synovial joints, 32, 90, 330–331, **545**
    fibrocartilage, 204
    kinesthetics, 337
Synovial membranes, *330–331*
Synthetase. *See* Ligase
Synthetic mRNA. *See* Messenger RNA
Syphilis, 161, 162, 163, 351
Systemic arteries, 30, 63, 544
Systemic capillaries, 434
Systemic circulation, 82, 109, 259, **545–546**
    coronary circulation and, 121
    ductus arteriosis, 152
    fetal, 201, 261
    pulmonary circulation and, 471
Systemic diseases, 58
Systemic lupus erythematosis, 33
Systemic sclerosis, 28
Systemic veins, 579
Systole. *See* Cardiac systole
Systolic pressure, 472
Szent-Györgyi, Albert, 276, **546–548**
Szent-Györgyi Foundation, 548

T

T3. *See* Triiodothyronine
T4. *See* Thyroxine
T-4 cells, 549
T-8 cells, 549–550
T cell receptor (TCR), 549
T lymphocytes, 14, 309, 310–312, 355, 366, **549–550**
    B lymphocytes and, 39
    Doherty, Peter C. on, 146
    hemopoiesis and, 264
    interferons from, 324–325
    Jerne, Niels Kaj on, 330
    Köhler, Georges on, 341
    thymus and, 559
    Zinkernagel, Rolf M. on, 607
T waves. *See* Cardiac rhythms
Table sugar. *See* Sucrose
Tabulae anatomicae sex (Vesalius), 581
Tachycardia, 148, 166
Takamine, Jokichi, 276
Tamoxifen, for cancer, 103
Tarsal plates, 193
Tarsals, 29, 521
Tartar, 563
Taste, 225, **550–551**
    saliva and, 502
    structures, **251**
Taste buds, 225, 251, 550
Tatum, Edward Lawrie, 158
Tau, 17

Dale, Henry Hallett on, 132
in erythema, 64
Furchgott, Robert F. on, 217
nitric oxide as, 420
Vasomotion, in capillaries, 24–25
Vasopressins. *See* Antidiuretic hormone (ADH)
Vegetative functions, 359, **579**
Vein walls, 109
Veins, 109–110, 578, **579–580**
abdominal, **2**
blood pressure, 63
in coronary circulation, 121
Galen on, 219
heart development, 261
Herophilus on, 266
study of, 24–25
in systemic circulation, 545
thoracic, **558**
varicose, 110, 579
Vena cava, 2, 109, 259, 579
cardiac cycle, 82
coronary circulation, 121
development of, 577, 578
fetal circulation, 201, 261
thoracic veins and, 558
Venous sinuses, 533
Venous system. *See* Veins
Ventilation. *See* Respiration
Ventilators, 254
Ventral. *See* Anatomical nomenclature
Ventral tegmental area, 149
Ventricles. *See* Cerebral ventricles; Coronary ventricles
Ventricular contraction
ECG and, 166
regulation, 262
Ventricular fibrillation, 123
Ventricular hypertrophy, 166
Ventricular pressure, 82, 259
Ventricular septal defect, 260
Ventricular system
cerebrospinal fluid, 101
coronary circulation, 121
Ventromedial nucleus, 359
Venules, 24, 25, 109, 580
Vermiform appendix. *See* Appendix
Vermis, 99, 100
Vernix caseosa, 202, 324
Vertebrae, 519, **580**
cervical, 98, 520, 580
fibrocartilage, 204
lumbar, 520
notochord and, 424
sacral, 519
spinal cord and, 532
thoracic, 520, 580
Vertebral arteries, 108
Vertebral column, 461, 519, **580–581**
Vesalius, Andreas, **581**
on anatomy, 22, 273, 274
Fallopius, Gabriel and, 198
Vesicle, 96
fluid transport, 95, 213

Katz, Bernard on, 334
Palade, George Emil on, 438
Vestibular disorientation, 35
Vestibular nuclei, 71, 507
Vestibular system, 155, 337, 507, 512
Bárány, Robert on, 44
embryonic development, 508
in space, 527
Vestibule, 482
Vestibulocochlear nerve. *See* Auditory nerve
Vestigial structures, 29, 115, 281, 540, **581–582**
Vibration. *See* Sound vibration
Vibrissae, 375
Vierordt, Karl, 276
Vieussens, Raymond, 274, **582–583**
Villi
arachnoid, 381
intestinal, 226, 326
placental, 452–453
Vinblastine, for cancer, 103
Vinci, Leonardo da, 273, **583–584**
Vincristine, for cancer, 103
Viral capsid, 585
Viral cultures, 76
Viral genes, 585–586
Viral infections
diarrhea from, 142
Doherty, Peter C. on, 146
fever from, 202
hemorrhagic fever, **264–265**
interferons for, 324–325
Kornberg, Arthur on, 343
nails, 409
responses, **584–587**
Viral meningitis, 98, 381
Viral replication, 585, 586, 587
Virchow, Rudolph Carl, 97, 241, 275, 514, **584**, 595
Virology
Burnet, Frank Macfarlane on, 76
Claude, Alfred on, 112
Virulence, of bacteria, 41
Virus-host relationship, 585
Virus Hunters (Williams), 555
Viruses, **584–587**
Burnet, Frank Macfarlane on, 76
cancer as, 80, 497–498
Carrel, Alexis on, 88, 89
Claude, Albert on, 112
diarrhea from, 142
Doherty, Peter C. on, 146
drugs for, 103, 170
fever and, 202
HIV, 12–14
See also names of specific viruses
Viscera, 533
Visceral arteries, 31, 64
Visceral pericardium. *See* Pericardium
Visceral pleura. *See* Pleura
Visceral processes, **526–527**
Visceral reflexes, 418
Vision, 194, 509–510, **587–588**
Granit, Ragnar Arthur on, 244
Hartline, Haldan Keffer on, 255

X

Y

Z